Emerging Technologies for Electric and Hybrid Vehicles

Emerging Technologies for Electric and Hybrid Vehicles

Selected Articles Published by MDPI

MDPI • Basel • Beijing • Wuhan • Barcelona • Belgrade

MDPI

Editorial Office
MDPI
St. Alban-Anlage 66
Basel, Switzerland

This is a reprint of articles published online by the open access publisher MDPI from 2016 to 2017 (available at: www.mdpi.com/journal/energies/special_issues/Vehicles_collection).

For citation purposes, cite each article independently as indicated on the article page online and as indicated below:

LastName, A.A.; LastName, B.B.; LastName, C.C. Article Title. *Journal Name* **Year**, *Article Number*, Page Range.

ISBN 978-3-03897-190-0 (Pbk)
ISBN 978-3-03897-191-7 (PDF)

Contents

Preface to "Emerging Technologies for Electric and Hybrid Vehicles"

Electric and hybrid vehicles are probably the cleanest and greenest road transportation, and are currently superseding internal combustion engine vehicles. The purpose of this book is to collect the wisdom of contributors with expertise in various technologies for electric and hybrid vehicles. Hence, the book will consolidate emerging technologies for electric and hybrid vehicles, with an emphasis on three main themes—the energy systems, the propulsion systems and the auxiliary systems.

In this book, fifteen outstanding papers are collected. Firstly, it starts with general reviews of electric vehicle technologies as well as their challenges, developments and applications. Secondly, some emerging technologies within the field of energy systems for electric and hybrid vehicles, including their battery costing, modeling, fault diagnosis and parameter estimation, are presented. Apart from these on-board energy systems, off-board energy systems for electric vehicles are discussed, which include the optimal siting of charging stations, grid-aware peak shaving and the bidirectional converter interface with DC microgrids. Thirdly, two emerging technologies in the field of propulsion systems are elaborated, namely open-end winding permanent magnet synchronous motor driving for electric vehicles and power split type hybrid drivetrains for plug-in hybrid electric vehicles. Finally, three emerging technologies in the field of auxiliary systems, namely the optimal energy management strategy for hybrid electric vehicles, wireless charging systems for electric vehicles and heat pump air conditioning systems for electric vehicles are discussed.

There is a Chinese idiom, "when you drink water, think of its source". Hence, I have to express my special thanks to all contributors to this book. Another Chinese idiom, "collective wisdom reaps wide benefits" assures that the expertise of various contributors will gather together to derive solid knowledge of electric and hybrid vehicles, which is beneficial not only to technological advancement but also to knowledge exchange. Without their contributions, this book would not have seen the light of day.

While electric and hybrid vehicles are a driving force of a better environment, my family is the propulsive force for my work on electric and hybrid vehicles. I would like to make use of this chance to express my heartfelt gratitude to Aten Man-Ho and Joan Wai-Yi for their heartfelt support throughout.

K.T. Chau
The University of Hong Kong
Hong Kong

energies

MDPI

Review

A Comprehensive Study of Key Electric Vehicle (EV) Components, Technologies, Challenges, Impacts, and Future Direction of Development

Fuad Un-Noor [1], Sanjeevikumar Padmanaban [2,*], Lucian Mihet-Popa [3], Mohammad Nurunnabi Mollah [1] and Eklas Hossain [4,*]

[1] Department of Electrical and Electronic Engineering, Khulna University of Engineering and Technology, Khulna 9203, Bangladesh; fuad9304@gmail.com (F.U.-N.); nurunnabim12@gmail.com (M.N.M.)
[2] Department of Electrical and Electronics Engineering, University of Johannesburg, Auckland Park 2006, South Africa
[3] Faculty of Engineering, Østfold University College, Kobberslagerstredet 5, 1671 Kråkeroy-Fredrikstad, Norway; lucian.mihet@hiof.no
[4] Department of Electrical Engineering & Renewable Energy, Oregon Tech, Klamath Falls, OR 97601, USA
* Correspondence: sanjeevi_12@yahoo.co.in (S.P.); eklas.hossain@oit.edu (E.H.); Tel.: +27-79-219-9845 (S.P.); +1-541-885-1516 (E.H.)

Academic Editor: Sergio Saponara
Received: 8 May 2017; Accepted: 21 July 2017; Published: 17 August 2017

Abstract: Electric vehicles (EV), including Battery Electric Vehicle (BEV), Hybrid Electric Vehicle (HEV), Plug-in Hybrid Electric Vehicle (PHEV), Fuel Cell Electric Vehicle (FCEV), are becoming more commonplace in the transportation sector in recent times. As the present trend suggests, this mode of transport is likely to replace internal combustion engine (ICE) vehicles in the near future. Each of the main EV components has a number of technologies that are currently in use or can become prominent in the future. EVs can cause significant impacts on the environment, power system, and other related sectors. The present power system could face huge instabilities with enough EV penetration, but with proper management and coordination, EVs can be turned into a major contributor to the successful implementation of the smart grid concept. There are possibilities of immense environmental benefits as well, as the EVs can extensively reduce the greenhouse gas emissions produced by the transportation sector. However, there are some major obstacles for EVs to overcome before totally replacing ICE vehicles. This paper is focused on reviewing all the useful data available on EV configurations, battery energy sources, electrical machines, charging techniques, optimization techniques, impacts, trends, and possible directions of future developments. Its objective is to provide an overall picture of the current EV technology and ways of future development to assist in future researches in this sector.

Keywords: electric vehicle; energy sources; motors; charging technologies; effects of EVs; limitations of EVs; energy management; control algorithms; global EV sales; trends and future developments

1. Introduction

In recent times, electric vehicles (EV) are gaining popularity, and the reasons behind this are many. The most eminent one is their contribution in reducing greenhouse gas (GHG) emissions. In 2009, the transportation sector emitted 25% of the GHGs produced by energy related sectors [1]. EVs, with enough penetration in the transportation sector, are expected to reduce that figure, but this is not the only reason bringing this century old and once dead concept back to life, this time as a commercially viable and available product. As a vehicle, an EV is quiet, easy to operate, and does not have the fuel costs associated with conventional vehicles. As an urban transport mode, it is highly useful. It does not use any stored energy or cause any emission while idling, is capable of frequent start-stop driving, provides the total torque from the startup, and does not require trips to the gas station. It does not contribute either to any of the smog making the city air highly polluted. The instant torque makes it highly preferable for motor sports. The quietness and low infrared signature makes it useful in military use as well. The power sector is going through a changing phase where renewable sources are gaining momentum. The next generation power grid, called 'smart grid' is also being developed. EVs are being considered a major contributor to this new power system comprised of renewable generating facilities and advanced grid systems [2,3]. All these have led to a renewed interest and development in this mode of transport.

The idea to employ electric motors to drive a vehicle surfaced after the innovation of the motor itself. From 1897 to 1900, EVs became 28% of the total vehicles and were preferred over the internal combustion engine (ICE) ones [1]. But the ICE types gained momentum afterwards, and with very low oil prices, they soon conquered the market, became much more mature and advanced, and EVs got lost into oblivion. A chance of resurrection appeared in the form of the EV1 concept from General Motors, which was launched in 1996, and quickly became very popular. Other leading carmakers, including Ford, Toyota, and Honda brought out their own EVs as well. Toyota's highly successful Prius, the first commercial hybrid electric vehicle (HEV), was launched in Japan in 1997, with 18,000 units sold in the first year of production [1]. Today, almost none of those twentieth century EVs exist; an exception can be Toyota Prius, still going strong in a better and evolved form. Now the market is dominated by Nissan Leaf, Chevrolet Volt, and Tesla Model S; whereas the Chinese market is in the grip of BYD Auto Co., Ltd. (X'an National Hi-tech Industrial Development Zone, Xi'an, China).

EVs can be considered as a combination of different subsystems. Each of these systems interact with each other to make the EV work, and there are multiple technologies that can be employed to operate the subsystems. In Figure 1, key parts of these subsystems and their contribution to the total system is demonstrated. Some of these parts have to work extensively with some of the others, whereas some have to interact very less. Whatever the case may be, it is the combined work of all these systems that make an EV operate.

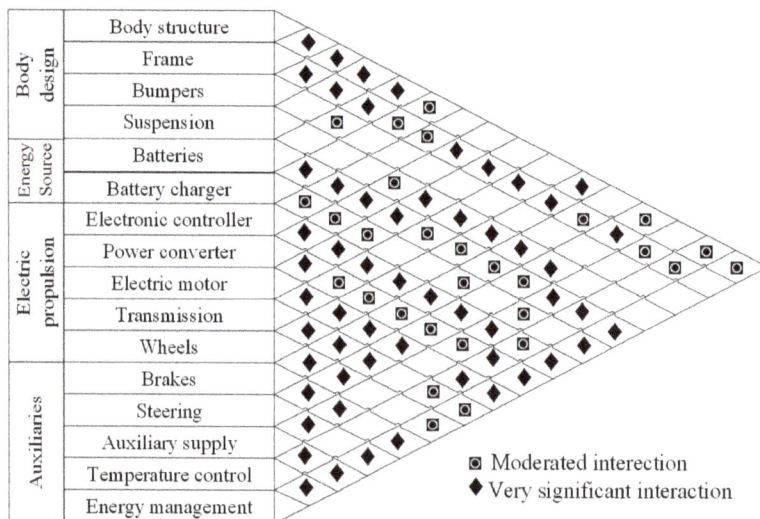

Figure 1. Major EV subsystems and their interactions. Some of the subsystems are very closely related while some others have moderated interactions. Data from [4].

There are quite a few configurations and options to build an EV with. EVs can be solely driven with stored electrical power, some can generate this energy from an ICE, and there are also some vehicles that employ both the ICE and the electrical motors together. The general classification is discussed in Section 2, whereas different configurations are described in Section 3. EVs use different types of energy storage to store their power. Though batteries are the most used ones, ultracapacitors, flywheels and fuel cells are also up and coming as potential energy storage systems (ESS). Section 4 is dedicated to these energy sources. The types of motors that have been used in EVs and can be used in future are discussed in Section 5. Different charging voltages and charger configurations can be used in charging the vehicles. Wireless charging is also being examined and experimented with to increase convenience. These charger standards, configurations and power conversion systems are demonstrated in Sections 6–8 discusses the effects EVs create in different sectors. Being a developing technology, EVs still have many limitations that have to be overcome to enable them to penetrate deeper into the market. These limitations are pointed out in Section 9 along with probable solutions. Section 10 summed up some strategies used in EVs to enable proper use of the available power. Section 11 presented different types of control algorithms used for better driving assistance, energy management, and charging. The current state of the global EV market is briefly presented in Section 12, followed by Section 13 containing the trends and sectors that may get developed in the future. Finally, the ultimate outcomes of this paper is presented in Section 14. The topics covered in this paper have been discussed in different literatures. Over the years, a number of publications have been made discussing different aspects of EV technology. This paper was created as an effort to sum up all these works to demonstrate the state-of-the art of the system and to position different technologies side by side to find out their merits and demerits, and in some cases, which one of them can make its way to the future EVs.

2. EV Types

EVs can run solely on electric propulsion or they can have an ICE working alongside it. Having only batteries as energy source constitutes the basic kind of EV, but there are kinds that can employ other energy source modes. These can be called hybrid EVs (HEVs). The International Electrotechnical Commission's Technical Committee 69 (Electric Road Vehicles) proposed that vehicles using two or more types of energy source, storage or converters can be called as an HEV as long as at least one of those provide electrical energy [4]. This definition makes a lot of combinations possible for HEVs like ICE and battery, battery and flywheel, battery and capacitor, battery and fuel cell, etc. Therefore, the common population and specialists both started calling vehicles with an ICE and electric motor combination HEVs, battery and capacitor ones as ultra-capacitor-assisted EVs, and the ones with battery and fuel cell FCEVs [2–4]. These terminologies have become widely accepted and according to this norm, EVs can be categorized as follows:

(1) Battery Electric Vehicle (BEV)
(2) Hybrid Electric Vehicle (HEV)
(3) Plug-in Hybrid Electric Vehicle (PHEV)
(4) Fuel Cell Electric Vehicle (FCEV)

2.1. Battery Electric Vehicle (BEV)

EVs with only batteries to provide power to the drive train are known as BEVs. BEVs have to rely solely on the energy stored in their battery packs; therefore the range of such vehicles depends directly on the battery capacity. Typically they can cover 100 km–250 km on one charge [5], whereas the top-tier models can go a lot further, from 300 km to 500 km [5]. These ranges depend on driving condition and style, vehicle configurations, road conditions, climate, battery type and age. Once depleted, charging the battery pack takes quite a lot of time compared to refueling a conventional ICE vehicle. It can take as long as 36 h completely replenish the batteries [6,7], there are far less time consuming ones as well, but none is comparable to the little time required to refill a fuel tank. Charging time depends on the charger configuration, its infrastructure and operating power level. Advantages of BEVs are their simple construction, operation and convenience. These do not produce any greenhouse gas (GHG), do not create any noise and therefore beneficial to the environment. Electric propulsion provides instant and high torques, even at low speeds. These advantages, coupled with their limitation of range, makes them the perfect vehicle to use in urban areas; as depicted in Figure 2, urban driving requires running at slow or medium speeds, and these ranges demand a lot of torque. Nissan Leaf and Teslas are some high-selling BEVs these days, along with some Chinese vehicles. Figure 3 shows basic configuration for BEVs: the wheels are driven by electric motor(s) which is run by batteries through a power converter circuit.

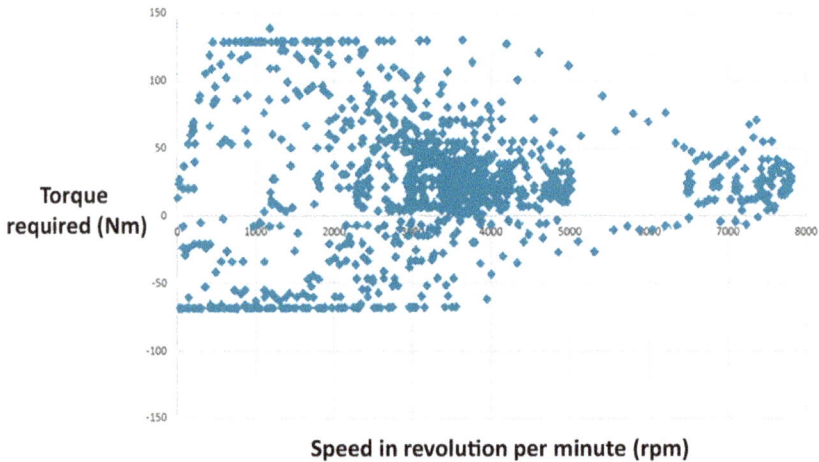

Figure 2. Federal Urban Driving Schedule torque-speed requirements. Most of the driving is done in the 2200 to 4800 rpm range with significant amount of torque. Lower rpms require torques as high as 125 Nm; urban vehicles have to operate in this region regularly as they face frequent start-stops. Data from [4].

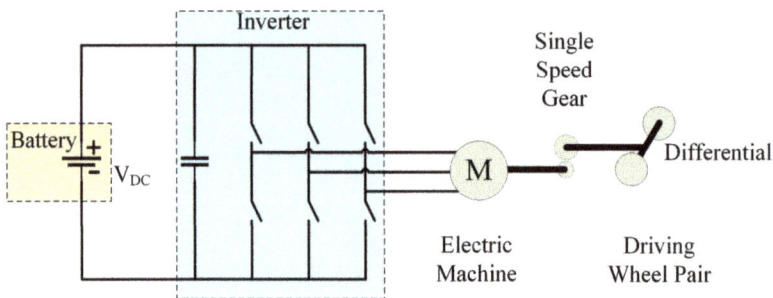

Figure 3. BEV configuration. The battery's DC power is converted to AC by the inverter to run the motor. Adapted from [5].

2.2. Hybrid Electric Vehicle (HEV)

HEVs employ both an ICE and an electrical power train to power the vehicle. The combination of these two can come in different forms which are discussed later. An HEV uses the electric propulsion system when the power demand is low. It is a great advantage in low speed conditions like urban areas;

it also reduces the fuel consumption as the engine stays totally off during idling periods, for example, traffic jams. This feature also reduces the GHG emission. When higher speed is needed, the HEV switches to the ICE. The two drive trains can also work together to improve the performance. Hybrid power systems are used extensively to reduce or to completely remove turbo lag in turbocharged cars, like the Acura NSX. It also enhances performance by filling the gaps between gear shifts and providing speed boosts when required. The ICE can charge up the batteries, HEVs can also retrieve energy by means of regenerative braking. Therefore, HEVs are primarily ICE driven cars that use an electrical drive train to improve mileage or for performance enhancement. To attain these features, HEV configurations are being widely adopted by car manufacturers. Figure 4 shows the energy flows in a basic HEV. While starting the vehicle, the ICE may run the motor as a generator to produce some power and store it in the battery. Passing needs a boost in speed, therefore the ICE and the motor both drives the power train. During braking the power train runs the motor as generator to charge the battery by regenerative braking. While cruising, ICE runs the both the vehicle and the motor as generator, which charges the battery. The power flow is stopped once the vehicle stops. Figure 5 shows an example of energy management systems used in HEVs. The one demonstrated here splits power between the ICE and the electric motor (EM) by considering the vehicle speed, driver's input, state of charge (SOC) of battery, and the motor speed to attain maximum fuel efficiency.

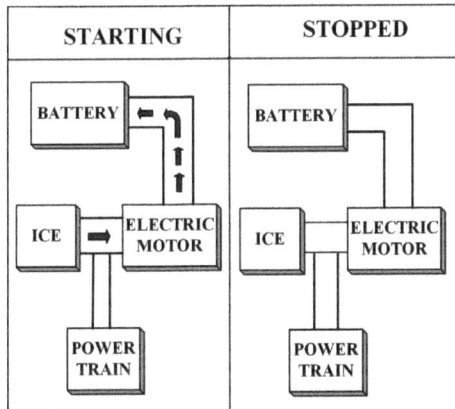

(a) Direction of power flow during starting and when stopped.

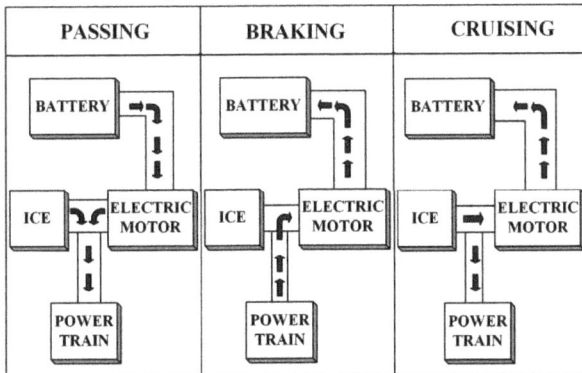

(b) Direction of power flow during passing, braking and cruising.

Figure 4. Power flow among the basic building blocks of an HEV during various stages of a drive cycle. Adapted from [8].

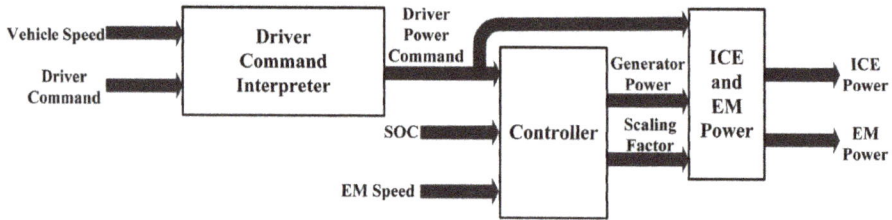

Figure 5. Example of energy management strategy used in HEV. The controller splits power between the ICE and the motor by considering different input parameters. Adapted from [8].

2.3. Plug-In Hybrid Electric Vehicle (PHEV)

The PHEV concept arose to extend the all-electric range of HEVs [9–14]. It uses both an ICE and an electrical power train, like a HEV, but the difference between them is that the PHEV uses electric propulsion as the main driving force, so these vehicles require a bigger battery capacity than HEVs. PHEVs start in 'all electric' mode, runs on electricity and when the batteries are low in charge, it calls on the ICE to provide a boost or to charge up the battery pack. The ICE is used here to extend the range. PHEVs can charge their batteries directly from the grid (which HEVs cannot); they also have the facility to utilize regenerative braking. PHEVs' ability to run solely on electricity for most of the time makes its carbon footprint smaller than the HEVs. They consume less fuel as well and thus reduce the associated cost. The vehicle market is now quite populated with these, Chevrolet Volt and Toyota Prius sales show their popularity as well.

2.4. Fuel Cell Electric Vehicle (FCEV)

FCEVs also go by the name Fuel Cell Vehicle (FCV). They got the name because the heart of such vehicles is fuel cells that use chemical reactions to produce electricity [15]. Hydrogen is the fuel of choice for FCVs to carry out this reaction, so they are often called 'hydrogen fuel cell vehicles'. FCVs carry the hydrogen in special high pressure tanks, another ingredient for the power generating process is oxygen, which it acquires from the air sucked in from the environment. Electricity generated from the fuel cells goes to an electric motor which drives the wheels. Excess energy is stored in storage systems like batteries or supercapacitors [2,3,16–18]. Commercially available FCVs like the Toyota Mirai or Honda Clarity use batteries for this purpose. FCVs only produce water as a byproduct of its power generating process which is ejected out of the car through the tailpipes. The configuration of an FCV is shown in Figure 6. An advantage of such vehicles is they can produce their own electricity which emits no carbon, enabling it to reduce its carbon footprint further than any other EV. Another major advantage of these are, and maybe the most important one right now, refilling these vehicles takes the same amount of time required to fill a conventional vehicle at a gas pump. This makes adoption of these vehicles more likely in the near future [2–4,19]. A major current obstacle in adopting this technology is the scarcity of hydrogen fuel stations, but then again, BEV or PHEV charging stations were not a common scenario even a few years back. A report to the U.S. Department of Energy (DOE) pointed to another disadvantage which is the high cost of fuel cells, that cost more than $200 per kW, which is far greater than ICE (less than $50 per kW) [20,21]. There are also concerns regarding safety in case of flammable hydrogen leaking out of the tanks. If these obstacles were eliminated, FCVs could really represent the future of cars. The possibilities of using this technology in supercars is shown by Pininfarina's H2 Speed (Figure 7). Reference [22] compared BEVs and FCEVs in different aspects, where FCEVs appeared to be better than BEVs in many ways; this comparison is shown in Figure 8. In this figure, different costs and cost associated issues of BEV and FCEV: weight, required storage volume, initial GHG emission, required natural gas energy, required wind energy, incremental costs, fueling infrastructure cost per car, fuel cost per kilometer, and incremental life cycle cost are all compared for 320 km (colored blue) and 480 km (colored green) ranges. The horizontal axis shows the attribute ratio of BEV to FCEV. As having a less value in these attributes indicates an advantage, any value higher than one in the horizontal axis will declare FCEVs superior to BEVs in that attribute. That being said, BEVs only appear better in the fields of required wind energy and fuel cost per kilometer. Fuel cost still appears to be one of the major drawbacks of FCEVs, as a cheap, sustainable and environment-friendly way of producing hydrogen is still lacking, and the refueling infrastructure lags behind that of BEVs; but these problems may no longer prevail in the near future.

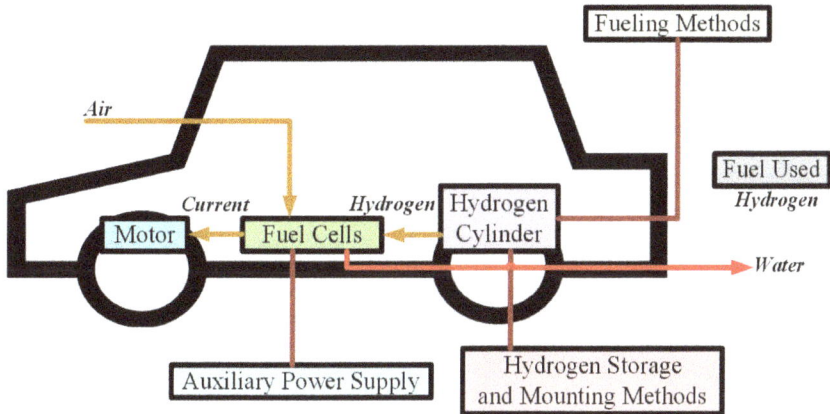

Figure 6. FCEV configuration. Oxygen from air and hydrogen from the cylinders react in fuel cells to produce electricity that runs the motor. Only water is produced as by-product which is released in the environment.

Figure 7. Pininfarina H2 Speed, a supercar employing hydrogen fuel cells.

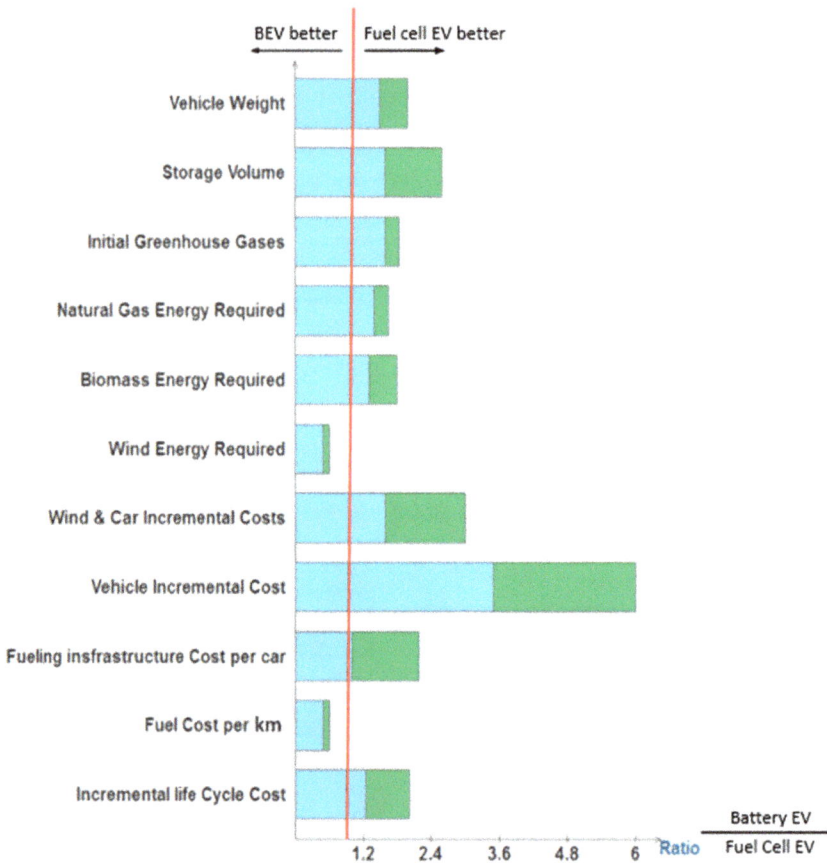

Figure 8. Advanced battery EV attribute and fuel cell EV attribute ratio for 320 km (colored blue) and 480 km (colored green) ranges, with assumptions of average US grid mix in 2010–2020 time-range and all hydrogen made from natural gas (values greater than one indicate a fuel cell EV advantage over the battery EV). Data from [22].

Rajashekara predicted a slightly different future for FCVs in [23]. He showed a plug-in fuel cell vehicle (PFCV) with a larger battery and smaller fuel cell, which makes it battery-dominant car. According to [23], if hydrogen for such vehicles can be made from renewable sources to run the fuel cells and the energy to charge the batteries comes from green sources as well, these PFCVs will be the future of vehicles. The FCVs we see today will not have much appeal other than some niche markets. Figure 9 shows a basic PFCV configuration. Table 1 compares the different vehicle types in terms of driving component, energy source, features, and limitations.

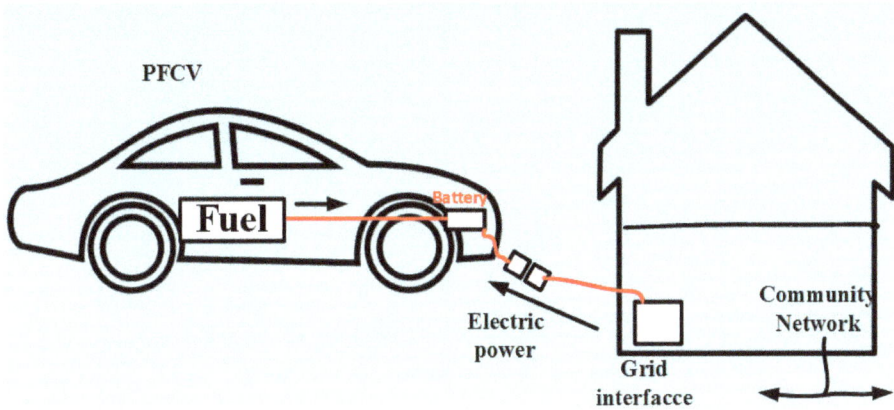

Figure 9. PFCV configuration. In addition to the fuel cells, this arrangement can directly charge the battery from a power outlet.

Table 1. Comparison of different vehicle types. Adapted from [4].

EV Type	Driving Component	Energy Source	Features	Problems
BEV	• Electric motor	• Battery • Ultracapacitor	• No emission • Not dependent on oil • Range depends largely on the type of battery used • Available commercially	• Battery price and capacity • Range • Charging time • Availability of charging stations • High price
HEV	• Electric motor • ICE	• Battery • Ultracapacitor • ICE	• Very little emission • Long range • Can get power from both electric supply and fuel • Complex structure having both electrical and mechanical drivetrains • Available commercially	• Management of the energy sources • Battery and engine size optimization
FCEV	• Electric motor	• Fuel cell	• Very little or no emission • High efficiency • Not dependent on supply of electricity • High price • Available commercially	• Cost of fuel cell • Feasible way to produce fuel • Availability of fueling facilities

3. EV Configurations

An electric vehicle, unlike its ICE counterparts, is quite flexible [4]. This is because of the absence of intricate mechanical arrangements that are required to run a conventional vehicle. In an EV, there is only one moving part, the motor. It can be controlled by different control arrangements and techniques. The motor needs a power supply to run which can be from an array of sources. These two components can be placed at different locations on the vehicle and as long as they are connected through electrical wires, the vehicle will work. Then again, an EV can run solely on electricity, but an ICE and electric motor can also work in conjunction to turn the wheels. Because of such flexibility, different configurations emerged which are adopted according to the type of vehicle. An EV can be considered as a system incorporating three different subsystems [4]: energy source, propulsion and auxiliary. The energy source subsystem includes the source, its refueling system and energy management system. The propulsion subsystem has the electric motor, power converter, controller, transmission and the driving wheels as its components. The auxiliary subsystem is comprised of auxiliary power supply, temperature control system and the power steering unit. These subsystems are shown in Figure 10.

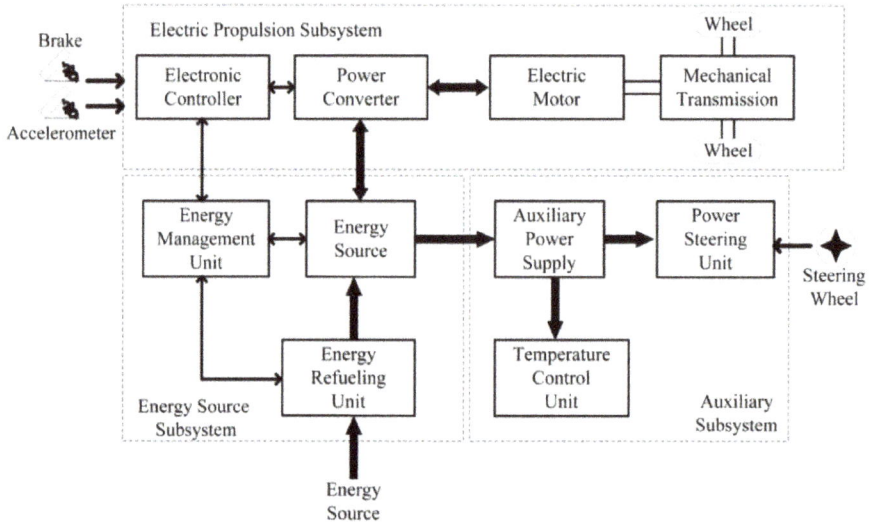

Figure 10. EV subsystems. Adapted from [4].

The arrows indicate the flow of the entities in question. A backward flow of power can be created by regenerative actions like regenerative braking. The energy source has to be receptive to store the energy sent back by regenerative actions. Most of the EV batteries along with capacitors/flywheels (CFs) are compatible with such energy regeneration techniques [4].

3.1. General EV Setup

EVs can have different configurations as shown in [4]. Figure 11a shows a front-engine front-wheel drive vehicle with just the ICE replaced by an electric motor. It has a gearbox and clutch that allows high torque at low speeds and low torque at high speeds. There is a differential as well that allows the wheels to rotate at different speeds. Figure 11b shows a configuration with the clutch omitted. It has a fixed gear in place of the gearbox which removes the chance of getting the desired torque-speed characteristics. The configuration of Figure 11c has the motor, gear and differential as a single unit that drives both the wheels. The Nissan Leaf, as well as the Chevrolet Spark, uses an electric motor mounted at the front to drive the front axle. In Figure 11d,e, configurations to obtain differential action by using two motors for the two wheels are shown. Mechanical interaction can be further reduced by placing the motors inside the wheels to produce an 'in-wheel drive'. A planetary gear system is employed here because advantages like high speed reduction ratio and inline arrangement of input and output shafts. Mechanical gear system is totally removed in the last configuration (Figure 11f) by mounting a low-speed motor with an outer rotor configuration on the wheel rim. Controlling the motor speed thus controls the wheel speed and the vehicle speed.

EVs can be built with rear wheel drive configuration as well. The single motor version of the Tesla Model S uses this configuration (Figure 12). The Nissan Blade Glider is a rear wheel drive EV with in-wheel motor arrangement. The use of in-wheel motors enables it to apply different amount of torques at each of the two rear wheels to allow better cornering.

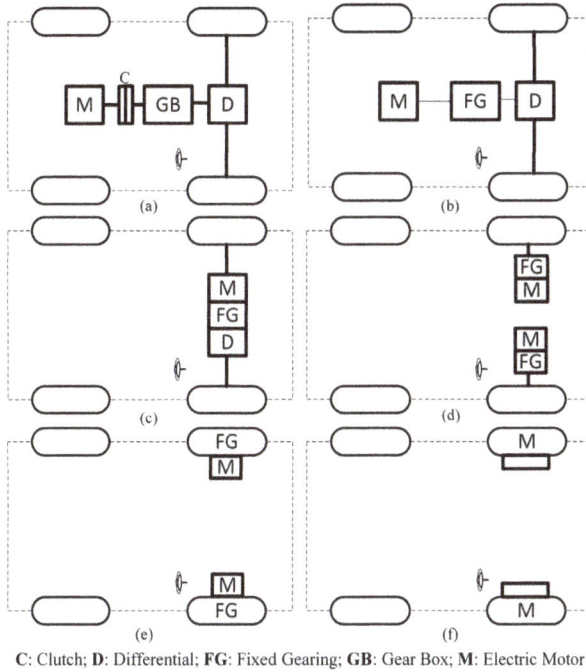

C: Clutch; **D**: Differential; **FG**: Fixed Gearing; **GB**: Gear Box; **M**: Electric Motor

Figure 11. Different front wheel drive EV configurations. (**a**) Front-wheel drive vehicle with the ICE replaced by an electric motor; (**b**) Vehicle configuration with the clutch omitted; (**c**) Configuration with motor, gear and differential combined as a single unit to drive the front wheels; (**d**) Configuration with individual motors with fixed fearing for the front wheels to obtain differential action; (**e**) Modified configuration of Figure 11d with the fixed gearing arrangement placed within the wheels; (**f**) Configuration with the mechanical gear system removed by mounting a low-speed motor on the wheel rim. Adapted from [4].

Figure 12. Tesla Model S, rear wheel drive configuration [22,24]. (Reprint with permission [24]; 2017, Tesla).

For more control and power, all-wheel drive (AWD) configurations can also be used, though it comes with added cost, weight and complexity. In this case, two motors can be used to drive the front and the rear axles. An all-wheel drive configuration is shown in Figure 13. AWD configurations are useful to provide better traction in slippery conditions, they can also use torque vectoring for better cornering performance and handling. AWD configuration can also be realized for in-wheel motor systems. It can prove quite useful for city cars like the Hiriko Fold (Figure 14) which has steering actuator, suspension, brakes and a motor all integrated in each wheel. Such arrangements can provide efficient all wheel driving, all wheel steering along with ease of parking and cornering.

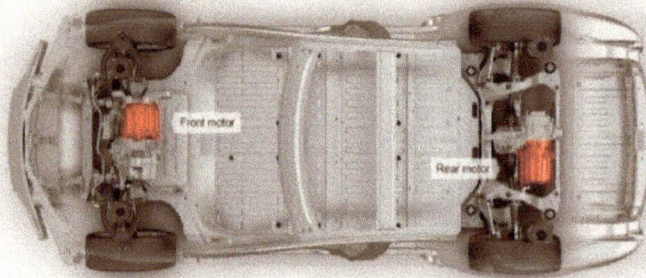

Figure 13. Tesla Model S, all-wheel drive configuration [24]. (Reprint with permission [24]; 2017, Tesla.)

Figure 14. Hiriko Fold—a vehicle employing in-wheel motors.

In-wheel motor configurations are quite convenient in the sense that they reduce the weight of the drive train by removing the central motor, related transmission, differential, universal joints and drive shaft [25]. They also provide more control, better turning capabilities and more space for batteries, fuel cells or cargo, but in this case the motor is connected to the power and control systems through wires that can get damaged because of the harsh environment, vibration and acceleration, thus causing serious trouble. Sato et al., proposed a wireless in-wheel motor system (W-IWM) in [26] which they had implemented in an experimental vehicle (shown in Figure 15). Simply put, the wires are replaced by two coils which are able to transfer power in-between them. Because of vibrations caused by road conditions, the motor and the vehicle can be misaligned and can cause variation in the secondary side voltage. In-wheel motor configurations are shown in Figure 16, whereas the efficiencies at different stages of such a system are shown in Figure 17. In conditions like this, magnetic resonance coupling is preferred for wireless power transfer [27] as it can overcome the problems associated with such misalignments [28]. The use of a hysteresis comparator and applying the secondary inverter power to a controller to counter the change in secondary voltage was also proposed in [28]. Wireless power transfer (WPT) employing magnetic resonance coupling in a series-parallel arrangement can provide a transmitting efficiency of 90% in both directions at 2 kW [29]. Therefore, W-IWM is compliant with regenerative braking as well.

Figure 15. Experimental vehicle with W-IWM system by Sato et al. [26]. (Reprint with permission [26]; 2015, IEEE.)

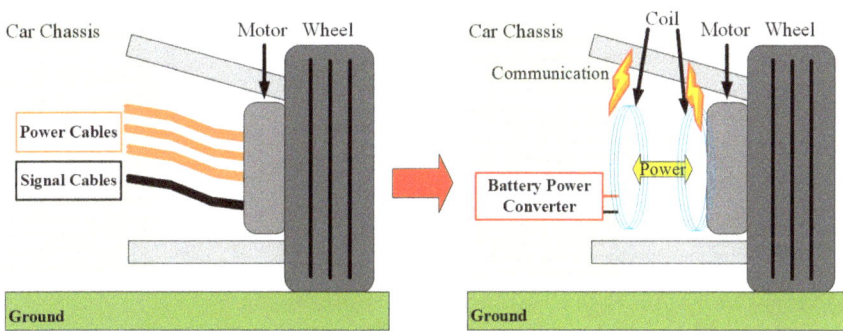

Figure 16. Conventional and wireless IWM. In the wireless setup, coils are used instead of wires to transfer power from battery to the motor. Adapted from [26].

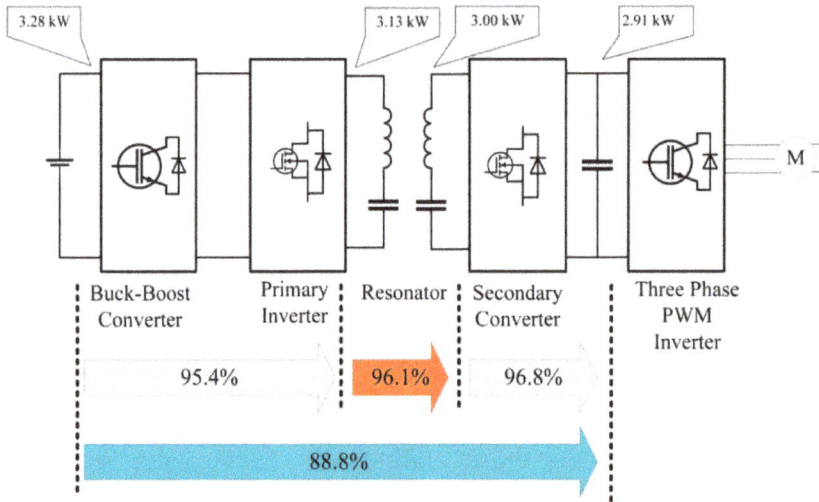

Figure 17. W-IWM setup showing efficiency at 100% torque reference. Adapted from [26].

3.2. HEV Setup

HEVs use both an electrical propulsion system and an ICE. Various ways in which these two can be set up to spin the wheels creates different configurations that can be summed up in four categories [4]:

(1) Series hybrid
(2) Parallel hybrid
(3) Series-parallel hybrid
(4) Complex hybrid

3.2.1. Series Hybrid

This configuration is the simplest one to make an HEV. Only the motor is connected to the wheels here, the engine is used to run a generator which provides the electrical power. It can be put as an EV that is assisted by an ICE generator [4]. Series hybrid drive train is shown in Figure 18. Table 2 shows the merits and demerits of this configuration.

Table 2. Advantages and limitations of series hybrid configuration. Adapted from [8].

Advantages	Efficient and optimized power-plant
	Possibilities for modular power-plant
	Optimized drive line
	Possibility of swift 'black box' service exchange
	Long lifetime
	Mature technology
	Fast response
	Capable of attaining zero emission
Limitations	Large traction drive system
	Requirement of proper algorithms
	Multiple energy conversion steps

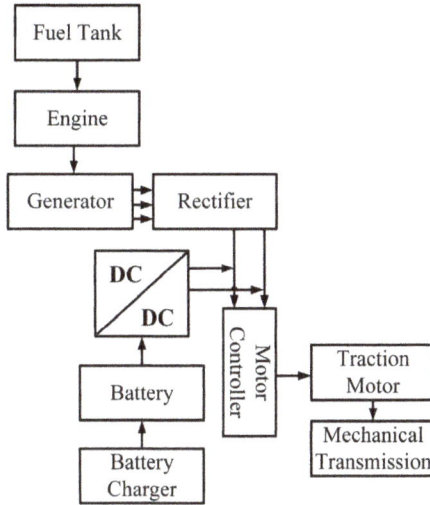

Figure 18. Drive train of series hybrid system. The engine is used to generate electricity only and supply to the motor through a rectifier. Power from the battery goes to the motor through a DC-DC converter [30].

3.2.2. Parallel Hybrid

This configuration connects both the ICE and the motor in parallel to the wheels. Either one of them or both take part in delivering the power. It can be considered as an IC engine vehicle with electric assistance [4]. The energy storages in such a vehicle can be charged by the electric motor by means of regenerative braking or by the ICE when it produces more than the power required to drive the wheels. Parallel hybrid drive train is shown in Figure 19. Table 3 shows the merits and demerits of this configuration, while Table 4 compares the series and the parallel systems.

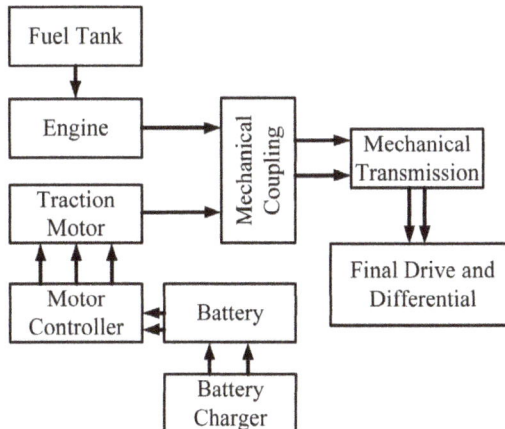

Figure 19. Drive train of parallel hybrid system. The engine and the motor both can run the can through the mechanical coupling [30].

Table 3. Advantages and limitations of parallel hybrid configuration. Adapted from [30].

Advantages	Capable of attaining zero emission Economic gain More flexibility
Limitations	Expensive Complex control Requirement of proper algorithms Need of high voltage to ensure efficiency

Table 4. Comparison of parallel and series hybrid configurations. Data from [8].

Parameters	Parallel HEV	Series HEV
Voltage	14 V, 42 V, 144 V, 300 V	216 V, 274 V, 300 V, 350 V, 550 V, 900 V
Power requirement	3 KW–40 KW	>50 KW
Relative gain in fuel economy (%)	5–40	>75

3.2.3. Series-Parallel Hybrid

In an effort to combine the series and the parallel configuration, this system acquires an additional mechanical link compared to the series type, or an extra generator when compared to the parallel type. It provides the advantages of both the systems but is more costly and complicated nonetheless. Complications in drive train are caused to some extent by the presence of a planetary gear unit [30]. Figure 20 shows a planetary gear arrangement: the sun gear is connected to the generator, the output shaft of the motor is connected to the ring gear, the ICE is coupled to the planetary carrier, and the pinion gears keep the whole system connected. A less complex alternative to this system is to use a transmotor, which is a floating-stator electric machine. In this system the engine is attached to the stator, and the rotor stays connected to the drive train wheel through the gears. The motor speed is the relative speed between the rotor and the stator and controlling it adjusts the engine speed for any particular vehicle speed [30]. Series-parallel hybrid drive train with planetary gear system is shown in Figure 21; Figure 22 shows the system with a transmotor.

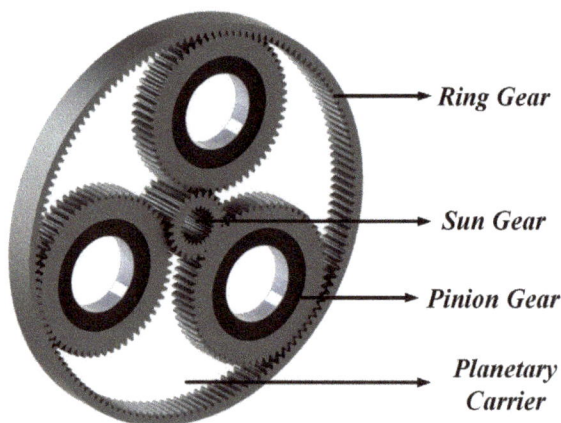

Ring Gear

Sun Gear

Pinion Gear

Planetary Carrier

Figure 20. Planetary gear system [31].

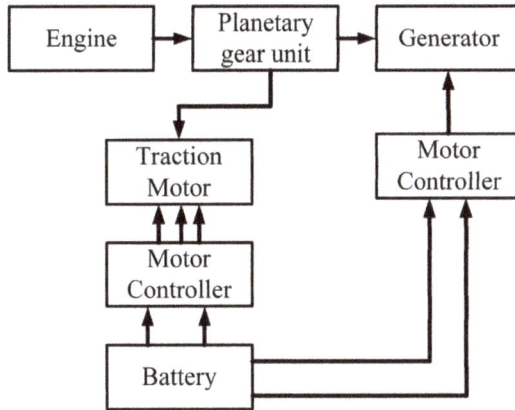

Figure 21. Drive train of series-parallel hybrid system using planetary gear unit. The planetary gear unit combines the engine, the generator and the motor [30].

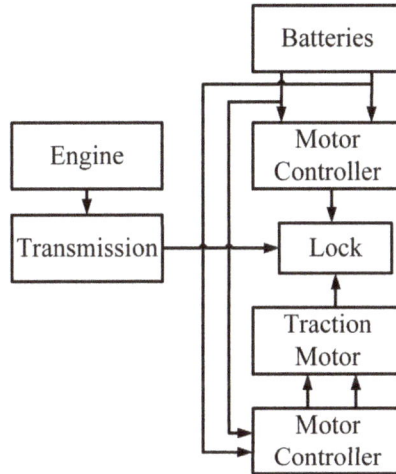

Figure 22. Drive train of series-parallel hybrid system using transmotor. The planetary gear system is absent in this arrangement [30].

3.2.4. Complex Hybrid

This system has one major difference with the series-parallel system, that is, it allows bidirectional flow of power whereas the series-parallel can provide only unidirectional power flow. However, using current market terminologies, this configuration is denoted as series-parallel system too. High complexity and cost are drawbacks of this system, but it is adopted by some vehicles to use dual-axle propulsion [4]. Constantly variable transmission (CVT) can be used for power splitting in a complex hybrid system or choosing between the power sources to drive the wheels. Electric arrangements can be used for such processes and this is dubbed as e-CVT, which has been developed and introduced by Toyota Motor Co. (Toyota City, Aichi Prefecture 471-8571, Japan). CVTs can be implemented hydraulically, mechanically, hydro-mechanically or electromechanically [32]. Two methods of power splitting—input splitting and complex splitting are shown in [32]. Input splitting got the name as it has a power split device placed at the transmission input. This system is used by certain Toyota and Ford models [32]. Reference [32] also showed different modes of these two splitting mechanisms and provided descriptions of e-CVT systems adopted by different manufacturers which are shown in Figures 23 and 24. Such power-split HEVs require two electric machines, wheels, an engine and a planetary gear (PG), combining all of them can be done in twenty-four different ways. If another PG is used, that number gets greater than one thousand. An optimal design incorporating a single PG is proposed in [31]. Four-wheel drive (4WD) configurations can benefit from using a two-motor hybrid configuration as it nullifies the need of a power transmission system to the back wheels (as they get their own motor) and provides the advantage of energy reproduction by means of regenerative braking [33]. Four-wheel drive HEV structure is shown in Figure 25. A stability enhancement scheme for such a configuration by controlling the rear motor is shown in [33].

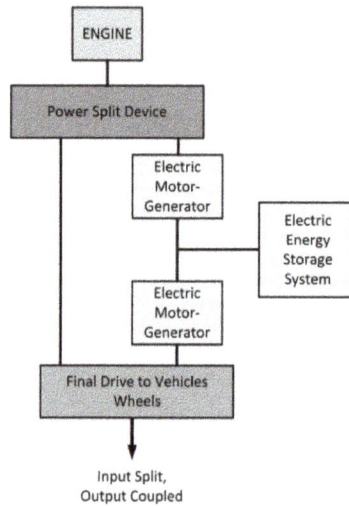

Figure 23. Input split e-CVT system. Adapted from [32].

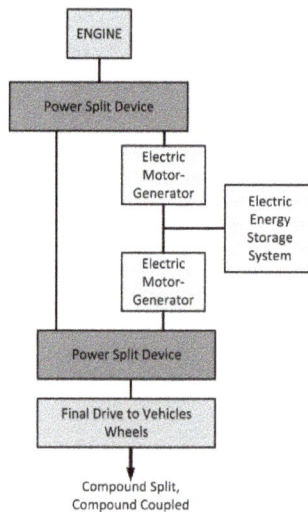

Figure 24. Compound split e-CVT system. Adapted from [32].

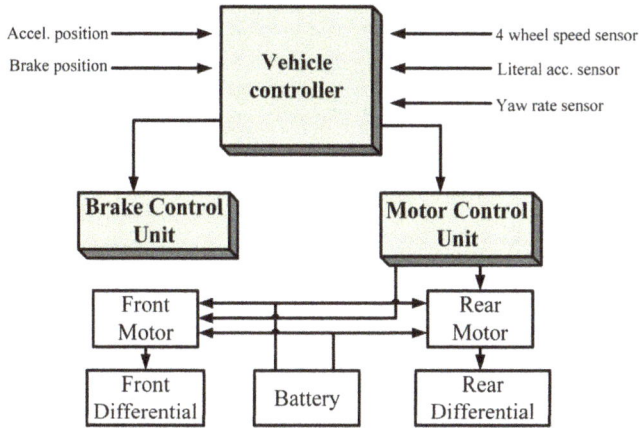

Figure 25. Structure for four-wheel drive HEV [32]. This particular system uses a vehicle controller which employs a number of sensors to perceive the driving condition and keeps the vehicle stable by controlling the brake control and the motor control units.

4. Energy Sources

EVs can get the energy required to run from different sources. The criteria such sources have to satisfy are mentioned in [4], high energy density and high power density being two of the most important ones [30]. There are other characteristics that are sought after to make a perfect energy source, fast charging, long service and cycle life, less cost and maintenance being a few of them. High specific energy is required from a source to provide a long driving range whereas high specific power helps to increase the acceleration. Because of the diverse characteristics that are required for the perfect source, quite a few sources or energy storage systems (ESS) come into discussion; they are also used in different combinations to provide desired power and energy requirements [4].

4.1. Battery

Batteries have been the major energy source for EVs for a long time; though of course, was time has gone by, different battery technologies have been invented and adopted and this process is still going on to attain the desired performance goals. Table 5 shows the desired performance for EV batteries set by the U.S. Advanced Battery Consortium (USABC).

Table 5. Performance goal of EV batteries as set by USABC. Data from [4].

	Parameters	Mid-Term	Long-Term
Primary goals	Energy density (C/3 discharge rate) (Wh/L)	135	300
	Specific energy (C/3 discharge rate) (Wh/kg)	80 (Desired: 100)	200
	Power density (W/l)	250	600
	Specific power (80% DOD/30 s) (W/kg)	150 (Desired: 200)	400
	Lifetime (year)	5	10
	Cycle life (80% DOD) (cycles)	600	1000
	Price (USD/kWh)	<150	<100
	Operating temperature (°C)	−30 to 65	−40 to 84
	Recharging time (hour)	<6	3 to 6
	Fast recharging time (40% to 80% SOC) (hour)	0.25	
Secondary goals	Self-discharge (%)	<15 (48 h)	<15 (month)
	Efficiency (C/3 discharge, 6 h charge) (%)	75	80
	Maintenance	No maintenance	No maintenance
	Resistance to abuse	Tolerance	Tolerance
	Thermal loss	3.2 W/kWh	3.2 W/kWh

Some of the prominent battery types are: lead-acid, Ni-Cd, Ni-Zn, Zn/air, Ni-MH, Na/S, Li-polymer and Li-ion batteries. Yong et al., also showed a battery made out of graphene for EV use whose advantages, structural model and application is described in [34]. Different battery types have their own pros and cons, and while selecting one, these things have to be kept in mind. In [35], Khaligh et al., provided key features of some known batteries which are demonstrated in Table 6. In Table 7, common battery types are juxtaposed to relative advantage of one battery type over the others.

Table 6. Common battery types, their basic construction components, advantages and disadvantages. Data from [35–44].

Battery Type	Components	Advantage	Disadvantage
Lead-acid	• Negative active material: spongy lead • Positive active material: lead oxide • Electrolyte: diluted sulfuric acid	• Available in production volume • Comparatively low in cost • Mature technology as used for over fifty years	• Cannot discharge more than 20% of its capacity • Has a limited life cycle if operated on a deep rate of SOC (state of charge) • Low energy and power density • Heavier • May need maintenance
NiMH (Nickel-Metal Hydride)	• Electrolyte: alkaline solution • Positive electrode: nickel hydroxide • Negative electrode: alloy of nickel, titanium, vanadium and other metals.	• Double energy density compared to lead-acid • Harmless to the environment • Recyclable • Safe operation at high voltage • Can store volumetric power and energy • Cycle life is longer • Operating temperature range is long • Resistant to over-charge and discharge	• Reduced lifetime of around 200–300 cycles if discharged rapidly on high load currents • Reduced usable power because of memory effect
Li-Ion (Lithium-Ion)	• Positive electrode: oxidized cobalt material • Negative electrode: carbon material • Electrolyte: lithium salt solution in an organic solvent	• High energy density, twice of NiMH • Good performance at high temperature • Recyclable • Low memory effect • High specific power • High specific energy • Long battery life, around 1000 cycles	• High cost • Recharging still takes quite a long time, though better than most batteries
Ni-Zn (Nickel-Zinc)	• Positive electrode: nickel oxyhydroxide • Negative electrode: zinc	• High energy density • High power density • Uses low cost material • Capable of deep cycle • Friendly to environment • Usable in a wide temperature range from $-10\,^{\circ}C$ to $50\,^{\circ}C$	• Fast growth of dendrite, preventing use in vehicles
Ni-Cd (Nickel-Cadmium)	• Positive electrode: nickel hydroxide • Negative electrode: cadmium	• Long lifetime • Can discharge fully without being damaged • Recyclable	• Cadmium can cause pollution in case of not being properly disposed of • Costly for vehicular application

Table 7. Cross comparison of different battery types to show relative advantages. Adapted from [45].

Advantages Over	Lead-Acid	Ni-Cd (Nickel-Cadmium)	NiMH (Nickel-Metal Hydride)	Li-Ion (Lithium-Ion)	
				Conventional	Polymer
Lead-acid		Volumetric energy density • Gravimetric energy density • Range of operating temperature • Rate of self-discharge reliability	Volumetric energy density • Gravimetric energy density • Rate of self-discharge	Volumetric energy density • Gravimetric energy density • Rate of self-discharge	Volumetric energy density • Gravimetric energy density • Rate of self-discharge • Design features
Ni-Cd (Nickel-Cadmium)	Output voltage • Cost • Higher cyclability		Volumetric energy density • Gravimetric energy density	Volumetric energy density • Gravimetric energy density • Rate of self-discharge • Output voltage	Volumetric energy density • Gravimetric energy density • Rate of self-discharge • Design features
NiMH (Nickel-Metal Hydride)	Output voltage • Cost • Higher cyclability	Range of operating temperature • Cost • Higher cyclability • Rate of self-discharge		Volumetric energy density • Gravimetric energy density • Range of operating temperature	Volumetric energy density • Gravimetric energy density • Range of operating temperature • Rate of self-discharge • Design features
Li-Ion (conventional)	Cost • Safety • Higher cyclability • Re-cyclability	Range of operating temperature • Cost • Safety • Higher cyclability • Recyclability	Cost • Safety • Rate of discharge • Re-cyclability		Volumetric energy density • Gravimetric energy density (potential) • Cost • Design features • Safety
Li-Ion (polymer)	Cost • Higher cyclability	Range of operating temperature • Higher cyclability • Cost	Volumetric energy density • Cost • Higher cyclability	Range of operating temperature • Higher cyclability	
Absolute advantages	Cost • Higher cyclability	Cost • Range of operating temperature	Volumetric energy density	Volumetric energy density • Gravimetric energy density • Range of operating temperature • Rate of self-discharge • Output voltage	Volumetric energy density • Gravimetric energy density • Range of operating temperature • Rate of self-discharge • Output voltage • Design features

The battery packs used in EVs are made of numerous battery cells (Figure 26). The Tesla Model S, for example, has 7104 Li-Ion cells in the 85 kWh pack. All these cells are desired to have the same SOC at all times to have the same degradation rate and same capacity over the lifetime, preventing premature end of life (EOL) [46]. A power electronic control device, called a cell voltage equalizer, can achieve this feat by taking active measures to equalize the SOC and voltage of each cell. The equalizers can be of different types according to their construction and working principle. Resistive equalizers keep all the cells at the same voltage level by burning up the extra power at cells with higher voltages. Capacitive equalizers, on the other hand, transfers energy from the higher energy cells to the lower energy ones by switching capacitors. Inductive capacitors can be of different configurations: basic, Cuk, and single of multiple transformer based; but all of them transfer energy from higher energy cells to the ones with lower energy by using inductors [46–52]. All these configurations have their own merits and demerits, which are shown in Table 8; the schematics are shown in Figures 27 and 28. Table 9 shows comparisons between the equalizer types.

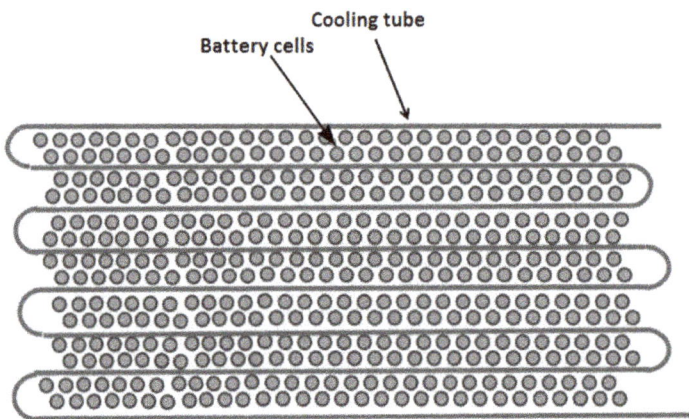

Figure 26. Battery cell arrangement in a battery pack. Cooling tubes are used to dissipate the heat generated in the battery cells.

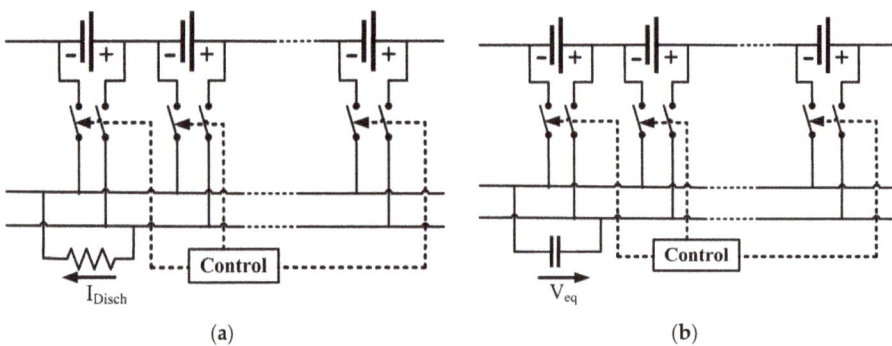

Figure 27. Equalizer configurations: (**a**) Resistive equalizer, extra power from any cell is burned up in the resistance; (**b**) Capacitive equalizer, excess energy is transferred to lower energy cells by switching of capacitors.

Table 8. Advantages and disadvantages of different equalizer types. Data from [46–52].

Equalizer Type	Advantage	Disadvantage
Resistive	• Cheapest, widely utilized for laptop batteries	• Inherent heating problem • Low equalizing current (300–500) mA • Only usable in the last stages of charging and flotation • Efficiency is almost 0% • All equalizing current transforms into heat for EV application, therefore not recommended
Capacitive	• Better current capabilities than resistive equalizers • No control issue • Simple implementation	• Unable to control inrush current • Potentially harmful current ripples can flow for big cell voltage differences • Cannot provide any required voltage difference which is essential for SOC equalization
Basic Inductive	• Relatively simple • Capable of transporting high amount of energy • Can handle complex control schemes like voltage difference control and current limitation • Can compensate for internal resistance of cells • Increased equalizing current • Not dependent on cell voltage	• Requires additional components to prevent ripple currents • Needs two switches in addition to drivers and controls in each cell • Current distribution is highly concentrated in neighboring cells because of switching loss
Cuk Inductive	• Has all the advantages of inductive equalizres • Can accommodate complex control and withstand high current	• Additional cost of higher voltage and current rated switches, power capacitors • Subjected to loss caused by series capacitor • A little less efficient than typical inductive equalizers • Faces problems during distributing equalizing currents all over the cell string • May need additional processing power
Transformer based Inductive	• Theoretically permits proper current distribution in all cells without addition control or loss	• Complex transformer with multiple secondary, which is very much challenging to mass produce • Not an option for EV packs • Cannot handle complex control algorithms
Multiple transformer based Inductive	• Separate transformers are used which are easier for mass production	• Still difficult to build with commercial inductors without facing voltage and current imbalance

Table 9. Comparison of equalizers; a ↑ sign indicates an advantage whereas the ↓ signs indicate drawbacks. Adapted from [46].

Equalizer Type	Equalizer Current	Current Distribution	Current Control	Current Ripple	Manufacture	Cost	Control
Resistive	↓↓	N/A	↑	↑↑↑	↑↑↑	↑↑↑	↑↑↑
Capacitive	↓	↑	↓↓	↓↓	↑↑	↑↑	↑↑
Basic Inductive	↑↑	↑	↑	↑↑	↑	↓	↓
Cuk	↑↑	↑	↑	↑↑↑	↓	↓↓	↓
Transformer	↑	↑↑↑	↓↓	↓↓	↓↓	↓↓	↑↑

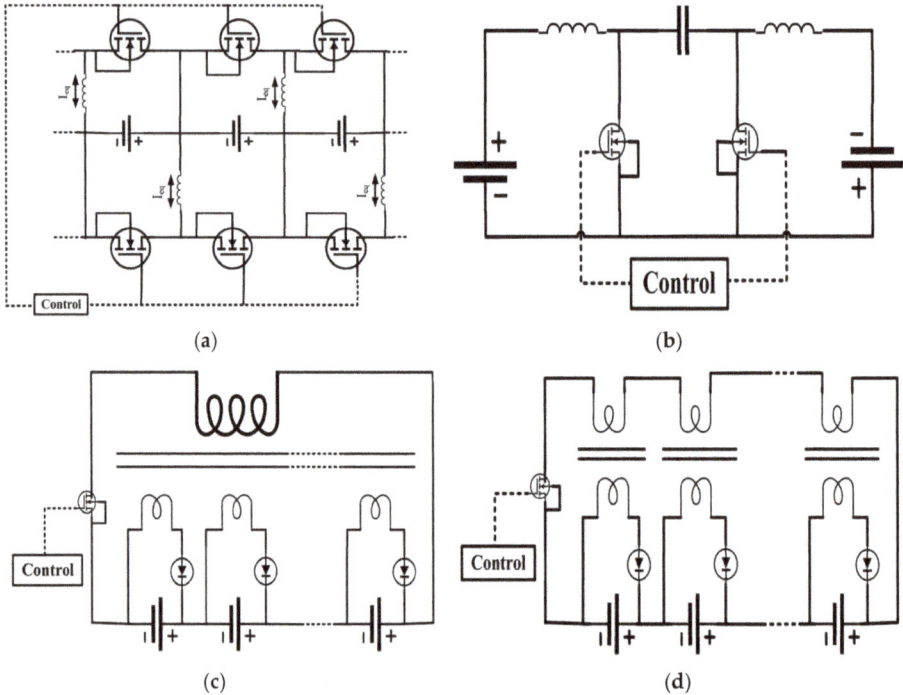

Figure 28. Inductive equalizer configurations: (**a**) Basic; (**b**) Cuk; (**c**) Transformer based; (**d**) Multiple transformers based. Excess energy is transferred to lower energy cells by using inductors.

Lithium-ion batteries are being used everywhere these days. It has replaced the lead-acid counterpart and became a mature technology itself. Their popularity can be justified by the fact that best-selling EVs, for example, Nissan Leaf and Tesla Model S—all use these batteries [53,54]. Battery parameters of some current EVs are shown in Table 10. Lithium batteries also have lots of scope to improve [55]. Better battery technologies have been discovered already, but they are not being pursued because of the exorbitant costs associated with their research and development, so it can be said that, lithium batteries will dominate the EV scene for quite some time to come.

Table 10. Battery parameters of some current EVs. Data from [5].

Model	Total Energy (kWh)	Usable Energy (kWh)	Usable Energy (%)
i3	22	18.8	85
C30	24	22.7	95
B-Class	36	28	78
e6	61.4	57	93
RAV4	41.8	35	84

4.2. Ultracapacitors (UCs)

UCs have two electrodes separated by an ion-enriched liquid dielectric. When a potential is applied, the positive electrode attracts the negative ions and the negative electrode gathers the positive ones. The charges get stored physically stored on electrodes this way and provide a considerably high power density. As no chemical reactions take place on the electrodes, ultra- capacitors tend to have a long cycle life; but the absence of any chemical reaction also makes them low in energy density [35].

The internal resistance is low too, making it highly efficient, but it also causes high output current if charged at a state of extremely low SOC [56,57]. A UC's terminal voltage is directly proportional to its SOC; so it can also operate all through its voltage range [35]. Basic construction of an UC cell is shown in Figure 29. EVs go through start/stop conditions quite a lot, especially in urban driving situations. This makes the battery discharge rate highly changeable. The average power required from batteries is low, but during acceleration or conditions like hill-climb a high power is required in a short duration of time [4,35]. The peak power required in a high-performance electric vehicle can be up to sixteen times the average power [4]. UCs fit in perfectly in such a scenario as it can provide high power for short durations. It is also fast in capturing the energy generated by regenerative braking [2,35]. A combined battery-UC system (as shown in Figure 30) negates each other's shortcomings and provides an efficient and reliable energy system. The low cost, load leveling capability, temperature adaptability and long service life of UCs make them a likable option as well [4,30].

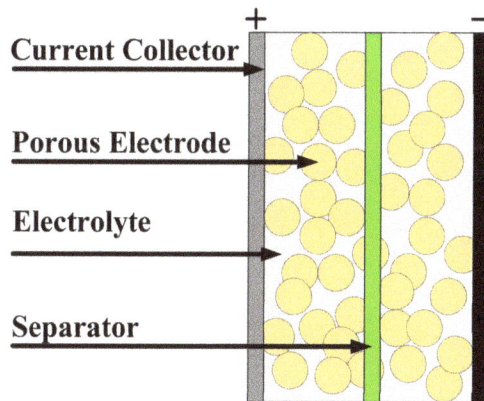

Figure 29. An UC cell; a separator keeps the two electrodes apart [58].

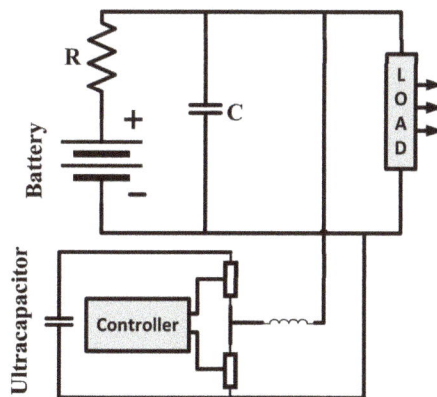

Figure 30. Combination of battery and UC to complement each-other's shortcomings [59].

4.3. Fuel Cell (FC)

Fuel cells generate electricity by electrochemical reaction. An FC has an anode (A), a cathode (C) and an electrolyte (E) between them. Fuel is introduced to the anode, gets oxidized there, the ions created travel through the electrolyte to the cathode and combine with the other reactant introduced there. The electrons produced by oxidation at the anode produce the electricity. Hydrogen is used in FCEVs because of its high energy content, and the facts it is non-polluting (producing only water as exhaust) and abundant in Nature in the form of different compounds such as hydrocarbons [4]. Hydrogen can be stored in different methods for use in EVs [4]; commercially available FCVs like the Toyota Mirai use cylinders to store it. The operating principle of a general fuel cell is demonstrated in Figure 31, while Figure 32 shows a hydrogen fuel cell. According to the material used, fuel cells can be classified into different types. A comparison among them is shown in Table 11. The chemical reaction governing the working of a fuel cell is stated below:

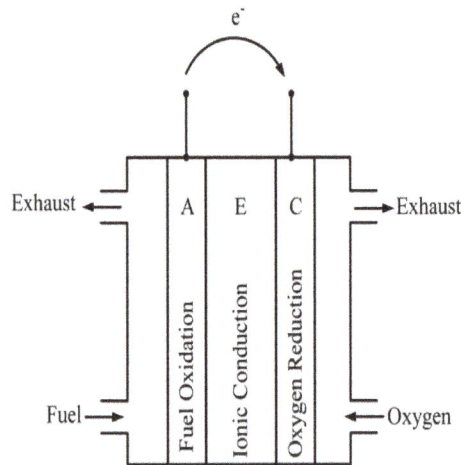

$$2H_2 + O_2 = 2H_2O \tag{1}$$

Figure 31. Working principle of fuel cell. Fuel and oxygen is taken in, exhaust and current is generated as the products of chemical reaction. Adapted from [4].

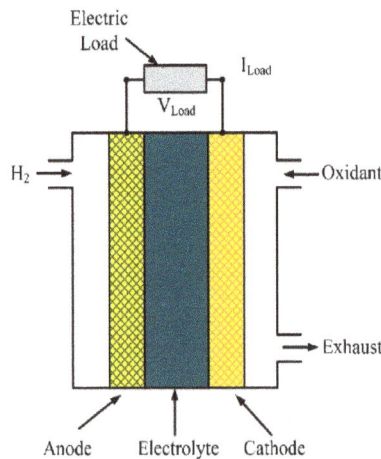

Figure 32. Hydrogen fuel cell configuration. Hydrogen is used as the fuel which reacts with oxygen and produces water and current as products. Adapted from [35].

Table 11. Comparison of different fuel cell configurations. Data from [2].

	PAFC	AFC	MCFC	SOFC	SPFC	DMFC
Working temp. (°C)	150–210	60–100	600–700	900–1000	50–100	50–100
Power density (W/cm^2)	0.2–0.25	0.2–0.3	0.1–0.2	0.24–0.3	0.35–0.6	0.04–0.25
Estimated life (kh)	40	10	40	40	40	10
Estimated cost (USD/kW)	1000	200	1000	1500	200	200

PAFC: Phosphoric acid fuel cell; AFC: Alkaline fuel cell; SOFC: Solid oxide fuel cell; SPFC: Solid polymer fuel cell, also known as proton exchange membrane fuel cell.

Fuel cells have many advantages for EV use like efficient production of electricity from fuel, noiseless operation, fast refueling, no or low emissions, durability and the ability to provide high density current output [24,60]. A main drawback of this technology is the high price. Hydrogen also have lower energy density compared to petroleum derived fuel, therefore larger fuel tanks are required for FCEVs, these tanks also have to capable enough to contain the hydrogen properly and to minimize risk of any explosion in case of an accident. FC's efficiency depends on the power it is supplying; efficiency generally decreases if more power is drawn. Voltage drop in internal resistances cause most of the losses. Response time of FCs is comparatively higher to UCs or batteries [35]. Because of these reasons, storage like batteries or UCs is used alongside FCs. The Toyota Mirai uses batteries to power its motor and the FC is used to charge the batteries. The batteries receive the power reproduced by regenerative braking as well. This combination provides more flexibility as the batteries do not need to be charged, only the fuel for the FC has to be replenished and it takes far less time than recharging the batteries.

4.4. Flywheel

Flywheels are used as energy storage by using the energy to spin the flywheel which keeps on spinning because of inertia. The flywheel acts as a motor during the storage stage. When the energy is needed to be recovered, the flywheel's kinetic energy can be used to rotate a generator to produce power. Advanced flywheels can have their rotors made out of sophisticated materials like carbon composites and are placed in a vacuum chamber suspended by magnetic bearings. Figure 33 shows a flywheel used in the Formula One (F1) racing kinetic energy recovery system (KERS). The major components of a flywheel are demonstrated in Figure 34. Flywheels offer a lot of advantages over other storage forms for EV use as they are lighter, faster and more efficient at absorbing power from regenerative braking, faster at supplying a huge amount of power in a short time when rapid acceleration is needed and can go through a lot of charge-discharge cycles over their lifetime. They are especially favored for hybrid racecars which go through a lot of abrupt braking and acceleration, which are also at much higher g-force than normal commuter cars. Storage systems like batteries or UCs cannot capture the energy generated by regenerative braking in situations like this properly. Flywheels, on the other hand, because of their fast response, have a better efficiency in similar scenarios, by making use of regenerative braking more effectively; it reduces pressure on the brake pads as well. The Porsche 911GT3R hybrid made use of this technology. Flywheels can be made with different materials, each with their own merits and demerits. Characteristics of some these materials are shown in Table 12; among the ones displayed in the table, carbon T1000 offers the highest amount of energy density, but it is much costlier than the others. Therefore, there remains a trade-off between cost and performance.

Figure 33. A flywheel used in the Formula One racing kinetic energy recovery system (KERS).

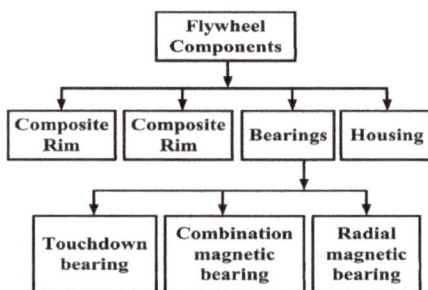

Figure 34. Basic flywheel components. The flywheel is suspended in tis hosing by bearings, and is connected to a motor-generator to store and supply energy [61].

Table 12. Characteristics of different materials used for flywheels [62].

Material		Density (kg/m^3)	Tensile Strength (mpa)	Max Energy Density (mj/kg)	Cost (USD/kg)
Monolithic material	4340 steel	7700	1520	0.19	1
Composites	E-glass	2000	100	0.05	11
	S2-glass	1920	1470	0.76	24.6
	Carbon T1000	1520	1950	1.28	101.8
	Carbon AS4C	1510	1650	1.1	31.3

Currently, no single energy source can provide the ideal characteristics, i.e., high value of both power and energy density. Table 13 shows a relative comparison of the energy storages to demonstrate this fact. Hybrid energy storages can be used to counter this problem by employing one source for high energy density and another for high power density. Different combinations are possible to create this hybrid system. It can be a combination of battery and ultracapacitor, battery and flywheel, or fuel cell and battery [4]. Table 14 shows the storage systems used by some current vehicles.

Table 13. Relative energy and power densities of different energy storage systems [63].

Storage	Energy Density	Power Density
Battery	High	Low
Ultracapacitor	Low	High
Fuel cell	High	Low
Flywheel	Low	High

Table 14. Vehicles using different storage systems.

Storage System	Vehicles Using the System
Battery	Tesla Model S, Nissan Leaf
Fuel cell + battery	Toyota Mirai, Honda Clarity
Flywheel	Porsche 911GT3R Hybrid

5. Motors Used

The propulsion system is the heart of an EV [64–69], and the electric motor sits right in the core of the system. The motor converts electrical energy that it gets from the battery into mechanical energy which enables the vehicle to move. It also acts as a generator during regenerative action which sends energy back to the energy source. Based on their requirement, EVs can have different numbers of motors: the Toyota Prius has one, the Acura NSX has three—the choice depends on the type of the vehicle and the functions it is supposed to provide. References [4,23] listed the requirements for a motor for EV use which includes high power, high torque, wide speed range, high efficiency, reliability, robustness, reasonable cost, low noise and small size. Direct current (DC) motor drives demonstrate some required properties needed for EV application, but their lack in efficiency, bulky structure, lack in reliability because of the commutator or brushes present in them and associated maintenance requirement made them less attractive [4,30]. With the advance of power electronics and control systems, different motor types emerged to meet the needs of the automotive sector, induction and permanent magnet (PM) types being the most favored ones [23,30,70].

5.1. Brushed DC Motor

These motors have permanent magnets (PM) to make the stator; rotors have brushes to provide supply to the stator. Advantages of these motors can be the ability to provide maximum torque in low speed. The disadvantages, on the other hand, are its bulky structure, low efficiency, heat generated because of the brushes and associated drop in efficiency. The heat is also difficult to remove as it is generated in the center of the rotor. Because of these reasons, brushed DC motors are not used in EVs any more [70].

5.2. Permanent Magnet Brushless DC Motor (BLDC)

The rotor of this motor is made of PM (most commonly NdFeB [4]), the stator is provided an alternating current (AC) supply from a DC source through an inverter. As there are no windings in the rotor, there is no rotor copper loss, which makes it more efficient than induction motors. This motor is also lighter, smaller, better at dissipating heat (as it is generated in the stator), more reliable, has more torque density and specific power [4]. But because of its restrained field-weakening ability, the constant power range is quite short. The torque also decreases with increased speed because of back EMF generated in the stator windings. The use of PM increases the cost as well [30,70]. However, enhancement of speed range and better overall efficiency is possible with additional field windings [4,71]. Such arrangements are often dubbed PM hybrid motors because of the presence of both PM and field windings. But such arrangements too are restrained by complexity of structure; the speed ratio is not enough to meet the needs of EV use, specifically in off-roaders [30]. PM hybrid motors can also be constructed using a combination of reluctance motor and PM motor. Controlling the conduction angle of the power converter can improve the efficiency of PM BLDCs as well as speed range, reaching as high as four times the base speed, though the efficiency may decrease at very high speed resulting from demagnetization of PM [4]. Other than the PM hybrid configurations, PM BLDCs can be buried magnet mounted—which can provide more air gap flux density, or surface magnet mounted—which require less amount of magnet. BLDCs are useful for use in small cars requiring a maximum 60 kW of power [72]. The characteristics of PM BLDCs are shown in Figure 35.

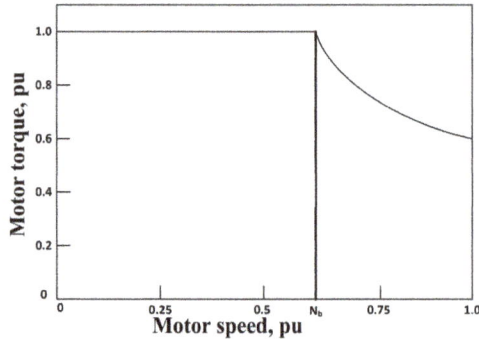

Figure 35. Characteristics of a Permanent Magnet Brushless DC Motor. The torque remains constant at the maximum right from the start, but starts to decrease exponentially for speeds over the base speed.

5.3. Permanent Magnet Synchronous Motor (PMSM)

These machines are one of the most advanced ones, capable of being operated at a range of speeds without the need of any gear system. This feature makes these motors more efficient and compact. This configuration is also very suitable for in-wheel applications, as it is capable of providing high torque, even at very low speeds. PMSMs with an outer rotor are also possible to construct without the need of bearings for the rotor. But these machines' only notable disadvantage also comes in during in-wheel operations where a huge iron loss is faced at high speeds, making the system unstable [73]. NdFeB PMs are used for PMSMs for high energy density. The flux linkages in the air-gap are sinusoidal in nature; therefore, these motors are controllable by sinusoidal voltage supplies and vector control [70]. PMSM is the most used motor in the BEVs available currently; at least 26 vehicle models use this motor technology [5].

5.4. Induction Motor (IM)

Induction motors are used in early EVs like the GM EV1 [23] as well as current models like the Teslas [54,74]. Among the different commutatorless motor drive systems, this is the most mature one [2]. Vector control is useful to make IM drives capable of meeting the needs of EV systems. Such a system with the ability to minimize loss at any load condition is demonstrated in [75]. Field orientation control can make an IM act like a separately excited DC motor by decoupling its field control and torque control. Flux weakening can extend the speed range over the base speed while keeping the power constant [30], field orientation control can achieve a range three to five times the base speed with an IM that is properly designed [76]. Three phase, four pole AC motors with copper rotors are seen to be employed in current EVs. Characteristics of IM are shown in Figure 36.

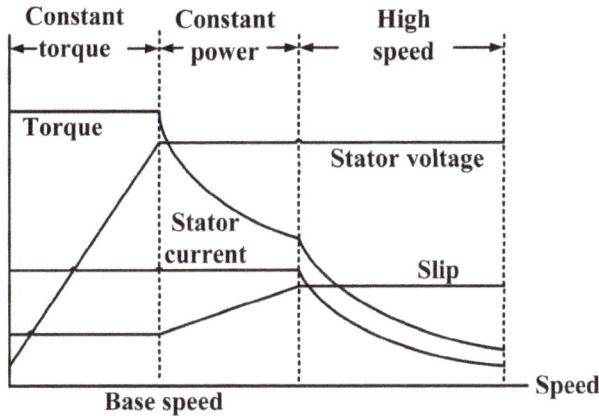

Figure 36. Induction motor drive characteristics. Maximum torque is maintained till base speed, and then decreases exponentially. Adapted from [4].

5.5. Switched Reluctance Motor (SRM)

SRMs, also known as doubly salient motor (because of having salient poles both in the stator and the rotor) are synchronous motors driven by unipolar inverter-generated current. They demonstrate simple and robust mechanical construction, low cost, high-speed, less chance of hazards, inherent long constant power range and high power density useful for EV applications. PM is not required for such motors and that facilitates enhanced reliability along with fault tolerance. On the downside, they are very noisy because of the variable torque nature, have low efficiency, and are larger in size and weight when compared to PM machines. Though such machines have a simple construction, their design and control are not easy resulting from fringe effect of slots and poles and high saturation of the pole-tips [4,23,30,70]. Because of such drawbacks, these machines did not advance as much as the PM or induction machines. However, because of the high cost rare-rare earth materials needed in PM machines, interest in SRMs are increasing. Advanced SRMs like the one demonstrated by Nidec in 2012 had almost interior permanent machine (IPM)-like performance, with a low cost. Reducing the noise and torque ripple are the main concerns in researches associated with SRMs [23]. One of the configurations that came out of these researches uses a dual stator system, which provides low inertia and noise, superior torque density and increased speed-range compared to conventional SRMs [77,78]. Design by finite element analysis can be employed to reduce the total loss [79], control by fuzzy sliding mode can also be employed to reduce control chattering and motor nonlinearity management [80].

5.6. Synchronous Reluctance Motor (SynRM)

A Synchronous Reluctance Motor runs at a synchronous speed while combining the advantages of both PM and induction motors. They are robust and fault tolerant like an IM, efficient and small like a PM motor, and do not have the drawbacks of PM systems. They have a control strategy similar to that of PM motors. The problems with SynRM can be pointed as the ones associated with controllability, manufacturing and low power factor which hinder its use in EVs. However, researches have been going on and some progress is made as well, the main area of concern being the rotor design. One way to improve this motor is by increasing the saliency which provides a higher power factor. It can be achieved by axially or transversally laminated rotor structures, such an arrangement is shown in Figure 37. Improved design techniques, control systems and advanced manufacturing can help it make its way into EV applications [23].

Figure 37. SynRM with axially laminated rotor [23].

5.7. PM Assisted Synchronous Reluctance Motor

Greater power factors can be achieved from SynRMs by integrating some PMs in the rotor, creating a PM assisted Synchronous Reluctance Motor. Though it is similar to an IPM, the PMs used are fewer in amount and the flux linkages from them are less too. PMs added in the right amount to the core of the rotor increase the efficiency with negligible back EMF and little change to the stator. This concept is free from the problems associated with demagnetization resulting from overloading and high temperature observed in IPMs. With a proper efficiency optimization technique, this motor can have the performance similar to IPM motors. A PM-assisted SynRM suitable for EV use was demonstrated by BRUSA Elektronik AG (Sennwald, Switzerland). Like the SynRM, PM-assisted SynRMs can also get better with improved design techniques, control systems and advanced manufacturing systems [23]. A demonstration of the rotor of PM-assisted SynRM is shown in Figure 38.

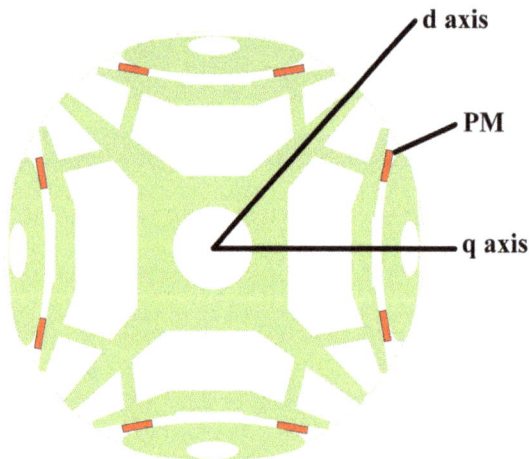

Figure 38. Permanent magnet (PM) assisted SynRM. Permanent magnets are embedded in the rotor [23].

5.8. Axial Flux Ironless Permanent Magnet Motor

According to [70], this motor is the most advanced one to be used in EVs. It has an outer rotor with no slot; use of iron is avoided here as well. The stator core is absent too, reducing the weight of the machine. The air gap here is radial field type, providing better power density. This motor is a variable speed one too. One noteworthy advantage of this machine is that the rotors can be fitted on lateral sides of wheels, placing the stator windings on the axle centrally. The slot-less design also improves the efficiency by minimizing copper loss as there is more space available [70].

Power comparison of three different motor types is conducted in Table 15. Table 16 compares torque densities of three motors. Table 17 summarizes the advantages and disadvantages of different motor types, and shows some vehicles using different motor technologies.

Table 15. Power comparison of different motors having the same size. Data from [72].

Motor Type	Power (kW)		Base Speed	Maximum Speed
	HEV	BEV		
IM	57	93	3000	12,000
SRM	42	77	2000	12,000
BLDC	75	110	4000	9000

Table 16. Typical torque density values of some motors. Data from [30].

Motor Type	Torque/Volume (Nm/m^3)	Torque/Cu Mass (Nm/kg Cu)
PM motor	28,860	28.7–48
IM	4170	6.6
SRM	6780	6.1

Table 17. Advantages, disadvantages and usage of different motor types.

Motor Type	Advantage	Disadvantage	Vehicles Used In
Brushed DC Motor	• Maximum torque at low speed	• Bulky structure • Low efficiency • Heat generation at brushes	Fiat Panda Elettra (Series DC motor), Conceptor G-Van (Separately excited DC motor)
Permanent Magnet Brushless DC Motor (BLDC)	• No rotor copper loss • More efficiency than induction motors • Lighter • Smaller • Better heat dissipation • More reliability • More torque density • More specific power	• Short constant power range • Decreased torque with increase in speed • High cost because of PM	Toyota Prius (2005)
Permanent Magnet Synchronous Motor (PMSM)	• Operable in different speed ranges without using gear systems • Efficient • Compact • Suitable for in-wheel application • High torque even at very low speeds	• Huge iron loss at high speeds during in-wheel operation	Toyota Prius, Nissan Leaf, Soul EV
Induction Motor (IM)	• The most mature commutatorless motor drive system • Can be operated like a separately excited DC motor by employing field orientation control		Tesla Model S, Tesla Model X, Toyota RAV4, GM EV1

Table 17. *Cont.*

Motor Type	Advantage	Disadvantage	Vehicles Used In
Switched Reluctance Motor (SRM)	• Simple and robust construction • Low cost • High speed • Less chance of hazard • Long constant power range • High power density	• Very noisy • Low efficiency • Larger and heavier than PM machines • Complex design and control	Chloride Lucas
Synchronous Reluctance Motor (SynRM)	• Robust • Fault tolerant • Efficient • Small	• Problems in controllability and manufacturing • Low power factor	
PM assisted Synchronous Reluctance Motor	• Greater power factor than SynRMs • Free from demagnetizing problems observed in IPM		BMW i3
Axial Flux Ironless Permanent Magnet Motor	• No iron used in outer rotor • No stator core • Lightweight • Better power density • Minimized copper loss • Better efficiency • Variable speed machine • Rotor is capable of being fitted to the lateral side of the wheel		Renovo Coupe

6. Charging Systems

For charging of EVs, DC or AC systems can be used. There are different current and voltage configurations for charging, generally denoted as 'levels'. The time required for a full charge depends on the level being employed. Wireless charging has also been tested and researched for quite a long time. It has different configurations as well. The charging standards are shown in Table 18. The safety standards that should be complied by the chargers are the following [46]:

- SAE J2929: Electric and Hybrid Vehicle Propulsion Battery System Safety Standard
- ISO 26262: Road Vehicles—Functional safety
- ISO 6469-3: Electric Road Vehicles—Safety Specifications—Part 3: Protection of Persons Against Electric Hazards
- ECE R100: Protection against Electric Shock
- IEC 61000: Electromagnetic Compatibility (EMC)
- IEC 61851-21: Electric Vehicle Conductive Charging system—Part 21: Electric Vehicle Requirements for Conductive Connection to an AC/DC Supply
- IEC 60950: Safety of Information Technology Equipment
- UL 2202: Electric Vehicle (EV) Charging System Equipment
- FCC Part 15 Class B: The Federal Code of Regulation (CFR) FCC Part 15 for EMC Emission Measurement Services for Information Technology Equipment.
- IP6K9K, IP6K7 protection class
- $-40\,°C$ to $105\,°C$ ambient air temperature

6.1. AC Charging

AC charging system provides an AC supply that is converted into DC to charge the batteries. This system needs an AC-DC converter. According to the SAE EV AC Charging Power Levels, they can be classified as below:

- Level 1: The maximum voltage is 120 V, the current can be 12 A or 16 A depending on the circuit ratings. This system can be used with standard 110 V household outlets without requiring any special arrangement, using on-board chargers. Charging a small EV with this arrangement can take 0.5–12.5 h. These characteristics make this system suitable for overnight charging [5,46,81].
- Level 2: Level 2 charging uses a direct connection to the grid through an Electric Vehicle Service Equipment (EVSE). On-board charger is used for this system. Maximum system ratings are 240 V, 60 A and 14.4 kW. This system is used as a primary charging method for EVs [46,81].
- Level 3: This system uses a permanently wired supply dedicated for EV charging, with power ratings greater than 14.4 kW. 'Fast chargers'—which recharge an average EV battery pack in no more than 30 min, can be considered level 3 chargers. All level 3 chargers are not fast chargers though [46,82]. Table 19 shows the AC charging characteristics defined by Society of Automotive Engineers (SAE).

Table 18. Charging standards. Data from [81].

Standard		Scope
IEC 61851: Conductive charging system	IEC 61851-1	Defines plugs and cables setup
	IEC 61851-23	Explains electrical safety, grid connection, harmonics, and communication architecture for DCFC station (DCFCS)
	IEC 61851-24	Describes digital communication for controlling DC charging
IEC 62196: Socket outlets, plugs, vehicle inlets and connectors	IEC 62196-1	Defines general requirements of EV connectors
	IEC 62196-2	Explains coupler classifications for different modes of charging
	IEC 62196-3	Describes inlets and connectors for DCFCS
IEC 60309: Socket outlets, plugs, and couplers	IEC 60309-1	Describes CS general requirements
	IEC 60309-2	Explains sockets and plugs sizes having different number of pins determined by current supply and number of phases, defines connector color codes according to voltage range and frequency.
IEC 60364		Explains electrical installations for buildings
SAE J1772: Conductive charging systems		Defines AC charging connectors and new Combo connector for DCFCS
SAE J2847: Communication	SAE J2847-1	Explains communication medium and criteria for connecting EV to utility for AC level 1&2 charging
	SAE J2847-2	Defines messages for DC charging
SAE J2293	SAE J2293-1	Explains total EV energy transfer system, defines requirements for EVSE for different system architectures
SAE J2344		Defines EV safety guidelines
SAE J2954: Inductive charging		Being developed

Table 19. SAE (Society of Automotive Engineers) AC charging characteristics. Data from [44,80].

AC Charging System	Supply Voltage (V)	Maximum Current (A)	Branch Circuit Breaker Rating (A)	Output Power Level (kW)
Level 1	120 V, 1-phase	12	15	1.08
	120 V, 1-phase	16	20	1.44
Level 2	208 to 240 V, 1-phase	16	20	3.3
	208 to 240 V, 1-phase	32	40	6.6
	208 to 240 V, 1-phase	≤80	Per NEC 635	≤14.4
Level 3	208/480/600 V	150–400	150	3

6.2. DC Charging

DC systems require dedicated wiring and installations and can be mounted at garages or charging stations. They have more power than the AC systems and can charge EVs faster. As the output is DC, the voltage has to be changed for different vehicles to suit the battery packs. Modern stations have the capability to do it automatically [46]. All DC charging systems has a permanently connected Electric Vehicle Service Equipment (EVSE) that incorporates the charger. Their classification is done depending on the power levels they supply to the battery:

- Level 1: The rated voltage is 450 V with 80 A of current. The system is capable of providing power up to 36 kW.
- Level 2: It has the same voltage rating as the level 1 system; the current rating is increased to 200 A and the power to 90 kW.
- Level 3: Voltage in this system is rated to 600 V. Maximum current is 400 A with a power rating of 240 kW. Table 20 shows the DC charging characteristics defined by Society of Automotive Engineers (SAE).

Table 20. SAE (Society of Automotive Engineers) DC charging characteristics. Data from [46].

DC Charging System	DC Voltage Range (V)	Maximum Current (A)	Power (kW)
Level 1	200–450	≤80	≤36
Level 2	200–450	≤200	≤90
Level 3	200–600	≤400	≤240

6.3. Wireless Charging

Wireless charging or wireless power transfer (WPT) enjoys significant interest because of the conveniences it offers. This system does not require the plugs and cables required in wired charging systems, there is no need of attaching the cable to the car, low risk of sparks and shocks in dirty or wet environment and less chance of vandalism. Forerunners in WPT research include R&D centers and government organizations like Phillips Research Europe, Energy Dynamic Laboratory (EDL), US DOT, DOE; universities including the University of Tennessee, the University of British Columbia, Korea Advance Institute of Science and Technology (KAIST); automobile manufacturers including Daimler, Toyota, BMW, GM and Chrysler. The suppliers of such technology include Witricity, LG, Evatran, HaloIPT (owned by Qualcomm), Momentum Dynamics and Conductix-Wampfler [27]. However, this technology is not currently available for commercial EVs because of the health and safety concerns associated with the current technology. The specifications are determined by different standardization organizations in different countries: Canadian Safety Code 6 in Canada [83], IEEE C95.1 in the USA [84], ICNIRP in Europe [85] and ARPANSA in Australia [86]. There are different technologies that are being considered to provide WPT facilities. They differ in the operating frequency, efficiency, associated electromagnetic interference (EMI), and other factors.

Inductive power transfer (IPT) is a mature technology, but it is only contactless, not wireless. Capacitive power transfer (CPT) has significant advantage at lower power levels because of low cost and size, but not suitable for higher power applications like EV charging. Permanent magnet coupling power transfer (PMPT) is low in efficiency, other factors are not favorable as well. Resonant inductive power transfer (RIPT) as well as On-line inductive power transfer (OLPT) appears to be the most promising ones, but their infrastructuret may not allow them to be a viable solution. Resonant antennae power transfer (RAPT) is made on a similar concept as RIPT, but the resonant frequency in this case is in MHz range, which is capable of damage to humans if not shielded properly. The shielding is likely to hinder range and performance; generation of such high frequencies is also a challenge for power electronics [87]. Table 21 compares different wireless charging systems in terms of performance, cost, size, complexity, and power level. Wireless charging for personal vehicles is unlikely to be available soon because of health, fire and safety hazards, misalignment problems and range. Roads with WPT systems embedded into them for charging passing vehicles also face major cost issues [27]. Only a few wireless systems are available now, and those too are in trial stage. WiTricity is working with Delphi Electronics, Toyota, Honda and Mitsubishi Motors. Evatran is collaborating with Nissan and GM for providing wireless facilities for Nissan Leaf and Chevrolet Volt models. However, with significant advance in the technology, wireless charging is likely to be integrated in the EV scenario, the conveniences it offers are too appealing to overlook.

Table 21. Comparison of wireless charging systems.

Wireless Charging System	Performance			Cost	Volume/Size	Complexity	Power Level
	Efficiency	EMI	Frequency				
Inductive power transfer (IPT)	Medium	Medium	10–50 kHz	Medium	Medium	Medium	Medium/High
Capacitive power transfer (CPT)	Low	Medium	100–500 kHz	Low	Low	Medium	Low
Permanent magnet coupling power transfer (PMPT)	Low	High	100–500 kHz	High	High	High	Medium/Low
Resonant inductive power transfer (RIPT)	Medium	Low	1–20 MHz	Medium	Medium	Medium	Medium/Low
On-line inductive power transfer (OLPT)	Medium	Medium	10–50 kHz	High	High	Medium	High
Resonant antennae power transfer (RAPT)	Medium	Medium	100–500 kHz	Medium	Medium	Medium	Medium/Low

For the current EV systems, on-board AC systems are used for the lowest power levels, for higher power, DC systems are used. DC systems currently have three existing standards [16]:

- Combined Charging System (CCS)
- CHAdeMO (CHArge de MOve, meaning: 'move by charge')
- Supercharger (for Tesla vehicles)

The powers offered by CCS and CHAdeMO are 50 kW and 120 kW for the Supercharger system [88,89]. CCS and CHAdeMO are also capable of providing fast charging, dynamic charging and vehicle to infrastructure (V2X) facilities [6,90]. Most of the EV charging stations at this time provides level 2 AC charging facilities. Level 3 DC charging network, which is being increased rapidly, is also available for Tesla cars. The stations may provide the CHAdeMO standard or the CCS, therefore, a vehicle has to be compatible with the configuration provided to be charged from the station. The CHAdeMO system is favored by the Japanese manufacturers like Nissan, Toyota and Honda whereas the European and US automakers, including Volkswagen, BMW, General Motors and Ford, prefer the CCS standard. Reference [5] discusses the charging systems used by current EVs along with the time required to get them fully charged.

7. Power Conversion Techniques

Batteries or ultracapacitors (UC) store energy as a DC charge. Normally they have to obtain that energy from AC lines connected to the grid, and this process can be wired or wireless. To deliver this energy to the motors, it has to be converted back again. These processes work in the reverse direction as well i.e., power being fed back to the batteries (regenerative braking) or getting supplied to grid when the vehicle in idle (V2G) [91]. Typical placement of different converters in an EV is shown in Figure 39 along with the power flow directions. This conversion can be DC-DC or DC-AC. For all this conversion work required to fill up the energy storage of EVs and then to use them to propel the vehicle, power converters are required [72], and they come in different forms. A detailed description of power electronics converters is provided in [92]. Further classification of AC-AC converters is shown in [93]. A detailed classification of converters is shown in Figure 40.

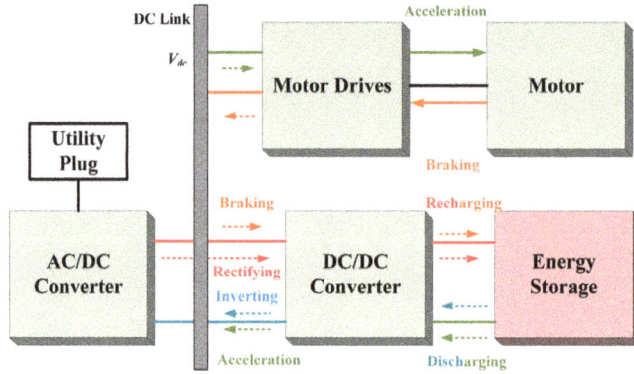

Figure 39. Typical placements of different converters in an EV. AC-DC converter transforms the power from grid to be stored in the storage through another stage of DC-DC conversion. Power is supplied to the motor from the storage through the DC-DC converter and the motor drives [72].

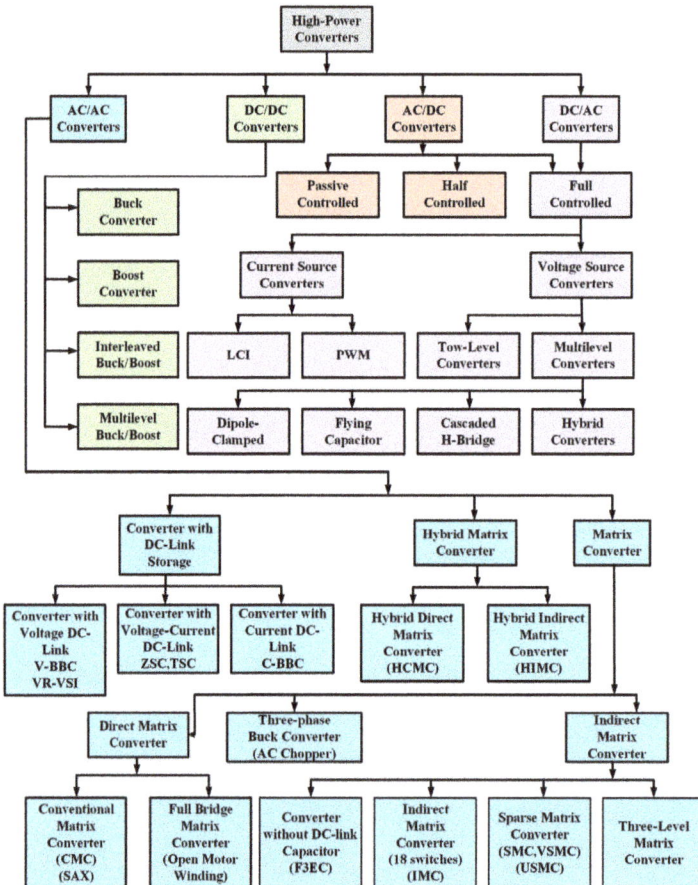

Figure 40. Detailed classification of converters. Data from [92,93].

7.1. Converters for Wired Charging

DC-DC boost converter is used to drive DC motors by increasing the battery voltage up to the operating level [72]. DC-DC converters are useful to combine a power source with a complementing energy source [94]. Figure 41 shows a universal DC-DC converter used for DC-DC conversion. It can be used as a boost converter for battery to DC link power flow and as a buck converter when the flow is reversed. The operating conditions and associated switching configuration is presented in Table 22. DC-DC boost converters can also use a digital signal processor [95].

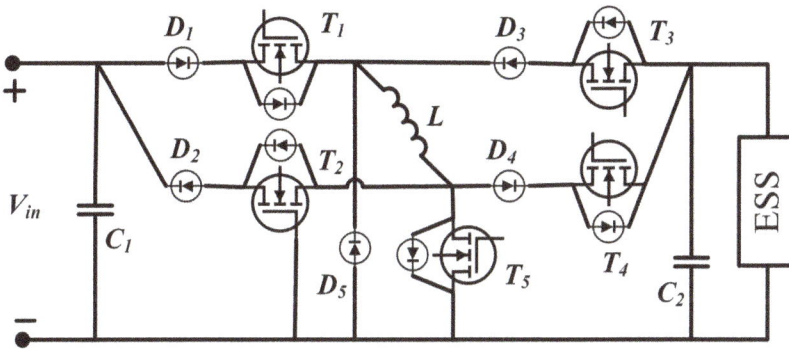

Figure 41. Universal DC-DC converter [72].

Table 22. Operating conditions for universal DC-DC converter. Adapted from [88].

Direction	Mode	T_1	T_2	T_3	T_4	T_5
V_{dc} to V_{batt}	Boost	On	Off	Off	On	PWM
V_{dc} to V_{batt}	Buck	PWM	Off	Off	On	Off
V_{batt} to V_{dc}	Boost	Off	On	On	Off	PWM
V_{batt} to V_{dc}	Buck	Off	On	PWM	Off	Off

According to [72], dual inverter is the most updated technology to drive AC motors like permanent magnet synchronous motors (PMSMs), shown in Figure 42. For dual voltage source applications, the system of Figure 43 is used [96]. These inverters operate on space vector PWM. For use on both PMSMs and induction motors (IMs), a bidirectional stacked matrix inverter can be used; such a system is shown in Figure 44.

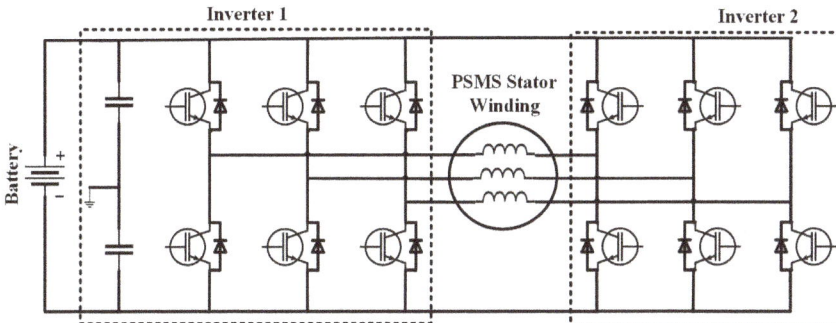

Figure 42. Dual inverter for single source [72].

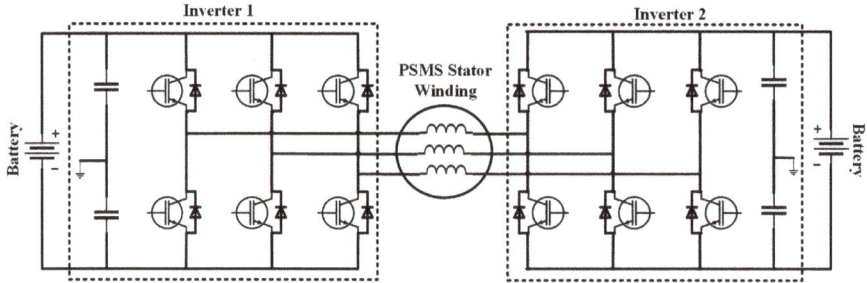

Figure 43. Dual inverter with dual sources [72].

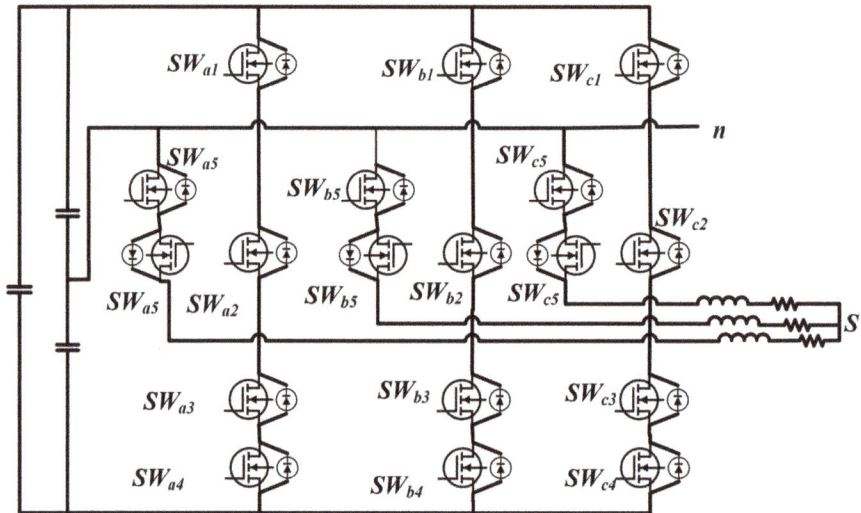

Figure 44. Novel stacked matrix inverter as shown in [97].

Some notable conventional DC-DC converters are: phase-shift full-bridge (PSFB), inductor-inductor-capacitor (LLC), and series resonant converter (SRC). A comparison of components used in these three converters is presented in [98], which is demonstrated here in Table 23. The DC-DC converters used are required to have low cost, weight and size for being used in automobiles [99]. Interleaved converters are a preferable option regarding these considerations, it offers some other advantages as well [100–103], though using it may increase the weight and volume of the inductors compared to the customary single-phase boost converters [99]. To solve this problem, Close-Coupled Inductor (CCI) and Loosely-Coupled Inductor (LCI) integrated interleaved converters have been proposed in [99]. In [48] converters for AC level-1 and level-2 chargers are shown by Williamson et al., who stated that Power Factor Correction (PFC) is a must to acquire high power density and efficiency. Two types of PFC technique are shown here: single-stage approach and two-stage approach. The first one suits for low-power use and charge only lead-acid batteries because of high low frequency ripple. To avoid these problems, the second technique is used.

Table 23. Comparison of components used in PSFB, LLC and SRC converter. Adapted from [98].

Item	PSFB	LLC	SRC
Number of switch blocks	4	4	4
Number of diode blocks	4	4	4
Number of transformers	1	1	2
Number of inductors	1	0	0
Additional capacitor	Blocking capacitor	-	-
Output filter size	Small	-	Large

In [34], Yong et al., presented the front end AC-DC converters. The Interleaved Boost PFC Converter (Figure 45) has a couple of boost converters connected in parallel and working in 180° out of phase [104–106]. The ripple currents of the inductors cancel each other. This configuration also provides twice the effective switching frequency and provides a lower ripple in input current, resulting in a relatively small EMI filter [103,107]. In Bridgeless/Dual Boost PFC Converter (Figure 46), the gating signals are made identical here by tying the power-train switches. The MOSFET gates are not made decoupled. Rectifier input bridge is not needed here. The Bridgeless Interleaved Boost PFC Converter (Figure 47) is proposed to operate above the 3.5 kW level. It has two MOSFETS and uses two fast diodes; the gating signals have a phase difference of 180°.

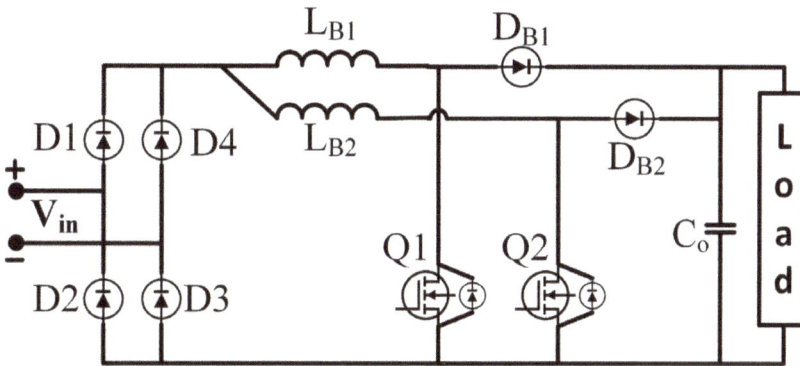

Figure 45. Interleaved Boost PFC Converter [46].

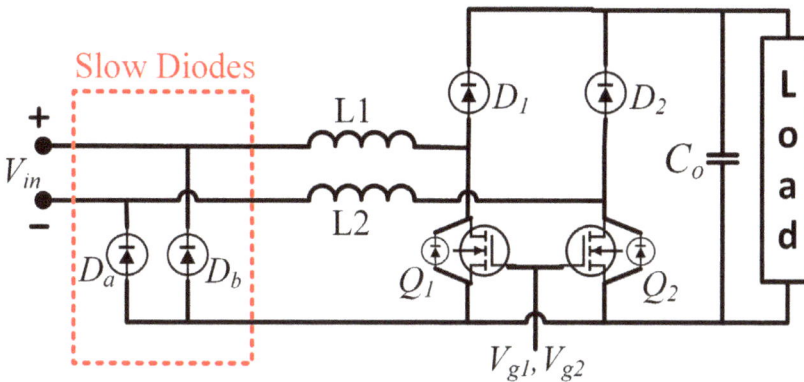

Figure 46. Bridgeless/Dual Boost PFC Converter. Adapted from [46].

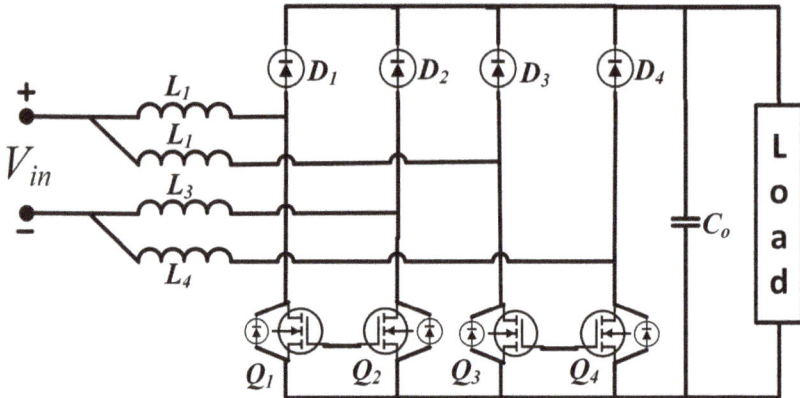

Figure 47. Bridgeless Interleaved Boost PFC Converter [46].

Williamson et al., presented some isolated DC-DC converter topologies in [44]. The ZVS FB Converter with Capacitive Output Filter (Figure 48) can achieve high efficiency as it uses zero voltage switching (ZVS) along with the capacitive output filters which reduces the ringing of diode rectifiers. The trailing edge PWM full-bridge system proposed in [107]. The Interleaved ZVS FB Converter with Voltage Doubler (Figure 49) further reduces the voltage stress and ripple current on the capacitive output filter, it reduces the cost too. Interleaving allows equal power and thermal loss distribution in each cell. The number of secondary diodes is reduced significantly by the voltage doubler rectifier at the output [34]. Among its operating modes, DCM (discontinuous conduction mode) and BCM (boundary conduction mode) are preferable. The Full Bridge LLC Resonant Converter (Figure 50) is widely used in telecom industry for the benefits like high efficiency at resonant frequency. But unlike the telecom sector, EV applications require a wide operating range. Reference [41] shows a design procedure for such configurations for these applications.

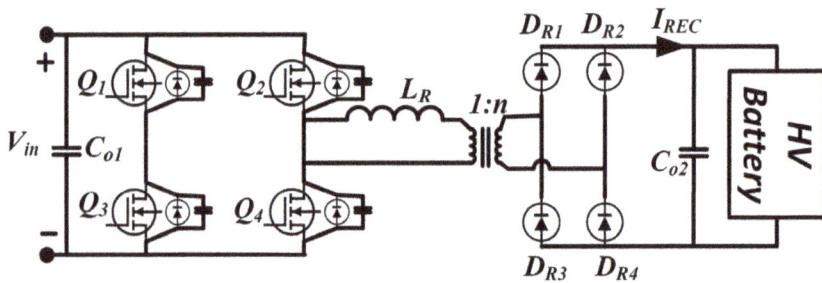

Figure 48. ZVS FB Converter with Capacitive Output Filter [46].

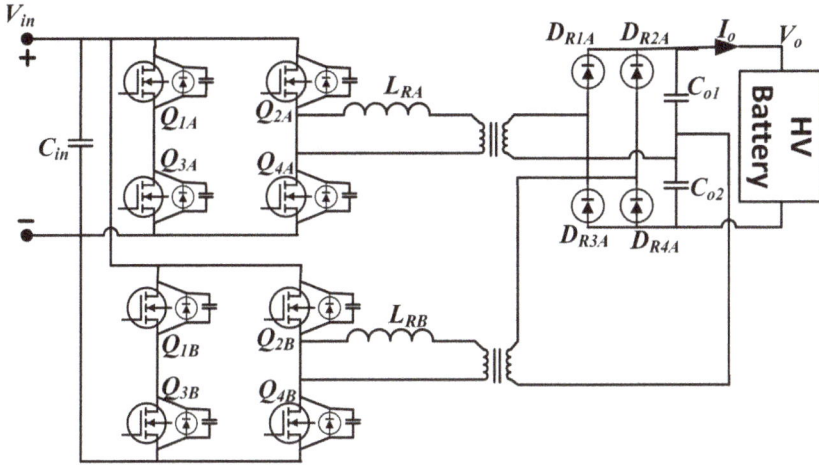

Figure 49. Interleaved ZVS FB Converter with Voltage Doubler [46].

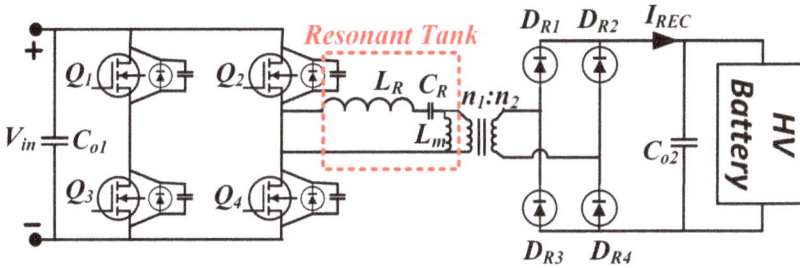

Figure 50. Full Bridge LLC Resonant Converter. Adapted from [46].

Balch et al., showed converter configurations that are used in different types of EVs in [42]. In Figure 51, a converter arrangement for a BEV is shown. An AC-DC charger is used for charging the battery pack here while a two-quadrant DC-DC converter is used for power delivery to the DC bus form the battery pack. This particular example included an ultracapacitor as well. An almost similar arrangement was shown in [42] for PHEVs (Figure 52) where a bidirectional DC-DC converter was used between the DC bus and the battery pack to facilitate regeneration. Use of integrated converter in PHEV is shown in Figure 53. Figure 54 shows converter arrangement for a PFCV; this configuration is quite similar to one shown for BEV, but it contains an additional boost converter to adjust the power produced by the fuel cell stack to be sent to the DC bus.

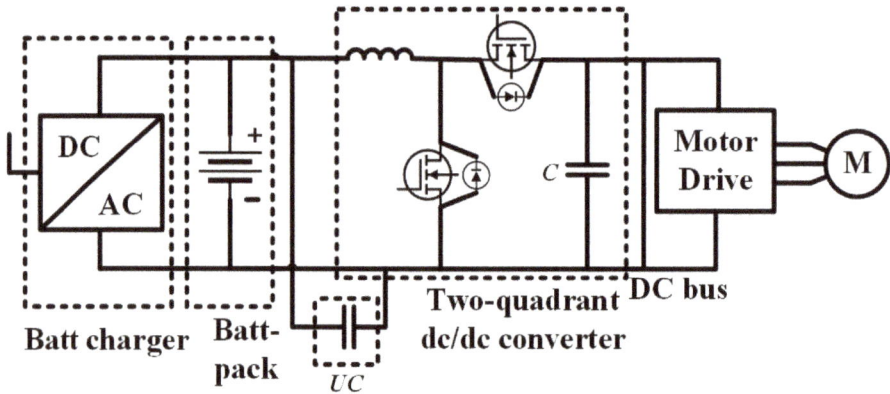

Figure 51. Converter placement in a pure EV [35]. The charger has an AC-DC converter to supply DC to the battery from the grid, whereas the DC-DC converter converts the battery voltage into a value required to drive the motor.

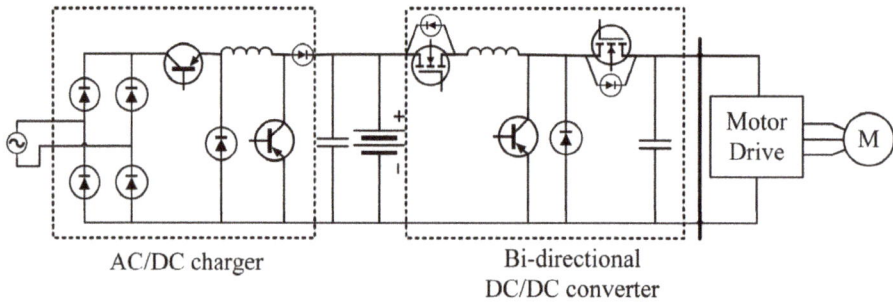

Figure 52. Cascaded converter to use in PHEV. Adapted from [35]. A bidirectional DC-DC converter is used between the DC bus and the battery pack to allow regenerated energy to flow back to the battery from the motor.

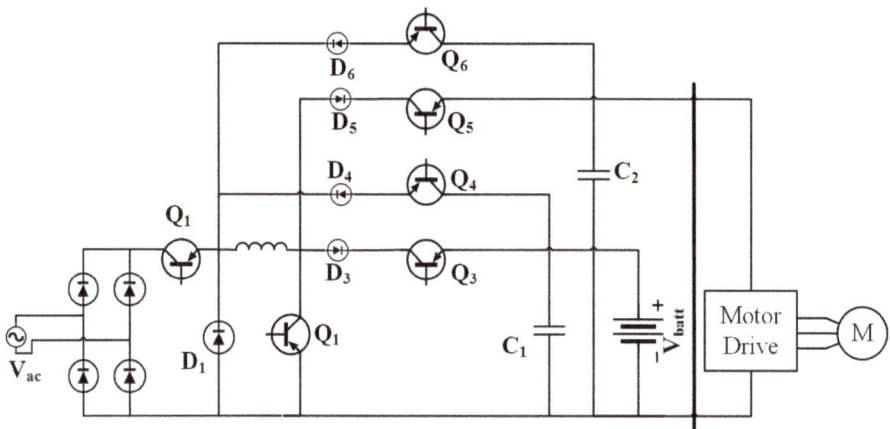

Figure 53. Integrated converter used in PHEV [35].

44

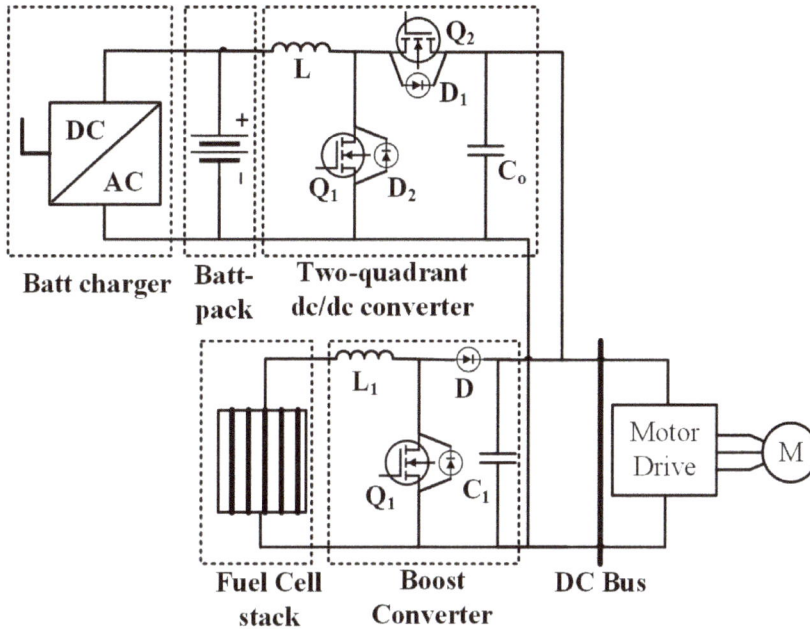

Figure 54. Converter arrangement in PFCV. Adapted from [35]. An AC-DC converter is used to convert the power from the grid; DC-DC converter is used for power exchange between the DC bus and battery; boost converter is used to make the voltage generated from the fuel cell stack suitable for the DC bus.

Bidirectional converters allow transmission of power from the motors to the energy sources and also from vehicle to grid. Novel topologies for bidirectional AC/DC-DC/DC converters to be used in PHEVs are being researched [103,108–112], such a configuration in shown in Figure 55. Kok et al., showed different DC-DC converter arrangements for EVs using multiple energy sources in [94] which are presented in Figure 56. The first system has both battery and ultracapacitor added in cascade, while the second one has them connected in parallel. The third one shows a system employing fuel cells, and battery for backup. In [113], Koushki et al., classified bidirectional AC-DC converters into two main groups: Low frequency AC-High frequency AC-DC (Figure 57), and Low frequency AC-DC- High frequency AC-DC (Figure 58). The first kind can also be called single-stage converters where the latter may be described as two-stage, which can be justified from their topologies. Converters employed for EV application are compiled in Table 24. From this table, it is evident that step down converters are required for charging the batteries from a higher voltage grid voltage, bidirectional converters are needed for providing power flow in both directions, and specialized converters such as the last three, are needed for better charging performances.

AC-DC converters are used to charge the batteries from AC supply-lines; DC-DC converters are required for sending power to the motors from the batteries. The power flow can be reversed in case of regenerative actions or V2G. Bidirectional converters are required in such cases. Different converter configurations have different advantages and shortcomings which engendered a lot of research and proliferation of hybrid converter topologies.

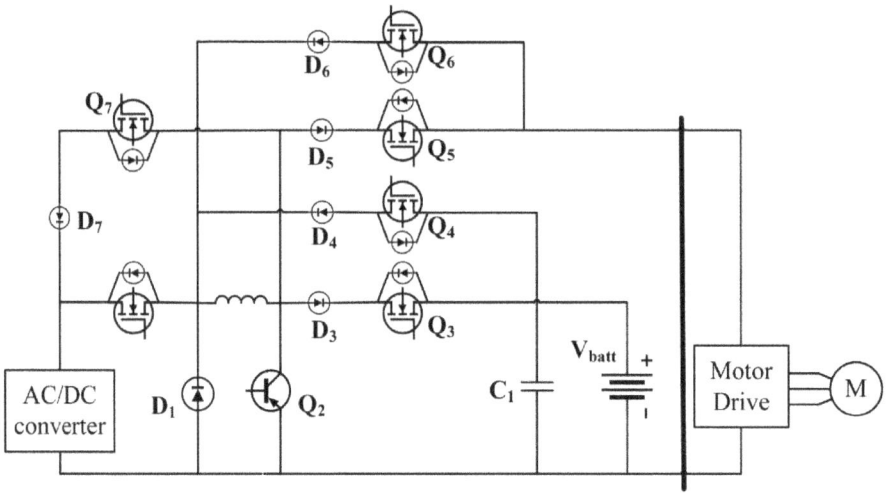

Figure 55. Integrated bidirectional AC/DC-DC/DC converter [33].

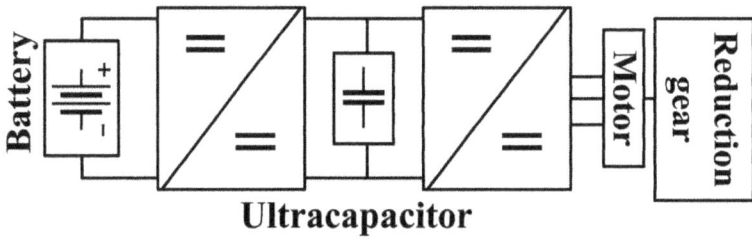

(a)

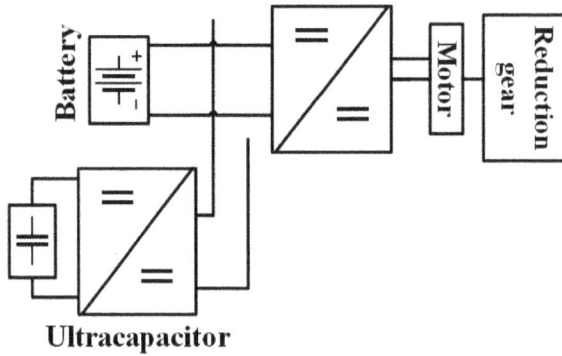

(b)

Figure 56. *Cont.*

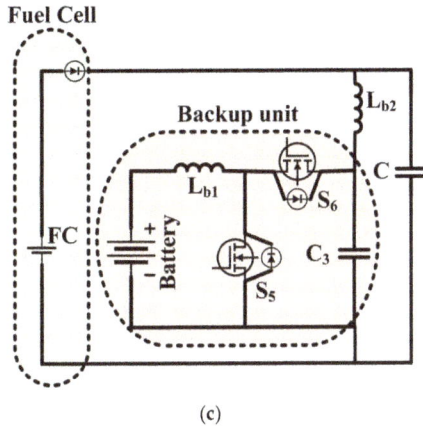

(c)

Figure 56. Converter arrangements as shown in [94]: (**a**) Cascaded connection; (**b**) Parallel connection; (**c**) Fuel cell with battery backup. Adapted from [94].

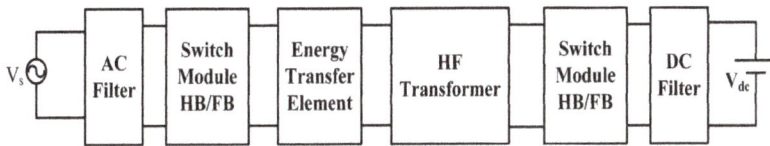

Figure 57. Low frequency AC-High frequency AC-DC converter, also called single-stage converter [113].

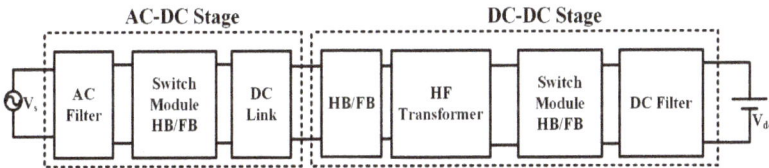

Figure 58. Low frequency AC-DC-High frequency AC-DC converter, also called two-stage converter. Adapted from [113].

Table 24. Converters with EV application displaying their key features and uses in EVs.

Configuration	Reference	Operation	Key Features	Application in EV
Buck converter	Bose [92]	Step down	Can operate in continuous or discontinuous mode	Sending power to the battery
Buck-Boost converter	Bose [92]	Step up and step down	Two quadrant operation of chopper	Regenerative action
Interleaved Boost PFC converter	Williamson et al. [46]	Step up with power factor correction	Relatively small input EMI filter	Charging
Bridgeless/Dual Boost PFC Converter	Williamson et al. [46]	Step up with power factor correction	Does not require rectifier input bridge	Charging
ZVS FB Converter with Capacitive Output Filter	Williamson et al. [46]	AC-DC conversion	Zero voltage switching	Charging

7.2. Systems for Wireless Charging

Wireless charging or wireless power transfer (WPT) uses a principle similar to transformer. There is a primary circuit at the charger end, from where the energy is transferred to the secondary circuit located at the vehicle. In case of inductive coupling, the voltage obtained at the secondary side is:

$$v_2 = L_2(di_2/dt) + M(di_1/dt) \tag{2}$$

M is the mutual inductance and can be calculated by:

$$M = k\sqrt{(L_1 L_2)} \tag{3}$$

The term k here is the coupling co-efficient; L_1 and L_2 are the inductances of primary and secondary circuit. Figure 59 shows the 'double D' arrangement for WPT which demonstrates the basic principle of wireless power transfer by means of flux linkages. A variety of configurations can be employed for wireless power transfer; some of them meet a few desired properties to charge vehicles. Inductive WPT, shown in Figure 60a, is the most rudimentary type, transfer power from one coil to another just like the double D system. Capacitive WPT (Figure 60b) uses a similar structure as the inductive system, but it has two coupling transformers at its core. Low frequency permanent magnet coupling power transfer (PMPT) is shown in Figure 60c; it uses a permanent magnet rotor to transmit power, another rotor placed in the vehicle acts as the receiver. Resonant antennae power transfer (RAPT) (Figure 60d) uses resonant antennas for wireless transfer of power. Resonant inductive power transfer (RIPT), shown in Figure 60e, uses resonance circuits for power transfer. Online power transfer (OLPT) has a similar working principle as RIPT, it can be used in realizing roadways that can charge vehicles wirelessly by integrating the transmitter with the roadway (pilot projects using similar technology placed them just beneath the road surface), and equipping vehicles with receivers to collect power from there. Schematic for this system is shown in Figure 60f. Characteristics of these systems are shown in Table 25.

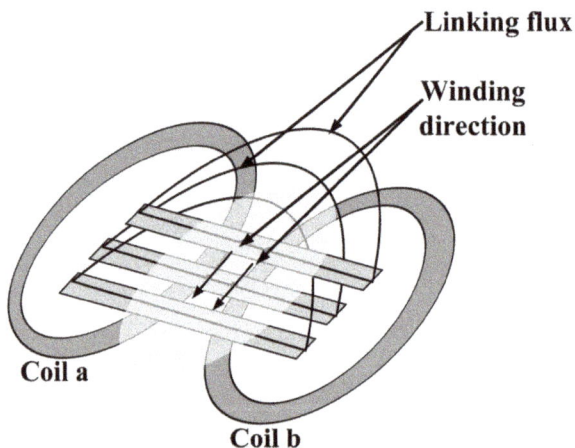

Figure 59. Double D arrangement for WPT. Fluxes generated in one coil cut the other one and induces a voltage there, enabling power transfer between the coils without any wired connection [27].

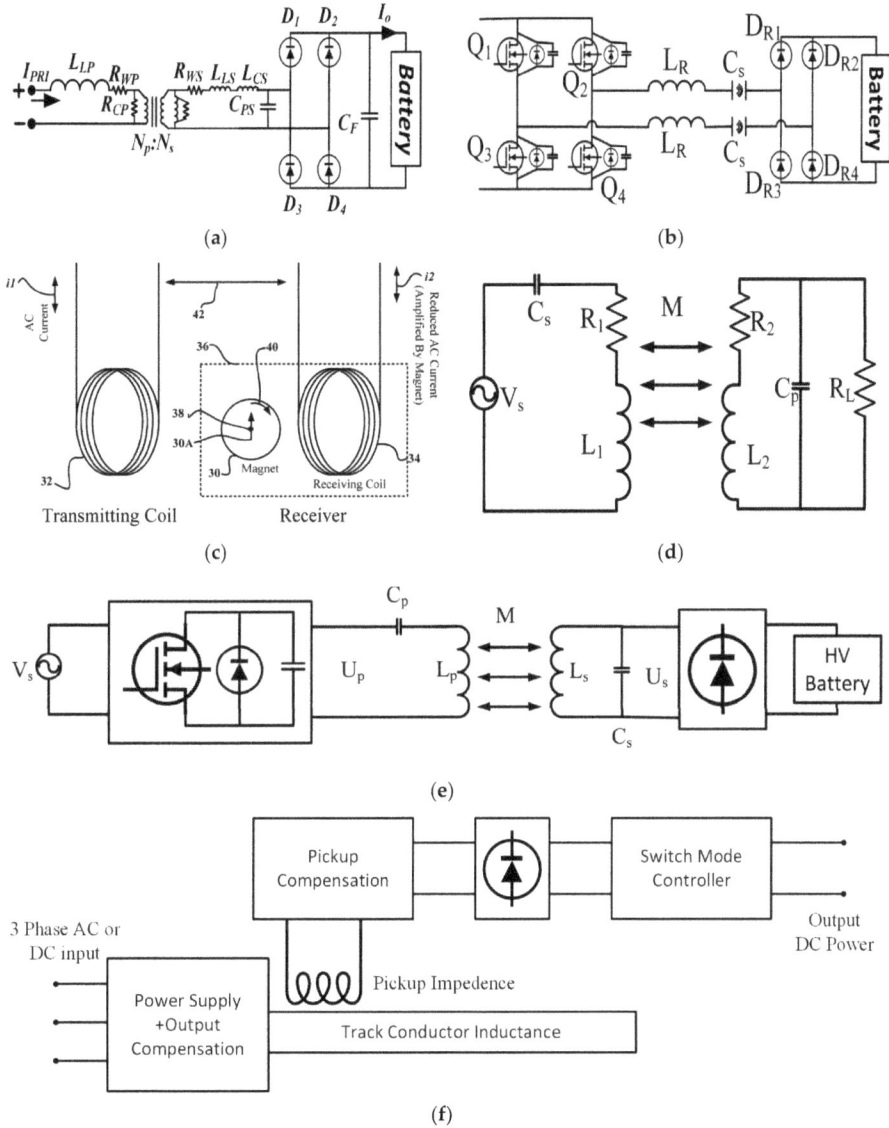

Figure 60. Different configurations used for wireless power transfer over the years: (**a**) Inductive WPT; (**b**) Capacitive WPT; (**c**) Low frequency permanent magnet coupling power transfer (PMPT); (**d**) Resonant antennae power transfer (RAPT); (**e**) Resonant inductive power transfer (RIPT); (**f**) Online power transfer (OLPT).

Table 25. Characteristics of wireless charging systems [87].

Technology	Characteristics
Inductive WPT	• It is not actually wireless, just does not require any connection. • Primary and secondary coils are sealed in epoxy. • Can provide power of either 6.6 kW or 50 kW. • Coaxial winding transformer can be used to place all the transformer core materials off-board. • Losses including geometric effects, eddy current loss, EMI are mainly caused by nonlinear flux distribution. • A piecewise assembly of ferrite core and dividing the secondary winding symmetrically can help minimizing the losses.
Capacitive WPT	• Capacitive power transfer or CPT interface is built with two coupling transformers at the center; the rest of the system is similar to inductive WPT. • Capacitive interface is helpful in reducing the size and cost of the required galvanic isolating parts. • Cheaper and smaller for lower power applications, but not preferred for high power usage. • Useful in consumer electronics, may not be sufficient for EV charging.
Low frequency permanent magnet coupling power transfer (PMPT)	• The transmitter is a cylinder-shaped, permanent magnet rotor driven by static windings placed on the rotor, inside it if the rotor is hollow, or outside the motor, separated by an air-gap. • The receiver is placed on the vehicle, similar to the transmitter in construction. • Transmitter and receiver have to be within 150 mm for charging. • Because of magnetic gear effect, the receiver rotor rotates at the same speed as the transmitter and energy is transferred. • The disadvantages may be the vibration, noise and lifetime associated with the mechanical components used.
Resonant inductive power transfer (RIPT)	• Most popular WPT system. • Uses two tuned resonant tanks or more, operating in the same frequency in resonance. • Resonant circuits enable maximum transfer of power, efficiency optimization, impedance matching, compensation of magnetic coupling and magnetizing current variation. • Can couple power for a distance of up to 40 cm. • Advantages include extended range, reduced EMI, operation at high frequency and high efficiency.
Online power transfer (OLPT)	• Has a similar concept like RIPT, but uses a lower resonant frequency. • Can be used for high power applications. • This system is proposed to be applied in public transport system in [87]. • The primary circuit—a combination of the input of resonant converter and distributed primary windings is integrated in the roadway. This primary side is called the 'track'. • The secondary is placed in vehicles and is called the 'pickup coil'. • Supply of this system is high voltage DC or 3-phase AC. • It can provide frequent charging of the vehicles while they are on the move, reducing the required battery capacity, which will reduce the cost and weight of the cars. • The costs associated with such arrangement may also make its implementation unlikely.
Resonant antennae power transfer (RAPT)	• This system uses two resonant antennas, or more, with integrated resonant inductances and capacitances. The antennas are tuned to identical frequencies. • Large WPT coils are often used as antennas; resonant capacitance is obtained there by controlled separation in the helical structure. • The frequencies used are in MHz range. • Can transfer power efficiently for distances up to 10 m. • The radiations emitted by most of such systems exceed the basic limits on human exposure and are difficult to shield without affecting the range and performance. • Generating frequencies in the MHz range is also challenging and costly with present power electronics technologies.

8. Effects of EVs

Vehicles may serve the purpose of transportation, but they affect a lot of other areas. Therefore, the shift in the vehicle world created by EVs impacts the environment, the economy, and being electric, the electrical systems to a great extent. EVs are gaining popularity because of the benefits they provide in all these areas, but with them, there come some problems as well. Figure 61 illustrates the impacts of EVs on the power grid, environment and economy.

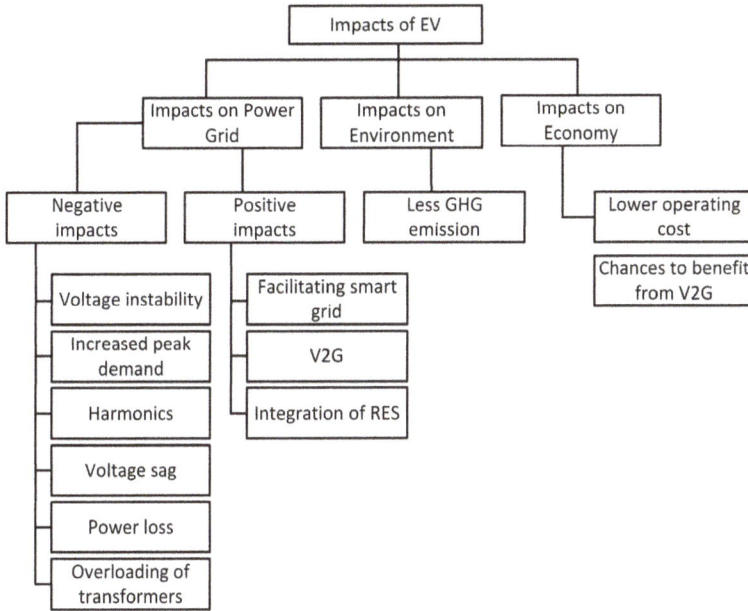

Figure 61. A short list of the impacts of EVs on the power grid, environment and economy.

8.1. Impact on the Power Grid

8.1.1. Negative Impacts

EVs are considered to be high power loads [114] and they affect the power distribution system directly; the distribution transformers, cables and fuses are affected by it the most [115,116]. A Nissan Leaf with a 24 kWh battery pack can consume power similar to a single European household. A 3.3 kW charger in a 220 V, 15 A system can raise the current demand by 17% to 25% [117]. The situation gets quite alarming if charging is done during peak hours, leading to overload on the system, damage of the system equipment, tripping of protection relays, and subsequently, an increase in the infrastructure cost [117]. Charging without any concern to the time of drawing power from the grid is denoted as uncoordinated charging, uncontrolled charging or dumb charging [117,118]. This can lead to the addition of EV load in peak hours which can cause load unbalance, shortage of energy, instability, and decrease in reliability and degradation of power quality [116,119]. In case of the modified IEEE 23 kV distribution system, penetration of EVs can deviate voltage below the 0.9 p.u. level up to 0.83 p.u., with increased power losses and generation cost [118]. Level 1 charging from an 110 V outlet does not affect the power system much, but problems arise as the charging voltage increases. Adding an EV for fast charging can be equivalent to adding several households to the grid. The grid is likely to be capable of withstanding it, but distribution networks are designed with specific numbers of households kept into mind, sudden addition of such huge loads can often lead to problems. Reducing the charging time to distinguish their vehicles in the EV market has become the current norm among the manufacturers, and it requires higher voltages than ever. Therefore, mitigating the adverse effects is not likely by employing low charging voltages.

To avoid these effects, and to provide efficient charging with the available infrastructure, coordinated charging (also called controlled or smart charging) has to be adopted. In this scheme, the EVs are charged during the time periods when the demand is low, for example, after midnight. Such schemes are beneficial in a lot of ways. It not only prevents addition of extra load during peak hours, but also increases the load in valley areas of the load curve, facilitating proper use of the power plants with better efficiency. In [116], Richardson et al., showed that a controlled charging rate can make high EV penetration possible in the current residential power network with only a few upgrades in the infrastructure. Geng et al., proposed a charging strategy in [120] comprising of two stages aimed at providing satisfactory charging for all connected EVs while shifting the loads on the transformers. On the consumer side, it can reduce the electricity bill as the electricity is consumed by the EVs during off peak hours, which generally have a cheaper unit rate than peak hours. According to [121], smart charging systems can reduce the increase investment cost in distribution system by 60–70%. The major problems that are faced in the power systems because of EVs can be charted as following:

- Voltage instability: Normally power systems are operated close to their stability limit. Voltage instabilities in such systems can occur because of load characteristics, and that instability can lead to blackouts. EV loads have nonlinear characteristics, which are different than the general industrial or domestic loads, and draw large quantities power in a short time period [81,122]. Reference [123] corroborated to the fact that EVs cause serious voltage instability in power systems. If the EVs have constant impedance load characteristics, then it is possible for the grid to support a lot of vehicles without facing any instability [81]. However, the EV loads cannot be assumed beforehand and thus their power consumptions stay unpredictable; addition of a lot of EVs at a time therefore can lead to violation of distribution constraints. To anticipate these loads properly, appropriate modeling methods are required. Reference [124] suggested tackling the instabilities by damping the oscillations caused by charging and discharging of EV batteries using a wide area control method. The situation can also be handled by changing the tap settings of transformers [125], by a properly planned charging system, and also by using control systems like fuzzy logic controllers to calculate voltages and SOCs of batteries [81].
- Harmonics: The EV charger characteristics, being nonlinear, gives raise high frequency components of current and voltage, known as harmonics. The amount of harmonics in a system can be expressed by the parameters total current harmonic distortion (THD_i) and total voltage harmonic distortion (THD_v):

$$THD_i = \frac{\sqrt{\sum_{h=2}^{H} I_h^2}}{I_1} \times 100\% \tag{4}$$

$$THD_v = \frac{\sqrt{\sum_{h=2}^{H} V_h^2}}{V_1} \times 100\% \tag{5}$$

Harmonics distort the voltage and current waveforms, thus can reduce the power quality. It also causes stress in the power system equipment like cables and fuses [122]. The present cabling is capable of withstanding 25% EV penetration if slow charging is used, in case of rapid charging, the amount comes down to 15% [126]. Voltage imbalance and harmonics can also give rise to current flow in the neutral wire [127,128]. Different approaches have been adopted to determine the effects of harmonics due to EV penetration. Reference [127] simulated the effects of harmonics using Monte Carlo analysis to determine the power quality. In [129] the authors showed that THD_v can reach 11.4% if a few number of EVs are fast charging. This is alarming as the safety limit of THD_v is 8%. According to

Melo et al. [130], *THD*$_i$ also becomes high, in the range of 12% to 14%, in case of fast charging, though it remains in the safe limit during times of slow charging. Studies conducted in [131] show the modern EVs generate less *THD*$_i$ than the conventional ones, though their *THD*$_v$ values are higher. However, with increased number of EVs, there are chances of harmonics cancellation because of different load patterns [132,133]. Different EV chargers can produce different phase angles and magnitudes which can lead to such cancellations [133]. It is also possible to reduce, even eliminate harmonics by applying pulse width modulation in the EV chargers [132]. High *THD*$_i$ can be avoided by using filtering equipment at the supply system [134].

- Voltage sag: A decrease in the RMS value of voltage for half a cycle or 1 min is denoted as voltage sag. It can be caused by overload or during the starting of electric machines. Simulation modeled with an EV charger and a power converter in [135] stated 20% EV penetration can exceed the voltage sag limit. Reference [136] stated that 60% EV penetration is possible without any negative impact is possible if controlled charging is employed. The amount, however, plummets to 10% in case of uncontrolled charging. Leemput et al., conducted a test employing voltage droop charging and peak shaving by EV charging [137]. This study exhibited considerable decrease in voltage sag with application of voltage droop charging. Application of smart grid can help in great extents in mitigating the sag [138].
- Power loss: The extra loss of power caused by EV charging can be formulated as:

$$PL_E = PL_{EV} - PL_{original} \qquad (6)$$

PL$_{original}$ is the loss occurred when the EVs are not connected to the grid and PLEV is the loss with EVs connected. Reference [121] charted the increased power loss as high as 40% in off peak hours considering 60% of the UK PEVs to be connected to distribution system. Uncoordinated charging, therefore, can increase the amount of loss furthermore. Taking that into account, a coordinated charging scheme, based on objective function, to mitigate the losses was proposed in [139]. Coordinated charging is also favored by [140,141] to reduce power losses significantly. Power generated in the near vicinity can also help minimizing the losses [142], and distributed generation can be quite helpful in this prospect, with the vehicle owners using energy generated at their home (by PV cells, CHP plants, etc.) to charge the vehicles.

- Overloading of transformers: EV charging directly affects the distribution transformers [81]. The extra heat generated by EV loads can lead to increased aging rate of the transformers, but it also depends on the ambient temperature. In places with generally cold weather like Vermont, the aging due to temperature is negligible [81]. Estimation of the lifetime of a transformer is done in [143], where factors taken into account are the rate of EV penetration, starting time of charging and the ambient temperature. It stated that transformers can withstand 10% EV penetration without getting any decrease in lifetime. The effect of level 1 charging, is in fact, has negligible effect on this lifetime, but significant increase in level 2 charging can lead to the failure of transformers [144]. Elnozahy et al., stated that overloading of transformer can happen with 20% PHEV penetration for level 1 charging, whereas level 2 does it with 10% penetration [145]. According to [122], charging that takes place right after an EV being plugged in can be detrimental to the transformers.
- Power quality degradation: The increased amount of harmonics and imbalance in voltage will degrade the power quality in case of massive scale EV penetration to the grid.

8.1.2. Positive Impacts

On the plus side, EVs can prove to be quite useful to the power systems in a number of ways:

- Smart grid: In the smart grid system, intelligent communication and decision making is incorporated with the grid architecture. Smart grid is highly regarded as the future of power grids and offers a vast array of advantages to offer reliable power supply and advanced control. In such a system, the much coveted coordinated charging is easily achievable as interaction with the grid system becomes very much convenient even from the user end. The interaction of EVs and smart grid can facilitate opportunities like V2G and better integration of renewable energy. In fact, EV is one the eight priorities listed to create an efficient smart grid [117].

- V2G: V2G or vehicle to grid is a method where the EV can provide power to the grid. In this system, the vehicles act as loads when they are drawing energy, and then can become dynamic energy storages by feeding back the energy to the grid. In coordinated charging, the EV loads are applied in the valley points of the load curve, in V2G; EVs can act as power sources to provide during peak hours. V2G is realizable with the smart grid system. By making use of the functionalities of smart grid, EVs can be used as dynamic loads or dynamic storage systems. The power flow in this system can be unidirectional or bidirectional. The unidirectional system is analogous to the coordinated charging scheme, the vehicles are charged when the load is low, but the time to charge the vehicles is decided automatically by the system. Vehicles using this scheme can simply be plugged in anytime and put there; the system will choose a suitable time and charge it. Smart meters are required for enabling this system. With a driver variable charging scheme, the peak power demand can be reduced by 56% [117]. Sortomme et al., found this system particularly attractive as it required little up gradation of the existing infrastructure; creating a communication system in-between the grid and the EVs is all that is needed [146]. The bidirectional system allows vehicles to provide power back to the grid. In this scenario, vehicles using this scheme will supply energy to the grid from their storage when it is required. This method has several appealing aspects. With ever increasing integration of renewable energy sources (RES) to the grid, energy storages are becoming essential to overcome their intermittency, but the storages have a very high price. EVs have energy storages, and in many cases, they are not used for a long time. Example for this point can be the cars in the parking lots of an office block, where they stay unused till the office hour is over, or vehicles that are used in a specific time of the year, like a beach buggy. Studies also revealed that, vehicles stay parked 95% of the time [117]. These potential storages can be used when there is excess generation or low demand and when the energy is needed, it is taken back to the grid. The vehicle owners can also get economically beneficial by selling this energy to the grid. In [147], Clement-Nyns et al., concluded that a combination of PHEVs can prove beneficial to distributed generation sources by providing storage for the excess generation, and releasing that to the grid later. Bidirectional charging, however, needs chargers capable of providing power flow in both directions. It also needs smart meters to keep track of the units consumed and sold, and advanced metering architecture (AMI) to learn about the unit charges in real time to get actual cost associated with the charging or discharging at the exact time of the day. The AMI system can shift 54% of the demand to off-peak periods, and can reduce peak consumption by 36% [117]. The bidirectional system, in fact, can provide 12.3% more annual revenue than the unidirectional one. But taking the metering and protections systems required in the bidirectional method, this revenue is nullified and indicates the unidirectional system is more practical. Frequent charging and discharging caused by bidirectional charging can also reduce battery life and increase energy losses from the conversion processes [81,117]. In a V2G scenario, operators with a vehicle fleet are likely to reduce their cost of operation by 26.5% [117]. Another concept is produced using the smart grid and the EVs, called virtual power plant (VPP), where a cluster of vehicles is considered as a power plant and dealt like one in the system. VPP architecture and control is shown in Figure 62. Table 26 shows the characteristics of unidirectional and bidirectional V2G.

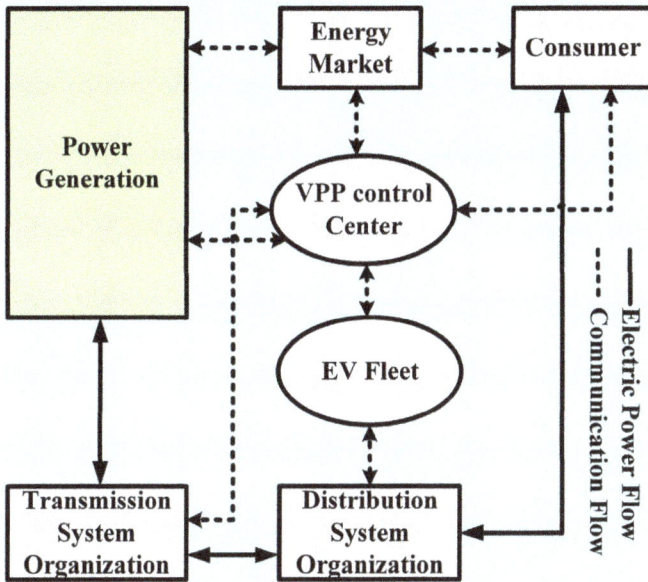

Figure 62. VPP architecture and control [117].

Table 26. Unidirectional and bidirectional V2G characteristics. Adapted from [1].

V2G System	Description	Services	Advantages	Limitations
Unidirectional	Controls EV charging rate with a unidirectional power flow directed from grid to EV based on incentive systems and energy scheduling	• Ancillary service—load levelling	• Maximized profit • Minimized power loss • Minimized operation cost • Minimized emission	• Limited service range
Bidirectional	Bidirectional power flow between grid and EV to attain a range of benefits	• Ancillary service—spinning reserve • Load leveling • Peak power shaving • Active power support • Reactive power support/Power factor correction • Voltage regulation • Harmonic filtering • Support for integration of renewable	• Maximized profit • Minimized power loss • Minimized operation cost • Minimized emission • Prevention of grid overloading • Failure recovery • Improved load profile • Maximization of renewable energy generation	• Fast battery degradation • Complex hardware • High capital cost • Social barriers

- Integration of renewable energy sources: Renewable energy usage becomes more promising with EVs integrated into the picture. EV owners can use RES to generate power locally to charge their EVs. Parking lot roofs have high potential for the placement of PV panels which can charge the vehicles parked underneath as well as supplying the grid in case of excess generation [148–150], thus serving the increase of commercial RES deployment. The V2G structure is further helpful to integrate RES for charging of EVs, and to the grid as well, as it enables the selling of energy to the grid when there is surplus, for example, when vehicles are parked and the system knows the user will not need the vehicle before a certain time. V2G can also enable increased penetration of wind energy (41%–59%) in the grid in an isolated system [121]. References [151–154] worked with different architectures to observe the integration scenario of wind energy with EV assistance. Figure 63 demonstrates integration of wind and solar farm with conventional coal and nuclear power grid with EV charging station employing bidirectional V2G. Table 27 shows the types of assistance EVs can provide for integrating renewable energy sources to the grid.

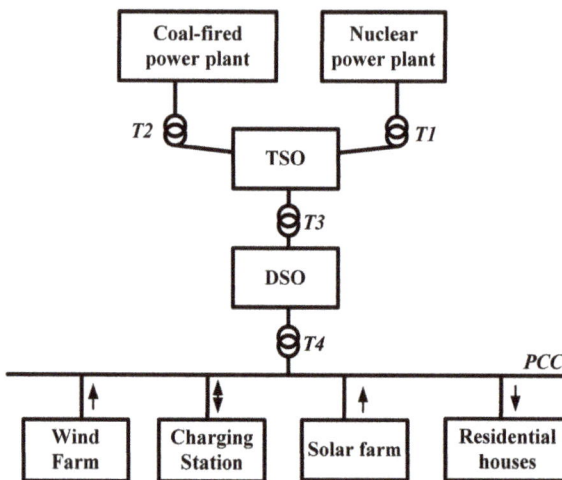

Figure 63. Wind and solar integration in the grid with the help of EV in V2G system. TSO stands for transmission system organization; DSO for distribution system organization; T1 to T4 represent the transformers coupling the generation, transmission, and distribution stages [117].

Table 27. Scopes of assisting renewable energy source (RES) integration using EV. Adapted from [1].

Interaction with RES	Field of Application	Contribution
Solar PV	Smart home	• Implementation of PV and EV in smart home to reduce emission • Development of stand-alone home EV charger based on solar PV system • Development of future home with uninterruptable power by implementing V2G with solar PV
	Parking lot	• Analysis of EV charging using solar PV at parking lots • Scheduling of charging and discharging for intelligent parking lot
	Grid distribution network	• Assessment of power system performance with integration of grid connected EV and solar PV • Development of EV charging control strategy for grid connected solar PV based charging station • Development of optimization algorithm to coordinate V2G services
	Micro grid	• Development of generation scheduling for micro grid consisting of EV and solar PV
Wind turbine	Grid distribution network	• Determination of EV interaction potential with wind energy generation • Development of V2G systems to overcome wind intermittency problems
	Micro grid	• Development of coordinating algorithm for energy dispatching of V2G and wind generation

Table 27. *Cont.*

Interaction with RES	Field of Application	Contribution
Solar PV and wind turbine	Smart home	• Development of control strategy for smart homes with grid-interactive EV and renewable sources
	Parking lot	• Design of intelligent optimization framework for integrating renewable sources and EVs
	Grid distribution network	• Potential analysis of grid connected EVs for balancing intermittency of renewable sources • Emission analysis of EVs associated with renewable generation • Development of optimized algorithm to integrate EVs and renewable sources to the grid
	Micro grid	• Development of V2G control for maximized renewable integration in micro grid

8.2. Impact on Environment

One of the main factors that propelled the increase of EVs' popularity is their contribution to reduce the greenhouse gas (GHG) emissions. Conventional internal combustion engine (ICE) vehicles burn fuels directly and thus produce harmful gases, including carbon dioxide and carbon monoxide. Though HEVs and PHEVs have IC engines, their emissions are less than the conventional vehicles. But there are also theories that the electrical energy consumed by the EVs can give rise to GHG emission from the power plants which have to produce more because of the extra load added in form of EVs. This theory can be justified by the fact that the peak load power plants are likely to be ICE type, or can use gas or coal for power generation. If EVs add excess load during peak hours, it will lead to the operation of such plants and will give rise to CO_2 emission [155]. Reference [156] also stated that power generation from coal and natural gas will produce more CO_2 from EV penetration than ICEs. However, all the power is not generated from such resources. There are many other power generating technologies that produce less GHG. With those considered, the GHG production from power plants because of EV penetration is less than the amount produced by equivalent power generation from ICE vehicles. The power plants also produce energy in bulk, thus minimizing the per unit emission. With renewable sources integrated properly, which the EVs can support strongly, the emission from both power generation and transportation sector can be reduced [115]. Over the lifetime, EVs cause less emission than conventional vehicles. This parameter can be denoted as well-to-wheel emission and it has a lower value for EVs [157]. In [158], well-to-wheel and production phases are taken into account to calculate the impact of EVs on the environment. This approach stated the EVs to be the least carbon intensive among the vehicles. Denmark managed to reduce 85% CO_2 emission from transportation by combining EVs and electric power. EVs also produce far less noise, which can highly reduce sound pollution, mostly in urban areas. The recycling of the batteries raises serious concerns though, as there are few organizations capable of recycling the lithium-ion batteries fully. However, like the previous nickel-metal and lead-acid ones, lithium-ion cells are not made of caustic chemicals, and their reuse can reduce 'peak lithium' or 'peak oil' demands [81].

8.3. Impact on Economy

From the perspective of the EV owners, EVs provide less operating cost because of their superior efficiency [22]; it can be up to 70% where ICE vehicles have efficiencies in the range of 60% to 70% [159]. The current high cost of EVs is likely to come down from mass production and better energy policies [3] which will further increase the economic gains of the owners. V2G also allows the owners to obtain a financial benefit from their vehicles by providing service to the grid [160]. The power service providers benefit from EV integration mainly by implementing coordinated charging and V2G. It allows them to adopt better peak shaving strategies as well as to integrate renewable sources. EV fleets can lead to $200 to $300 savings in cost per vehicle per year [161,162].

8.4. Impacts on Motor Sports

Hybrid technologies are not used extensively in motor sports to enhance the performance of the vehicles. Electric vehicles now have their own formula racing series named 'Formula E' [163] which started in Beijing in September 2014. Autonomous EVs are also being planned to take part in a segment of this series called 'Roborace'.

9. Barriers to EV Adoption

Although electric vehicles offer a lot of promises, they are still not widely adopted, and the reasons behind that are quite serious as well.

9.1. Technological Problems

The main obstacles that have frustrated EVs' domination are the drawbacks of the related technology. Batteries are the main area of concern as their contribution to the weight of the car is significant. Range and charging period also depend on the battery. These factors, along with a few others, are demonstrated below:

9.1.1. Limited Range

EVs are held back by the capacity of their batteries [4]. They have a certain amount of energy stored there, and can travel a distance that the stored energy allows. The range also depends on the speed of the vehicle, driving style, cargo the vehicle is carrying, the terrain it is being driven on, and the energy consuming services running in the car, for example air conditioning. This causes 'range anxiety' among the users [81], which indicates the concern about finding a charging station before the battery drains out. People are found to be willing to spend up to $75 extra for an extra range of one mile [164]. Though even the current BEVs are capable of traversing equivalent or more distance than a conventional vehicle can travel with a full tank (Tesla Model S 100D has a range of almost 564 km on 19″ wheels when the temperature is 70 °C and the air conditioning is off [24], the Chevrolet Bolt's range is 238 miles or 383 km [165]), range anxiety remains a major obstacle for EVs to overcome. This does not affect the use of EVs for urban areas though, as in most cases this range is enough for daily commutation inside city limits. Range extenders, which produce electricity from fuel, are also available with models like BMW i3 as an option. Vehicles with such facilities are currently being called as Extended Range Electric Vehicles (EREV).

9.1.2. Long Charging Period

Another major downside of EVs is the long time they need to get charged. Depending on the type of charger and battery pack, charging can take from a few minutes to hours; this truly makes EVs incompetent against the ICE vehicles which only take a few minutes to get refueled. Hidrue et al., found out that, to have an hour decreased from the charging time; people are willing to pay $425–$3250 [164]. A way to make the charging time faster is to increase the voltage level and employment of better chargers. Some fast charging facilities are available at present, and more are being studied. There are also the fuel cell vehicles that do not require charging like other EVs. Filling up the hydrogen tank is all that has to be done in case of these vehicles, which is as convenient as filling up a fuel tank, but FCVs need sufficient hydrogen refueling stations and a feasible way to produce the hydrogen in order to thrive.

9.1.3. Safety Concerns

The concerns about safety are rising mainly about the FCVs nowadays. There are speculations that, if hydrogen escapes the tanks it is kept into, can cause serious harm, as it is highly flammable. It has no color either, making a leak hard to notice. There is also the chance of the tanks to explode in case of a collision. To counter these problems, the automakers have taken measures to ensure the integrity of the tanks; they are wrapped with carbon fibers in case of the Toyota Mirai. In this car, the hydrogen handling parts are placed outside the cabin, allowing the gas to disperse easily in case of any leak, there are also arrangements to seal the tank outlet in case of high-speed collision [166].

9.2. Social Problems

9.2.1. Social Acceptance

The acceptance of a new and immature technology, along with its consequences, takes some time in the society as it means change of certain habits [167]. Using an EV instead of a conventional vehicle means change of driving patters, refueling habits, preparedness to use an alternative transport in case of low battery, and these are not easy to adopt.

9.2.2. Insufficient Charging Stations

Though public charging stations have increased a lot in number, still they are not enough. Coupled with the lengthy charging time, this acts as a major deterrent against EV penetration. Not all the public charging stations are compatible with every car as well; therefore it also becomes a challenge to find a proper charging point when it is required to replete the battery. There is also the risk of getting a fully occupied charging station with no room for another car. But, the manufacturers are working on to mitigate this problem. Tesla and Nissan have been expanding their own charging networks, as it, in turn means they can sell more of their EVs. Hydrogen refueling stations are not abundant yet as well. It is necessary as well to increase the adoption of FCVs. In [168], a placement strategy for hydrogen refueling stations in California is discussed. It stated that a total of sixty-eight such stations will be sufficient to provide service to FCVs in the area. To get the better out of the remaining stations, there are different trip planning applications, both web based and manufacturer provided, which helps to obtain a route so that there are enough charging facilities to reach the destination.

9.3. Economic Problems

High Price

The price of the EVs is quite high compared to their ICE counterparts. This is because of the high cost of batteries [81] and fuel cells. To make people overlook this factor, governments in different countries including the UK and Germany, have provided incentives and tax breaks which provide the buyers of EVs with subsidies. Mass production and technological advancements will lead to a decrease in the prices of batteries as well as fuel cells. Affordable EVs with a long range like the Chevrolet Bolt has already appeared in the market, while another vehicle with the same promises (the Tesla Model 3) is anticipated to arrive soon. Figure 64 shows the limitations of EVs in the three sectors. Table 29 demonstrates the drawbacks in key factors, while Table 28 suggests some solutions for the existing limitations.

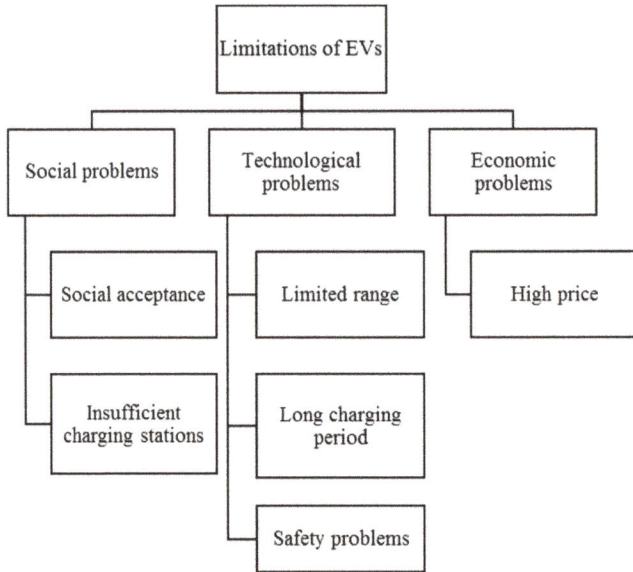

Figure 64. Social, technological, and economic problems faced by EVs.

Table 28. Tentative solutions of current limitations of EVs.

Limitation	Probable Solution
Limited range	Better energy source and energy management technology
Long charging period	Better charging technology
Safety problems	Advanced manufacturing scheme and build quality
Insufficient charging stations	Placement of sufficient charging stations capable of providing services to all kinds of vehicles
High price	Mass production, advanced technology, government incentives

10. Optimization Techniques

To make the best out of the available energy, EVs apply various aerodynamics and mass reduction techniques, lightweight materials are used to decrease the body weight as well. Regenerative braking is used to restore energy lost in braking. The restored energy can be stored in different ways. It can be stored directly in the ESS, or it can be stored by compressing air by means of hydraulic motor, springs can also be employed to store this energy in form of gravitational energy [169].

Table 29. Hurdles in key EV factors. Adapted from [170].

Factor	Hurdles
Recharging	Weight of charger, durability, cost, recycling, size, charging time
Hybrid EV	Battery, durability, weight, cost
Hydrogen fuel cell	Cost, hydrogen production, infrastructure, storage, durability, reliability
Auxiliary power unit	Size, cost, weight, durability, safety, reliability, cooling, efficiency

Formula One vehicles employ kinetic energy recovery systems (KERSs) to use the energy gathered during braking to provide extra power during accelerating. The Porsche 911 GT3R hybrid uses a flywheel energy storage system to store this energy. The energy consuming accessories on a car include power steering, air conditioning, lights, infotainment systems etc. Operating these in an energy efficient way or turning some of these off can increase the range of a vehicle. LEDs can be used for lighting because of their high efficiency [169]. Table 30 shows different methods of recovering the energy lost during braking.

Table 30. Different methods of recovering energy during braking [169].

Storage System	Energy Converter	Recovered Energy	Application
Electric storage	Electric motor/generator	~50%	BEV, HEV
Compressed gas storage	Hydraulic motor	>70%	Heavy-duty vehicles
Flywheel	Rotational kinetic energy	>70%	Formula One (F1) racing
Gravitational energy storage	Spring storage system	-	Train

Aerodynamic techniques are used in vehicles to reduce the drag coefficient, which reduces the required power. Power needed to overcome the drag force is:

$$P_d = \frac{1}{2}\rho v^3 A C_d \tag{7}$$

Here C_d is the drag coefficient, the power to overcome the drag increases if the drag coefficient's value increases. The Toyota Prius claims a drag coefficient of 0.24 for the 2017 model, the same as the Tesla Model S. The 2012 Nissan Leaf SL had this value set at 0.28 [171].

To ensure efficient use of the available energy, different energy management schemes can be employed [6]. Presented different control strategies for energy management which included systems using fuzzy logic, deterministic rule and optimization based schemes. Geng et al., worked on a plug-in series hybrid FCV. The objective of their control system was to consume the minimum amount of hydrogen while preserving the health of the proton exchange membrane fuel cell (PEMFC) [172]. The control system was comprised of two stages; the first stage determined the SOC and control references, whereas the second stage determined the PEMFC health parameters. This method proved to be capable of reducing the hydrogen consumption while increasing the life-time to the fuel cell. Another intelligent management system is examined in [173] by Murphey et al., which used machine learning combined with dynamic programming to determine energy optimization strategies for roadway and traffic-congestion scenarios for real-time energy flow control of a hybrid EV. Their system is simulated using a Ford Escape Hybrid model; it revealed the system was effective in finding out congestion level, optimal battery power and optimal speed. Geng et al., proposed a control mechanism for energy management for a PHEV employing batteries and a micro turbine in [174]. In this work, they introduced a new parameter, named the "energy ratio", to produce the equivalent factor (EF) which was used in the popular Equivalent Consumption Minimization Strategy (ECMS) to deduce the minimum driving cost by applying Pontryagin's minimum principle. This method claimed to reduce the cost by 7.7–21.6%. In [175], Moura et al., explored efficient ways to split power demand among different power sources of mid-sized sedan PHEVs. They used a number of drive cycles, rather than a single one, assessed the potential of depleting charge in a controlled manner, and considered relative pricing of fuel and electricity for optimal power management of the vehicle.

11. Control Algorithms

Control systems are crucial for proper functioning of EVs and associated systems. Sophisticated control mechanisms are required for providing a smooth and satisfactory ride quality, for providing the enough power when required, estimating the energy available from the on-board sources and using them properly to cover the maximum distance, charging in a satisfactory time without causing burden on the grid, and associated tasks. Different algorithms are used in these areas, and as the EV culture is becoming more mainstream, need for better algorithms are on the rise.

Driving control systems are required to assist the driver in keeping the vehicle in control, especially at high speeds and in adverse conditions such as slippery surfaces caused by rain or snow. Driving control systems such as traction control, cruise control, and different driving modes have been being applied in conventional vehicles for a long time. Application of such systems appeared more efficient in EVs as the driving forces of EVs can be controlled with more ease, with less conversion required in-between the mechanical and the electrical domains. In any condition, forces act on a vehicle at different directions; for a driving control system, if is essential to perfectly perceive these forces, along with other sensory inputs, and provide torques to the wheels to maintain desired stability. In Figure 65, the forces in different direction acting on each wheel of a car is shown in a horizontal plane. In [176], Magallan et al., proposed and simulated a control system to utilize the maximum torque in a rear-wheel-drive EV without causing the tires to skid. The model they worked on had independent driving systems for the two rear wheels. A sliding mode system, based on a *LuGre* dynamic friction model, was used to estimate the vehicle's velocity and wheel slip on unknown road surfaces. Utilizing these data, the control algorithm determined the maximum allowable traction force, which was applied to the road by torque controlling of the two rear motors. Juyong Kang et al., presented an algorithm aimed at driving control systems for four-wheel-drive EVs in [177]. Their vehicle model had two motors driving the front and the rear shafts. The algorithm had three parts: a supervisory level for determine the desirable dynamics and control mode, an upper level computing the yaw moment and traction force inputs, and a lower level determining the motor and braking commands. This system proved useful for enhancing lateral stability, maneuverability, and reducing rollover. Figure 66 shows the acting components of this system on a vehicle model while Figure 67 shows a detailed diagram of the system with the inputs, controller levels, and actuators. Tahami et al., introduced a stability system for driving assistance for all-wheel drive EVs in [25]. They trained a neural network to produce a reference yaw rate. A fuzzy logic controller dictated independent wheel torques; a similar controller was used for controlling wheel slip. This system is shown in Figure 68. In [178], Wang et al., showed a system to assist steering using differential drive for in-wheel drive system. A proportional integral (PI) closed loop control system was used here to monitor the reference steering position. It was achieved by distributing torque at the front wheels. Direct yaw moment control and traction control were also employed to make the differential drive system better. This approach maintained the lateral stability of the vehicle, and improved stability at high speeds. The structure of this system is shown in Figure 69. In a separate study conducted by Nam et al., lateral stability of an in-wheel drive EV was attained by estimating the sideslip angle of the vehicle employing sensors to measure lateral tire forces [179]. In this study, a state observer was proposed which was derived from extended-Kalman-filtering (EKF) method and was evaluated by implementing in an experimental EV alongside Matlab/Simulink-Carsim simulations.

Energy management is a big issue for EVs. Proper measurement of the available energy is crucial for calculating the range and plans the driving strategy thereafter. For vehicles with multiple energy sources (e.g., HEVs), efficient energy management algorithms are required to make proper use of the energy on-board. Zhou et al., proposed a battery state-of-charge (SOC) measuring algorithm for lithium polymer batteries which made use of a combination of particle filter and multi-model data fusion technique to produce results real time and is not affected by measurement noise [180]. They used different battery models and presented the tuning strategies for each model as well. Their multi-model approach proved to be more effective than single model methods for providing real time results. Working principle of this system is shown in Figure 70. Moura et al., explored efficient ways to split power demand among different power sources of mid-sized sedan PHEVs in [175], which can be used for other vehicle configurations as well. Their method made use of different drive cycles, rather than using a single one; assessed the potential of depleting charge in a controlled manner; and considered relative pricing of fuel and electricity to optimally manage the power of the vehicle. In [181], Hui et al., presented a novel hybrid vehicle using parallel hybrid architecture which employed a hydraulic/electric synergy configuration to mitigate the drawbacks faced by heavy hybrid vehicles using a single energy source. Transition among the operating modes of such a vehicle is shown in Figure 71. They developed an algorithm to optimize the key parameters and adopted a logic threshold approach to attain desired performance, stable SOC at the rational operating range constantly, and maximized fuel economy. The operating principle of this system is shown in Figure 72. Chen et al., proposed an energy management algorithm in [182] to effectively control battery current, and thus reduce fuel usage by allowing the engine operate more effectively. Quadratic programming was used here to calculate the optimum battery current. In [183], Li et al., used fuzzy logic to create a new quantity: battery working state or BWS which was used in an energy management system run by fuzzy logic to provide proper power division between the engine and the battery. Simulation results proved this approach to be effective in making the engine operate in the region of maximum fuel efficiency while keeping the battery away from excess discharging. Yuan et al., compared Dynamic Programming and Pontryagin's Minimum Principle (PMP) for energy management in parallel HEVs using Automatic Manual Transmission. The PMP method proved better as it was more efficient to implement, required considerably less computational time, and both of the systems provided almost similar results [184]. In [185], Bernard et al., proposed a real time control system to reduce hydrogen consumption in FCEVs by efficiently sharing power between the fuel cell arrangement and the energy buffer (ultracapacitor or battery). This control system was created from an optimal control theory based non-causal optimization algorithm. It was eventually implemented in a hardware arrangement built around a 600 W fuel cell arrangement. In an attempt to create an energy management system for a still-not-commercialized PHEV employing a micro turbine, Geng et al., used an equivalent consumption minimization strategy (ECMS) in [174] to estimate the optimum driving cost. Their system used the battery SOC and the vehicle telemetry to produce the results, which were available in real time and provided driving cost reductions of up to 21.6%.

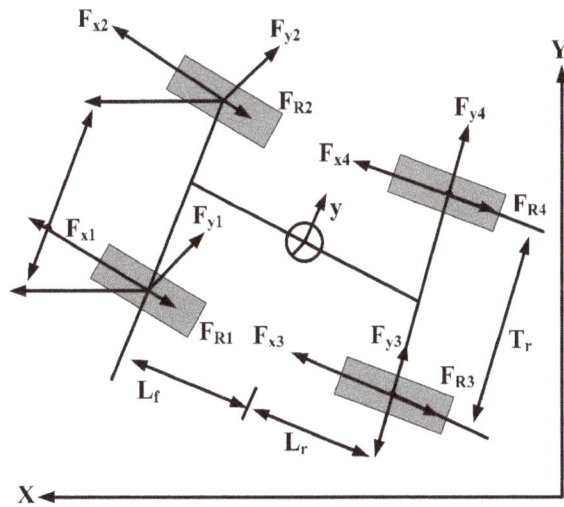

Figure 65. Forces acting on the wheels of a car. Each of the wheels experience forces in all three directions, marked with the 'F' vectors. L_f and L_r show the distances of front and rear axles from the center of the vehicle, while T_r shows the distance between the wheels of an axle. Adapted from [25].

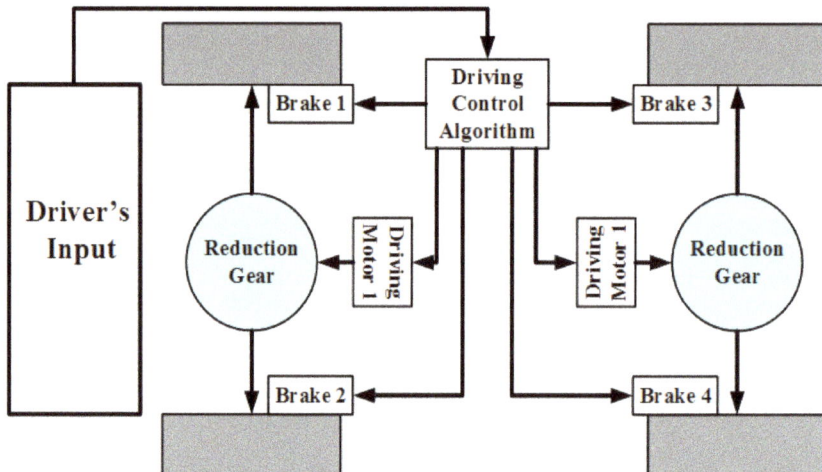

Figure 66. Main working components of the driving control system for four-wheel-drive EVs proposed by Juyong Kang et al. The driving control algorithm takes the driver's inputs, and then determines the actions of the brakes and the motors according to the control mode [177].

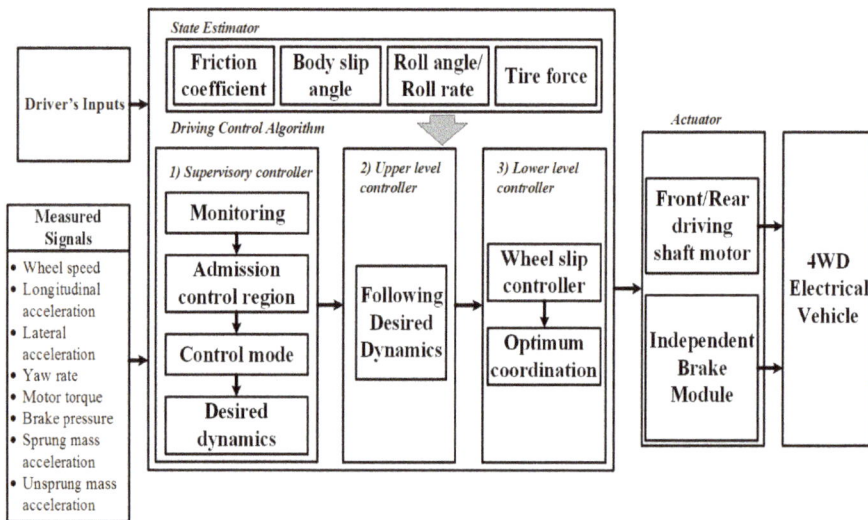

Figure 67. Working principle of the control system proposed by Kang et al. The system uses both the driver's commands and sensor measurements as inputs, and then drives the actuators as determined by the three level control algorithms. Adapted from [177].

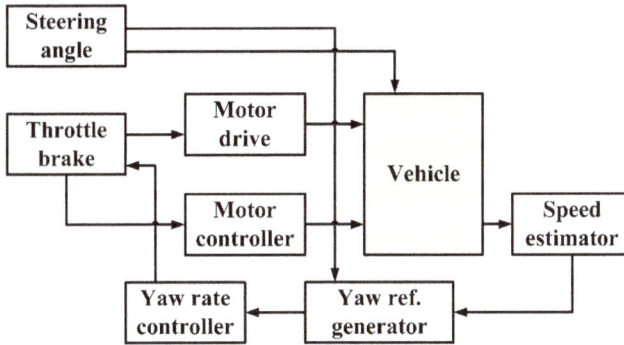

Figure 68. Working principle of vehicle stability system proposed by Tahami et al. A neural network was used in the yaw reference generator [25].

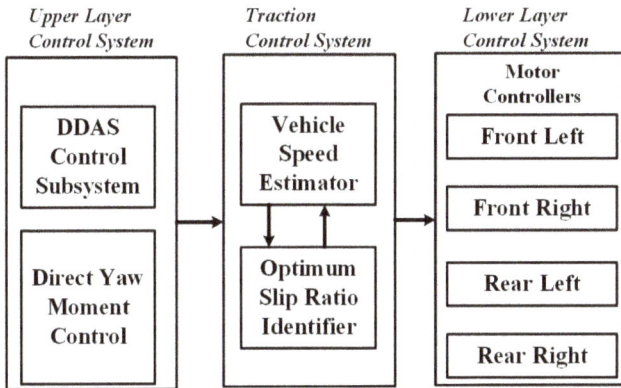

Figure 69. Independent torque control system proposed by Wang et al., Differential drive assisted steering (DDAS) subsystem and direct yaw moment control subsystem creates the upper layer. The traction control subsystem processes the inputs, and the controlling is done through the lower layer [178].

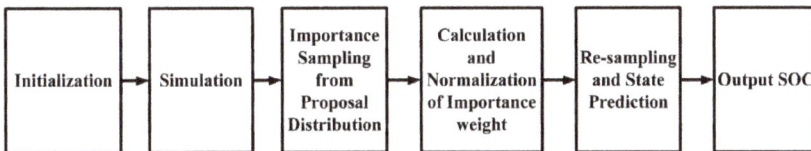

Figure 70. Working principle of the SOC measuring algorithm proposed by Zhou et al. [180].

As pointed out in Section 8, the grid is facing some serious problems with the current rise in EV penetration. Reducing the charging time of the vehicles while creating minimal pressure on the grid has become difficult goal to achieve. However, ample research has already been done on this matter and a number of charging system algorithms have been proposed to attain satisfactory charging performance.

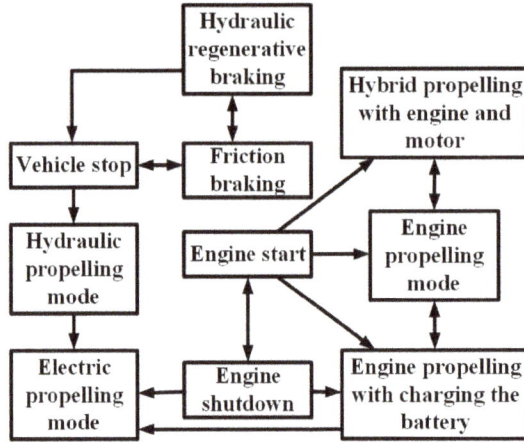

Figure 71. Transition of the operating modes of the vehicle used in [181] by Hui et al. From engine start to shutdown through stops, the vehicle can use either the hydraulic or the electric system, or it can use both.

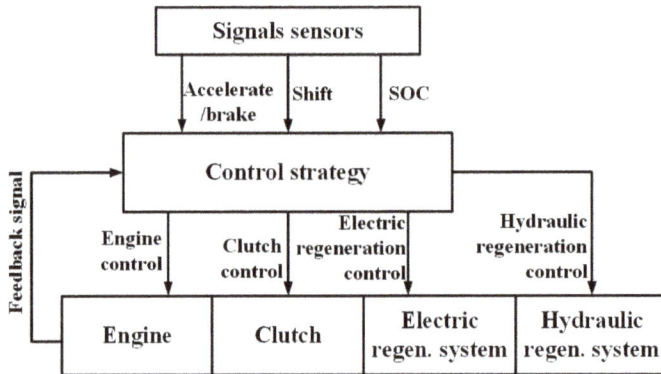

Figure 72. Operating principle of the control system proposed by Hui et al. The control strategy drives the actuating systems according to the decisions made from the sensor inputs. Adapted from [181].

In [186], Su et al., presented an algorithm (shown in Figure 73) capable of providing charge intelligently to a large fleet of PHEVs docked at a municipal charging station. This algorithm—which used the estimation of distribution (EDA) algorithm—considered real-world factors such as remaining charging time, remaining battery capacity, and energy price. The load management system proposed by Deilami et al., in [140] considered market energy prices that vary with time, time zones preferred by EV owners by priority selection, and random plugging-in of EVs—for providing coordinated charging in a smart grid system. It then used the maximum sensitivities selection (MSS) optimization technique to enable EVs charge as soon as possible depending on the priority time zones while maintaining the operation criteria of the grid such as voltage profile, limits of generation, and losses. This system was simulated using an IEEE 23 kV distribution system modified for this purpose. Mohamed et al., designed an energy management algorithm to be applied in EV charging parks incorporating renewable generation such as PV systems [187]. The system they developed used a fuzzy controller to manage the charging/discharging times of the connected EVs, power sharing among them, and V2G services.

The goal of this system was to minimize the charging cost while reducing the impact on the grid as well as contributing to peak shaving. The flowchart associated to this system is shown in Figure 74.

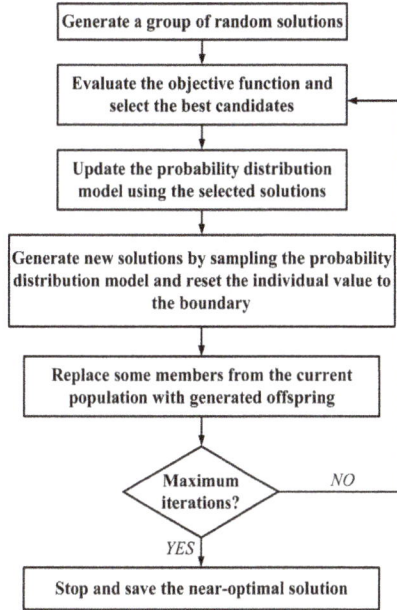

Figure 73. Intelligent charging algorithm proposed by Su et al., for a municipal charging station [186].

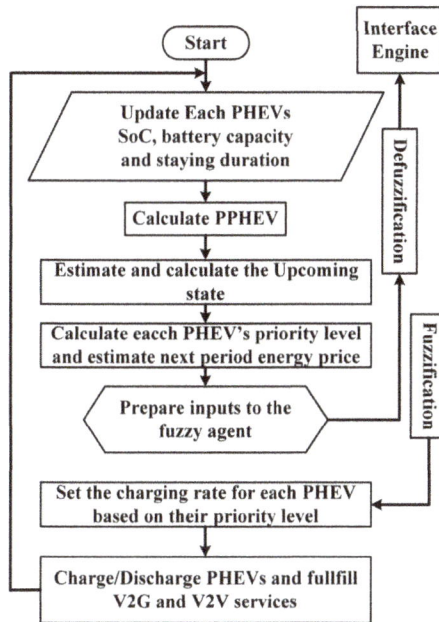

Figure 74. Flowchart of the management system proposed by Mohamed et al. [187].

To alleviate the problems at the distribution stage of the grid—which is highly affected by EV penetration—Geng et al., proposed a charging strategy comprising of two stages aimed at providing satisfactory charging for all connected EVs while shifting the loads on the transformers [120]. The first stage utilized Pontryagin's minimum principle and was based on the concept of dynamic aggregator; it derived the optimal charging power for all the EVs in the system. The second stage used fuzzy logic to distribute the power calculated in the first stage among the EVs. According to the authors, the system was feasible to be implemented practically [120]. In [116], Richardson et al., employed a linear programming based technique to calculate the optimal rate of charging for each EV connected in a distribution network to enable maximized power delivery to the vehicles while maintaining the network limits. This approach can provide high EV penetration possible in existing residential power systems with no or a little upgrade. Sortomme et al., developed an algorithm to maximize profit from EV charging in a unidirectional V2G system where an aggregator is present to manage the charging [146]. Table 31 summarizes the algorithms presented in this section.

Table 31. Summary of the control algorithms presented.

References	Algorithm Based on	Application
Magallan et al. [176]	*LuGre* dynamic friction model	Driving control system in rear-wheel-drive EV
Kang et al. [177]	Optimization-based control allocation strategy	Driving control system in four-wheel-drive EV
Tahami et al. [25]	Fuzzy logic	Driving control system in all-wheel-drive EV
Wang et al. [178]	Proportional-integral (PI) closed loop control system	Driving control system in in-wheel-drive EV
Nam et al. [179]	Extended Kalman filtering (EKF) method	Driving control system in in-wheel-drive EV
Zhou et al. [180]	Particle filter and multi-model data fusion	SOC measurement for lithium polymer batteries
Moura et al. [175]	Markov process	Power splitting in mid-sized sedan PHEV
Hui et al. [181]	Torque control strategy	Heavy hybrid vehicles using a single energy source
Chen et al. [182]	Quadratic programming	Reduction of fuel consumption by effective battery current control
Li et al. [183]	Fuzzy logic	Attaining maximum fuel efficiency without excess discharging of battery
Yuan et al. [184]	Dynamic Programming and Pontryagin's Minimum Principle	Efficient energy management in parallel HEV using Automatic Manual Transmission or AMT
Bernard et al. [185]	Non-causal optimization algorithm	Reduction of hydrogen consumption in FCEV
Geng et al. [174]	Equivalent consumption minimization strategy (ECMS)	Energy management in PHEV employing microturbine
Su et al. [186]	Estimation of distribution (EDA) algorithm	Intelligent charging of large fleet of PHEVs docked at a municipal charging station
Deilami et al. [140]	Maximum sensitivities selection (MSS) optimization	Load management system for intelligent charging
Mohamed et al. [187]	Fuzzy controller	V2G system for EV charging parks incorporating renewable generation
Geng et al. [120]	Pontryagin's minimum principle, fuzzy logic	Load shifting while charging EVs in the distribution network
Richardson et al. [116]	Linear programming	Enabling high EV penetration in existing residential power system network
Sortomme et al [146]	Preferred operating point (POP) algorithm	Maximizing profit from EV charging through an aggregator

12. Global EV Sales Figures

The electric vehicle market is growing much faster than the conventional vehicle market, and in some regions EVs are catching up with ICE vehicles in terms of the number of units sold. China has become the largest market for EVs, its market claiming 35.4% of the worldwide EV scene in 2017, an exorbitant rise from the mere 6.3% in 2013 [188]. Chinese consumers bought a world-topping 24.38 million passenger electric vehicles in 2016. China has the greatest number of manufacturers, led by BYD autos, which sold 96,000 EVs in 2016. This drive in China is fueled by government initiatives adopted to promote EV use to mitigate the country's serious air pollution. However, the majority of Chinese vehicles are in the $36,000 range and offers limited range, but high-end vehicles manufacturing is on the rise in China too. This huge market has attracted major carmakers all over the world—Ford, Volkswagen, Volvo, and General Motors—who have their own EVs in the Chinese market and are poised to introduce more models in the coming years [189]. Figure 75 shows the ten highest selling EVs in China in 2016.

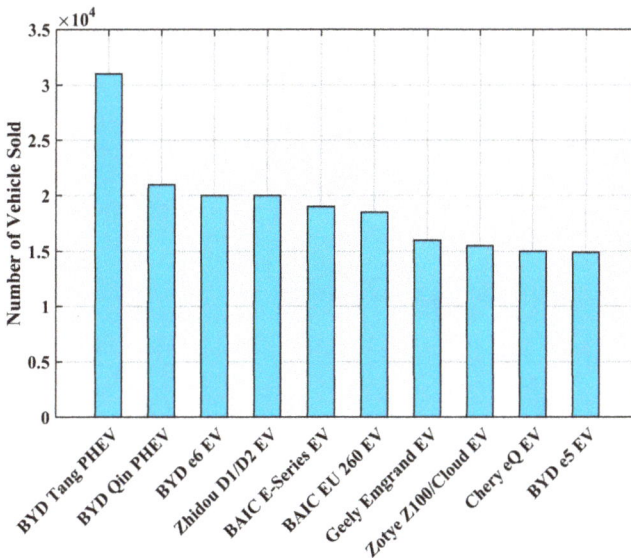

Figure 75. Top ten EVs in China in 2016 according to the number of units sold. Data from [190].

From a global perspective, sales of EV grew by 36% in the USA; Europe saw a growth of 13%, while Japan observed a decrease of 11% in the same period. BYD dominated the global market with a 13.2% share, followed by Tesla in second place (9.9%); the other major contributors can be listed as Volkswagen Group, BMW Group, Nissan, BAIC, and Zoyte. However, the Tesla Model S remained the best-selling EV in 2016 with 50,935 units sold, followed by the Nissan Leaf EV with 49,818 units [191]. The top ten best-selling vehicles around the globe in shown in Figure 76.

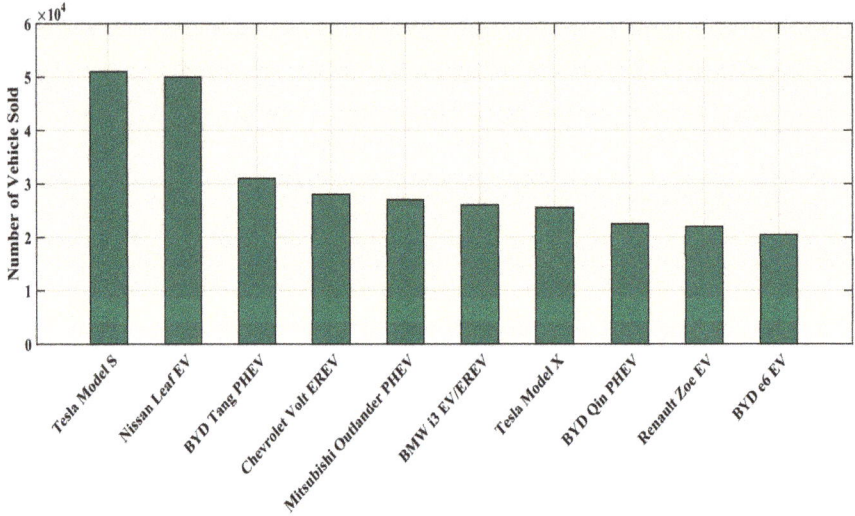

Figure 76. Top ten best-selling EVs globally in 2016. Data from [191].

The American market was dominated predictably by the Tesla Model S in 2016, 28,821 of these were sold; Chevrolet Volt EREV sold 24,739 units, thus securing the second place. The third place was achieved by another Tesla, the Model X; 18,192 of these SUVs were sold in 2016 [192]. The ten best-selling EVs in the USA in 2016 are shown in Figure 77.

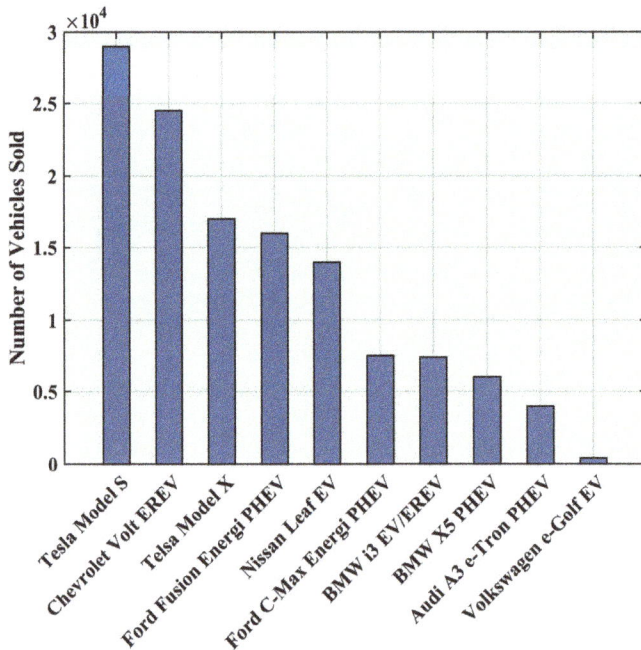

Figure 77. Top ten best-selling EVs in the USA in 2016. Data from [192].

The Renault Zoe was the best-selling BEV in Europe in 2016, with 21,338 units sold, followed by the Nissan Leaf with 18,614 units. In the PHEV segment, the Mitsubishi Outlander PHEV was the market leader in Europe in 2016, with 21,333 units sold; the Volkswagen Passat GTE held the second position with 13,330 units [193]. Figures 78 and 79 shows the BEV and PHEV market shares in Europe in 2016.

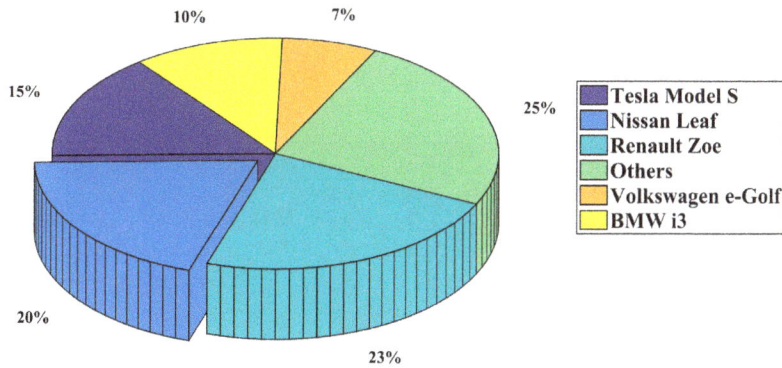

Figure 78. BEV market shares in Europe in 2016. Data from [193].

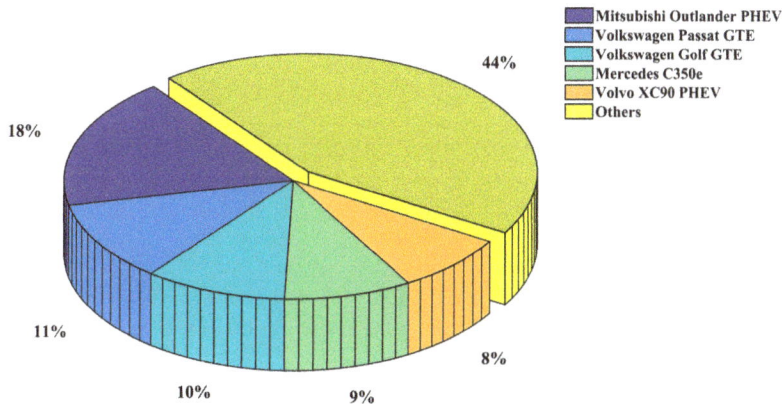

Figure 79. PHEV market shares in Europe in 2016. Data from [193].

13. Trends and Future Developments

The adoption of EVs has opened doors for new possibilities and ways to improve both the vehicles and the systems associated with it, the power system, for example. EVs are being considered as the future of vehicles, whereas the smart grid appears to be the grid of the future [194,195]. V2G is the link between these two technologies and both get benefitted from it. With V2G comes other essential systems required for a sustainable EV scenario—charge scheduling, VPP, smart metering etc. The existing charging technologies have to improve a lot to make EVs widely accepted. The charging time has to be decreased extensively for making EVs more flexible. At the same time, chargers and EVSEs have to able to communicate with the grid for facilitating V2G, smart metering, and if needed, bidirectional charging [23]. Better batteries are a must to take the EV technology further. There is a need for batteries that use non-toxic materials and have higher power density, less cost and weight, more capacity, and needs less time to recharge. Though technologies better than Li-ion have been discovered already, they are not being pursued industrially because of the huge costs associated with creating a working version. Besides, Li-ion technology has the potential to be improved a lot more. Li-air batteries could be a good option to increase the range of EVs [23]. EVs are likely to move away from using permanent magnet motors which use rare-earth materials. The motors of choice can be induction motor, synchronous reluctance motor, and switched reluctance motor [23]. Tesla is using an induction motor in its models at present. Motors with internal permanent magnet may stay in use [23]. Wireless power transfer systems are likely to replace the current cabled charging system. Concepts revealed by major automakers adopted this feature to highlight their usefulness and convenience. The Rolls-Royce 103EX and the Vision Mercedes-Maybach 6 can be taken as example for that. Electric roads for wireless charging of vehicles may appear as well. Though this is not still viable, the situation may change in the future. Recent works in this sector includes the work of Electrode, an Israeli startup, which claims to be able to achieve this feat in an economic way. Vehicles that follow a designated route along the highway, like trucks, can get their power from overhead lines like trains or trams. It will allow them to gather energy as long as their route resides with the power lines, then carry on with energy from on-board sources. Such a system has been tested by Siemens using diesel-hybrid trucks from Scania on a highway in Sweden [196]. New ways of recovering energy from the vehicle may appear. Goodyear has demonstrated a tire that can harvest energy from the heat generated there using thermo-piezoelectric material. There are also chances of solar-powered vehicles. Until now, these have not appeared useful as installed solar cells only manage to convert up to 20% of the input power [70]. Much research is going on to make the electronics and sensors in EVs more compact, rugged and cheaper—which in many cases are leading to advanced solid state devices that can achieve these goals with promises of cheaper products if they can be mass-produced. Some examples can be the works on gas sensors [197], smart LED drivers [198], smart drivers for automotive alternators [199], advanced gearboxes [200], and compact and smart power switches to weather harsh conditions [201]. The findings of [202–208] may prove helpful for studies regarding fail-proof on-board power supplies for EVs. The future research topics will of course, revolve around making the EV technology more efficient, affordable, and convenient. A great deal of research has already been conducted on making EVs more affordable and capable of covering more distance: energy management, materials used for construction, different energy sources etc. More of such researches are likely to go on emphasizing on better battery technologies, ultracapacitors, fuel cells, flywheels, turbines, and other individual and hybrid configurations. FCVs may get significant attention in military and utility-based studies, whereas the in-wheel drive configuration for BEVs may be appealing to researchers focusing on better urban transport systems. Better charging technologies will remain a crucial research topic in near future. This is one of the areas the EV technology is lacking very badly; wireless charging technologies are very likely to attract more researchers' attention. A lot of research has already been done incorporating EVs and the grid: the challenges and possibilities that the EVs bring with them to the existing grid and also to the grid of the future. With more implementation of smart grids, distributed generation, and renewable energy sources, researches in these fields are likely to increase. And as researches in the entire aforementioned field's increase, exploration for better algorithms to run the systems is bound to rise. Figure 80 shows the major trends and sectors for future developments for EVs.

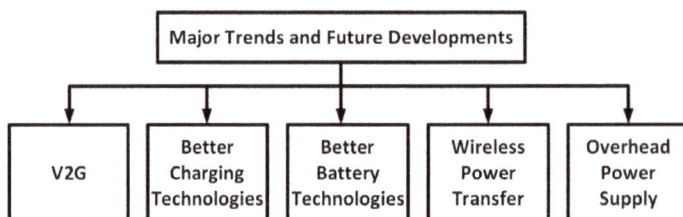

Figure 80. Major trends and sectors for future developments for EV.

14. Outcomes

The goal of this paper is to focus on the key components of EV. Major technologies in different sections are reviewed and the future trends of these sectors are speculated. The key findings of this paper can be summarized as follows:

- EVs can be classified as BEV, HEV, PHEV, and FCEV. BEVs and PHEVs are the current trends. FCEVs can become mainstream in future. Low cost fuel cells are the main prerequisite for that and there is need of more research to make that happen. There are also strong chances for BEVs to be the market dominators with ample advancement in key technologies; energy storage and charging systems being two main factors. Currently FCVs appear to have little chance to become ubiquitous, these may find popularity in niche markets, for example, the military and utility vehicles.
- EVs can be front wheel drive, rear wheel drive, even all-wheel drive. Different configurations are applied depending on the application of the vehicle. The motor can also be placed inside the wheel of the vehicle which offers distinct advantages. This configuration is not commercially abundant now, and has scopes for more study to turn it into a viable product.
- The main HEV configurations are classified as series, parallel, and series-parallel. Current vehicles are using the series-parallel system mainly as it can operate in both battery-only and ICE-only modes, providing more efficiency and less fuel consumption than the other two systems.
- Currently EVs use batteries as the main energy source. Battery technology has gone through significant changes, the lead-acid technology is long gone, as is the NiMH type. Li-ion batteries are currently in use, but even they are not capable enough to provide the amount of energy required to appease the consumers suffering from 'range anxiety' in most cases. Therefore the main focus of research in this area has to be creating batteries with more capacity, and also with better power densities. Metal-air batteries can be the direction where the EV makers will head towards. Lithium-sulfur battery and advanced rechargeable zinc batteries also have potential provide better EVs. Nevertheless, low cost energy sources will be sought after always as ESS cost is one of the major contributors to high EV cost.
- Ultracapacitors are considered as auxiliary power sources because of their high power densities. If coupled with batteries, ultracapacitors produce a hybrid ESS that can satisfy some requirements demanded from an ideal source. Flywheels are also being used, especially because of their compact build and capability to store and discharge power on demand. Fuel cells can also be used more in the future if FCVs gain popularity.
- Different types of motors can be employed for EV use. The prominent ones can be listed as induction motor, permanent magnet synchronous motor, and synchronous reluctance motor. Induction motors are being extensively these days, they can also dominate in future because of their independence on rare-earth material permanent magnets.
- EVs can be charged with AC or DC supply. There are different voltage levels and they are designated accordingly. Higher voltage levels provide faster charging. DC supplies negate the need of rectification from AC, which reduces delay and loss. However, with increased voltage level, the pressure on the grid increases and can give rise to harmonics as well as voltage imbalance in an unsupervised system. Therefore, there are ample chances of research in the field of mitigating the problems associated with high-voltage charging.
- Two charger configurations are mainly available now: CCS and CHAdeMO. These two systems are not compatible with each other and each has a number of automakers supporting them. Tesla also brought their own 'supercharger' system, which provides a faster charging facility. It is not possible to determine now which one of these will prevail, or if both will co-exist, technical study is needed to find out the most useful one of these configurations or ways to make them compatible with each other.

- Whatever the charging system is, the charging time is still very long. This is a major disadvantage that is thwarting the growth of the EV market. Extensive research is needed in this sector to provide better technologies that can provide much faster charging and can be compatible with the small time required to refill an ICE vehicle. Wireless charging is also something in need of research. With all the conveniences it promises, it is still not in a viable form to commercialize.

- EV impacts the environment, power system, and economy alongside the transportation sector. It shows promises to reduce the GHG emissions as well as efficient and economical transport solutions. At the same time, it can cause serious problems in the power system including voltage instability, harmonics, and voltage sag, but these shortcomings may be short-lived if smart grid technologies are employed. There are prospects of research in the areas of V2G, smart metering, integration of RES, and system stability associated with EV penetration.

- EVs employ different techniques to reduce energy loss and increase efficiency. Reducing the drag coefficient, weight reduction, regenerative braking, and intelligent energy management are some of these optimization techniques. Further research directions can be better aerodynamic body designs, new materials with less weight and desired strength, ways to generate and restore the lost energy.

- Different control algorithms have been developed for driving assist, energy management, and charging. There is lots of room left for more research into charging and energy management algorithms. With increased EV penetration in the future, demands for efficient algorithms are bound to increase.

15. Conclusions

EVs have great potential of becoming the future of transport while saving this planet from imminent calamities caused by global warming. They are a viable alternative to conventional vehicles that depend directly on the diminishing fossil fuel reserves. The EV types, configurations, energy sources, motors, power conversion and charging technologies for EVs have been discussed in detail in this paper. The key technologies of each section have been reviewed and their characteristics have been presented. The impacts EVs cause in different sectors have been discussed as well, along with the huge possibilities they hold to promote a better and greener energy system by collaborating with smart grid and facilitating the integration of renewable sources. Limitations of current EVs have been listed along with probable solutions to overcome these shortcomings. The current optimization techniques and control algorithms have also been included. A brief overview of the current EV market has been presented. Finally, trends and ways of future developments have been assessed followed by the outcomes of this paper to summarize the whole text, providing a clear picture of this sector and the areas in need of further research.

Acknowledgments: No funding has been received in support of this research work.

Author Contributions: All authors contributed for bringing the manuscript in its current state. Their contributions include detailed survey of the literatures and state of art which were essential for the completion of this review paper.

Conflicts of Interest: The authors declare no conflict of interest.

References

1. Yong, J.Y.; Ramachandaramurthy, V.K.; Tan, K.M.; Mithulananthan, N. A review on the state-of-the-art technologies of electric vehicle, its impacts and prospects. *Renew. Sustain. Energy Rev.* **2015**, *49*, 365–385. [CrossRef]
2. Camacho, O.M.F.; Nørgård, P.B.; Rao, N.; Mihet-Popa, L. Electrical Vehicle Batteries Testing in a Distribution Network using Sustainable Energy. *IEEE Trans. Smart Grid* **2014**, *5*, 1033–1042. [CrossRef]
3. Camacho, O.M.F.; Mihet-Popa, L. Fast Charging and Smart Charging Tests for Electric Vehicles Batteries using Renewable Energy. *Oil Gas Sci. Technol.* **2016**, *71*, 13–25. [CrossRef]

4. Chan, C.C. The state of the art of electric and hybrid vehicles. *Proc. IEEE* **2002**, *90*, 247–275. [CrossRef]
5. Grunditz, E.A.; Thiringer, T. Performance Analysis of Current BEVs Based on a Comprehensive Review of Specifications. *IEEE Trans. Transp. Electr.* **2016**, *2*, 270–289. [CrossRef]
6. SAE International. SAE Electric Vehicle and Plug-in Hybrid Electric Vehicle Conductive Charge Coupler. In *SAE Standard J1772*; Society of Automotive Engineers (SAE): Warrendale, PA, USA, 2010.
7. Yilmaz, M.; Krein, P.T. Review of battery charger topologies, charging power levels, and infrastructure for plug-in electric and hybrid vehicles. *IEEE Trans. Power Electr.* **2013**, *28*, 2151–2169. [CrossRef]
8. Bayindir, K.Ç.; Gözüküçük, M.A.; Teke, A. A comprehensive overview of hybrid electric vehicle: Powertrain configurations, powertrain control techniques and electronic control units. *Energy Convers. Manag.* **2011**, *52*, 1305–1313. [CrossRef]
9. Marchesoni, M.; Vacca, C. New DC–DC converter for energy storage system interfacing in fuel cell hybrid electric vehicles. *IEEE Trans. Power Electron.* **2007**, *22*, 301–308. [CrossRef]
10. Schaltz, E.; Khaligh, A.; Rasmussen, P.O. Influence of battery/ultracapacitor energy-storage sizing on battery lifetime in a fuel cell hybrid electric vehicle. *IEEE Trans. Veh. Technol.* **2009**, *58*, 3882–3891. [CrossRef]
11. Kramer, B.; Chakraborty, S.; Kroposki, B. A review of plug-in vehicles and vehicle-to-grid capability. In Proceedings of the 34th IEEE Industrial Electronics Annual Conference, Orlando, FL, USA, 10–13 November 2008; pp. 2278–2283.
12. Williamson, S.S. Electric drive train efficiency analysis based on varied energy storage system usage for plug-in hybrid electric vehicle applications. In Proceedings of the IEEE Power Electronics Specialists Conference, Orlando, FL, USA, 17–21 June 2007; pp. 1515–1520.
13. Wirasingha, S.G.; Schofield, N.; Emadi, A. Plug-in hybrid electric vehicle developments in the US: Trends, barriers, and economic feasibility. In Proceedings of the IEEE Vehicle Power and Propulsion Conference, Harbin, China, 3–5 September 2008; pp. 1–8.
14. Gao, Y.; Ehsani, M. Design and control methodology of plug-in hybrid electric vehicles. *IEEE Trans. Ind. Electron.* **2010**, *57*, 633–640.
15. EG&G Technical Services, Inc. *The Fuel Cell Handbook*, 6th ed.; U.S. Department of Energy: Morgantown, WV, USA, 2002.
16. Miller, J.F.; Webster, C.E.; Tummillo, A.F.; DeLuca, W.H. Testing and evaluation of batteries for a fuel cell powered hybrid bus. In Proceedings of the Energy Conversion Engineering Conference, Honolulu, HI, USA, 27 July–1 August 1997; Volume 2, pp. 894–898.
17. Rodatz, P.; Garcia, O.; Guzzella, L.; Büchi, F.; Bärtschi, M.; Tsukada, A.; Dietrich, P.; Kötz, R.; Scherer, G.; Wokaun, A. Performance and operational characteristics of a hybrid vehicle powered by fuel cells and supercapacitors. In Proceedings of the SAE 2003 World Congress and Exhibition, Detroit, MI, USA, 3 March 2003; Volume 112, pp. 692–703.
18. Thounthong, P.; Raël, S.; Davat, B. Utilizing fuel cell and supercapacitors for automotive hybrid electrical system. In Proceedings of the Applied Power Electronics Conference and Exposition, Austin, TX, USA, 6–10 March 2005; Volume 1, pp. 90–96.
19. Why the Automotive Future Will Be Dominated by Fuel Cells—IEEE Spectrum. Available online: http://spectrum.ieee.org/green-tech/fuel-cells/why-the-automotive-future-will-be-dominated-by-fuel-cells (accessed on 8 May 2017).
20. Rose, R. *Questions and Answers about Hydrogen and Fuel Cells; Report Style*; U.S. Department of Energy: Washington, DC, USA, 2005.
21. *U.S. Climate Technology Program: Technology Options for the Near and Long Term (Report Style)*; U.S. Climate Change Technology Program: Washington, DC, USA, 2005.
22. Thomas, C.E. Fuel cell and battery electric vehicles compared. *Int. J. Hydrogen Energy* **2009**, *34*, 6005–6020. [CrossRef]
23. Rajashekara, K. Present status and future trends in electric vehicle propulsion technologies. *IEEE J. Emerg. Sel. Top. Power Electron.* **2013**, *1*, 3–10. [CrossRef]
24. Model S | Tesla. Available online: https://www.tesla.com/models (accessed on 8 May 2017).
25. Tahami, F.; Kazemi, R.; Farhanghi, S. A novel driver assist stability system for all-wheel-drive electric vehicles. *IEEE Trans. Veh. Technol.* **2003**, *52*, 683–692. [CrossRef]
26. Sato, M.; Yamamoto, G.; Gunji, D.; Imura, T.; Fujimoto, H. Development of Wireless In-Wheel Motor Using Magnetic Resonance Coupling. *IEEE Trans. Power Electron.* **2016**, *31*, 5270–5278. [CrossRef]

27. Kurs, A.; Karalis, A.; Moffatt, R.; Joannopoulos, J.D.; Fisher, P.; Soljačić, M. Wireless power transfer via strongly coupled magnetic resonances. *Science* **2007**, *317*, 83–86. [CrossRef] [PubMed]

28. Imura, I.; Uchida, T.; Hori, Y. Flexibility of contactless power transfer using magnetic resonance coupling to air gap and misalignment for EV. *World Electr. Veh. J.* **2009**, *3*, 24–34.

29. Nakadachi, S.; Mochizuki, S.; Sakaino, S.; Kaneko, Y.; Abe, S.; Yasuda, T. Bidirectional contactless power transfer system expandable from unidirectional system. In Proceedings of the 2013 IEEE Energy Conversion Congress and Exposition, Denver, CO, USA, 15–19 September 2013; pp. 3651–3657.

30. Gao, Y.; Ehsani, M.; Miller, J.M. Hybrid Electric Vehicle: Overview and State of the Art. In Proceedings of the IEEE International Symposium on Industrial Electronics, Dubrovnik, Croatia, 20-23 June 2005; pp. 307–316.

31. Kim, H.; Kum, D. Comprehensive Design Methodology of Input- and Output-Split Hybrid Electric Vehicles: In Search of Optimal Configuration. *IEEE/ASME Trans. Mechatron.* **2016**, *21*, 2912–2923. [CrossRef]

32. Miller, J.M. Hybrid electric vehicle propulsion system architectures of the e-CVT type. *IEEE Trans. Power Electron.* **2006**, *21*, 756–767. [CrossRef]

33. Kim, D.; Hwang, S.; Kim, H. Vehicle Stability Enhancement of Four-Wheel-Drive Hybrid Electric Vehicle Using Rear Motor Control. *IEEE Trans. Veh. Technol.* **2008**, *57*, 727–735.

34. Li, Y.; Yang, J.; Song, J. Nano energy system model and nanoscale effect of graphene battery in renewable energy electric vehicle. *Renew. Sustain. Energy Rev.* **2017**, *69*, 652–663. [CrossRef]

35. Khaligh, A.; Li, Z. Battery, ultracapacitor, fuel cell, and hybrid energy storage systems for electric, hybrid electric, fuel cell, and plug-in hybrid electric vehicles: State of the art. *IEEE Trans. Veh. Technol.* **2010**, *59*, 2806–2814. [CrossRef]

36. Olson, J.B.; Sexton, E.D. Operation of lead–acid batteries for HEV applications. In Proceedings of the 15th Battery Conference on Applications and Advances, Long Beach, CA, USA, 11–14 January 2000; pp. 205–210.

37. Edwards, D.B.; Kinney, C. Advanced lead acid battery designs for hybrid electric vehicles. In Proceedings of the 16th Battery Conference on Applications and Advances, Long Beach, CA, USA, 12 January 2001; pp. 207–212.

38. Cooper, A.; Moseley, P. Progress in the development of lead–acid batteries for hybrid electric vehicles. In Proceedings of the IEEE Vehicle Power and Propulsion Conference, Windsor, UK, 6–8 September 2006; pp. 1–6.

39. Fetcenko, M.A.; Fetcenko, M.A.; Ovshinsky, S.R.; Reichman, B.; Young, K.; Fierro, C.; Koch, J.; Zallen, A.; Mays, W.; Ouchi, T. Recent advances in NiMH battery technology. *J. Power Sources* **2007**, *165*, 544–551. [CrossRef]

40. Li, H.; Liao, C.; Wang, L. Research on state-of-charge estimation of battery pack used on hybrid electric vehicle. In Proceedings of the Asia-Pacific Power and Energy Engineering Conference, Wuhan, China, 27–31 March 2009; pp. 1–4.

41. Chalk, S.G.; Miller, J.F. Key challenges and recent progress in batteries, fuel cells, and hydrogen storage for clean energy systems. *J. Power Sources* **2006**, *159*, 73–80. [CrossRef]

42. Balch, R.C.; Burke, A.; Frank, A.A. The affect of battery pack technology and size choices on hybrid electric vehicle performance and fuel economy. In Proceedings of the 16th IEEE Annual Battery Conference on Applications and Advances, Long Beach, CA, USA, 12 January 2001; pp. 31–36.

43. Viera, J.C.; Gonzalez, M.; Anton, J.C.; Campo, J.C.; Ferrero, F.J.; Valledor, M. NiMH vs. NiCd batteries under high charging rates. In Proceedings of the 28th Annual Telecommunications Energy Conference, Providence, RI, USA, 10–14 September 2006; pp. 1–6.

44. Gao, Y.; Ehsani, M. Investigation of battery technologies for the army's hybrid vehicle application. In Proceedings of the 56th IEEE Vehicular Technology Conference, Vancouver, BC, Canada, 24–28 September 2002; pp. 1505–1509.

45. Pilot, C. The Rechargeable Battery Market and Main Trends 2014–2025. Available online: http://www.avicenne.com/pdf/Fort_Lauderdale_Tutorial_C_Pillot_March2015.pdf (accessed on 29 July 2017).

46. Williamson, S.S.; Rathore, A.K.; Musavi, F. Industrial electronics for electric transportation: Current state-of-the-art and future challenges. *IEEE Trans. Ind. Electron.* **2015**, *62*, 3021–3032. [CrossRef]

47. Cassani, P.A.; Williamson, S.S. Feasibility analysis of a novel cell equalizer topology for plug-in hybrid electric vehicle energy-storage systems. *IEEE Trans. Veh. Technol.* **2009**, *58*, 3938–3946. [CrossRef]

48. Baughman, A.C.; Ferdowsi, M. Double-tiered switched-capacitor battery charge equalization technique. *IEEE Trans. Ind. Electron.* **2008**, *55*, 2277–2285. [CrossRef]

49. Nishijima, K.; Sakamoto, H.; Harada, K. A PWM controlled simple and high performance battery balancing system. In Proceedings of the IEEE Power Electronics Specialists Conference, Galway, Ireland, 23 June 2000; Volume 1, pp. 517–520.

50. Cassani, P.A.; Williamson, S.S. Design, testing, and validation of a simplified control scheme for a novel plug-in hybrid electric vehicle battery cell equalizer. *IEEE Trans. Ind. Electron.* **2010**, *57*, 3956–3962. [CrossRef]

51. Lee, Y.S.; Cheng, M.W. Intelligent control battery equalization for series connected lithium-ion battery strings. *IEEE Trans. Ind. Electron.* **2005**, *52*, 1297–1307. [CrossRef]

52. Lee, Y.S.; Cheng, M.W.; Yang, S.C.; Hsu, C.L. Individual cell equalization for series connected lithium-ion batteries. *IEICE Trans. Commun.* **2006**, *E89-B*, 2596–2607. [CrossRef]

53. 2017 Nissan LEAF® Electric Car Specs. Available online: https://www.nissanusa.com/electric-cars/leaf/versions-specs/ (accessed on 8 May 2017).

54. Model S Specifications | Tesla. Available online: https://www.tesla.com/support/model-s-specifications (accessed on 8 May 2017).

55. Why We Still Don't Have Better Batteries—MIT Technology Review. Available online: https://www.technologyreview.com/s/602245/why-we-still-dont-have-better-batteries/ (accessed on 8 May 2017).

56. Ribeiro, P.F.; Johnson, B.K.; Crow, M.L.; Arsoy, A.; Liu, Y. Energy storage systems for advanced power applications. *Proc. IEEE* **2001**, *89*, 1744–1756. [CrossRef]

57. Bartley, T. Ultracapacitors and batteries for energy storage in heavy-duty hybrid-electric vehicles. In Proceedings of the 22nd International Battery Seminar & Exhibit, Fort Lauderdale, FL, USA, 14–17 March 2005.

58. Gigaom | How Ultracapacitors Work (and Why They Fall Short). Available online: https://gigaom.com/2011/07/12/how-ultracapacitors-work-and-why-they-fall-short/ (accessed on 8 May 2017).

59. Singh, A.; Karandikar, P.B. A broad review on desulfation of lead-acid battery for electric hybrid vehicle. *Microsyst. Technol.* **2017**, *23*, 1–11. [CrossRef]

60. Chiu, H.J.; Lin, L.W. A bidirectional DC-DC converter for fuel cell electric vehicle driving system. *IEEE Trans. Power Electron.* **2006**, *21*, 950–958. [CrossRef]

61. Mahlia, T.M.I.; Saktisahdan, T.J.; Jannifar, A.; Hasan, M.H.; Matseelar, H.S.C. A review of available methods and development on energy storage; technology update. *Renew. Sustain. Energy Rev.* **2014**, *33*, 532–545. [CrossRef]

62. Bolund, B.; Bernhoff, H.; Leijon, M. Flywheel energy and power storage systems. *Renew. Sustain. Energy Rev.* **2007**, *11*, 235–258. [CrossRef]

63. Luo, X.; Wang, J.; Dooner, M.; Clarke, J. Overview of current development in electrical energy storage technologies and the application potential in power system operation. *Appl. Energy* **2015**, *137*, 511–536. [CrossRef]

64. Chan, C.C.; Chau, K.T. An overview of power electronics in electric vehicles. *IEEE Trans. Ind. Electron.* **1997**, *44*, 3–13. [CrossRef]

65. Chan, C.C.; Chau, K.T.; Jiang, J.Z.; Xia, W.A.X.W.; Zhu, M.; Zhang, R. Novel permanent magnet motor drives for electric vehicles. *IEEE Trans. Ind. Electron.* **1996**, *43*, 331–339. [CrossRef]

66. Chan, C.C.; Chau, K.T.; Yao, J. Soft-switching vector control for resonant snubber based inverters. In Proceedings of the IEEE International Conference Industrial Electronics, New Orleans, LA, USA, 14 November 1997; pp. 605–610.

67. Chan, C.C.; Jiang, J.Z.; Chen, G.H.; Chau, K.T. Computer simulation and analysis of a new polyphase multipole motor drive. *IEEE Trans. Ind. Electron.* **1993**, *40*, 570–576. [CrossRef]

68. Chan, C.C.; Jiang, J.Z.; Chen, G.H.; Wang, X.Y.; Chau, K.T. A novel polyphase multipole square-wave permanent magnet motor drive for electric vehicles. *IEEE Trans. Ind. Appl.* **1994**, *30*, 1258–1266. [CrossRef]

69. Chan, C.C.; Jiang, J.Z.; Xia, W.; Chan, K.T. Novel wide range speed control of permanent magnet brushless motor drives. *IEEE Trans. Power Electron.* **1995**, *10*, 539–546. [CrossRef]

70. Jose, C.P.; Meikandasivam, S. A Review on the Trends and Developments in Hybrid Electric Vehicles. In *Innovative Design and Development Practices in Aerospace and Automotive Engineering*; Springer: Singapore, 2017; pp. 211–229.

71. Chan, C.C.; Chau, K.T. *Morden Elcetric Vehicle Technology*; Oxford University Press, Inc.: New York, NY, USA, 2001; pp. 122–133.

72. Lulhe, A.M.; Date, T.N. A technology review paper for drives used in electrical vehicle (EV) & hybrid electrical vehicles (HEV). In Proceedings of the 2015 International Conference on Control, Instrumentation, Communication and Computational Technologies (ICCICCT), Kumaracoil, India, 18–19 December 2015.

73. Magnussen, F. On design and analysis of synchronous permanent magnet for field—Weakening operation. Ph.D. Thesis, KTH Royal Institute of Technology, Sweden, 2004.

74. Model X Specifications | Tesla. Available online: https://www.tesla.com/support/model-x-specifications (accessed on 8 May 2017).

75. Yamada, K.; Watanabe, K.; Kodama, T.; Matsuda, I.; Kobayashi, T. An efficiency maximizing induction motor drive system for transmissionless electric vehicle. In Proceedings of the 13th International Electric Vehicle Symposium, Osaka, Japan, 13–16 Octobor 1996; Volume II, pp. 529–536.

76. Boglietti, A.; Ferraris, P.; Lazzari, M.; Profumo, F. A new design criteria for spindles induction motors controlled by field oriented technique. *Electr. Mach. Power Syst.* **1993**, *21*, 171–182. [CrossRef]

77. Abbasian, M.; Moallem, M.; Fahimi, B. Double-stator switched reluctance machines (DSSRM): Fundamentals and magnetic force analysis. *IEEE Trans. Energy Convers.* **2010**, *25*, 589–597. [CrossRef]

78. Cameron, D.E.; Lang, J.H.; Umans, S.D. The origin and reduction of acoustic noise in doubly salient variable-reluctance motors. *IEEE Trans. Ind. Appl.* **1992**, *28*, 1250–1255. [CrossRef]

79. Chan, C.C.; Jiang, Q.; Zhan, Y.J.; Chau, K.T. A high-performance switched reluctance drive for P-star EV project. In Proceedings of the 13th International Electric Vehicle Symposium, Osaka, Japan, 13–16 Octobor 1996; Volume II, pp. 78–83.

80. Zhan, Y.J.; Chan, C.C.; Chau, K.T. A novel sliding-mode observer for indirect position sensing of switched reluctance motor drives. *IEEE Trans. Ind. Electron.* **1999**, *46*, 390–397. [CrossRef]

81. Shareef, H.; Islam, M.M.; Mohamed, A. A review of the stage-of-the-art charging technologies, placement methodologies, and impacts of electric vehicles. *Renew. Sustain. Energy Rev.* **2016**, *64*, 403–420. [CrossRef]

82. Yu, X.E.; Xue, Y.; Sirouspour, S.; Emadi, A. Microgrid and transportation electrification: A review. In Proceedings of the 2012 IEEE Transportation Electrification Conference and Expo (ITEC), Dearborn, MI, USA, 18–20 June 2012.

83. Consumer and Clinical Radiation Protection Bureau; Environmental and Radiation Health Sciences Directorate; Healthy Environments and Consumer Safety Branch; Health Canada. Limits of human exposure to radiofrequency electromagnetic energy in the frequency range from 3 kHz to 300 GHz. *Health Can. Saf. Code* **2009**, *6*, 10–11.

84. *IEEE Standard for Safety Levels with Respect to Human Exposure to Radio Frequency Electromagnetic Fields, 3 kHz to 300 GHz*; IEEE Std C95.1; IEEE: New York, NY, USA, 1999.

85. Ahlbom, A.; Bergqvist, U.; Bernhardt, J.H.; Cesarini, J.P.; Court, L.A.; Grandolfo, M.; Hietanen, M.; McKinlay, A.F.; Repacholi, M.H.; Sliney, D.H. Guidelines: For limiting exposure to time-varying electric, magnetic and electromagnetic fields (up to 300 GHz). *Health Phys.* **1998**, *74*, 494–521.

86. Australian Radiation Protection and Nuclear Safety Agency (ARPANSA). *Radiation Protection Standard: Maximum Exposure Levels to Radiofrequency Fields—3 kHz to 300 GHz*; Radiation Protection Series Publication No. 3; ARPANSA: Melbourne, Australia, 2002.

87. Musavi, F.; Eberle, W. Overview of wireless power transfer technologies for electric vehicle battery charging. *IET Power Electron.* **2014**, *7*, 60–66. [CrossRef]

88. Chademo-Ceritifed Chrager List. Available online: www.chademo.com (accessed on 6 July 2015).

89. Supercharger. Available online: www.chademo.com (accessed on 7 July 2015).

90. International Electrotechnical Commission. *Standard IEC 62196—Plugs, Socket-Outlets, Vehicle Couplers and Vehicle Inlets—Conductive Charging of Electric Vehicles*; The International Electrotechnical Commission (IEC): Geneva, Switzerland, 2003.

91. Onar, O.C.; Kobayashi, J.; Khaligh, A. A Fully Directional Universal Power Electronic Interface for EV, HEV, and PHEV Applications. *IEEE Trans. Power Electron.* **2013**, *28*, 5489–5498. [CrossRef]

92. Bose, B.K. Power electronics—A technology review. *Proc. IEEE* **1992**, *80*, 1303–1334. [CrossRef]

93. Yaramasu, V.; Wu, B.; Sen, P.C.; Kouro, S.; Narimani, M. High-power wind energy conversion systems: State-of-the-art and emerging technologies. *Proc. IEEE* **2015**, *103*, 740–788. [CrossRef]

94. Kok, D.; Morris, A.; Knowles, M. Novel EV drive train topology—A review of the current topologies and proposal for a model for improved drivability. In Proceedings of the 2013 15th European Conference on Power Electronics and Applications (EPE), Lille, France, 2–6 September 2013.

95. Hegazy, O.; Van Mierlo, J.; Lataire, P. Analysis, control and comparison of DC/DC boost converter topologies for fuel cell hybrid electric vehicle applications. In Proceedings of the 14th European Conference on Power Electronics and Applications (EPE 2011), Birmingham, UK, 30 August–1 September 2011.

96. Hong, J.; Lee, H.; Nam, K. Charging Method for the Secondary Battery in Dual-Inverter Drive Systems for Electric Vehicles. *IEEE Trans. Power Electron.* **2015**, *30*, 909–921. [CrossRef]

97. Sangdehi, S.M.M.; Hamidifar, S.; Kar, N.C. A novel bidirectional DC/AC stacked matrix converter design for electrified vehicle applications. *IEEE Trans. Veh. Technol.* **2014**, *63*, 3038–3050. [CrossRef]

98. Kim, Y.J.; Lee, J.Y. Full-Bridge+ SRT Hybrid DC/DC Converter for a 6.6-kW EV On-Board Charger. *IEEE Trans. Veh. Technol.* **2016**, *65*, 4419–4428. [CrossRef]

99. Kimura, S.; Itoh, Y.; Martinez, W.; Yamamoto, M.; Imaoka, J. Downsizing Effects of Integrated Magnetic Components in High Power Density DC–DC Converters for EV and HEV Applications. *IEEE Trans. Ind. Appl.* **2016**, *52*, 3294–3305. [CrossRef]

100. Schroeder, J.C.; Fuchs, F.W. Detailed Characterization of Coupled Inductors in Interleaved Converters Regarding the Demand for Additional Filtering. In Proceedings of the 2012 IEEE Energy Energy Conversion Congress and Exposition (ECCE), Raleigh, NC, USA, 15–20 September 2012; pp. 759–766.

101. Imaoka, J.; Yamamoto, M.; Nakamura, Y.; Kawashima, T. Analysis of output capacitor voltage ripple in multi-phase transformer-linked boost chopper circuit. *IEEE J. Ind. Appl.* **2013**, *2*, 252–260. [CrossRef]

102. Zhu, J.; Pratt, A. Capacitor Ripple Current in an interleaved PFC Converter. *IEEE Trans. Power Electron.* **2009**, *24*, 1506–1514. [CrossRef]

103. Wang, C.; Xu, M.; Lee, F.C.; Lu, B. EMI Study for the Interleaved Multi-Channel PFC. In Proceedings of the IEEE Power Electronics Specialists Conference (PESC), Orlando, FL, USA, 17–21 June 2007; pp. 1336–1342.

104. O'Loughlin, M. An Interleaved PFC Preregulator for High-Power Converters. Available online: http://www.ti.com/download/trng/docs/seminar/Topic5MO.pdf (accessed on 7 August 2017).

105. Balogh, L.; Redl, R. Power-factor correction with interleaved boost converters in continuous-inductor-current mode. In Proceedings of the IEEE Applied Power Electronics Conference and Exposition, San Diego, CA, USA, 7–11 March 1993; pp. 168–174.

106. Jang, Y.; Jovanovic, M.M. Interleaved boost converter with intrinsic voltage-doubler characteristic for universal-line PFC front end. *IEEE Trans. Power Electron.* **2007**, *22*, 1394–1401. [CrossRef]

107. Kong, P.; Wang, S.; Lee, F.C.; Wang, C. Common-mode EMI study and reduction technique for the interleaved multichannel PFC converter. *IEEE Trans. Power Electron.* **2008**, *23*, 2576–2584. [CrossRef]

108. Gautam, D.S.; Musavi, F.; Eberle, W.; Dunford, W.G. A zero voltage switching full-bridge dc-dc converter with capacitive output filter for a plug-in-hybrid electric vehicle battery charger. In Proceedings of the IEEE Applied Power Electronics Conference and Exposition, Orlando, FL, USA, 5–9 February 2012; pp. 1381–1386.

109. Musavi, F.; Craciun, M.; Gautam, D.S.; Eberle, W.; Dunford, W.G. An LLC resonant DC-DC Converter for wide output voltage range battery charging applications. *IEEE Trans. Power Electron.* **2013**, *28*, 5437–5445. [CrossRef]

110. Gurkaynak, Y.; Li, Z.; Khaligh, A. A novel grid-tied, solar powered residential home with plug-in hybrid electric vehicle (PHEV) loads. In Proceedings of the 5th Annual IEEE Vehicle Power and Propulsion Conference, Dearborn, MI, USA, 7–10 September 2009; pp. 813–816.

111. Onar, O. *Bi-Directional Rectifier/Inverter and Bi-Directional DC/DC Converters Integration for Plug-in Hybrid Electric Vehicles with Hybrid Battery/Ultra-Capacitors Energy Storage Systems*; Illinois Institute of Technology: Chicago, IL, USA, 2009.

112. Rashid, M.H. *Power Electronics Handbook: Devices, Circuits and Applications*; Elsevier: Amsterdam, The Netherlands, 2010.

113. Koushki, B.; Safaee, A.; Jain, P.; Bakhshai, A. Review and comparison of bi-directional AC-DC converters with V2G capability for on-board EV and HEV. In Proceedings of the 2014 IEEE Transportation Electrification Conference and Expo (ITEC), Dearborn, MI, USA, 15–18 June 2014.

114. Yao, L.; Lim, W.H.; Tsai, T.S. A Real-Time Charging Scheme for Demand Response in Electric Vehicle Parking Station. *IEEE Trans. Smart Grid* **2017**, *8*, 52–62. [CrossRef]

115. Kütt, L.; Saarijärvi, E.; Lehtonen, M.; Mõlder, H.; Niitsoo, J. A review of the harmonic and unbalance effects in electrical distribution networks due to EV charging. In Proceedings of the 2013 12th International Conference on Environment and Electrical Engineering (EEEIC), Wroclaw, Poland, 5–8 May 2013.

116. Richardson, P.; Flynn, D.; Keane, A. Optimal charging of electric vehicles in low-voltage distribution systems. *IEEE Trans. Power Syst.* **2012**, *27*, 268–279. [CrossRef]

117. Mwasilu, F.; Justo, J.J.; Kim, E.K.; Do, T.D.; Jung, J.W. Electric vehicles and smart grid interaction: A review on vehicle to grid and renewable energy sources integration. *Renew. Sustain. Energy Rev.* **2014**, *34*, 501–516. [CrossRef]

118. Green, R.C.; Wang, L.; Alam, M. The impact of plug-in hybrid electric vehicles on distribution networks: A review and outlook. *Renew. Sustain. Energy Rev.* **2011**, *15*, 544–553. [CrossRef]

119. Qian, K.; Zhou, C.; Allan, M.; Yuan, Y. Modeling of load demand due to EV battery charging in distribution systems. *IEEE Trans. Power Syst.* **2011**, *26*, 802–810. [CrossRef]

120. Geng, B.; Mills, J.K.; Sun, D. Two-stage charging strategy for plug-in electric vehicles at the residential transformer level. *IEEE Trans. Smart Grid* **2013**, *4*, 1442–1452. [CrossRef]

121. Fernandez, L.P.; San Román, T.G.; Cossent, R.; Domingo, C.M.; Frias, P. Assessment of the impact of plug-in electric vehicles on distribution networks. *IEEE Trans. Power Syst.* **2011**, *26*, 206–213. [CrossRef]

122. Gómez, J.C.; Morcos, M.M. Impact of EV battery chargers on the power quality of distribution systems. *IEEE Trans. Power Deliv.* **2003**, *18*, 975–981. [CrossRef]

123. Dharmakeerthi, C.H.; Mithulananthan, N.; Saha, T.K. Impact of electric vehicle fast charging on power system voltage stability. *Int. J. Electr. Power Energy Syst.* **2014**, *57*, 241–249. [CrossRef]

124. Mitra, P.; Venayagamoorthy, G.K. Wide area control for improving stability of a power system with plug-in electric vehicles. *IET Gener. Transm. Distrib.* **2010**, *4*, 1151–1163. [CrossRef]

125. Rajakaruna, S.; Shahnia, F.; Ghosh, A. *Plug in Electric Vehicles in Smart Grids*, 1st ed.; Springer Science and Business Media Singapore Pte Ltd.: Singapore, 2015.

126. Akhavan-Rezai, E.; Shaaban, M.F.; El-Saadany, E.F.; Zidan, A. Uncoordinated charging impacts of electric vehicles on electric distribution grids: Normal and fast charging comparison. In Proceedings of the IEEE Power and Energy Society General Meeting, San Diego, CA, USA, 22–26 July 2012; pp. 1–7.

127. Jiang, C.; Torquato, R.; Salles, D.; Xu, W. Method to assess the power-quality impact of plug-in electric vehicles. *IEEE Trans. Power Deliv.* **2014**, *29*, 958–965.

128. Desmet, J.J.M.; Sweertvaegher, I.; Vanalme, G.; Stockman, K.; Belmans, R.J.M. Analysis of the neutral conductor current in a three- phase supplied network with nonlinear single-phase loads. *IEEE Trans. Ind. Appl.* **2003**, *39*, 587–593. [CrossRef]

129. Nguyen, V.L.; Tuan, T.Q.; Bacha, S. Harmonic distortion mitigation for electric vehicle fast charging systems. In Proceedings of the 2013 IEEE Grenoble PowerTech (POWERTECH), Grenoble, France, 16–20 June 2013; pp. 1–6.

130. Melo, N.; Mira, F.; De Almeida, A.; Delgado, J. Integration of PEV in Portuguese distribution grid: Analysis of harmonic current emissions in charging points. In Proceedings of the International Conference on Electrical Power Quality and Utilization, Lisbon, Portugal, 17–19 October 2011; pp. 791–796.

131. Zamri, M.; Wanik, C.; Siam, M.F.; Ayob, A.; Mohamed, A.; Hanifahazit, A.; Sulaiman, S.; Ali, M.A.M.; Hussein, Z.F.; MatHussin, A.K. Harmonic measurement and analysis during electric vehicle charging. *Engineering* **2013**, *5*, 215–220.

132. Bentley, E.C.; Suwanapingkarl, P.; Weerasinghe, S.; Jiang, T.; Putrus, G.A.; Johnston, D. The interactive effects of multiple EV chargers within a distribution network. In Proceedings of the IEEE Vehicle Power and Propulsion Conference (VPPC), Lille, France, 1–3 September 2010; pp. 1–6.

133. Staats, P.T.; Grady, W.M.; Arapostathis, A.; Thallam, R.S. A statistical analysis of the effect of electric vehicle battery charging on distribution system harmonic voltages. *IEEE Trans. Power Deliv.* **1998**, *13*, 640–646. [CrossRef]

134. Balcells, J.; García, J. Impact of plug-in electric vehicles on the supply grid. In Proceedings of the IEEE Vehicle Power and Propulsion Conference (VPPC), Lille, France, 1–3 September 2010; pp. 1–4.

135. Lee, S.J.; Kim, J.H.; Kim, D.U.; Go, H.S.; Kim, C.H.; Kim, E.S.; Kim, S.K. Evaluation of voltage sag and unbalance due to the system connection of electric vehicles on distribution system. *J. Electr. Eng. Technol.* **2014**, *9*, 452–460. [CrossRef]

136. Tie, C.H.; Gan, C.K.; Ibrahim, K.A. The impact of electric vehicle charging on a residential low voltage distribution network in Malaysia. In Proceedings of the 2014 IEEE Innovative Smart Grid Technologies—Asia (ISGT Asia), Kuala Lumpur, Malaysia, 20–23 May 2014; pp. 272–277.

137. Leemput, N.; Geth, F.; Van Roy, J.; Delnooz, A.; Buscher, J.; Driesen, J. Impact of electric vehicle on board single-phase charging strategies on a Flemish residential grid. *IEEE Trans. Smart Grid* **2014**, *5*, 1815–1822. [CrossRef]

138. Masoum, M.A.S.; Moses, P.S.; Deilami, S. Load management in smart grids considering harmonic distortion and transformer derating. In Proceedings of the IEEE Innovative Smart Grid Technologies Europe (ISGT Europe), Gaithersburg, MD, USA, 19–21 January 2010; pp. 1–7.

139. Nyns, K.C.; Haesen, E.; Driesen, J. The impact of charging plug-in hybrid electric vehicles on a residential distribution grid. *IEEE Trans. Power Syst.* **2010**, *25*, 371–380. [CrossRef]

140. Deilami, S.; Masoum, A.S.; Moses, P.S.; Masoum, M.A.S. Real-time coordination of plug-in electric vehicle charging in smart grids to minimize power losses and improve voltage profile. *IEEE Trans. Smart Grid* **2011**, *2*, 456–467. [CrossRef]

141. Sortomme, E.; Hindi, E.M.M.; MacPherson, S.D.J.; Venkata, S.S. Coordinated charging of plug-in hybrid electric vehicles to minimize distribution system losses. *IEEE Trans. Smart Grid* **2011**, *2*, 198–205. [CrossRef]

142. Sadeghi-Barzani, P.; Rajabi-Ghahnavieh, A.; Kazemi-Karegar, H. Optimal fast charging station placing and sizing. *Appl. Energy* **2014**, *125*, 289–299. [CrossRef]

143. Qian, K.; Zhou, C.; Yuan, Y. Impacts of high penetration level of fully electric vehicles charging loads on the thermal ageing of power transformers. *Int. J. Electr. Power Energy Syst.* **2015**, *65*, 102–112. [CrossRef]

144. Razeghi, G.; Zhang, L.; Brown, T.; Samuelsen, S. Impacts of plug-in hybrid electric vehicles on a residential transformer using stochastic and empirical analysis. *J. Power Sources* **2014**, *252*, 277–285. [CrossRef]

145. Elnozahy, M.S.; Salama, M.M. A comprehensive study of the impacts of PHEVs on residential distribution networks. *IEEE Trans. Sustain. Energy* **2014**, *5*, 332–342. [CrossRef]

146. Sortomme, E.; El-Sharkawi, M.A. Optimal charging strategies for unidirectional vehicle-to-grid. *IEEE Trans. Smart Grid* **2011**, *2*, 131–138. [CrossRef]

147. Clement-Nyns, K.; Haesen, E.; Driesen, J. The impact of vehicle-to-grid on the distribution grid. *Electr. Power Syst. Res.* **2011**, *81*, 185–192. [CrossRef]

148. Tulpule, P.; Marano, V.; Yurkovich, S.; Rizzoni, G. Economic and environmental impacts of a PV powered workplace parking garage charging station. *Appl. Energy* **2013**, *108*, 323–332. [CrossRef]

149. Birnie, D.P. Solar-to-vehicle (S2V) systems for powering commuters of the future. *J. Power Sources* **2009**, *186*, 539–542. [CrossRef]

150. Derakhshandeh, S.Y.; Masoum, A.S.; Deilami, S.; Masoum, M.A.; Golshan, M.H. Coordination of generation scheduling with PEVs charging in industrial microgrids. *IEEE Trans. Power Syst.* **2013**, *28*, 3451–3461. [CrossRef]

151. Pillai, R.J.; Heussen, K.; Østergaard, P.A. Comparative analysis of hourly and dynamic power balancing models for validating future energy scenarios. *Energy* **2011**, *36*, 3233–3243. [CrossRef]

152. Borba, B.S.M.; Szklo, A.; Schaeffer, R. Plug-in hybrid electric vehicles as a way to maximize the integration of variable renewable energy in power systems: The case of wind generation in northeastern Brazil. *Energy* **2012**, *37*, 469–481. [CrossRef]

153. Wu, T.; Yang, Q.; Bao, Z.; Yan, W. Coordinated energy dispatching in microgrid with wind power generation and plug-in electric vehicles. *IEEE Trans. Smart Grid* **2013**, *4*, 1453–1463. [CrossRef]

154. Liu, C.; Wang, J.; Botterud, A.; Zhou, Y.; Vyas, A. Assessment of impacts of PHEV charging patterns on wind-thermal scheduling by stochastic unit commitment. *IEEE Trans. Smart Grid* **2012**, *3*, 675–683. [CrossRef]

155. Ma, H.; Balthser, F.; Tait, N.; Riera-Palou, X.; Harrison, A. A new comparison between the life cycle greenhouse gas emissions of battery electric vehicles and internal combustion vehicles. *Energy Policy* **2012**, *44*, 160–173. [CrossRef]

156. Sioshansi, R.; Miller, J. Plug-in hybrid electric vehicles can be clean and economical in dirty power systems. *Energy Policy* **2011**, *39*, 6151–6161. [CrossRef]

157. Donateo, T.; Ingrosso, F.; Licci, F.; Laforgia, D. A method to estimate the environmental impact of an electric city car during six months of testing in an Italian city. *J. Power Sources* **2014**, *270*, 487–498. [CrossRef]

158. Onat, N.C.; Kucukvar, M.; Tatari, O. Conventional, hybrid, plug-in hybrid or electric vehicles? State-based comparative carbon and energy footprint analysis in the United States. *Appl. Energy* **2015**, *150*, 36–49. [CrossRef]

159. Jorgensen, K. Technologies for electric, hybrid and hydrogen vehicles: Electricity from renewable energy sources in transport. *Util. Policy* **2008**, *16*, 72–79. [CrossRef]

160. Kempton, W.; Letendrem, S. Electric vehicles as a new power source for electric utilities. *Transp. Res. Part D* **1997**, *2*, 157–175. [CrossRef]
161. Peterson, S.; Whitacre, J.; Apt, J. The economics of using plug-in hybrid electric vehicles battery packs for grid storage. *J. Power Sources* **2010**, *195*, 2377–2384. [CrossRef]
162. Sioshansi, R.; Denholm, P. The value of plug-in hybrid electric vehicles as grid resources. *Energy J.* **2010**, *31*, 1–16. [CrossRef]
163. Formula, E. Available online: http://www.fiaformulae.com/en (accessed on 8 May 2017).
164. Hidrue, M.K.; Parsons, G.R.; Kempton, W.; Gardner, M.P. Willingness to pay for electric vehicles and their attributes. *Resour. Energy Econ.* **2011**, *33*, 686–705. [CrossRef]
165. 2017 Bolt EV: All-Electric Vehicle | Chevrolet. Available online: http://www.chevrolet.com/bolt-ev-electric-vehicle.html (accessed on 8 May 2017).
166. Hydrogen Fuel Cell Car | Toyota Mirai. Available online: https://ssl.toyota.com/mirai/fcv.html (accessed on 8 May 2017).
167. Wolsink, M. The research agenda on social acceptance of distributed generation in smart grids: Renewable as common pool resources. *Renew. Sustain. Energy Rev.* **2012**, *16*, 822–835. [CrossRef]
168. Kang, J.E.; Brown, T.; Recker, W.W.; Samuelsen, G.S. Refueling hydrogen fuel cell vehicles with 68 proposed refueling stations in California: Measuring deviations from daily travel patterns. *Int. J. Hydrogen Energy* **2014**, *39*, 3444–3449. [CrossRef]
169. Tie, S.F.; Tan, C.W. A review of energy sources and energy management system in electric vehicles. *Renew. Sustain. Energy Rev.* **2013**, *20*, 82–102. [CrossRef]
170. Chan, C.C.; Wong, Y.S. Electric vehicles charge forward. *IEEE Power Energy Mag.* **2004**, *2*, 24–33. [CrossRef]
171. Five Slippery Cars Enter a Wind Tunnel, One Slinks Out a Winner. Available online: https://www.tesla.com/sites/default/files/blog_attachments/the-slipperiest-car-on-the-road.pdf (accessed on 8 May 2017).
172. Geng, B.; Mills, J.K.; Sun, D. Two-stage energy management control of fuel cell plug-in hybrid electric vehicles considering fuel cell longevity. *IEEE Trans. Veh. Technol.* **2012**, *61*, 498–508. [CrossRef]
173. Murphey, Y.L.; Park, J.; Chen, Z.; Kuang, M.L.; Masrur, M.A.; Phillips, A.M. Intelligent hybrid vehicle power control—Part I: Machine learning of optimal vehicle power. *IEEE Trans. Veh. Technol.* **2012**, *61*, 3519–3530. [CrossRef]
174. Geng, B.; Mills, J.K.; Sun, D. Energy management control of microturbine powered plug-in hybrid electric vehicles using telemetry equivalent consumption minimization strategy. *IEEE Trans. Veh. Technol.* **2011**, *60*, 4238–4248. [CrossRef]
175. Moura, S.J.; Fathy, H.K.; Callaway, D.S.; Stein, J.L. A stochastic optimal control approach for power management in plug-in hybrid electric vehicles. *IEEE Trans. Control Syst. Technol.* **2011**, *19*, 545–555. [CrossRef]
176. Magallan, G.A.; De Angelo, C.H.; Garcia, G.O. Maximization of the traction forces in a 2WD electric vehicle. *IEEE Trans. Veh. Technol.* **2011**, *60*, 369–380. [CrossRef]
177. Kang, J.; Yoo, J.; Yi, K. Driving control algorithm for maneuverability, lateral stability, and rollover prevention of 4WD electric vehicles with independently driven front and rear wheels. *IEEE Trans. Veh. Technol.* **2011**, *60*, 2987–3001. [CrossRef]
178. Wang, J.N.; Wang, Q.N.; Jin, L.Q.; Song, C.X. Independent wheel torque control of 4WD electric vehicle for differential drive assisted steering. *Mechatronics* **2011**, *21*, 63–76. [CrossRef]
179. Nam, K.; Fujimoto, H.; Hori, Y. Lateral stability control of in-wheel-motor-driven electric vehicles based on sideslip angle estimation using lateral tire force sensors. *IEEE Trans. Veh. Technol.* **2012**, *61*, 1972–1985.
180. Zhou, D.; Ravey, A.; Gao, F.; Miraoui, A.; Zhang, K. Online Estimation of Lithium Polymer Batteries State-of-Charge Using Particle Filter-Based Data Fusion with Multimodels Approach. *IEEE Trans. Ind. Appl.* **2016**, *52*, 2582–2595. [CrossRef]
181. Hui, S.; Lifu, Y.; Junqing, J.; Yanling, L. Control strategy of hydraulic/electric synergy system in heavy hybrid vehicles. *Energy Convers. Manag.* **2011**, *52*, 668–674. [CrossRef]
182. Chen, Z.; Mi, C.C.; Xiong, R.; Xu, J.; You, C. Energy management of a power-split plug-in hybrid electric vehicle based on genetic algorithm and quadratic programming. *J. Power Sources* **2014**, *248*, 416–426. [CrossRef]
183. Li, S.G.; Sharkh, S.M.; Walsh, F.C.; Zhang, C.N. Energy and battery management of a plug-in series hybrid electric vehicle using fuzzy logic. *IEEE Trans. Veh. Technol.* **2011**, *60*, 3571–3585. [CrossRef]

184. Yuan, Z.; Teng, L.; Fengchun, S.; Peng, H. Comparative study of dynamic programming and Pontryagin's minimum principle on energy management for a parallel hybrid electric vehicle. *Energies* **2013**, *6*, 2305–2318. [CrossRef]

185. Bernard, J.; Delprat, S.; Guerra, T.M.; Büchi, F.N. Fuel efficient power management strategy for fuel cell hybrid powertrains. *Control Eng. Pract.* **2010**, *18*, 408–417. [CrossRef]

186. Su, W.; Chow, M.Y. Performance evaluation of an EDA-based large-scale plug-in hybrid electric vehicle charging algorithm. *IEEE Trans. Smart Grid* **2012**, *3*, 308–315. [CrossRef]

187. Mohamed, A.; Salehi, V.; Ma, T.; Mohammed, O. Real-time energy management algorithm for plug-in hybrid electric vehicle charging parks involving sustainable energy. *IEEE Trans. Sustain. Energy* **2014**, *5*, 577–586. [CrossRef]

188. Worldwide EV Sales Are on The Move. Available online: http://evercharge.net/blog/infographic-worldwide-ev-sales-are-on-the-move/ (accessed on 8 May 2017).

189. China's Quota Threat Charges Up Electric Car Market | The Daily Star. Available online: http://www.thedailystar.net/business/chinas-quota-threat-charges-electric-car-market-1396066 (accessed on 8 May 2017).

190. EV-Volumes—The Electric Vehicle World Sales Database. Available online: http://www.ev-volumes.com/news/china-plug-in-sales-2016-q4-and-full-year/ (accessed on 8 May 2017).

191. EV-Volumes—The Electric Vehicle World Sales Database. Available online: http://www.ev-volumes.com/country/total-world-plug-in-vehicle-volumes/ (accessed on 8 May 2017).

192. EV-Volumes—The Electric Vehicle World Sales Database. Available online: http://www.ev-volumes.com/country/usa/ (accessed on 8 May 2017).

193. EAFO. Available online: http://www.eafo.eu/vehicle-statistics/m1 (accessed on 8 May 2017).

194. Hossain, E.; Kabalci, E.; Bayindir, R.; Perez, R. Microgrid testbeds around the world: State of art. *Energy Convers. Manag.* **2014**, *86*, 132–153. [CrossRef]

195. Bayindir, R.; Hossain, E.; Kabalci, E.; Perez, R. A comprehensive study on microgrid technology. *Int. J. Renew. Energy Res.* **2014**, *4*, 1094–1107.

196. eHighway—Siemens Global Website. Available online: https://www.siemens.com/press/en/feature/2015/mobility/2015-06-eHighway.php?content[]=MO (accessed on 8 May 2017).

197. Saponara, S.; Petri, E.; Fanucci, L.; Terreni, P. Sensor modeling, low-complexity fusion algorithms, and mixed-signal IC prototyping for gas measures in low-emission vehicles. *IEEE Trans. Instrum. Meas.* **2011**, *60*, 372–384. [CrossRef]

198. Saponara, S.; Pasetti, G.; Costantino, N.; Tinfena, F.; D'Abramo, P.; Fanucci, L. A flexible LED driver for automotive lighting applications: IC design and experimental characterization. *IEEE Trans. Power Electron.* **2012**, *27*, 1071–1075. [CrossRef]

199. Saponara, S.; Pasetti, G.; Tinfena, F.; Fanucci, L.; D'Abramo, P. HV-CMOS design and characterization of a smart rotor coil driver for automotive alternators. *IEEE Trans. Ind. Electron.* **2013**, *60*, 2309–2317. [CrossRef]

200. Baronti, F.; Lazzeri, A.; Roncella, R.; Saletti, R.; Saponara, S. Design and characterization of a robotized gearbox system based on voice coil actuators for a Formula SAE Race Car. *IEEE/ASME Trans. Mechatron.* **2013**, *18*, 53–61. [CrossRef]

201. Costantino, N.; Serventi, R.; Tinfena, F.; D'Abramo, P.; Chassard, P.; Tisserand, P.; Saponara, S.; Fanucci, L. Design and test of an HV-CMOS intelligent power switch with integrated protections and self-diagnostic for harsh automotive applications. *IEEE Trans. Ind. Electron.* **2011**, *58*, 2715–2727. [CrossRef]

202. Saponara, S.; Fanucci, L.; Bernardo, F.; Falciani, A. Predictive diagnosis of high-power transformer faults by networking vibration measuring nodes with integrated signal processing. *IEEE Trans. Instrum. Meas.* **2016**, *65*, 1749–1760. [CrossRef]

203. Abhishek, A.; Karthikeyan, V.; Sanjeevikumar, P.; Rajasekar, S.; Blaabjerg, F.; Asheesh, K.S. Optimal Planning of Electric Vehicle Charging Station at the Distribution System Using Hybrid Optimization Algorithm. *Energy* **2017**, *133*, 70–78.

204. Febin Daya, J.L.; Sanjeevikumar, P.; Blaabjerg, F.; Wheeler, P.; Ojo, O.; Ahmet, H.E. Analysis of Wavelet Controller for Robustness in Electronic Differential of Electric Vehicles—An Investigation and Numerical Implementation. *Electr. Power Compon. Syst.* **2016**, *44*, 763–773. [CrossRef]

205. Febin Daya, J.L.; Sanjeevikumar, P.; Blaabjerg, F.; Wheeler, P.; Ojo, O. Implementation of Wavelet Based Robust Differential Control for Electric Vehicle Application. *IEEE Trans. Power Electron.* **2015**, *30*, 6510–6513. [CrossRef]
206. Sanjeevikumar, P.; Febin Daya, J.L.; Blaabjerg, F.; Mir-Nasiri, N.; Ahmet, H.E. Numerical Implementation of Wavelet and Fuzzy Transform IFOC for Three-Phase Induction Motor. *Eng. Sci. Technol. Int. J.* **2016**, *19*, 96–100.
207. Dragonas, F.A.; Nerrati, G.; Sanjeevikumar, P.; Grandi, G. High-Voltage High-Frequency Arbitrary Waveform Multilevel Generator for DBD Plasma Actuators. *IEEE Trans. Ind. Appl.* **2015**, *51*, 3334–3342. [CrossRef]
208. Mohan, K.; Febin Daya, J.L.; Sanjeevikumar, P.; Mihet-Popa, L. Real-time Analysis of a Modified State Observer for Sensorless Induction Motor Drive used in Electric Vehicle Applications. *Energies* **2017**, *10*, 1077.

![energies logo] *energies*

MDPI

Article

Electric Vehicles in Logistics and Transportation: A Survey on Emerging Environmental, Strategic, and Operational Challenges

Angel Alejandro Juan [1,*]**, Carlos Alberto Mendez** [2]**, Javier Faulin** [3]**, Jesica de Armas** [1]
and Scott Erwin Grasman [4]

[1] Computer Science Department, Internet Interdisciplinary Institute, Open University of Catalonia, 08018 Barcelona, Spain; jde_armasa@uoc.edu

[2] Instituto de Desarrollo Tecnológico para la Industria Química, Universidad Nacional del Litoral, CONICET, 3000 Santa Fe, Argentina; cmendez@intec.unl.edu.ar

[3] Statistics and Operations Research Department, Public University of Navarre, 31006 Pamplona, Spain; javier.faulin@unavarra.es

[4] Industrial and Systems Engineering, Rochester Institute of Technology, Rochester, NY 14623, USA; segeie@rit.edu

* Correspondence: ajuanp@uoc.edu; Tel.: +34-93-326-3600; Fax: +34-93-356-8822

Academic Editor: K. T. Chau
Received: 14 December 2015; Accepted: 25 January 2016; Published: 28 January 2016

Abstract: Current logistics and transportation (L&T) systems include heterogeneous fleets consisting of common internal combustion engine vehicles as well as other types of vehicles using "green" technologies, e.g., plug-in hybrid electric vehicles and electric vehicles (EVs). However, the incorporation of EVs in L&T activities also raise some additional challenges from the strategic, planning, and operational perspectives. For instance, smart cities are required to provide recharge stations for electric-based vehicles, meaning that investment decisions need to be made about the number, location, and capacity of these stations. Similarly, the limited driving-range capabilities of EVs, which are restricted by the amount of electricity stored in their batteries, impose non-trivial additional constraints when designing efficient distribution routes. Accordingly, this paper identifies and reviews several open research challenges related to the introduction of EVs in L&T activities, including: (a) environmental-related issues; and (b) strategic, planning and operational issues associated with "standard" EVs and with hydrogen-based EVs. The paper also analyzes how the introduction of EVs in L&T systems generates new variants of the well-known Vehicle Routing Problem, one of the most studied optimization problems in the L&T field, and proposes the use of metaheuristics and simheuristics as the most efficient way to deal with these complex optimization problems.

Keywords: electric vehicles; logistics and transportation; green vehicle routing problems

1. Introduction

Logistics and transportation (L&T) activities represent a key sector in worldwide economies, and are a significant contributor to social and economic progress in modern societies. The prevalence of the L&T industry is due to its constant growth and impact in terms of regional Gross Domestic Product (GDP). In particular, road L&T activities using motorized vehicles have significantly increased in response to the rise of globalization and commercial interchanges among countries. With the aim of making them more efficient, L&T systems have been widely studied by the Operations Research/Computer Science (OR/CS) communities for decades. Due to its potential applications to real-life operations, one of the most recurrent topics in the L&T literature is that of modeling and optimizing tour assignments of vehicles. This is known as the Vehicle Routing Problem (VRP) [1]. Numerous variants of this problem have been addressed over the last years. The basic variant is the so-called Capacitated VRP (CVRP), where each customer has a given demand that has to be satisfied without exceeding a maximum vehicle capacity. The VRP with Time Windows extends the CVRP by adding time windows to the depot and the customers. To account for additional real world aspects, the classical VRP has been redefined in various manners that are often called Rich VRPs [2] or Multi-attribute VRPs [3]. Despite the extensive literature in the VRP area, most of the existing contributions have assumed that the fleet to be managed comprises only internal combustion engine vehicles (ICEVs), which is not exactly the current picture.

A large percentage of the oil consumed in regions such as Europe or the USA is used in transport, while road transport accounts for an important percentage of CO_2 emissions of the overall transport activity. Furthermore, the whole transport sector causes about 28% of the total greenhouse gas (GHG) emissions in countries such as the USA. In order to mitigate this situation, one possibility is to incorporate emission costs as an objective to be minimized in routing models, thus trading off environmental and economic goals [4,5]. A different approach is the utilization of less polluting means of transport such as plug-in hybrid electric vehicles and electric vehicles (EVs), whose specific characteristics have to be included in adequate routing models. In effect, as part of the initiative to improve the local air quality, modern cities encourage fleets of vehicles to adopt alternative technologies, such as EVs. Several factors are promoting the use of these technologies, including: (i) companies receive incentives to reduce their carbon footprint; (ii) high variability of oil-based products and long-term cost risk associated with dependence on oil-based energy sources; (iii) availability of government subsidies to reduce acquisition cost; and (iv) advances in alternative energy technologies (such as EVs), which have potential for a more environmentally sustainable solutions at a cost that is starting to be competitive. From both an environmental and energy standpoints, the use of EVs should be a first priority for the reduction of primary energy consumption. Although higher concerns are the advantages of EVs in terms of efficiency and flexibility in the use of energy, the EV technology is currently facing several weak points, which can be summarized as follows: (i) the low energy density of batteries compared to the fuel of ICEVs; (ii) the long recharge times of EVs batteries compared to the relatively fast process of refueling a tank in ICEVs; and (iii) the scarcity of public and/or private charging stations for EV batteries. In earlier years, EVs failed because of excessive battery prices and very short driving ranges. As EVs have become one of the major research areas in the automotive sector, the magnitude of these problems has been notably diminished. Although the replacement of conventional ICEVs with EVs is not profitable under most operation scenarios given the current cost conditions, the availability of increasingly long-lived batteries, the trends for rising fuel costs, and lower EV purchase costs are likely to change the picture [6]. Figure 1 shows the noticeable increase experienced during the last years in the number of EV-related articles published in Scopus-indexed journals, which proves the growing interest that the use of EVs is arising among researchers and practitioners.

Accordingly, this paper identifies and reviews, from an OR/CS perspective, several open research challenges related to the introduction of EVs in L&T activities, including the following dimensions: (a) environmental-related issues; (b) strategic and planning challenges associated with "standard" EVs and with hydrogen-based EVs; and (c) emerging operational issues related to the use of EVs in VRPs. Table 1 summarizes the different research challenges that have been identified in our study and that will be conveniently described and reviewed in different sections of this manuscript. For a better understanding, these research issues have been classified in three dimensions: environmental, strategic and planning, and operational. The paper also analyzes in detail how the introduction of EVs in L&T systems generates new VRP variants, *i.e.*, in the context of the Green VRP, this work points out some of the most promising research lines yet to be fully explored. Finally, the paper also includes a discussion on which optimization approaches can better contribute to deal with these open and difficult research challenges.

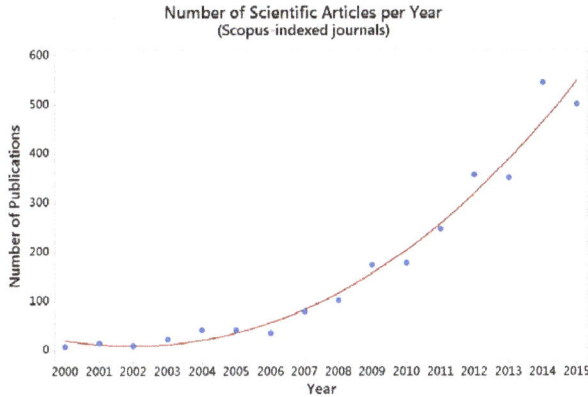

Figure 1. Evolution of electric vehicle (EV) related publications in Scopus-indexed journals.

Table 1. Open Operations Research/Computer Science (OR/CS) research challenges associated with the use of EVs.

Dimension	Research challenges
Environmental	(1) Including the cost of externalities (noise, air pollution, infrastructure wear, *etc.*) in L&T activities. (2) Analyzing how the increasing use of EVs reduces the environmental impact of L&T activities. Exploring new environmentally-sustainable yet efficient ways of doing freight deliveries in urban areas. In particular, considering energy cost and carbon footprint in Vehicle Routing Problems. Studying the environmental cost of manufacturing EVs as well as producing the energy needed to power them. (3) Measuring the effect of using small EVs (e.g., electric bikes, drones, *etc.*) to perform urban last mile distribution.
Strategic and Planning	(1) Analyzing different EV related technologies and infrastructures (e.g., standard EV *vs.* hydrogen vehicles). (2) Computing the necessary recharging stations, both for standard EVs as well as for hydrogen vehicles, and analyzing their integration in the transport network, *i.e.*, number and type of stations, location, capacity, *etc.* (3) Determining the optimal combination of EVs and internal combustion engine vehicles (fleet size and mix problem). In particular, developing new optimization approaches for the Fleet Size and Mix Vehicle Routing Problem. (4) Exploring potential uses of renewably-generated electricity to power hydrogen vehicles. (5) Quantifying the benefits of horizontal cooperation among stakeholders of EV fleets (e.g., fleet manager, auto manufacturer, electricity supplier, *etc.*).
Operational	(1) Analyzing the impact of EVs recharging times in Vehicle Routing Problems with time-related constraints. (2) Comparing battery swapping *vs.* battery recharging strategies, and proposing the right combination of both. In particular comparing these strategies in Vehicle Routing Problems with EVs. (3) Considering the new issues derived from the driving-range limitations of EVs. In particular, developing new optimization approaches for the Vehicle Routing Problem with multiple driving-range constraints.

The remainder of this paper is structured as follows: Section 2 reviews and analyzes some of the environmental issues related to the use of EVs. Section 3 builds on Section 2 and identifies the main strategic and planning challenges related to the introduction of EVs in "green" L&T activities. Section 4 extends Sections 2 and 3 by focusing on how the introduction of hybrid fleets with both ICEVs and EVs imposes new operational challenges and exploring opportunities on the popular VRP. Section 5 points out some new solving approaches that allow facing these challenges in an efficient way. Section 6 provides an overview of other related emerging issues, such as lifecycle cost analysis and the use of EVs in rural areas and to confront natural disasters. Finally, Section 7 summarizes the main contributions of this paper.

2. Environmental Issues Related to the Use of Electric Vehicles (EVs)

2.1. Environmental Aspects of Transportation

Transportation activities involve the side effects (externalities) of noise pollution, air pollution, and traffic congestion, which current city planning strategies hardly take into account. Given that the transport sector accounted for more than a 25% of world energy consumption, and producing energy increases air contamination, these externalities must be considered to ensure the sustainable growth of transportation worldwide [7]. A complete description of the problem of externalities in transportation would involve the introduction of the following sources of external costs: noise, air pollution, infrastructure wear, visual intrusion, flow congestion, traffic accidents, and so on. Nonetheless, the main environmental studies are performed on the noise and air pollution caused by transportation, due to the fact that they are very well-known externalities. Some reports and studies have tried to assess the economic impact and pricing of these externalities in Europe, but their results have not been conclusive so far. In effect, there is a great divergence in the cost estimation of externalities [8]. According to Korzhenevych *et al.* [9], external costs of transportation activities account for about 8.5% of the GDP in regions such as the European Union. These activities represent one of the largest sources of CO_2 emissions, and there is a strong interest in mitigating their effects.

Nowadays, there is a general agreement on the need to consider these negative externalities when formulating transportation policies and logistic strategies. For instance, the European Union has developed an infrastructure-use taxation system based on the "user and polluter pays" tenet. In extraordinary cases involving infrastructures in mountainous areas, the directives suggested the rise of toll charges. Some of these directives highlight that particular attention should be devoted to mountainous regions, such as the Alps or the Pyrenees, with the consequent apportionment of European Union taxpayers' money to its related projects. The suitable pricing of all the aforementioned externalities is essential for the success of any consideration of adequate payments in transportation policies. As stated before, two main types of traffic-related environmental pollution are considered: air pollution and noise pollution. The previous discussion about the importance of controlling different types of contaminating emissions explains the need for searching new technologies that allow reducing the environmental impact of freight transportation activities. In this sense, the increasing use of EVs in hybrid fleets constitutes a fundamental step in this direction [10]. Some basic figures will help to understand the potential contribution of EVs in promoting sustainability of freight-distribution operations: according to Figliozzi [11], while a diesel van delivering goods in a "standard" city releases about 1.0375 $kgCO_2$/mile, using an EV instead would produce just about 0.01915 $kgCO_2$/mile (notice that this number corresponds to the estimated emissions produced by the source of electrical power necessary to run the EV). In other words, a diesel van (ICEV) covering a mile produces about 54 times the CO_2 emissions released by an EV to do the same distribution activity. Of course, this huge advantage has to be considered along with the associated disadvantages, e.g., limited autonomy of the EVs, reduction of the load to carry per mile, *etc.* Further, there are different alternatives to the use of "pure" or "standard" EVs, among them hydraulic hybrid EVs [12] as well as hydrogen-based EVs [13], which would further reduce CO_2 emissions. From this simple example, it seems clear that the contribution of EVs to environmental sustainability can be significant.

2.2. Environmental Impact of Delivering Goods in Urban Areas

During the last decades, a large percentage of the world's population has moved into cities [14]. Therefore, all L&T issues related to procurement and supply management of cities are critical. City managers try to identify new strategies to increase the quality of life of their citizens while maintaining their economic competitiveness. For that reason, cities around the world are worried about designing sustainable yet efficient ways of doing freight deliveries in urban areas [15]. In Europe and USA, around 80% of the population lives in urban areas already. Since urban areas usually contain large populations, extensive commercial establishments, and an increasing demand of services and commodities, there is a need to increase the frequency of urban deliveries, which requires an intensive use of the existing infrastructure. According to the US Federal Highway Administration, the total vehicle miles of travel increased 21% in the urban areas between 1996 and 2006. In fact, according to Feng and Figliozzi [16], the proportion of freight vehicles crossing urban zones also increased from 4.8% to 5.2%. As discussed before, freight distribution in densely populated cities are related to negative transportation externalities, such as congestion, pollution (both gas emissions and noise), energy inefficiencies, decreasing road safety, infrastructures degradation, lack of roadway capacity and parking space, worse breathing conditions, *etc.* As pointed out by Russo and Comi [17], urban freight vehicles account for about 6%–18% of the total urban travel, for about 19% of the energy use, and for about 21% of the CO_2 pollution.

Some of the typical pollutants in urban districts are the following [18]: (i) mono-nitrogen oxides, which are produced by the combustion of fossil fuels and contribute to generate smog and acid rain; (ii) sulfur oxides; (iii) carbon monoxide; (iv) particulate matter; and (v) volatile organic compounds. All the previous gases have negative effects on people's health, as discussed in Bernard *et al.* [19]. As a consequence, there is a strong impulse at international, national, and local levels to mitigate them by switching to zero emissions technologies, the shift to EVs being one of the most promising policies. Also, this shift can be a good solution to relieve other problems related to urban distribution. Thus, for instance, Nüesch *et al.* [20] propose a method to minimize the fuel consumption using hybrid EVs while tracking a given reference trajectory for both emissions and the battery state of charge. Similarly, Collin *et al.* [21] design a generic methodology to incorporate environmental and battery-related constraints into on-line energy management strategies for different types of hybrid EVs, while Chen *et al.* [22] introduce an energy management approach to minimize total energy cost associated with the use of hybrid EVs. Finally, in the context of urban transportation networks, Hwang *et al.* [23] propose a stochastic model to minimize the expected total cost of freight truck activities, where these costs include total delivery time, different types of emissions, and a penalty for late or early arrival.

Urban transport usually involves vehicles operating with low loading levels, thus resulting in a non-efficient use of oil-based energy. Moreover, urban delivery vans have a low average driving speed, and electric engines are more efficient at low speeds. Likewise, the routes covered in urban distribution are quite similar from one day to another, which can facilitate the design of stable policies for battery recharging or battery swapping. It is clear then that a shift from a fossil fuel fleet to an electric-powered fleet is necessary in order to reduce pollutant emissions in cities. A conversion to EVs would imply the conjoint development of transportation and power generation sectors, and would shift GHG emissions from conventional vehicle tailpipes to big electric power plants.

On the one hand, EVs using electricity from the public grid will play a critical role in reducing GHG emissions and in mitigating negative transport externalities. Nevertheless, these reductions in emissions will be only possible within a scenario of low-carbon electricity production, *i.e.*, the replacement of ICEVs by EVs is only reasonable if the electricity generation has a low level of carbon production. Otherwise, one pollutant technology would be swapped by another pollutant technology (maybe less pollutant, but not really sustainable either). Additionally, EVs are ideal to make the distribution of light products with a low emission pollutants impact in city centers. That fact is due to: (i) the lack of gas releases in EVs; and (ii) the usually small size of EVs, which allows them to easily access high congested streets with limited parking space. Many cities allow EVs to use parking spaces for free. Thus, EVs are constrained to a lesser degree by the existence of congestion or lack of parking areas than ICEVs. This implies that the walking distances covered by the drivers of EVs are usually shorter than the ones performed by ICEVs drivers [24]. In fact, it is common to see conventional vehicles double-parked or parked in restricted areas to diminish the walking distance delivery, having to pay in many cases extra costs due to parking fines. Moreover, other advantages of EVs are related to the lower noise level produced by their use in comparison with the noise level of ICEVs [25]. Many papers have been published highlighting the good properties and advantages of EVs [26], but little work has been done so far to evaluate the environmental impact of the EVs production itself and the electricity power generation. In fact, to the best of our knowledge, there is no published carbon footprint comparison between companies which use ICEVs and those using EVs [27].

2.3. Decarbonizing the Last-Mile Delivery Process with the Use of EVs

The study and development of the EV Routing Problem, along with the variation in competitiveness due to the introduction of EVs in hybrid fleets, is a recent study area with many real applications. As noticed by Afroditi *et al.* [28], this study is especially interesting in the "last-mile" delivery process. The distribution process is usually critical in the last mile of the supply chain, where most of the difficult operational decisions to make are present. In effect, it is in this last mile where more details can affect the quality of the delivery service, where more routes are formed, and where the direct contact with the final customer makes a critical mix between L&T and marketing. This situation involves an exhaustive use of L&T resources to achieve the expecting quality of the delivery process. An exhaustive use of resources usually causes more negative externalities (congestion, emissions, and noise, among others). Therefore, the use of EVs in the last-mile activity can help to significantly reduce the level of the aforementioned externalities. This improvement has been clearly shown in many European cities such as Paris, London, or Vienna [29]. Thus, the EVs are revealed as a very useful tool to "decarbonize" the last mile delivery process, although their range limitation could be an important disadvantage in some cases.

The typical design of an EV is conceived in the shape of a small vehicle to take advantage of its capacity and its performance according to the electric power of its battery. Nevertheless, it is also possible to design EVs with the shape of a bike with a small electric engine hybridized with human power propulsion. These bike EVs are usually presented in the way of tricycles to provide them with more capacity. Delivery actions in the last-mile range using electric tricycles are becoming increasingly common, mainly in very congested cities [30]. These vehicles clearly benefit from the option of recharging batteries with the use of human power propulsion. Some companies showing this experience in freight deliveries are, among others: Ecopostale (Brussels, Belgium), B-line (Portland, OR, USA), La Petite Reine (Paris, France), or Txita (San Sebastian, Spain). Some studies performed by these companies range the savings in CO_2 emissions from the 8.5 tons per year (Ecopostale) to the 89.125 tons per year (La Petite Reine) of oil equivalent. Another interesting experience concerning the evaluation of results in last mile delivery optimization is depicted by Browne *et al.* [25], who described a trial of shipped goods from a suburban depot serving customers in London. In their study, the fleet of ICEVs performing deliveries in London was replaced by EVs and tricycles working in a consolidation center in the British capital. The normal use of those EVs is not interfered with by any fossil fuel consumption or greenhouse effect due to the fact that the electricity they use was exclusively generated by renewable sources. By making a direct comparison between the emissions with conventional ICEVs and with EVs, it is possible to conclude that CO_2 emissions fell by 20% if using standard EVs and by 54% if using tricycles. Moreover, GNewt Cargo, the operator of the micro-consolidation center, certified that it is possible to cut the CO_2 emissions by 62%. Similarly, Conway *et al.* [29] describe two case studies in New York (USA) where the use of electric cargo cycles involves a savings of 11–13 tons/year of CO_2 emissions and 2–2.5 lbs/year of particulate matter for the first case, and 8.3 tons/year and 1.6 lbs/year, respectively, for the second case.

As discussed in Bektas and Laporte [5], the inclusion of pollutant emissions in vehicle routing problems has allowed the design of new routing models and the development of new optimization algorithms. Likewise, it has generated an updated classification of pollution pricing models inside the VRP framework [31,32].

3. Strategic and Planning Issues Related to the Use of EVs

Due to the different aspects that distinguish EVs from traditional ICEVs, the L&T problems that have been addressed so far for conventional vehicles need to be rethought and reformulated to take into account the new features of EVs. Unlike conventional vehicles, EVs must refuel frequently due to the short length of their batteries in terms of travel distance, *i.e.*, their limited driving range [10]. Thus, users must consider how many miles can be covered before a recharge is needed. There is no doubt that this restricts their use as transport tools. Therefore, the provision of the necessary recharging stations and their integration in the transport network are important issues to address. The main issues to determine are: (i) the number and kind of refueling stations to establish; (ii) the location of these stations; and (iii) their optimal capacity. Moreover, companies need to assess the impact of the introduction of EVs in their fleet, so that they can choose the best size and mix of vehicles to use. Hence, the fleet size and mix is another important issue to analyze. The following subsections are devoted to review and describe the influence of some of these aspects in the L&T arena.

3.1. Different Kinds of Recharging Stations

As EVs are entering the market, there is a rising demand for public refueling stations. Nowadays, when the EV's batteries are depleted there are two possibilities: recharge them or exchange them. Charging stations can be divided into two categories: fast charging and slow charging. A fast refueling station can quickly recharge an EV in less than five minutes [33], but this kind of charging can significantly shorten the life of the batteries. Conversely, a slow refueling station needs a longer time to recharge an EV. At slow recharging stations of Level 1 or 2 (110–240 V), vehicles need to wait from 2 up to 8 h to fully charge their batteries. At recharge stations of Level 3 (480 V), charging a battery fully takes about 20–40 min. Therefore, recharge time has been a critical factor influencing public acceptance of EVs. A major solution could be to remove the existing battery that is nearly depleted and replace the battery with a fully charged one, as proposed by Li [34]. Such a method is called battery swapping. The main benefit associated with the swapping model is the speed. The whole operation could take less than 10 min, which is on par with conventional vehicles and much faster than even some fast recharging stations. Other noted features of battery swap stations include the following: (i) charging depleted batteries can be left for the night when the charging cost is low; (ii) the provision of grid-support service in a centralized charging and discharging manner; (iii) the ability for drivers to resume their journeys in minutes with a full-capacity battery; (iv) the charging of batteries in slow-charging mode to extend their lifetime; and (v) the savings in cost of EVs by providing batteries by operators. As pointed out by Yang *et al.* (2015) [35], a battery swapping model could be considered more appropriate than a battery recharging model since the former not only improves the productivity of vehicles but also lowers the charging cost. Due to the battery driving range limitation and the nature of battery swapping, distribution network optimization with a battery swapping infrastructure could be an important part of establishing any green L&T policy. However, companies can take this possibility, since the best battery swapping infrastructure ownership model is the company-owned business model, which indicates that the L&T companies establish and operate the battery swap stations for the EVs by themselves. This way, determining the ideal battery swap stations location strategy and vehicle routing plan for a distribution network is mainly a question of service level and operational cost for the logistics enterprises.

All in all, the major challenges encountered regarding the kind of EVs recharging stations are summarized next. On the one hand, EV consumers expect a short charging time just like refueling their current vehicles. This requirement makes fast charging stations more preferred, but this kind of charging can shorten the life of the batteries. Moreover, as clearly explained in [36], implementing centralized charging/discharging control under plug-in mode is very difficult, since EV users present a stochastic charging profile. In order to avoid uncontrolled charging, which may produce a significantly increase of the peak load and endanger power system security , some incentive strategies could be proposed. On the other hand, considering the aforementioned challenges, an alternative strategy based on a battery swap station has received increasing attention during recent years [37–39]. However, battery swap stations have the problem of the lack of unified battery standards for various EVs. As will be discussed in the next subsection, the majority of papers in the literature tackle transportation problems using EVs with charging stations. Nevertheless, there is an emerging number of works considering battery swap stations.

3.2. Recharging Station Location

As mentioned before, one of the main issues to be addressed regarding the EVs success is to determine the location of recharging stations. Therefore, it is important to develop methods that allow minimizing the costs of developing an alternative infrastructure. This "station location problem" can be considered a specific case of Facility Location Problem (FLP). The key questions commonly faced by facility planners include: (i) the number of facilities; (ii) the locations of these facilities; and (iii) the types of facilities (in terms of size, product variety and other design aspects). Most location models focus on either minimizing the average cost of travel (the median problem) or minimizing the maximum cost of travel (the center problem). In the particular case of optimally locating recharging stations, several location models have been proposed. These models can be divided in two main groups, node-based and flow-based, depending on their assumptions of refueling demand type [40]. The p-median model is a well-known node-based model that has been used in many articles to locate conventional gasoline or alternative-fuel vehicles refueling stations [41]. However, since the demand for vehicles is generally in the form of traffic flow that passes by the refueling facilities [42], the majority of papers in the literature are based on flow-based models. Specifically, the locations of recharging stations for EVs, which presents some peculiarities due to their limited driving range, is usually tacked using flow-based models. Thus, Hodgson [42] provided a basic theoretical framework for dealing with the problem of locating stations. However, this approach depends on the assumption that, if one station is sited on a node of a path, then all the related traffic flows will be captured. Unfortunately, this assumption cannot be applied to alternative fuel vehicles since these have a limited range and need a multi-stop system to extend their driving distance and carry out long-distance journeys.

In order to achieve the multi-stops needed for long-distance travel, Kuby and Lim [43] proposed a flow refueling location model (FRLM). The objective of the FRLM is to maximize the capture of the traffic flows on each path if a combination of stations sited on the paths can be successfully used to refuel vehicles, so that they can complete their trips. This model needs to be solved in two stages. The first stage is to find feasible combinations of candidate locations of stations to refuel the flows on each path, and the second stage is when these combinations are used as inputs to the model to determine the station locations. Due to the time-consuming process of generating combinations in first stage, Lim and Kuby [44] provided some heuristic algorithms to solve larger scale problems. Capar and Kuby [45] developed a new approach to solve the flow refueling location problem in one stage. Three locating logics were used to check whether a path could be refueled by the sited refueling stations. The first is if there is no station built at the origin then there should be at least one station built within half the vehicle range to the origin node, so that it can be reachable by half a tank of fuel or half a battery charge. The second is if there is a station built at a location, then the next built facility should be within the vehicle range, otherwise the vehicle cannot reach the next station. The third is if the vehicle range is greater than or equal to two times the path length, then a single station at any point can refuel the entire path. However, these logics are available only when the vehicle has regained its full fuel or charge level (for maximum range) after each period of refueling at the stations, e.g., via fuel-tank or battery exchange, which makes the newer approach difficult to apply with regard to multiple types of stations with different refueling or recharging efficiencies. In addition, this approach cannot solve the capacitated location problem, whereby each station has a limited number of demands to handle. Basically, such models do not consider the factors of refueling or recharging efficiency and time, and are limited to the location of a single kind of station for performing the battery exchange (or very fast refueling) to refill the vehicles.

For deploying battery swapping network infrastructure and battery management, Mak *et al.* [46] developed two distributionally robust optimization models for the battery swap station location problem under ambiguous information on demand distribution. A different flow-based model for economically siting fast-refueling stations, such as battery swapping or hydrogen refilling stations, was proposed by Wang and Lin [47]. The model was formulated based on vehicle refueling logic which can ensure the alternative fuel vehicle has sufficient fuel to move between the nodes, and a feasible path can then be achieved. The model can also be solved in one stage, *i.e.*, it does not need to pre-determine the feasible combination of stations, like the original FRLM does. Moreover, this approach does not need the fuel or charge level after each refueling or recharging to be full, and thus has more flexibility with regard to different situations. Wang and Wang [33] extended the aforementioned model and proposed a bi-objective model to simultaneously consider intercity (path flow demands) and intra-city travel (the nodal demands). The flow-based model was also extended to consider battery recharging efficiency and time to locate sufficient slow-recharging stations for electric scooters traveling in a destination area [48]. However, these previous models still adopt an approach for locating a single kind of refueling stations. In Wang and Lin [47], the authors extended the previous slow-recharging station location model by considering facility budget constraints, multiple kinds of recharging stations, and vehicle routing behavior. These authors also proposed more generalized models to locate multiple kinds of refueling stations for the (maximal) coverage of battery (or non-battery) powered EV journeys on each path. At each site along paths, multiple types of charging stations, including slow-recharging, fast-recharging, and battery exchange stations, would be candidates to locate stations based on consideration of the station locating cost, recharging efficiency and time, and vehicle routing behavior. Furthermore, the available refueling time (also the length of stay) at each site can be divided into three categories, including the sight-seeing or recreational time at attractions, the battery switching time at convenience stores, and the normal refueling time at common sites (similar to the refueling time at gasoline stations). This new proposed model was compared to that produced for siting a single type of recharging station.

You and Hsieh [49] developed another model to address the problem which simultaneously combines the locations and types of recharging stations. In this case, the objective was to find the optimal origin-destination trips and alternative-fuel vehicle kinds of stations such that the number of people who can complete round-trip itineraries is maximized. These authors proposed a hybrid heuristic approach to solve this model.

Regarding hydrogen-based EVs, Melaina [50] performed a preliminary analysis in order to estimate the number of initial hydrogen stations required by emulating the existing gasoline infrastructure. Nicholas *et al.* [51] used geographical information systems (GIS) to map stations to locations. Nicholas and Ogden [52] based the placement of stations on customer convenience, which is taken into consideration by the average travel time to the nearest station. Schwoon [53] combined agent-based trip modeling and GIS to construct various snapshots of the initial hydrogen filling station network along Germany highways, while Stiller *et al.* [54] analyzed hydrogen fueling stations in highly populated regions and corridors in Europe.

3.3. Capacity of Recharging Stations

The size or capacity of recharging stations for EVs affects the transportation planning. Usually, the capacity of these stations is limited and during a specified time, a station cannot serve more than its capacity, especially recharging stations. This means that only a small number of vehicles can be recharged simultaneously. Changing the departure times of vehicles belonging to a logistic company may require different times for recharging. Moreover, travelers who start their trips in different times may also reach a station at different times. In the specified time in which vehicles reach a station, if the station is occupied, the vehicles must wait in queues. The recharge time, capacity of stations, and waiting are important problems that have been mostly neglected in the EVs station location literature. Hosseini and MirHassani [55] is one of the few works in the literature that considers some of these issues. The objective addressed in their work is to establish a strategic plan in order to build recharging stations in such a way that minimize total costs. These total costs include stations-construction cost, waiting time cost, and refueling cost. Olivella-Rosell *et al.* [56] propose an agent-based simulation approach that allows forecasting the EV charging demand in a certain urban area, and they successfully test the efficiency of their model in the city of Barcelona, Spain.

In the specific case of battery swapping stations, when a vehicle arrives, it requests a fully charged battery pallet to replace the nearly depleted batteries it currently holds. The request could either be satisfied by a fully charged battery pallet from the station storage, or by a pallet that is just completing its charging. If the request is indeed satisfied, the vehicle in turn deposits a fully or partially spent pallet. If there are idle battery pallet chargers at the station, the spent battery pallet is placed on one of them and its recharging begins, otherwise it is kept in a queue until a battery pallet charger is available. If, instead, there is no fully charged battery available at the station, then the vehicle could leave and go to a different station. Alternatively, it could wait for a battery to fully charge, which may take some time. The vehicle could even take, if necessary, a replacement battery that is only partially charged and use that partially charged battery to travel to another battery swap station on its route. In this case, the vehicle will have to stop earlier than planned, and this influences the routes planning, which estimated some stops at stations and suddenly the vehicle is forced to perform other not covered stops. Depending on both the number of battery pallet chargers the station holds and number of battery pallets the station keeps on hand, the size and attendant cost of the station will change. The availability of charged battery pallets at any given time depends on the size of the station, the inventory of pallets, and the demand for charged pallets the station is experiencing. The station incurs an indirect cost from the unavailability of charged pallets when an EV arrives for an exchange because the driver will not have to pay for a battery swap, and there may be a loss of goodwill from the unserved customer. Models to evaluate total direct and indirect costs for possible decisions on station sizing and inventory holding would be very important in designing the battery swapping infrastructure.

In the literature related to battery-swap station size, Zheng *et al.* [36] proposed a method for locating and sizing battery swap stations in distribution systems, which are two determinants keys in the take-up of EVs as explained before. The problem is modeled as maximizing the net present value of the battery swap station project, where the battery swap station model, load type, network reinforcement, and reliability are taken into consideration.

In the case of hydrogen-based EVs, since the price of hydrogen exhibits an inverse feedback interaction with the adoption rate of fuel cell vehicles and corresponding demand for hydrogen, this behavior has a compounding cyclical effect [57]. Existing models often fall short with respect to incorporating this effect into capacity decisions. Further, the capacity decision depends on the demand that is unknown *a priori* [58]. Game theory may be required to determine the optimal timing of capacity investment. Thus, for example, Qin *et al.* [59] uses an option-based approach to demonstrate the behavior or optimal capacity decisions considering a variety of factors. Struben and Sterman [60] discusses requirements of sustained adoption of hydrogen EVs, and Gnann and Plötz [61] provide a review of integrated market and infrastructure models.

3.4. Fleet Size and Mix

EVs are likely to be used in delivery fleets with other kinds of vehicles. A well-studied branch of the VRP literature is precisely addressing the problem of heterogeneous fleets in delivery fleets [62]. As noticed in Lebeau *et al.* (2015) [63], merging the VRP research on electric vehicles with the fleet size and mix vehicle routing problem is therefore relevant to come with recommendations for logistics decision makers. One of the first attempts to investigate the specific characteristics of EVs as part of the fleet of a VRP was achieved by Gonçalves *et al.* [64]. They considered a VRP with pickup and delivery using a mixed fleet that consists of EVs and vehicles using internal-combustion engines. The objective is to minimize total costs, which consist of vehicle related fixed and variable costs. They consider time and capacity constraints and assume a time for recharging the EVs, which were calculated from the total distance travelled and the range using one battery charge. Vehicles can recharge anywhere during the routes, involving a time penalty. This way, several scenarios combining kinds of vehicles were evaluated, and finally the results showed that using EVs is a more costly alternative due to the high investment required for acquiring or converting these vehicles. According to Lebeau *et al.* (2015) [63], one of the problems of this model is that the locations of charging spots were not considered, meaning that EVs could virtually refuel anywhere on the delivery round once the battery was empty. Erdoğan and Miller-Hooks [65] improved the previous work by developing the green VRP with the possibility of refueling vehicles at the existing alternative fueling stations along the routes. The contributions of these two works were integrated by Schneider *et al.* [66] in their electric VRP with time windows. Charging locations and charging times are both considered in their model which approaches well the problem of EVs. A similar approach was also developed by Conrad and Figliozzi [67]. Based on a capacitated VRP with time windows constraints, they introduced the limited range and charging times in order to get the recharging vehicle routing problem. Their main difference is regarding charging locations, since these authors consider that charging is possible at some customer locations while the formulation in [66] is more flexible as other possible charging locations are possible in the network.

Bae *et al.* [68] also considered the EV and internal-combustion engines vehicles fleet size and mix problem as a two-player two-stage game. They focused on determination of the level of hybrid or alternative energy delivery fleet for a logistics and transportation company. In order to do that, they constructed a model of self-selection with heterogeneous consumers who value the firm's delivery service along two dimensions: the quality of delivery service and the relative reduction in emissions. These authors concluded that while subsidies may increase the operator's profit they may also result in higher prices for the customers. However, these customers will benefit from a reduction in pollution if more EVs are used. More recently, Van Duin *et al.* [69] dealt with the electrical vehicle fleet size and mix VRP with time windows. The aim is to determine an optimal fleet of EVs and delivery routes to offer a desired service level at minimal cost to a set of customers with delivery time windows. However, as noticed by Lebeau *et al.* (2015) [63], they approached the problem without considering the previous work on battery electric vehicles in VRP. As a result, the model involves similar weaknesses as in [64], *i.e.*, they do not consider the locations of charging points. An EV with a battery swapping system is modeled so that the range of this EV can be doubled. Nonetheless, the swapping system is not reflected in the constraints. It is in fact reflected in the range parameter of the vehicle which is simply doubled, meaning that the battery of the EV can be swapped virtually anywhere on the road. Hiermann *et al.* [70] developed that idea further to propose an electrical vehicle fleet size and mix VRP with time windows that also considers the decisions regarding the fleet composition and the choice of recharging times and locations. This work can be considered as the state of the art of delivery optimization with EVs. If vehicles cannot recharge or swap their batteries on the road, then another different problem can be discussed as in Juan *et al.* [10], who dealt with the VRP with multiple driving ranges, an extension of the classical routing problem where the total distance each vehicle can travel is limited and is not necessarily the same for all vehicles, *i.e.*, the fleet is heterogeneous with respect to maximum route lengths.

Regarding hydrogen-based EVs, additional papers highlight the use of fleet vehicles to take advantage of centralized fueling while hydrogen infrastructure is being developed. Thus, for instance, Mercuri *et al.* [71], Joffe *et al.* [72], O'Garra *et al.* [73], and Brey *et al.* [74] present examples in Italy, United Kingdom, and Spain, respectively. More recently, researchers have begun to include more complete systems into the scope of their models. For example, infrastructure models have been developed for China [75], Europe [76], Germany [77], Great Britain [78], South Korea [79], and the United States [80–82]. For a recent literature review aimed at optimizing hydrogen infrastructure see Agnolucci and McDowall [83].

4. Emerging Vehicle Routing Problem (VRP) Operational Issues Related to the Use of EVs

Novel emerging routing models for EVs have to include the most important practical constraints of logistics service providers that use EVs for last-mile deliveries. First, vehicle capacity restrictions have to be considered for a significant share of delivery operations. Second, many companies, e.g., in the small package shipping sector, face a high percentage of time-definite deliveries, which makes the integration of customer time windows into the routing model a necessity. The second aspect is especially interesting because recharging times for EVs cannot be assumed to be fixed but depend on the current battery charge of the vehicle when arriving at the recharging station. Moreover, recharging operations take a significant amount of time, especially compared to the relatively short customer service times of, e.g., small package shippers, and thus clearly affect the route planning.

Motivated by the current transportation circumstances, Electric Vehicle Management (EVM) has recently emerged as a new challenging problem, which is strongly related to the field of green logistics and has the purpose to expedite the establishment of a customer convenient, cost-effective, EVs infrastructure. Based on the increasing relevance of the problem in the last years, a number of research groups in this field have started working on some particular aspects of this new area. The next subsections analyze some typical strategic, tactical, and operational issues arising in EVM based on the new features of EVs.

4.1. Economic Issues of EVs

This is a strategic objective that tries to determine if the emerging novel EVs technology is sustainable for certain transportation activities. Many papers have been devoted to sustainable operations in the transportation area. However, research on the economic viability of EVs is limited in the production and operations management literature. In addition, the impact of EVs on the associated supply chain is yet to be examined. Research on EVs has been mostly focused on: (i) planning infrastructure deployment [46]; (ii) impact of integrating electric vehicles into the power system [84,85]; and (iii) using incentives to promote EVs [86]. Perspectives and insights on EV adoption are only recently being developed. Thus, in the Avci *et al.* [87] model, the interactions between the infrastructure provider and direct consumers in a principal-agent framework. Kleindorfer *et al.* [88] develop a framework to determine and value optimal fleet renewal strategies for the French postal service, La Poste, under two technology options (EVs and ICEVs), uncertain fuel costs, and uncertain battery prices. Wang and Lin [89] consider a firm's capacity adjustments over time given a portfolio of technology options when the demand and the fuel costs are uncertain. Chocteau *et al.* [90] use a game theory framework to study the value of cooperation between stakeholders such as fleet manager, auto manufacturer, and electricity supplier under multiple coalition settings. They also present conditions under which such cooperation can add value.

4.2. Fleet size and Mix Issues of EVs

A critical issue arising when EVs are incorporated into the set of vehicles to be managed gives place to the so-called VRPs with heterogeneous fleet. Contributions related to fleet size and mix considering EVs are recent and very limited. A Mixed Fleet or Heterogeneous VRP considers problems where different types of vehicles are available. It was first introduced in Golden *et al.* [91]. Subsequently, Baldacci *et al.* [92] identifies five major subclasses differing in the number of vehicles available (limited and unlimited), whether a fixed cost per vehicle is considered or not and if the routing cost depends on the vehicle type. The original formulation in [91] considers an unlimited number of vehicles with fixed acquisition costs and vehicle type independent routing costs, which can be classified as a Fleet Size and Mix VRP with Fixed costs (FSMF).

As described in [70], Liu and Shen [93] proposes the fleet size and mix VRP with time windows reformulating the FSMF to take into account time windows. The routing cost corresponds to the so-called *en route* time , which is the time between departing from and returning to the depot menus de cumulative service time at the customers in the respective route. This approach was tested using a new benchmark set based on the well-known Solomon instances for the VRP with time windows.

4.3. Charging Networks Issues of EVs

One major barrier to the success of EVs is the limited number of refueling stations. Due to the restricted range of batteries, the establishment of an infrastructure to facilitate recharging is a pressing concern. Two critical factors determine the need for infrastructure services such as battery swapping and recharging: daily driving distance and battery range. Due to the large capital costs involved in infrastructure investment, economic factors are very important in determining the number and location of stations. Therefore, studies must work to provide a theoretical basis for station deployment, such as with a facility location model, to economically and efficiently serve EV trips [33]. Location problems in general are spatial resource allocation problems dealing with one or more service facilities serving a spatially distributed set of demands. The objective is to locate facilities to optimize a spatially dependent objective such as the minimization of average traveling time or distance between demands and facilities. The most studied practical problem in this context concerns hydrogen station location. General criteria are proposed for identifying effective locations for early hydrogen stations: (i) close to areas with high traffic volume; (ii) in places to provide fuel during long distance trips; (iii) at high profile locations to increase public awareness; and (iv) in places that are accessible to individuals who are buying their first fuel-cell vehicle. These criteria are also needed to be taken into account in EVM in order to ensure consumer confidence in the reliability of the refueling network [33].

4.4. Routing Issues of EVs

Routing of EVs is a critical aspect of EVM, it consists of designing routes for maximizing the autonomy of vehicles. Efficient EV routing plays a major role for encouraging EV use. The energy shortest path problem and the energy routing problem and some relationships between these emerge as new challenges to face in the EVM. Restricted driving distance between battery charges is a fundamental impediment to increase consumer adoption of EVs. In the small-package shipping industry, several big companies, such as DHL, UPS and DPD have already started using EVs for last-mile deliveries, particularly in urban areas. Moreover, governments in all parts of the world promote the electrification trend and plan to provide the required infrastructure. As mentioned earlier, a successful transition from conventional vehicles to EVs requires the development of novel efficient route-planning techniques that take into account the specific features of EVs. Currently, the maximum driving range of most EVs is estimated to be about 100–150 miles [16], but it can be decreased significantly by cold temperatures and so-called range anxiety [94,95]. Thus, the available range is potentially not sufficient to perform the typical delivery tour of a logistics service provider in one run or to reach customers located far from the depot. Because reducing the number of deliveries performed by one vehicle is clearly not a profitable option, visits to recharging stations along the routes are required. These recharging visits have to be explicitly considered in the route planning to avoid inefficient vehicle routes with long detours, especially if the number of available recharging stations is scarce. In a recent work, Hung *et al.* [96] propose a queuing modeling framework to develop efficient routing strategies for EVs requesting charging at available stations. These authors show that the proposed routing strategies contribute to improving the throughput of the queuing system and also to reducing stopover times. In addition, Liu *et al.* [97] analyze a heterogeneous fleet version of the VRP in which the goal is to find a routing solution minimizing the carbon footprint. Similarly, Fang *et al.* [98] try to minimize the carbon footprint generated by bird watching tourist activities throughout optimal routing design supported by geographic information systems.

Table 2 summarizes some of the main decision variables, constraints, and objective functions related to the new VRP variants that emerge when considering heterogeneous fleets of ICEVs and EVs.

Table 2. Details of some new VRP variants related to EVs.

Variant	Decision Variables	Constraints	Objectives
Fleet size and mix	(1) Determine the number and type of EVs to be purchased. (2) Determine the ideal composition of the heterogeneous fleet.	(1) Environmental standards and price incentive to acquisition of EVs. (2) Fixed and variable charging times. (3) Limited budget to renew the fleet of vehicles.	(1) Minimize the acquisition and operating costs of new EVs. (2) Maximize the satisfaction of customer needs. (3) Minimize the environmental impact.
Charging networks	(1) Determine number and geographical position of recharging stations. (2) Determine capacity of recharging stations. (3) Determine technology of recharging stations (low or fast recharge). (4) Decide between swapping or recharging of batteries.	(1) Limited budget to install new recharging stations. (2) Needs of EVs to recharge or exchange batteries.	(1) Minimize the investment and operating costs of charging networks. (2) Maximize the level of service to customers.
Routing	(1) Determine the number of visits to recharging stations. (2) Determine the timing of visits to recharging stations. (3) Allocate available recharging resources to vehicles in recharging stations. (4) Select the option of recharging or swapping batteries.	(1) Geographical position of recharging stations. (2) Capacity of recharging stations. (3) Fixed or variable recharging/swapping times.	(1) Minimize routing cost considering recharging operations. (2) Minimize routing times considering recharging operations. (3) Minimize recharging and swapping costs.

5. Solving Approaches for VRPs with EVs

As discussed in the previous sections, the introduction of EVs in freight fleets imposes a number of strategic and operational challenges that must be efficiently addressed with the use of novel methods and approaches. This is particularly true for VRP models, which can be classified into three levels according to their degree of realism (Figure 2). The classical-basic VRP models are mainly theoretical models that allow the development of mathematical approaches, either if they use exact or approximate solving methods. These models are used to test solving methods in controlled environments, which allows assessing their performance before being used in practical applications. The classical-advanced VRP models are characterized by a higher level of realism, *i.e.*, large-scale problems, multi-objective functions, integrated routing and logistics. Examples of the latter are: VRPs combined with packing [99], allocation, inventory management [100], *etc.* More advanced and complex VRP variants are included in this category. Usually, these problems have been solved by metaheuristic approaches, such as Genetic Algorithms, Iterated Local Search, Ant Colony Optimization, Simulated Annealing, *etc.* Most of the existing work in the VRP literature so far deals with these two classical types of models. Recently, however, and due to both the maturity of existing exact and metaheuristic methods as well as to new business needs, new Rich VRP models are being considered. The solving methods for these models combine different exact and metaheuristic approaches (matheuristics) [101] or even simulation with metaheuristics (simheuristics) [102]. Simheuristics allow considering uncertainty both in the objective function and the constraints of a VRP model, thus making these models a more accurate representation of real-life routing distribution systems. These hybrid methods not only can deal with uncertainty and real-time decision making [103], but they can also consider aspects such as richer objective functions (including environmental costs), dynamism [104], diversity of vehicle driving ranges, multi-periodicity in the distribution activity, integration with other supply chain components, *etc.*

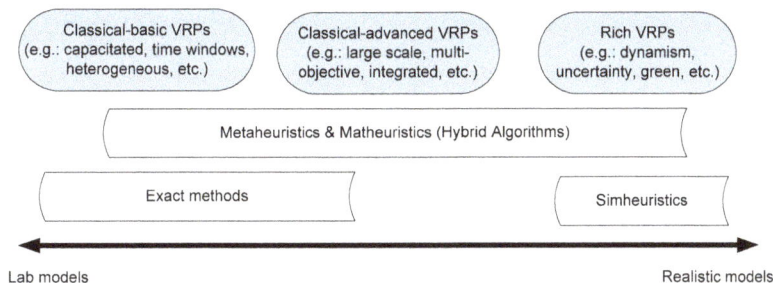

Figure 2. Classification of Vehicle Routing Problem (VRP) models according to their degree of realism.

6. Other Related and Emergent Issues

As some experts point out, life cycle cost analysis is a necessary step in order to properly assess the long-term benefits associated with substituting ICEVs by EVs. Thus, Aguirre *et al.* [105] perform a study to compare the lifecycle environmental costs (energy inputs and CO_2 emissions) of an ICEV, a hybrid vehicle, and an EV. According to their results, the hybrid vehicle is the most effective in terms of CO_2 emissions and also the one offering the lowest net present cost. However, the EV was the most efficient in terms of total environmental impact during its lifetime. Gao and Winfield [106] investigate the lifetime GHG emissions and energy use for different types of fuel-efficient vehicles, showing that all of them improve, in both dimensions, the values associated with ICEVs. They also conclude that all these advanced vehicles require more energy for production than ICEVs, mainly due to the additional power electronics and battery packs. Nevertheless, the energy savings in the fuel cycle for these advanced vehicles compensates the marginal energy required during the vehicle cycle (production stage). Li *et al.* [107] compare the vehicle cycle energy and gas emission impacts of both ICEVs and EVs in China. According to their analysis, when considering the entire life cycle EVs are the best choice in terms of energy consumption and gas emissions. However, these authors also remark the importance of solving some operational and technological challenges, e.g., charging facilities location and capacity, before massively adopting EVs as the standard solution. Finally, Noori *et al.* [108] analyze the life cycle cost and life cycle environmental emissions of ICEVs, hybrid electric vehicles, and three different types of EVs. According to their results, ICEVs are the most cost effective vehicle type in terms of life cycle cost. However, they also conclude that shifting towards EVs reduces the environmental damage costs when considering the vehicle lifetime. At the same time, they also notice that the use of EVs has a high impact on the water footprint due to upstream electricity generation and to the water consumption necessary for battery production.

Regarding the use of EVs in rural areas, Aultman-Hall *et al.* [109] discuss suitability and charging requirements in these environments. They conclude that, although hybrid vehicles will still have substantial utility in these areas, EVs are quickly becoming an attractive alternative for rural mobility demand, especially in those areas with an acceptable power supply and vehicle charging infrastructure. Newman *et al.* [110] support the idea that EVs can be extraordinarily useful in sub-urban and rural areas, especially as a complement to deficient public transport infrastructures. Nevertheless, they also notice that, quite often, habitants of rural areas have difficulties buying EVs due to their relatively low purchasing power. Wappelhorst *et al.* [111] recognize two of the main obstacles impeding the expansion of EVs: their cost and their driving range limitations. In order to partially overcome these problems, they propose the use of intermodal concepts and car-sharing practices. After some empirical studies, the authors conclude that car-sharing of EVs could have the same positive adoption level in rural areas as in the urban ones.

Interruption of power supply causes serious problems in civic life, especially during the evacuation of stricken areas. It impacts medical institutions, interrupts the supply chain, and causes serious damage to the economy. Therefore, in disaster cases it is essential to minimize the period of power failure. Focused in the vehicle industry, Kinomura *et al.* [112] describe the development of Toyota's electricity supply system, through which vehicles supply power directly to electrical devices, and may supply power at either a home site or at an evacuation center in the event of a natural disaster. Yamamura and Miwa [113] consider the design of an effective control method for store-carry-forward energy distribution after a disaster in which EVs and plug-in hybrid vehicles are mobile units having power generation and power storage capabilities. Finally, Yamagata *et al.* [114] propose and extend the concept of a community-based disaster resilient electricity sharing system. In this system, electricity generated from widely introduced photovoltaic panels is stored in the not-in-use cars.

7. Conclusions

This paper has reviewed some of the existing literature related to the introduction of electric vehicles in road transportation, paying special attention to environmental issues; the emerging strategic and operational challenges; the use of hydrogen electric vehicles as an alternative to other types of electric vehicles; and the new variants of the popular vehicle routing problem that arise as a consequence of introducing electric vehicles in the distribution fleets. From this analysis, it becomes evident that the use of sustainable energy sources in road logistics and transportation is more necessary than ever, and constitutes a critical factor for the evolution of an economically and environmentally stable world. The incorporation of electric vehicles in the road distribution activities, especially in urban areas, shows a promising trend yet to be explored in its full potential. However, the expanded use of electric vehicles raises a number of concerns and challenges that complicate planning efforts. From an operational research point of view, the following strategic and operational challenges can be highlighted: (a) the development of infrastructure networks for battery recharging/swapping, including the number, location, type, and capacity of the associated stations; (b) the size and mix composition of hybrid fleets with both traditional vehicles, hybrid electric vehicles, and pure electric vehicles; (c) the severe feasibility constraints imposed using heterogeneous fleets of vehicles with different driving ranges; (d) the additional time-window constraints related to short driving ranges; and (e) the economic impact of the introduction of electric vehicles over the entire supply chain. The development of new optimization and hybrid optimization-simulation methods to efficiently cope with these challenges, including dynamic scenarios and scenarios with uncertainty, is a necessary step to promote the desirable shift towards more sustainable energy sources in the logistics and transportation arena.

Acknowledgments: This work has been partially supported by the Spanish Ministry of Economy and Competitiveness (TRA2013-48180-C3-P and TRA2015-71883-REDT), FEDER, and the CYTED Program (CYTED2014-515RT0489). Likewise, we want to acknowledge the support received by the Catalan Government (2014-CTP-00001) and the CAN Foundation (CAN2014-3758 and CAN2015-70473).

Author Contributions: Angel Juan had the original idea for the study and coordinated the manuscript development. Javier Faulin was the main contributor of the section about environmental issues. Jesica de Armas analyzed the strategic and operational challenges related to the introduction of EVs. Scott E. Grasman provided much of the insight related to hydrogen-based EVs. Finally, Carlos Mendez was in charge of developing most of the parts related to the new EV-based variants for the Vehicle Routing Problem.

Conflicts of Interest: The authors declare no conflict of interest.

References

1. Toth, P.; Vigo, D. *Vehicle Routing: Problems, Methods, and Application*, 2nd ed.; Society for Industrial and Applied Mathematics (SIAM): Philadelphia, PA, USA, 2014.
2. Caceres, J.; Arias, P.; Guimarans, D.; Riera, D.; Juan, A. Rich Vehicle Routing Problem: Survey. *ACM Comput. Surv.* **2014**, *47*, 1–28. [CrossRef]
3. Vidal, T.; Crainic, T.G.; Gendreau, M.; Prins, C. Heuristics for multi-attribute vehicle routing problems: A survey and synthesis. *Eur. J. Oper. Res.* **2013**, *231*, 1–21. [CrossRef]

4. Figliozzi, M.A. Vehicle routing problem for emissions minimization. *Transp. Res. Rec. J. Transp. Res. Board* **2010**, *2197*, 1–7. [CrossRef]
5. Bektas, T.; Laporte, G. The pollution-routing problem. *Transp. Res. B* **2011**, *45*, 1232–1250. [CrossRef]
6. Davis, B.A.; Figliozzi, M.A. A methodology to evaluate the competitiveness of electric delivery trucks. *Transp. Res. E* **2013**, *49*, 8–23. [CrossRef]
7. Sinha, K.C.; Labi, S. *Transportation Decision Making: Principles of Project Evaluation and Programming*; John Wiley & Sons: New Jersey, NY, USA, 2007.
8. Piecyk, M.; McKinnon, A.; Allen, J. Evaluating and Internalizing the Environmental Costs of Logistics. In *Green Logistics: Improving the Environmental Sustainability of Logistics*; McKinnon, A., Browne, A., Whiteing, A., Eds.; Kogan Page: London, UK, 2012.
9. Korzhenevych, A.; Dehnen, N.; Bröcker, J.; Holtkamp, M.; Meier, H.; Gibson, G.; Varma, A.; Cox, V. Update of the Handbook on External Costs of Transport. Ricardo-AEA/R/ED57769. 2014. Available online: http://ec.europa.eu/transport/themes/sustainable/studies/doc/2014-handbook-external-costs-transport.pdf (accessed on 16 January 2016).
10. Juan, A.; Goentzel, J.; Bektaş, T. Routing fleets with multiple driving ranges: Is it possible to use greener fleet configurations? *Appl. Soft Comput.* **2014**, *21*, 84–94. [CrossRef]
11. Figliozzi, M.A. The impacts of congestion on commercial vehicle tour characteristics and costs. *Transp. Res. E Logist. Transp. Rev.* **2010**, *46*, 496–506. [CrossRef]
12. Chen, J. Energy efficiency comparison between hydraulic hybrid and hybrid electric vehicles. *Energies* **2015**, *8*, 4697–4723. [CrossRef]
13. Martin, K.B.; Grasman, S.E. An assessment of wind-hydrogen systems for light duty vehicles. *Int. J. Hydrog. Energy* **2009**, *34*, 6581–6588. [CrossRef]
14. Grimm, N.B.; Faeth, S.H.; Golubiewski, N.E.; Redman, C.L.; Wu, J.; Bai, X.; Briggs, J.M. Global change and the ecology of cities. *Science* **2008**, *319*, 756–760. [CrossRef] [PubMed]
15. Schliwa, G.; Armitage, R.; Aziz, S.; Evans, J.; Rhoades, J. Sustainable city logistics: Making cargo cycles viable for urban freight transport. *Res. Transp. Bus. Manag.* **2015**, *15*, 50–57. [CrossRef]
16. Feng, W.; Figliozzi, M.A. An economic and technological analysis of the key factors affecting the competitiveness of electric commercial vehicles: A case study from the USA market. *Transp. Res. C* **2013**, *26*, 135–145. [CrossRef]
17. Russo, F.; Comi, A. City characteristics and urban goods movements: A way to environmental transportation system in a sustainable city. *Procedia Soc. Behav. Sci.* **2012**, *39*, 61–73. [CrossRef]
18. Shahraeeni, M.; Ahmed, S.; Malek, K.; van Drimmelen, B.; Kjeang, E. Life cycle emissions and cost of transportation systems: Case study on diesel and natural gas for light duty trucks in municipal fleet operations. *J. Nat. Gas Sci. Eng.* **2015**, *24*, 26–34. [CrossRef]
19. Bernard, S.M.; Samet, J.M.; Grambsch, A.; Ebi, K.L.; Romieu, I. The potential impacts of climate variability and change on air pollution-related health effects in the United States. *Environ. Health Perspect.* **2001**, *109*, 199–209. [CrossRef] [PubMed]
20. Nüesch, T.; Cerofolini, A.; Mancini, G.; Cavina, N.; Onder, C.; Guzzella, L. Equivalent consumption minimization strategy for the control of real driving NO$_x$ emissions of a diesel hybrid electric vehicle. *Energies* **2014**, *7*, 3148–3178. [CrossRef]
21. Colin, G.; Chamaillard, Y.; Charlet, A.; Nelson-Gruel, D. Towards a friendly energy management strategy for hybrid electric vehicles with respect to pollution, battery and drivability. *Energies* **2014**, *7*, 6013–6030. [CrossRef]
22. Chen, Z.; Xiong, R.; Wang, K.; Jiao, B. Optimal energy management strategy of a plug-in hybrid electric vehicle based on a particle swarm optimization algorithm. *Energies* **2015**, *8*, 3661–3678. [CrossRef]
23. Hwang, T.; Ouyang, Y. Urban freight truck routing under stochastic congestion and emission considerations. *Sustainability* **2015**, *7*, 6610–6625. [CrossRef]
24. Jaller, M.; Holguín-Veras, J.; Hodge, S. Parking in the city: Challenges for freight traffic. *Transp. Res. Rec. J. Transp. Res. Board* **2013**, *2379*, 46–56. [CrossRef]
25. Browne, M.; Allen, J.; Leonardi, J. Evaluating the use of an urban consolidation centre and electric vehicles in central London. *IATSS Res.* **2011**, *35*, 1–6. [CrossRef]

26. Pelletier, S.; Jabali, O.; Laporte, G. Goods Distributions with Electric Vehicles: Review and Research Perspectives. CIRRELT Publications. CIRRELT-2014–44. 2014. Available online: https://www.cirrelt.ca/DocumentsTravail/CIRRELT-2014-43.pdf (accessed on 16 January 2016).

27. Feng, W.; Figliozzi, M.A. Conventional *vs* electric commercial vehicle fleets: A case study of economic and technological factors affecting the competitiveness of electric commercial vehicles in the USA. *Procedia Soc. Behav. Sci.* **2012**, *39*, 702–711. [CrossRef]

28. Afroditi, A.; Boile, M.; Theofanis, S.; Sdoukopoulos, E.; Margaritis, D. Electric Vehicle Routing Problem with industry constraints: Trends and insights for future research. *Transp. Res. Procedia* **2014**, *3*, 452–459. [CrossRef]

29. Conway, A.; Fatisson, P.E.; Eickemeyer, P.; Cheng, J.; Peters, D. Urban micro-consolidation and last mile goods delivery by freight-tricycle in Manhattan: Opportunities and challenges. In Proceedings of the 91st Transportation Research Board Annual Meeting, Washington, DC, USA, 22–26 January 2012.

30. Lenz, B.; Riehle, E. Bikes for urban freight? *Transp. Res. Rec. J. Transp. Res. Board* **2013**, *2379*, 39–45. [CrossRef]

31. Demir, E.; Bektas, T.; Laporte, G. A review of recent research on green road freight transportation. *Eur. J. Oper. Res.* **2014**, *237*, 775–793. [CrossRef]

32. Demir, E.; Huang, Y.; Scholts, S.; van Woensel, T. A selected review on the negative externalities of the freight transportation: Modeling and pricing. *Transp. Res. E* **2015**, *77*, 95–114. [CrossRef]

33. Wang, Y.; Wang, C. Locating passenger vehicle refueling stations. *Transp. Res. E* **2010**, *46*, 791–801. [CrossRef]

34. Li, J.Q. Transit bus scheduling with limited energy. *Transp. Sci.* **2014**, *48*, 521–539. [CrossRef]

35. Yang, J.; Sun, H. Battery swap station location-routing problem with capacitated electric vehicles. *Comput. Oper. Res.* **2015**, *55*, 217–232.

36. Zheng, Y.; Dong, Z.Y.; Xu, Y.; Meng, K.; Zhao, J.H.; Qiu, J. Electric vehicle battery charging/swap stations in distribution systems: Comparison study and optimal planning. *IEEE Trans. Power Syst.* **2014**, *29*, 221–229. [CrossRef]

37. Liu, Y.X.; Hui, F.H.; Xu, R.L.; Chen, T.; Xu, X.; Li, J. Investigation on the construction mode of charging and battery-exchange station. In Proceedings of the Power Energy Engineering Conference, Wuhan, China, 25–28 March 2011; pp. 1–2.

38. Lombardi, P.; Heuer, M.; Styczynski, Z. Battery switch station as storage system in an autonomous power system: Optimization issue. In Proceedings of the IEEE PES General Meeting, Minneapolis, MN, USA, 25–29 July 2010; pp. 1–6.

39. Zhou, F.Q.; Lian, Z.W.; Wan, X.L.; Yang, X.H.; Xu, Y.S. Discussion on operation mode to the electric vehicle charging station. *Power Syst. Prot. Control* **2010**, *1*, 65–72.

40. Upchurch, C.; Kuby, M. Comparing the p-median and flow refueling models for locating alternative-fuel stations. *J. Transp. Geogr.* **2010**, *18*, 750–758. [CrossRef]

41. Goodchild, M.F.; Noronha, V. *Location-Allocation and Impulsive Shopping the Case of Gasoline Retailing. Spatial Analysis and Location-Allocation Models*; Van Nostrand Reinhold: New York, NY, USA, 1987.

42. Hodgson, M. A flow-capturing location-allocation model. *Geogr. Anal.* **1990**, *22*, 270–279. [CrossRef]

43. Kuby, M.; Lim, S. The flow-refueling location problem for alternative-fuel vehicles. *Socio Econ. Plan. Sci.* **2005**, *39*, 125–145. [CrossRef]

44. Lim, S.; Kuby, M. Heuristic algorithms for siting alternative-fuel stations using the flow-refueling location model. *Eur. J. Oper. Res.* **2010**, *204*, 51–61. [CrossRef]

45. Capar, I.; Kuby, M. An efficient formulation of the flow refueling location model for alternative-fuel stations. *Inst. Ind. Eng. Trans.* **2012**, *44*, 622–636. [CrossRef]

46. Mak, H.Y.; Rong, Y.; Shen, Z.J.M. Infrastructure planning for electric vehicles with battery swapping. *Manag. Sci.* **2013**, *59*, 1557–1575. [CrossRef]

47. Wang, Y.W.; Lin, C.C. Locating road-vehicle refueling stations. *Transp. Res. E Logist. Transp. Rev.* **2009**, *45*, 821–829. [CrossRef]

48. Wang, Y.W. Locating flow-recharging stations at tourist destinations to serve recreational travelers. *Int. J. Sustain. Transp.* **2011**, *5*, 153–171. [CrossRef]

49. You, P.S.; Hsieh, Y.C. A hybrid heuristic approach to the problem of the location of vehicle charging stations. *Comput. Ind. Eng.* **2014**, *70*, 195–204. [CrossRef]

50. Melaina, M.W. Initiating hydrogen infrastructure: Preliminary analysis of a sufficient number of initial hydrogen stations in the US. *Int. J. Hydrog. Energy* **2003**, *28*, 743–755. [CrossRef]

51. Nicholas, M.A.; Handy, S.L.; Sperling, D. Using GIS to evaluate siting and networks of hydrogen stations. *Transp. Res. Rec. J. Transp. Res. Board* **2004**, *1880*, 126–134. [CrossRef]

52. Nicholas, M.; Ogden, J. Detailed analysis of urban station siting for California hydrogen highway network. *Transp. Res. Rec.* **2007**, 129–139. [CrossRef]

53. Schwoon, M. A tool to optimize the initial distribution of hydrogen fueling stations. *Transp. Res. D Transp. Environ.* **2007**, *12*, 70–82. [CrossRef]

54. Stiller, C.; Seydel, P.; Bünger, U.; Wietschel, M. Early hydrogen user centres and corridors as part of the European hydrogen energy roadmap (HyWays). *Int. J. Hydrog. Energy* **2008**, *33*, 4193–4208. [CrossRef]

55. Hosseini, M.; MirHassani, S.A. Selecting optimal location for electric recharging stations with queue. *KSCE J. Civ. Eng.* **2015**, *19*, 2271–2280. [CrossRef]

56. Olivella-Rosell, P.; Villafafila-Robles, R.; Sumper, A.; Bergas-Jané, J. Probabilistic agent-based model of electric vehicle charging demand to analyse the impact on distribution networks. *Energies* **2015**, *8*, 4160–4187. [CrossRef]

57. Keles, D.; Wietschel, M.; Most, D.; Rentz, O. Market penetration of fuel cell vehicles—Analysis based on agent behavior. *Int. J. Hydrog. Energy* **2008**, *33*, 4444–4455. [CrossRef]

58. Dangl, T. Investment and capacity choice under uncertain demand. *Eur. J. Oper. Res.* **1999**, *117*, 415–428. [CrossRef]

59. Qin, R.; Grasman, S.E.; Martin, K.B. An inventory modeling approach for hydrogen fueling station capacity considering and outside option. *Energy Syst.* **2013**, *4*, 195–217.

60. Struben, J.; Sterman, J. Transition challenges for alternative fuel vehicle and transportation systems. *Environ. Plan. B Plan. Des.* **2007**, *35*, 1070–1097. [CrossRef]

61. Gnann, T.; Plötz, P. A review of combined models for market diffusion of alternative fuel vehicles and refueling infrastructure. *Renew. Sustain. Energy Rev.* **2015**, *47*, 783–793. [CrossRef]

62. Lin, C.; Choy, K.L.; Ho, G.T.S.; Chung, S.H.; Lam, H.Y. Survey of green vehicle routing problem: Past and future trends. *Expert Syst. Appl.* **2014**, *3*, 1118–1138. [CrossRef]

63. Lebeau, P.; De Cauwer, C.; Van Mierlo, J.; Macharis, C.; Verbeke, W.; Coosemans, T. Conventional, Hybrid, or Electric Vehicles: Which Technology for an Urban Distribution Centre? *Sci. World J.* **2015**, 302867. [CrossRef] [PubMed]

64. Gonçalves, F.; Cardoso, S.R.; Relvas, S.; Barbosa-Povoa, A. Optimization of a distribution network using electric vehicles: A VRP problem. In Proceedings of the IO2011-15 Congresso da associação Portuguesa de Investigação Operacional, Coimbra, Portugal, 18–20 April 2011.

65. Erdoğan, S.; Miller-Hooks, E. A green vehicle routing problem. *Transp. Res. E Logist. Transp. Rev.* **2012**, *48*, 100–114. [CrossRef]

66. Schneider, M.; Stenger, A.; Goeke, D. The electric vehicle-routing problem with time windows and recharging stations. *Transp. Sci.* **2014**, *48*, 500–520. [CrossRef]

67. Conrad, R.G.; Figliozzi, M.A. The Recharging Vehicle Routing Problem. Available online: http://web.cecs.pdx.edu/~maf/Conference_Proceedings/2011_The_Recharging_Vehicle_Routing_Problem.pdf (accessed on 16 January 2016).

68. Bae, S.H.; Sarkis, J.; Yoo, C.S. Greening transportation fleets: Insights from a two-stage game theoretic model. *Transp. Res. E Logist. Transp. Rev.* **2011**, *47*, 793–807. [CrossRef]

69. Van Duin, J.H.R.; Tavasszy, L.A.; Quak, H.J. Towards E(lectric)-urban freight: First promising steps in the electric vehicle revolution. *Eur. Transp. Trasp. Eur.* **2013**, *54*, 1–19.

70. Hiermann, G.; Puchinger, J.; Hartl, R.F. The Electric Fleet Size and Mix Vehicle Routing Problem with Time Windows and Recharging Stations. 2015. Available online: http://prolog.univie.ac.at/research/publications/downloads/Hie_2015_638.pdf (accessed on 16 January 2016).

71. Mercuri, R.; Bauen, A.; Hart, D. Options for refueling hydrogen fuel cell vehicles in Italy. *J. Power Sources* **2002**, *106*, 353–363. [CrossRef]

72. Joffe, D.; Hart, D.; Bauen, A. Modelling of hydrogen infrastructure for vehicle refueling in London. *J. Power Sources* **2004**, *131*, 13–22. [CrossRef]

73. O'Garra, T.; Mourato, S.; Pearson, P. Analyzing awareness and acceptability of hydrogen vehicles: A London case study. *Int. J. Hydrog. Energy* **2005**, *30*, 649–659. [CrossRef]

74. Brey, J.J.; Brey, R.; Carazo, A.F.; Contreras, I.; Hernandez-Diaz, A.G.; Gallardo, V. Designing a gradual transition to a hydrogen economy on Spain. *J. Power Sources* **2006**, *159*, 1231–1240. [CrossRef]

75. Weinert, J.X.; Liu, S.; Ogden, J.M.; Ma, J. Hydrogen refueling station costs in Shanghai. *Int. J. Hydrog. Energy* **2007**, *32*, 4089–4100. [CrossRef]

76. Tzimas, E.; Castello, P.; Peteves, S. The evolution of size and cost of a hydrogen delivery infrastructure in Europe in the medium and long term. *Int. J. Hydrog. Energy* **2007**, *32*, 1355–1368. [CrossRef]

77. Ball, M.; Wietschel, M.; Rentz, O. Integration of a hydrogen economy into the German energy system: An optimizing modeling approach. *Int. J. Hydrog. Energy* **2007**, *32*, 1355–1368. [CrossRef]

78. Almansoori, A.; Shah, N. Design and operation of a future hydrogen supply chain: Snapshot model. *Chem. Eng. Res. Des.* **2006**, *84*, 423–438. [CrossRef]

79. Kim, J.; Moon, I. Strategic design of hydrogen infrastructure considering cost and safety using multiobjective optimization. *Int. J. Hydrog. Energy* **2008**, *33*, 5887–5896. [CrossRef]

80. Lin, Z.; Chen, C.; Ogden, J.; Fan, Y. The least-cost hydrogen for southern California. *Int. J. Hydrog. Energy* **2008**, *33*, 3009–3014. [CrossRef]

81. Johnson, N.; Yang, C.; Ogden, J. A GIS-based assessment of coal-based hydrogen infrastructure deployment in the State of Ohio. *Int. J. Hydrog. Energy* **2008**, *30*, 5287–5303. [CrossRef]

82. Ogden, J.M. Developing an infrastructure for hydrogen vehicles: A southern California case study. *Int. J. Hydrog. Energy* **1999**, *24*, 709–730. [CrossRef]

83. Agnolucci, P.; McDowall, W. Designing future hydrogen infrastructure: Insights from analysis at different spatial scales. *Int. J. Hydrog. Energy* **2013**, *38*, 5181–5191. [CrossRef]

84. Galus, M.D.; Zima, M.; Andersson, G. On integration of plug-in hybrid electric vehicles into existing power system structures. *Energy Policy* **2011**, *38*, 6736–6745. [CrossRef]

85. Sioshansi, R. OR Forum-Modeling the impacts of electricity tariffs on plug-in hybrid electric vehicle charging, costs, and emissions. *Oper. Res.* **2012**, *60*, 506–516. [CrossRef]

86. Huang, J.; Leng, M.; Liang, L.; Liu, J. Promoting electric automobiles: Supply chain analysis under a government's subsidy incentive scheme. *IIE Trans.* **2012**, *45*, 826–844. [CrossRef]

87. Avci, B.; Girortra, K.; Netessine, S. Electric vehicles with a battery switching station: Adoption and environmental impact. *Manag. Sci.* **2014**, *61*, 772–794. [CrossRef]

88. Kleindorfer, P.R.; Neboian, A.; Roset, A.; Spinler, S. Fleet renewal with electric vehicles at La Poste. *Interfaces* **2012**, *42*, 465–477. [CrossRef]

89. Wang, Y.W.; Lin, C.C. Locating multiple types of recharging stations for battery-powered electric vehicle transport. *Transp. Res. E Logist. Transp. Rev.* **2013**, *58*, 76–87. [CrossRef]

90. Chocteau, V.; Drake, D.; Kleindorfer, P.; Orsato, R.J.; Roset, A. Collaborative innovation for sustainable fleet operations: The Electric Vehicle Adoption Decision. 2011. Available online: http://www.insead.edu/facultyresearch/research/doc.cfm?did=47857 (accessed on 16 January 2016).

91. Golden, B.; Assad, A.; Levy, L.; Gheysens, F. The fleet size and mix vehicle routing problem. *Comput. Oper. Res.* **1984**, *11*, 49–66. [CrossRef]

92. Baldacci, R.; Battarra, M.; Vigo, D. Routing a Heterogeneous Fleet of Vehicles. In *The Vehicle Routing Problem: Latest Advances and New Challenges*; Golden, B., Raghavan, S., Wasil, E., Eds.; Springer: New York, NY, USA, 2008; pp. 3–27.

93. Liu, F.H.; Shen, S.Y. The fleet size and mix vehicle routing problem with time windows. *J. Oper. Res. Soc.* **1999**, *50*, 721–732. [CrossRef]

94. Tredeau, F.P.; Salameh, Z.M. Evaluation of lithium iron phosphate batteries for electric vehicles application. In Proceedings of the IEEE Vehicle Power and Propulsion Conference, Dearborn, MI, USA, 7–10 September 2009; pp. 1266–1270.

95. Botsford, C.; Szczepanek, A. Fast charging *vs.* slow charging: Pros and cons for the new age of electric vehicles. In Proceedings of the EVS24 International Battery, Hybrid and Fuel Cell Electric Vehicle Symposium, Stavanger, Norway, 13–16 May 2009.

96. Hung, Y.C.; Michailidis, G. Optimal routing for electric vehicle service systems. *Eur. J. Oper. Res.* **2015**, *247*, 515–524. [CrossRef]

97. Liu, W.Y.; Lin, C.C.; Chiu, C.R.; Tsao, Y.S.; Wang, Q. Minimizing the carbon footprint for the time-dependent heterogeneous-fleet vehicle routing problem with alternative paths. *Sustainability* **2014**, *6*, 4658–4684. [CrossRef]

98. Fang, W.T.; Huang, C.W.; Chou, J.Y.; Cheng, B.Y.; Shih, S.S. Low carbon footprint routes for bird watching. *Sustainability* **2015**, *7*, 3290–3310. [CrossRef]

99. Dominguez, O.; Juan, A.; Nuez, I.; Ouelhadj, D. An ILS-Biased Randomization algorithm for the two-dimensional loading HFVRP with sequential loading and items rotation. *J. Oper. Res. Soc.* **2015**. [CrossRef]
100. Juan, A.; Grasman, S.; Caceres, J.; Bektas, T. A Simheuristic Algorithm for the single-period stochastic inventory routing problem with stock-outs. *Simul. Model. Pract. Theory* **2014**, *46*, 40–52. [CrossRef]
101. Doerner, K.F.; Schmid, V. Survey: Matheuristics for Rich Vehicle Routing Problems. In *Hybrid Metaheuristics*; Blesa, M.J., Blum, C., Raidl, G., Roli, A., Sampels, M., Eds.; Springer: Berlin, Germany, 2010; pp. 206–221.
102. Juan, A.; Faulin, J.; Grasman, S.; Rabe, M.; Figueira, G. A review of Simheuristics: Extending metaheuristics to deal with stochastic optimization problems. *Oper. Res. Perspect.* **2015**, *2*, 62–72. [CrossRef]
103. Juan, A.; Faulin, J.; Jorba, J.; Caceres, J.; Marques, J. Using parallel & distributed computing for solving real-time vehicle routing problems with stochastic demands. *Ann. Oper. Res.* **2013**, *207*, 43–65.
104. Fikar, C.; Juan, A.; Martinez, E.; Hirsch, P. A discrete-event driven metaheuristic for dynamic home-service routing with synchronized trip sharing. *Eur. J. Ind. Eng.* **2016**, in press. [CrossRef]
105. Aguirre, K.; Eisenhardt, L.; Lim, C.; Nelson, B.; Norring, A.; Slowik, P.; Tu, N. Lifecycle Analysis Comparison of a Battery Electric Vehicle and a Conventional Gasoline vehicle. Available online: http://www.ioe.ucla.edu/perch/resources/files/batteryelectricvehiclelca2012.pdf (accessed on 16 January 2016).
106. Gao, L.; Winfield, Z.C. Life cycle assessment of environmental and economic impacts of advanced vehicles. *Energies* **2012**, *5*, 605–620. [CrossRef]
107. Li, S.; Li, J.; LI, N.; Gao, Y. *Vehicle Cycle Analysis Comparison of Battery Electric Vehicle and Conventional Vehicle in China*; SAE Technical Paper 2013-01-2581; SAE International: Warrendale, PA, USA, 2013.
108. Noori, M.; Gardner, S.; Tatari, O. Electric vehicle cost, emissions, and water footprint in the United States: Development of a regional optimization model. *Energy* **2015**, *89*, 610–625. [CrossRef]
109. Aultman-Hall, L.; Sears, J.; Dowds, J.; Hines, P. Travel demand and charging capacity for electric vehicles in rural states. *Transp. Res. Rec. J. Transp. Res. Board* **2012**, *2287*, 27–36. [CrossRef]
110. Newman, D.; Wells, P.; Donovan, C.; Nieuwenhuis, P.; Davies, H. Urban, sub-urban or rural: Where is the best place for electric vehicles? *Int. J. Autom. Technol. Manag.* **2014**, *14*, 306–323. [CrossRef]
111. Wappelhorst, S.; Sauer, M.; Hinkeldein, D.; Bocherding, A.; Glaß, T. Potential of electric carsharing in urban and rural areas. *Transp. Res. Procedia* **2014**, *4*, 374–386. [CrossRef]
112. Kinomura, S.; Kusafuka, H.; Kamichi, K.; Ono, T. *Development of Vehicle Power Connector Equipped with Outdoor Power Outlet Using Vehicle Inlet of Plug-in Hybrid Vehicle*; SAE Technical Papers 2013-01-1442; SAE International: Warrendale, PA, USA, 2013.
113. Yamamura, T.; Miwa, H. Store-carry-forward energy distribution method and routing control method for use in a disaster. In Proceedings of the 2014 International Conference on Intelligent Networking and Collaborative Systems, Salerno, Italy, 10–12 September 2014; pp. 289–295.
114. Yamagata, Y.; Seya, H.; Kuroda, S. Energy resilient smart community: Sharing green electricity using V2C technology. *Energy Procedia* **2014**, *61*, 84–87. [CrossRef]

Article

Cost Projection of State of the Art Lithium-Ion Batteries for Electric Vehicles Up to 2030

Gert Berckmans *, Maarten Messagie, Jelle Smekens, Noshin Omar, Lieselot Vanhaverbeke and Joeri Van Mierlo

MOBI Research Group, Vrije Universiteit Brussel, Pleinlaan 2, 1050 Brussels, Belgium;
maarten.messagie@vub.be (M.M.); jelle.smekens@vub.be (J.S.); noshomar@vub.ac.be (N.O.);
Lieselot.vanhaverbeke@vub.be (L.V.); joeri.van.mierlo@vub.be (J.V.M.)
* Correspondence: gert.berckmans@vub.be; Tel.: +32-2629-3399

Academic Editor: K.T. Chau
Received: 20 July 2017; Accepted: 24 August 2017; Published: 1 September 2017

Abstract: The negative impact of the automotive industry on climate change can be tackled by changing from fossil driven vehicles towards battery electric vehicles with no tailpipe emissions. However their adoption mainly depends on the willingness to pay for the extra cost of the traction battery. The goal of this paper is to predict the cost of a battery pack in 2030 when considering two aspects: firstly a decade of research will ensure an improvement in material sciences altering a battery's chemical composition. Secondly by considering the price erosion due to the production cost optimization, by maturing of the market and by evolving towards to a mass-manufacturing situation. The cost of a lithium Nickel Manganese Cobalt Oxide (NMC) battery (Cathode: NMC 6:2:2 ; Anode: graphite) as well as silicon based lithium-ion battery (Cathode: NMC 6:2:2 ; Anode: silicon alloy), expected to be on the market in 10 years, will be predicted to tackle the first aspect. The second aspect will be considered by combining process-based cost calculations with learning curves, which takes the increasing battery market into account. The 100 dollar/kWh sales barrier will be reached respectively between 2020–2025 for silicon based lithium-ion batteries and 2025–2030 for NMC batteries, which will give a boost to global electric vehicle adoption.

Keywords: process-based cost modeling; NMC battery; silicon lithium-ion battery; market prediction; learning curves

1. Introduction

Throughout the last decades, the emission of greenhouse gases have increased dramatically; however, their negative impact on the climate has been demonstrated [1,2]. To limit these adversary effects of climate change, several actions are undertaken on a worldwide scale, for example it has been agreed at COP21 in Paris to keep the temperature rise limited to maximum 2 °C [3]. Additionally, steps are undertaken by the European Commission to have a cleaner environment by setting new ambitious environmental targets. For example the EU target is to have a CO_2 reduction by 20% compared to the levels of 2008 as stated in their white paper [4]. Improving urban air quality and reducing its impact on climate change of transport comes down to (1) reducing the total consumption of kilometers by improving efficiency of the service and (2) providing the remainder of needed transport without fossil fuels. A technological option to substitute fossil based km is to use battery electric vehicles, powered by renewable fuels. To ensure a minimum of driving range a large, expensive battery is required for battery electric vehicles, explaining their high cost which is limiting its mass-adoption. The cost and performance of the battery, the most expensive component in a vehicle, is directly linked with the adaption of electric vehicles. The adoption towards battery electric vehicles mainly depends on the willingness to pay for the extra cost of the traction battery. Therefore will this paper study the price evolution of an automotive battery up to 2030 and answer the following questions. What is thus a reasonable price of a 1 kWh lithium battery in 2030? Can we expect an erosion of the price due to a production cost optimization in a mass-manufacturing situation?

In an initial phase, the current market of electric and hybrid vehicles is analyzed. Additionally, its sales up to 2030 are predicted based on historical data. Cost of a battery is inversely linked with the growth of the market of electric vehicles, since larger production quantities leads to lowers cost per unit. In this study this effect will not be taken into account, resulting in an underestimation or a very conservative estimation of the amount of EVs. In a second phase the cost and sales price of a battery are calculated and predicted up to 2030 based upon an innovative methodology. This innovative methodology will combine process-based cost modeling with learning curves to cope with the evolution from an immature to a mature battery market. Another innovative aspect is that current state of the art battery chemistries will be used alongside with battery chemistries which are believed to become the state of the art in 2030. A roadmap of future battery technologies will be presented, out of which a promising battery chemistry will be chosen.

2. Market and Technology Landscape of Electric Vehicles

This section will analyze the current global automotive market as well as the technological split between internal combustion driven vehicles, hybrid and electric vehicles. The current technological split is expected to change due to firstly the increased awareness regarding climate change and secondly the decreasing cost of electric vehicles. Based on historical sales figures a prediction of global sales of vehicles up to 2030 will be made, including the evolution of the technological split.

2.1. State of the Art—BEV

An overview of the most sold BEVs in the small and medium-large segment of 2016 are shown in Table 1, adapted from [5]. This is a non-exhaustive list, for example vehicles which do not reach 100 km/h are omitted as well as vehicles sold in low quantities. It can be seen that in the segment of small cars, which are mainly city cars, rather small batteries are used with an average energy content of 18.2 kWh and a range of 150 km. The average values are quite coherent since the median gives comparable results. In this segment rather small batteries are used due to two reasons. Firstly a battery represents 75% of an EVs powertrain cost [6], which means that implementing a bigger battery would significantly increase the overall cost. Secondly, because space is often limited in a city car to place a bigger battery. In the segment of medium to large cars it is clear from Table 1 that bigger batteries are used, namely on average a battery energy content of 36.2 kWh and of course a larger average range of 231 km. However, due to the large battery of the Tesla's the average can be misleading and the median gives a better representation of the current market, meaning an energy content of 24.2 kWh and a range of 190 km. This is consistent with the higher cost and size of these vehicles.

Table 1. Overview of electric vehicles commercially available in 2016, adapted from [5].

Vehicle Segment	Brand	Model	Model Year	Battery Energy Content (kWh)	Range (km)
	Smart	Fortwo	2014	17,6	160
	Toyota	iQ EV	2012	12	85
	Fiat	500e	2015	24	135
	Citroen	C-Zero	2014	14,5	150
	Peugeot	iOn	2014	14,5	150
Small	Mitsubitshi	i-MiEV	2014	16	160
	VW	e-up!	2013	18,7	160
	Chevrolet	Spark Ev	2015	18,4	130
	Bollore	Bluecar	2015	30	250
	Mitsubitshi	MinicabMiEV	2014	16	150
Average				**18.2**	**153**
Median				**16.8**	**150**
	BMW	i3	2014	22	190
	Renault	Zoe	2015	22	240
	Volvo	C30 Electric	2015	24	145
	VW	e-Golf	2016	24,2	190
	Nissan	Leaf (2016)	2014	30	250
	Honda	FIT EV	2012	20	130
	Renault	Fluence Z.E.	2015	22	185
Medium-Large	Ford	Focus EV	2015	23	162
	Kia	Soul Electric	2015	27	212
	Mercedes	B-class El.Dr.	2015	36	230
	BYD	e6	2015	61,4	205
	Nissan	e-NV200	2015	24	170
	Toyota	RAV 4 EV	2014	41,8	182
	Tesla	Model S	2015	75	480
	Tesla	Model X	2015	90	489
Average				**36.2**	**231**
Median				**24.2**	**190**

2.2. State of the Art—HEV

In Table 2 an overview of the 10 most European sold hybrid electric vehicles (HEV) in 2016 are given [7]. The list consists solely of high-end vehicles in which the electrical range is quite limited except the BMW i3. This is due to the current tax reductions for hybrid vehicles in several European countries, which use the electric power to reduce their average fuel consumption and emissions on which taxes are generally based. In this table the BMW i3 is also included, which has a high range and a large battery pack since it is a BEV with a range extender. Therefore it is more representable for the HEV category to use the median value to get a better insight in the battery energy contents and driving ranges used in the HEV segment. The average value does not give a good representation due to the influence of the BMW i3 and its large range and batteries since it is a BEV with a range extender. Therefore also the median is given. Small battery packs are used which can be seen from their low median battery energy content of 9 kWh and limited median driving range of 41 km.

The mass-adoption of BEVs and HEVs are somehow limited due to two reasons. Firstly the high initial cost of HEV and BEV, mainly because of the high purchasing cost of the battery pack [8]. A second problem with electric vehicles is range anxiety, meaning the fear of running out of fuel. Many research efforts are ongoing to improve both problems of which the first one will be more deeply discussed in Section 3. In literature [9–15] the range anxiety is identified as mainly a psychological barrier since most people drive less kilometers a day than the range of current EVs. This problem is enhanced by the long charging time of an EV as well as the lack of abundantly available charging stations for electric vehicles. Therefore, increasing the battery energy content to increase its range to about 500 km, so having a comparable battery capacity as a Tesla of about 75–90 kWh will significantly speed up the adoption of EVs. To implement such batteries their energy density has to increase significantly, since battery with such high energy content are to large to fit in small city cars.

Table 2. Overview of most sold hybrid electric vehicles in 2016 in Europe (10 most popular).

Vehicle Segment	Brand	Model	Model Year	Battery Energy Content (kWh)	Range (km)
	VW	Passat GTE	2015	9,9	50
	Mitsubishi	Outlander PHEV	2013	12	52
	Volvo	XC90 PHEV	2015	9,2	40
	Mercedes	GLC350e	2016	8,7	34
Medium-Large	BMW	225xe Active Tourer	2015	7,6	41
	Mercedes	C350e	2015	6,5	31
	BMW	330e	2015	7,6	40
	BMW	X5 40e	2015	9	31
	Audi	A3 e-Tron	2014	9	50
	BMW	i3 range extended	2013	22	320
Average				18.2	153
Median				9	41

2.3. Electric Vehicle Prediction Up to 2030

The following paragraph will make a prediction of the global sales of electric, hybrid, classical combustion engine vehicles as well as other types of vehicles such as compressed natural gas (CNG), liquefied petroleum gas (LPG), fuel cells vehicles, which will be combined in the category others. Based upon a literature review [16–21] a prediction of the technology split is shown in Figure 1. In 2015 the global sales are still dominated (99.3%) by the classical combustion engine(ICE) based vehicles, even by 2030 more ICE than electric vehicles are sold however its dominance decreases significantly. Due to the increasing effort of the automotive manufacturers more and more HEV and BEV models are available on the market, increasing the choice for consumers, which was quite limited up to in the past. In 2030 25% of all vehicles sold will be either fully electric or hybrid, requiring an enormous amount of batteries. To get a better idea of the quantities this analysis is expanded by combining the previous figure with the expected global sales of vehicles predictions worldwide, which can be seen in Figure 2. Only limited sources are available in literature [22–28] which predict the global sales up to 2030. Therefore it was opted to make a prediction based upon the sales in the past, more specifically the global sales between 2010 and 2015 were analyzed [28]. Since only a small period of time is analyzed a linear approximation is used, predicting in 2020, 2025, 2030 receptively 107, 122 and 138 million vehicles sold yearly. The linear approximation can be clearly seen in the increase of the total amount of vehicles sold, Figure 2. The peak of ICE vehicles will be reached in 2020 with a sales of more than 100 millions. By 2030 roughly 10 million BEVs and 20 million HEVs will be sold on a yearly. When assuming 75–90 kWh is needed for BEV, HEV will require half of the capacity roughly 23 billion kWh of battery are required yearly which is a very large potential market for battery manufactures.

PREDICTION TECHNOLOGY SPLIT

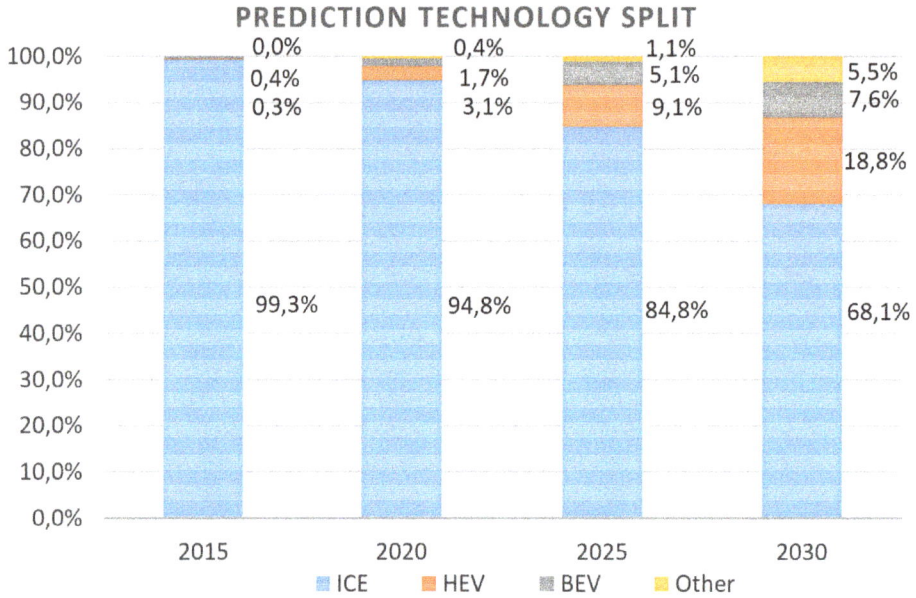

Figure 1. Evolution of the technology split between electric vehicles (**EV**), hybrid electric vehicles (**HEV**), internal combustion engines (**ICE**) and other up to 2030.

Global vehicle sales prediction including technological split

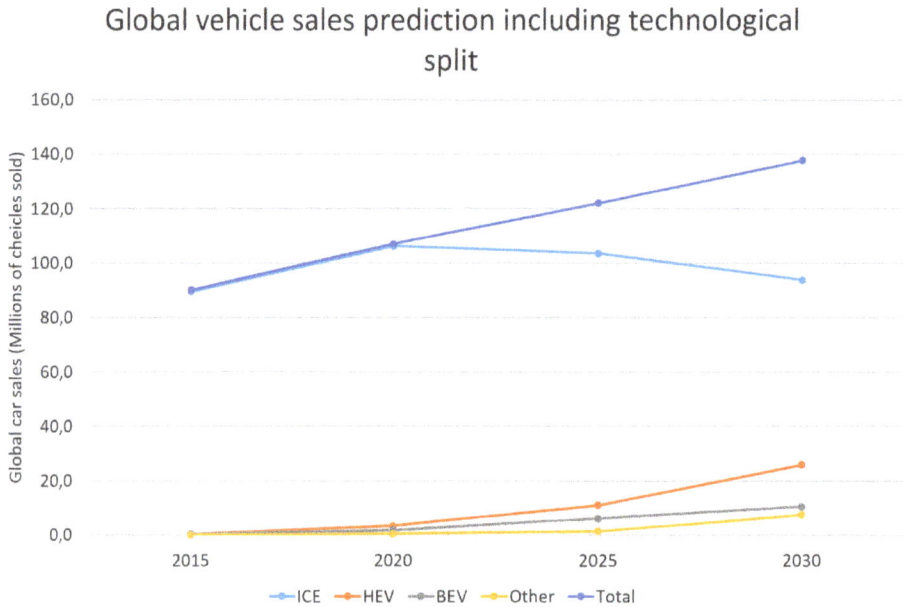

Figure 2. Global vehicles sales predictions up to 2030 including the evolution of the technological split between EV, HEV, ICE and others.

3. Battery Discussion

Batteries have become an indispensable product in society; they are being used in a variety of products ranging from cellphones up to electric vehicles. The most popular battery technologies are the lithium-ion batteries due to their high energy- and power-density as well as their high lifetime compared to other types [29].

The electrochemical storage of energy in a lithium-ion battery is achieved through intercalation in the positive and negative electrode, shown by Equation (1) [30].

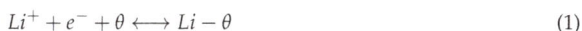

$$Li^+ + e^- + \theta \longleftrightarrow Li - \theta \tag{1}$$

With:

θ	The insertion material
$\theta - Li$	Lithium inserted in material θ
e^-	An electron
Li^+	A lithium-ion

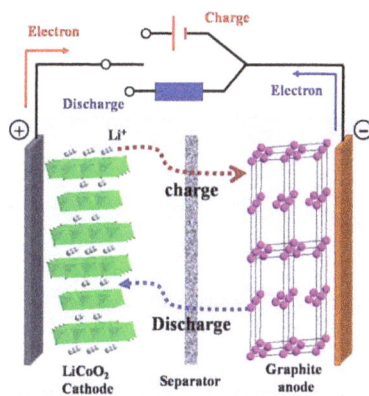

Figure 3. Schematic illustration of the charge/discharge process in a lithium-ion battery, reproduced from [31].

The functioning of a lithium-ion battery is summarized below; however, a more extended explanation can be found in literature, for example [32,33]. The intercalation of lithium-ions in the electrodes is the main reason for its long lifetime of this type of batteries. However, lithium undergoes numerous side-reactions, reducing the concentration of lithium available for intercalation, causing the battery to decrease in capacity over lifetime. A schematic illustration of a lithium-ion battery's functioning is shown in Figure 3. A battery consists of four main components: a negative electrode or often called an anode, a positive electrode or often called a cathode, an electrolyte and a separator. The variety and properties of several anodes and cathodes will be discussed in the next sections except the separator and electrolyte since in most case commercially available ones are used. The main property of the electrolyte is to transport ions from the anode to the cathode or vice-versa, while ensuring as little as possible side reactions with the Li-ions. Mostly it consists of water with some dissolved salts, lithium hexafluorophosphate most used, to ensure good ion conductivity. The purpose of the separator is to stop the transport of electrons while intervening in the rest of the processes as little as possible.

3.1. State of the Art—Anode

Twenty years ago graphite (372 mAh/g [34]) was first commercialized [35] as anode material in a lithium-ion battery and up to now it is still being used in most lithium-ion batteries. Its low cost, good electrochemical performance, low volume expansion during charging and discharging as well as that it is abundantly available, explains the widely accepted use of graphite as anode material [33,35,36]. Many research efforts allowed to optimize this material resulting it is almost reaching its maximum theoretical capacity and only incremental improvements can be expected [29]. However, by adding small amounts of metals with high theoretical energy densities, such as silicon (4200 mAh/g [37]), the overall energy density can be increased [38]. Adding high concentrations of these additional components cause numerous problems such as volume expansions up to 300% as well as reduced lifetime despite the current numerous research efforts, for example using silicon as nano-particles such as in [37].

Other often used anodes materials used are lithium alloyed metals with as most popular $Li_4Ti_5O_{12}$ LTO (175 mAh/g [35]). More noble metals are used resulting in a higher price than graphite. Other disadvantages are its lower energy capacity and reduced cell voltage compared to graphite. However its exceptional good stability over its lifetime makes it the ideal anodes in specific cases explaining its wide usage.

An overview of the two most used anode materials is shown in Table 3 [34,35].

Table 3. Anode materials—Overview including specific energy density, cost and lifetime.

Anode Material	Energy Density (mAh/g)	Cost	Lifetime
Graphite	372	Medium	Medium
$Li_4Ti_5O_{12}$ (LTO)	175	High	High

3.2. State of the Art—Cathode

The selection of the most suited cathode material is strongly dependent on the application itself. A selection has to be made of which key property is the most important for an application. The key properties of a battery are: energy density, power density, cost and lifetime. An overview of the most used cathode materials can be found in Table 4 [29,33,36,38–45].

Table 4. Cathode materials—Overview including energy density cost and lifetime.

Cathode Material	Energy Density (Wh/kg)	Cost	Lifetime
$LiCoO_2$ (LCO)	546	Medium	Medium
$LiMn_2O_4$ (LMO)	410–492	Low	Low
$LiNiMnCoO_2$ (NMC)	610–650	High	High
$LiFePO_4$ (LFP)	518–587	Medium	High
$LiNiCoAlO_2$ (NCA)	680–760	High	Medium

The oldest commercially used electrodes are $LiMn_2O_4$ (LMO) due to the low cost, however the lifetime is limited which is considered to be the biggest disadvantage but they are still frequently used. $LiCoO_2$ (LCO) another old electrode, characterized with a medium cost and high energy, has some safety drawbacks but is still used frequently. $LiNiMnCoO_2$ (NMC), a combination of LCO, LMO and nickel, is gaining popularity due to its high lifetime as well as its high energy density. The exact mixture of Ni, Mn and Co will define the property of the cathode of which a variety exist such as NMC (1:1:1), NMC (5:3:2), … The trend is to use Ni rich NMC since this gives an increased energy density. It is mainly used where cost is less important. $LiFePO_4$ (LFP) has excellent lifetime properties and is frequently used in combination with an LTO anode to get an excellent overall lifetime of the battery. LiNiCoAlO$_2$ (NCA), a relatively new cathode, has a very high energy density however it is potentially thermally unstable stability, meaning a reduced safety.

3.3. State of the Art—Roadmap

Demands on lifetime and energy/power density are ever increasing to extend the duration in which the battery can be used. Hence, there is a continuous need to further improve lithium-ion batteries [46]. This section will explain which future trends can be expected during the next decades, an overview is given in Figure 4, which is limited to lithium-ion batteries. Other types, such as sodium-ion, zinc air, lithium-air are still in a very early phase and thus omitted from this overview.

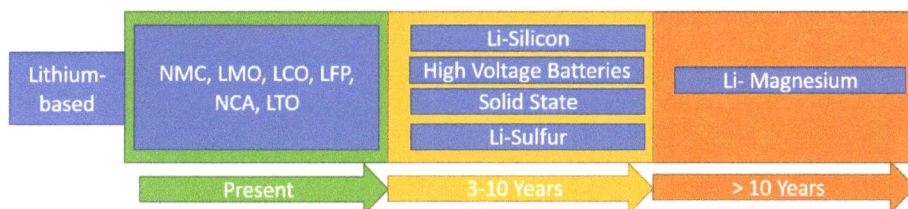

Figure 4. Roadmap of Lithium-ion based batteries from present up to >10 years.

The world of batteries is changing very rapidly which is the reason why it is very hard to predict the most promising battery chemistry. It can be disrupted very easily if a novel type/chemistry is discovered in material sciences with superior properties. However some trends are already visible when limiting to lithium-ion based batteries.

The first clear trend is to use different electrodes which have a significant higher theoretical capacity such as sulfur (1672 mAh/g [47]), silicon (4200 mAh/g [37]) and lithium metal (3860 mA/g [48]). This will inherently increase the energy density of the cell since the electrodes itself can store more energy. A second trend is to increase the voltage limit of a single cell to around 5 V since it is a harmonized voltage value used in the field of electronics. This trend will also increase the cells energy density due to its definition which can be simply represented as integral of the actual capacity multiplied with the actual voltage. The third trend is to go towards solid state electrolytes since using liquid electrolytes can cause safety problems when leaking. In general it can be concluded that in the near future the energy density and safety are the two key topics, in which significant improvement can be expected during the next decade [32].

To make predictions further than a decade is extremely difficult but lithium-magnesium is worth mentioning. It has superior energy density and is abundantly available, but is still in a very early phase [49].

In this paper, as already mentioned, the goal is to perform a price estimation up to 2030. The best overall chemistry now in 2015 is NMC (6:2:2) as cathode combined with graphite as anode due to their high energy density and lifetime. It can be seen in literature many research efforts or ongoing on silicon based cells. Therefore in 2030 namely a silicon-alloy anode combined a nickel rich cathode (NMC (6:2:2)) to maximize its energy content will be most likely on the market. An overview of the two battery chemistries, used this research as well as their pack energy density is shown in Table 5. Throughout these two battery types will be referred to as battery I and battery II.

Table 5. Overview of the cell chemistries used for cost calculations. The first one will be referred to as the NMC battery and the second one as the silicon based lithium-ion battery.

	Battery I	Battery II
Positive electrode	NMC (6:2:2)	NMC (6:2:2)
Negative electrode	Graphite	Silicon Alloy [50]
Pack energy density	155 Wh/kg	205 Wh/kg

3.4. Battery Cell Manufacturing

Several architectures and designs of battery cells exist such as cylindrical cells, pouch cells, hard casing with a variety of positions for the tabs [51]. However in all these designs three main processes can be identified as described in Table 6 [33]. Also in this table the material inputs (+) and outputs (−) are shown since material cost is the main cost of a battery, which will be demonstrated in Section 3.5.

The first step is electrode manufacturing in which the electrodes themselves are being prepared. The active material, conductive agents, solvents and binder are mixed to a slurry which is coated onto a current collector (aluminium for the positive tab and copper for the negative one). After which the cells are dried, in order to vaporise the solvents. To ensure a good electrical connection between the slurry and current collector the electrodes are calendared, which consists of pressing the two firmly together. As a last step in this electrode manufacturing the electrodes are cut to their correct size.

The second step is to make a cells assembly or a multilayer combination of a positive electrode, separator and negative electrode. These multilayers can be created through stacking or winding to create respectively pouch and cylindrical cells. Afterwards the cells are packaged (hard casing/soft casing/...) and temporarily sealed. The cells undergo a drying procedure to ensure no solvents remain after which they are filled with electrolyte and permanently sealed.

The battery cell is now ready to use, however to ensure stable and good quality of it its has to undergo some electrical formation cycles. These formation cycles are critical for the lifetime of the cell since its stabilizes the chemical structure of the cell. The final step is to test the cell's electrical performance to maintain a good quality control.

Table 6. Battery cell manufacturing process summary including material in- and outflow.

Manufacturing Process		Material
Electrode Manufacturing	Slurry Mixing	+ Active Material + Conductive agent + Solvents + Binder
	Coating	+ Al/Cu foil
	Drying	− Solvents
	Calendaring	
	Cutting	+ Remaining al/Cu foil
Cell Assembly	Stacking/Winding	+ Separator + Adhesive Tape + Al/Cu tabs
	Packaging (Pouch/Case)	+ Pouch Foil/casing
	Temporary sealing	+ Solvents
	Drying	
	Filling	− Remaining al/Cu foil
	Permanent Seal	
Formation	Formation	
	Cell Testing	

3.5. Process-Based Cost Modeling

This section will focus on cost and sales prices calculations and predictions by using process-based cost modeling of several battery chemistries. The methodology of process-based cost modeling is to calculate costs based on detailed process descriptions, which are well defined for batteries [52,53]. This methodology is being used in several application fields including battery cost calculations [54–58]. Battery production for automotive cells are still in an early phase and not yet in mass production. This means optimizations in the production process are possible by upscaling and by building more production plants. Process-based modeling can be combined with learning or dynamic curves, which are dependent on growth of the production capacity as demonstrated in other application fields [59]. This study [59] has performed a thorough analysis about the price evaluation of chemical products during a decade and linking the impact of increasing production capacities to the evolution of prices. For analysis of the price evolutions up to 2030 process-based cost modeling extended with learning curves will be used. The main drawbacks of this approach are that the exact process and composition of the battery chemistry has to be known. This can be overcome by combining recently published patents, which describe the processes in detail, as well as by performing an extended literature review.

The cost of two different cell chemistries will be analyzed, namely battery I (NMC(6:2:2) + Graphite) and battery II (NMC(6:2:2) + silicon alloy). NMC based batteries can be seen as the current state of the art batteries and silicon based ones as state of the art batteries in 10–15 years as shown by the roadmap in Figure 4. Process-based cost modeling is used in order to calculate the detailed material cost in dollar/kWh for each battery type. The methodology, including the key equations, are represented in Figure 5. In the first step the manufacturing procedure is split into logical substeps of which the material and energy in and outflows are analyzed, which is done in Section 3.4. In this step also the excess material, used during the manufacturing, should be taken into account. The next step will calculate the cost of goods sold, which entails the material, energy, labor and overhead costs. Two additional assumptions are made regarding the labor and overhead cost, which can be approximated by respectively 15% and 8% as demonstrated in [60]. When comparing with other calculations done in literature [54,56,61] similar assumptions were used. The sales price including the manufacturer's and retailer's profit are taken into account. A profit margin of 35% for the manufacturer is used, which is high but it entails novel products requiring a high profit margin [60]. When these batteries are not directly sold by the manufacturer and additional profit margin of 15% by the retailer is added. This leads to the final sales price. The prediction of the prices will be detailed in Section 3.6.

Throughout this paper several assumptions are made. Firstly all prices will be expressed in dollar/kWh in 2015. All prices will be expressed in price in 2015 to make it easier to compare the evolutions. This will imply that the price of a battery will be higher in 2030 than predicted due to the inflation, which will occur between 2015 and 2030.

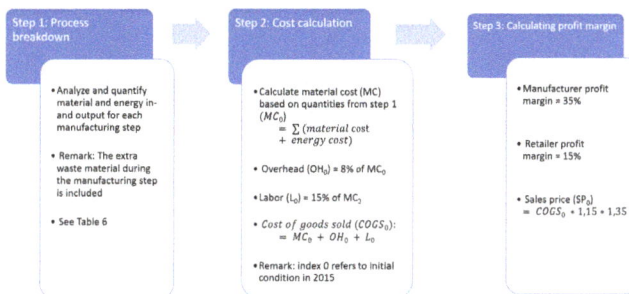

Figure 5. Overview of the used methodology including learning curves.

3.5.1. Battery I—Cost Calculation

The process of manufacturing of NMC lithium-ion batteries is well known; however, a variety of types are still possible. A NMC with ratio 6:2:2 will be used as a baseline for this calculations since this nickel rich type has a high energy density and is considered the current state of the art. It is also suited to increase the overall cell voltage which was described as one of the future trends. This NMC 6:2:2 cathode will be combined with standards binders, conductive agents and as anode graphite. The cost price calculations of a battery pack including a simple passive battery managements system (BMS) and casing are visually represented in Figure 6. The cost of 432 dollar/kWh is dominated by material cost representing 65% of the overall cost. Similar results can be found in literature [56,62] in which the material cost varies between 60–80%. The two most costly components are the positive and negative electrode. These require noble materials and are used in high quantities in a battery pack explaining their high cost. A detailed breakdown analysis of the negative electrode is also shown in Figure 6 in which it is clear that the active material is the main driver of the cost of the electrode (62% of the negative electrode cost). A similar trend can be seen in the positive electrode.

When including the profit margins of the producer and middle man a sales price of 670 dollar/kwh is reached. The breakdown is visualized in Figure 7. Also here the expensive electrodes represent a significant cost of 28%. Since these prices are valid for low production quantities, it is hard to compare them with the pricing of an automotive battery pack, however comparisons can be found varying from 700–1300 Dollar/kWh [63–65] of battery packs sold in lower quantities.

COST BREAKDOWN OF BATTERY I

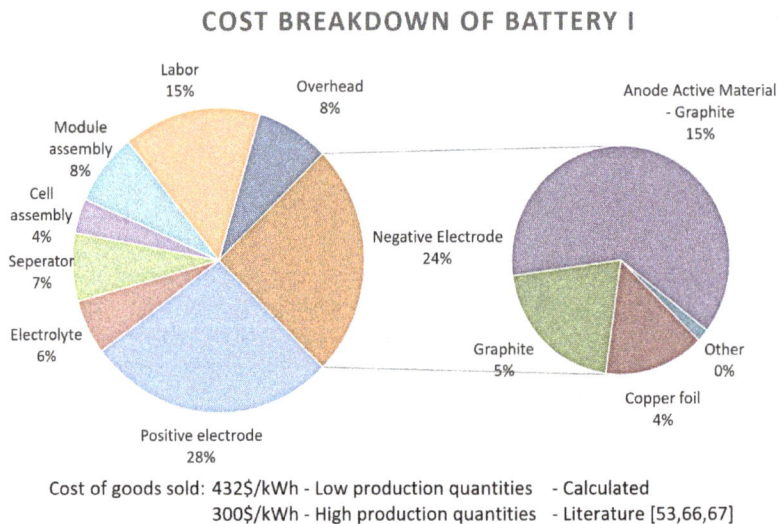

Cost of goods sold: 432$/kWh - Low production quantities - Calculated
300$/kWh - High production quantities - Literature [53,66,67]

Figure 6. Cost breakdown of battery I with a special focus on the anode composition.

SALES PRICE OF BATTERY I

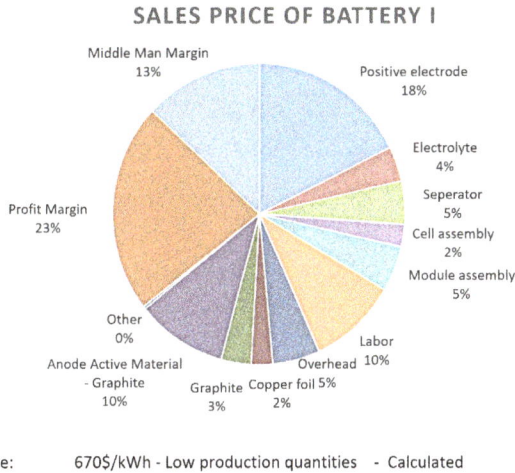

Sales price: 670$/kWh - Low production quantities - Calculated

Figure 7. Sales price of battery I.

3.5.2. Battery II—Cost Calculation

This section will quantify the impact of using silicon-based batteries, with a higher energy density, compared to the classical NMC combined with graphite. The difference between the two is mainly in the negative electrode which uses silicon alloy in stead of graphite. Many varieties of silicon based lithium-ion batteries are available, however the ones in which silicon is used as an alloy are the most promising. As a baseline the following patent WO2016089666 A1 [50] will be used for making a silicon alloy with high lifetime and high energy density. The exact calculation of the composition of this active material composed from raw materials is shown in Table 7. The other components such as binder, conductive agents are kept constant. In reality different materials are required, however in a similar quantity and at a comparable cost explaining why these cost are kept con.

To calculate the total cost per kWh of battery II the increased energy density of the pack should be taken into account. For battery I and battery II receptively the following energy densities are used 155 Wh/kg and 205 Wh/kg. This is an increase of 33% which can be expected from material calculations. Silicon has a 10 times higher theoretical capacity than graphite, but a mixture is used as shown by its composition. The cost breakdown is visualized in Figure 8 in which the negative electrode cost is decreased from 24% for battery I to 19% for battery II. The main impact however comes from the higher energy density explaining a significantly lower cost of silicon based batteries compared top NMC. This makes a total sales price of 456 Dollar/kWh compared to 431 dollar/kWh from NMC, visualized in Figure 9. This is large price reduction of 30%. However it should be stated that this battery is not yet a commercial product and some extra research should be done to increase its lifetime.

Table 7. Anode active material battery II—detailed.

Material	Amount (kg)	Price in 2015 (Dollar)
$Si_{73}Fe_{17}C_{10}$	0.6	2.76
Graphite	0.128	1.25
Carbon nanotubes	0.16	4.17
Carboxy methyl cellulose	0.032	2.96
LiPAA	0.08	11.25
Total	**1**	**22.39**

COST BREAKDOWN OF BATTERY II

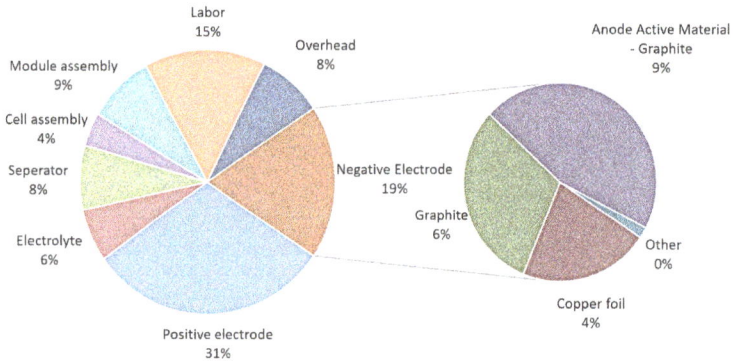

Cost of goods sold: 293$/kWh - Low production quantities - Calculated
Not commercially available in 2015

Figure 8. Cost breakdown of battery II with a special focus on the anode composition.

SALES PRICE OF BATTERY II

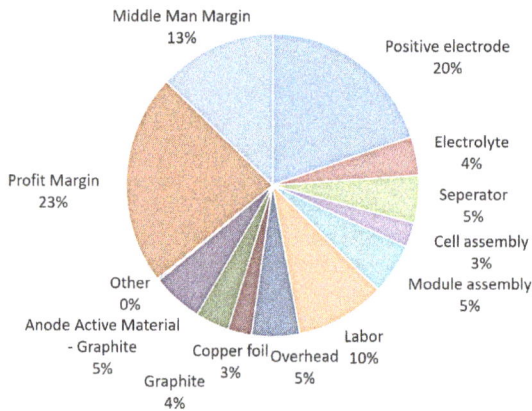

Sales price: 456$/kWh - Low production quantities   - Calculated
Not commercially available in 2015

Figure 9. Sales price of battery II.

3.6. Evolution of Cost in Time

The prediction of cost up to 2030 for automotive batteries based upon battery I and battery II will be detailed in this section. The cost and prices calculated in previous sections are only valid for small production quantities. Therefore the current cost of goods sold for automotive NMC battery packs will be used as a baseline, which is around 300 dollar/kWh according to literature [54,66,67]. Adding the assumed profit margin as stipulated in Figure 5 results in a sales price of 466 dollar/kWh. Also the relative amounts, calculated in the previous sections will be kept constant meaning labor 15%, overhead 8% and material cost 76% of the costs of goods sold (300 dollar/kWh).

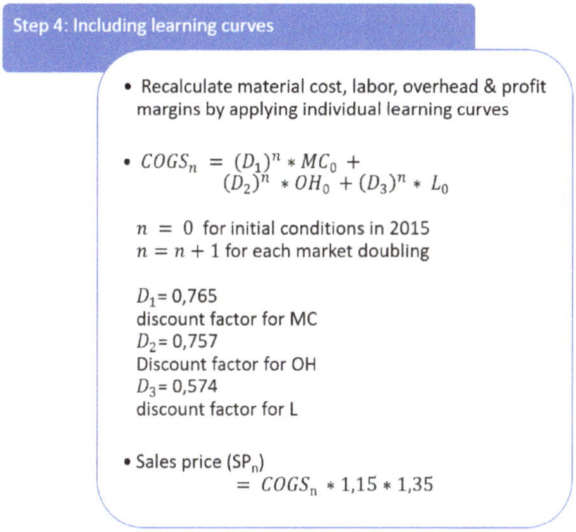

Step 4: Including learning curves

- Recalculate material cost, labor, overhead & profit margins by applying individual learning curves

- $COGS_n = (D_1)^n * MC_0 +$
  $\quad\quad\quad (D_2)^n * OH_0 + (D_3)^n * L_0$

  $n = 0$ for initial conditions in 2015
  $n = n + 1$ for each market doubling

  $D_1 = 0{,}765$
  discount factor for MC
  $D_2 = 0{,}757$
  Discount factor for OH
  $D_3 = 0{,}574$
  discount factor for L

- Sales price (SP_n)
  $\quad\quad\quad = COGS_n * 1{,}15 * 1{,}35$

Figure 10. Overview of the used methodology including its key equations.

As previously mentioned there is a clear link in doubling the production capacity of the industry and price reductions as shown by [59]. Three different types of cost can be analyzed with their own learning/discount rate per doubling capacity of the industry. The calculations of the costs of goods sold and the sales price calculations for predictions up to 2030 are shown in Figure 10. Combining process based modeling with these learning curves are a new approach to tackle the price predictions of batteries. Firstly the material cost is analyzed by Lieberman, [59] and a reduction of 23.5% can be found per doubling production capacity, corresponding to a discount factor of 0.765 of material cost. While for overhead and labor cost receptively the classical "six-tenths-rule" and "two-tenths-rule" [59] are used. The six-tenths-rule gives a relationship between size and cost as shown by following formula [68]:

$$\frac{Cost_2}{Cost_1} = \left(\frac{Scale_2}{Scale_1}\right)^{0.6}$$

This shows that for a size doubling the total cost increases to 151.4% instead of doubling. This can be achieved by for example building a new plant. When this total cost is redistributed over the two plants, the cost for each plant reduced to 75.7% of its original cost. This translate thus in a reduction of 24.3% per plant or a discount rate of 0.757 for overhead cost. Labor cost follows the two-tenths-rule meaning thus in the previous equation 0.6 should be replaced by 0.2. This results in a discount factor of 0.574 or a reduction by 42.6% for each size doubling for labor costs. For completeness the market growth for EV and BEV from 2015 up to 2030 are summarized in Table 8, based upon the predictions made in Section 2.3.

In Figure 11 the prediction of sales cost for battery II is shown. By 2020 a significant price reduction is expected (more than half of the total sales price) due to the rapid growth of the battery market, linked to the EVs one. The threshold of 100 dollar/kWh will be reached between 2025 and 2030.

The same approach is used for battery II. The price in 2015 is calculated using the same proportion of the sales price between NMC and silicon for the previous sections. After which the profit margins are added, resulting in a sales price of 317 dollar/kWh. The prediction of its price evolution is shown in Figure 12. One important remark is that silicon based lithium-ion batteries are not yet on the market but will be in 10 years. If comparing battery I this with battery II the threshold of 100 dollar/kWh will be reached much sooner in 2020–2025.

Table 8. BEV and EV market growth.

Year	Amount of BEV (Millions)	Amount of HEV (Millions)	Market growth
2015	0.4	0.3	1 (Baseline)
2020	1.8	3.4	7
2025	6.2	11.1	25
2030	10.5	25.8	52

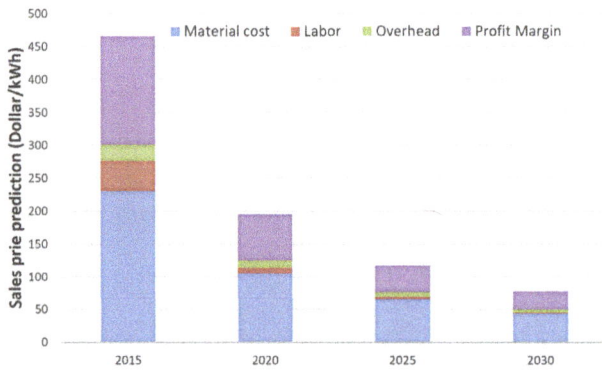

Figure 11. Prediction of sales price of battery I up to 2030.

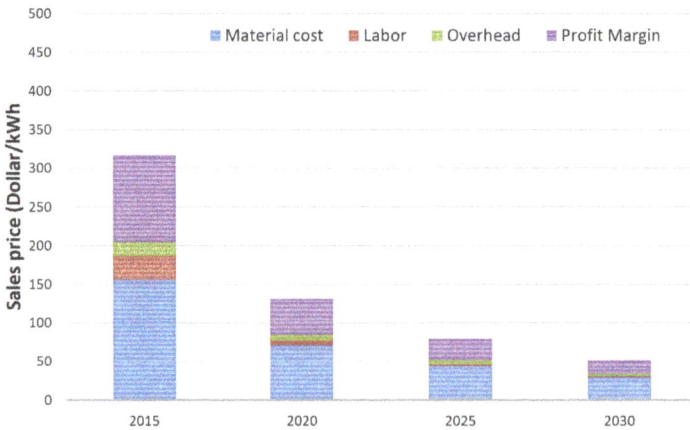

Figure 12. Prediction of sales price of battery II up to 2030.

3.7. Comparison

This section will compare the price predictions available in literature [69–75] of lithium-ion batteries in 2020, visualized in Figure 13. In this figure it can be seen that the prices are ranked from low to high, which also coincides with when the price was predicted. The Boston Consulting Group [71] predicted in 2010 an average price of 300 dollar/kWh, which is the oldest and highest price in the listed comparison. In 2012 and 2015 receptively Roland and Berger [73,74] and Avicenne [72] predicted a price of 275 and 250 dollar/kWh. The predictions in 2016 made by P3 Consulting [75] and GTM Research [69] even shows lower costs of 210 and 217 dollar/kWh. The most recent prediction made by Tesla [70] even 150 dollar/kWh can be expected in 2015. The predictions made by this research gives result of 195 and 131 dollar/kWh, which is in the same line as the recent predictions. However silicon based batteries can have a beneficial impact on the cost it is unlikely that it is already commercially available in 2020 but most likely by 2025.

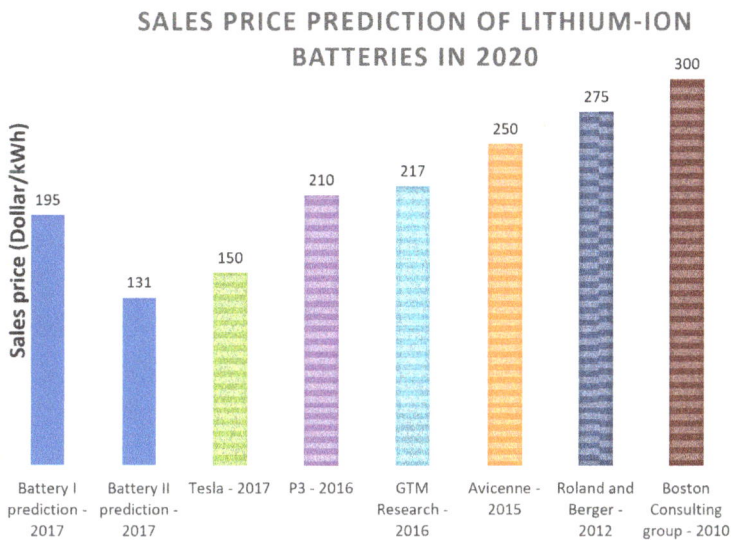

Figure 13. Sales price prediction of lithium-ion batteries in 2020.

4. Conclusions

From the previous analysis it is clear that the electric automotive market is still in its innovators/introduction stage [76]. However in the near future a step towards mass-adoption/ growth is expected. One of the indicators is to analyze the publications and patents about the electric automotive industry which is currently booming. BEVs still have a small driving range however a large effort was done by the manufacturers to significantly extend the choice of electric vehicles during the last years. The most popular HEVs are the higher priced vehicles in which cost plays is less important. Currently the market is still dominated by classical combustion engine vehicles however its dominance will decrease from 99% in 2015 to 68% in 2030. The market of electric vehicles (HEVs and BEVs) will have to increase by a factor of 52, which means a huge investment in battery manufacturing will be required to cope with this increase. This mass production will be one of the driving forces of the decreasing cost of battery pack.

The trends deduced from the roadmap of lithium-ion batteries show that within the next decade improvements regarding energy density and safety can be expected. To anticipate on this trend a innovative approach is used in which process-based cost calculations are used on two types of battery

Energies **2017**, *10*, 1314

chemistries, one which can be considered as state of the art in 2015, namely NMC and one which will be considered as state of the art in 2030, namely silicon based lithium-ion batteries. Additionally, this methodology is combined with learning curves to which will include the maturing of the battery market. Material costs represent the majority of costs in a battery pack (66%) of which the active material, responsible for the intercalation of li-ions, is the most costly component. By using silicon based batteries a cost reduction per kWh of 30%. The limit of 100 Dollar/kWh will be reached in 2020–2025 for silicon based batteries and in 2025–2030 for NMC batteries. This low price will have a significant impact on the overall price of an electric vehicles since the battery represents the largest cost. This price reduction will aide in the mass adoption of electric vehicles.

Author Contributions: Gert Berckmans performed the research and wrote the paper. The supervisors: Lieselot Vanhaverbeke and Joeri Van Mierlo as well as Maarten Messagie, Jelle Smekens and Noshin Omar provided interesting disucssions adn interesting insights inS the research. All authors read and approved the final manuscript.

Conflicts of Interest: The authors declare not conflicts of interest.

References

1. Hooftman, N.; Oliveira, L.; Messagie, M.; Coosemans, T.; Mierlo, J.V. Environmental Analysis of Petrol, Diesel and Electric Passenger Cars in a Belgian Urban Setting. *Energies* **2016**, *9*, 84.
2. Oliveira, L.; Messagie, M.; Rangaraju, S.; Sanfelix, J.; Rivas, M.H.; Mierlo, J.V. Key issues of lithium-ion batteries e from resource depletion to environmental performance indicators. *J. Clean. Prod.* **2015**, *108*, 354–362.
3. COP21. Available online: http://www.cop21paris.org/about/cop21/ (accessed on 19 April 2017).
4. European Commission. *White Paper: European Transport Policy for 2010: Time to Decide*; Technical Report; European Commission: Luxemburg, 2001.
5. Grunditz, E.A.; Thiringer, T. Performance Analysis of Current BEVs—Based on a Comprehensive Review of Specifications. *IEEE Trans. Transp. Electr.* **2016**, *2*, 270–289.
6. Wolfram, A.P.; Lutsey, N. *Electric Vehicles: Literature Review of Technology Costs and Carbon Emissions*; The International Council on Clean Transportation: Washington, DC, USA, 2016; pp. 1–23.
7. European Alternative Fuels Observatory. Available online: http://www.eafo.eu/vehicle-statistics/m1 (accessed on 15 March 2017).
8. Delucchi, M.E.; Lipman, T. An Analysis of the Retail and Lifecycle Cost of Barrery-Powered Electric Vehicles. *Transp. Res. Part D* **2001**, *6*, 371–404.
9. Neubauer, J.; Wood, E. The impact of range anxiety and home, workplace, and public charging infrastructure on simulated battery electric vehicle lifetime utility. *J. Power Source* **2014**, *257*, 12–20.
10. Rauh, N.; Franke, T.; Krems, J.F. Understanding the Impact of Electric Vehicle Driving Experience on Range Anxiety. *J. Hum. Factors Ergon. Soc.* **2015**, *57*, 177–187.
11. Franke, T.; Neumann, I.; Bühler, F.; Cocron, P.; Krems, J.F. Experiencing Range in an Electric Vehicle: Understanding Psychological Barriers. *Appl. Psychol.* **2012**, *61*, 368–391.
12. Franke, T.; Krems, J.F. What drives range preferences in electric vehicle users? *Transp. Policy* **2013**, *30*, 56–62.
13. Pearre, N.S.; Kempton, W.; Guensler, R.L.; Elango, V.V. Electric vehicles: How much range is required for a day's driving? *Transp. Res. Part C Emerg. Technol.* **2011**, *19*, 1171–1184.
14. Dong, J.; Liu, C.; Lin, Z. Charging infrastructure planning for promoting battery electric vehicles: An activity-based approach using multiday travel data. *Transp. Res. Part C Emerg. Technol.* **2014**, *38*, 44–55.
15. Bakker, J. Contesting Range Anxiety: The Role of Electric Vehicle Charging Infrastructure in the Transportation Transition. Master's Thesis, Eindhoven University of Technology, Eindhoven, The Netherlands, 2011.
16. International Energy Agency (IEA). *Technology Roadmap: Electric and Plug-in Hybrid Electric Vehicles*; Technical Report; International Energy Agency (IEA): Paris, France, 2011.
17. Klynveld Peat Marwick Goerdeler KPMG's Global Automotive Executive Survey. Available online: https://assets.kpmg.com/content/dam/kpmg/xx/pdf/2017/01/global-automotive-executive-survey-2017.pdf (accessed on 16 May 2017).

18. Lazard & Roland Berger. Global Automotive Supplier Study 2013. Available online: https://www.rolandberger. com/en/Publications/pub_global_automotive_supplier_study_by_roland_berger_and_lazard.html (accessed on 10 May 2017).

19. International Energy Agency. *Global EV Outlook 2016 Electric Vehicles Initiative*; IEA: Lisbon, Portugal, 2016; pp. 1–51.

20. Mac Donald, J. Electric Vehicles to Be 35% of Global New Car Sales by 2040. Available online: https://about.bnef.com/blog/electric-vehicles-to-be-35-of-global-new-car-sales-by-2040/ (accessed on 4 October 2017).

21. *Global Trends to 2025: A Transformed World*; Lukoil: Moscow, Russia, 2015.

22. King, N. Global Light Vehicle Sales Forecast to Exceed 100 Million Units in 2019. Available online: http://blog.euromonitor.com/2015/07/global-light-vehicle-sales-forecast-to-exceed-100-million-units-in-2019.html (accessed on 25 May 2017).

23. Goldman Sachs. Cars 2025. Available online: http://www.goldmansachs.com/our-thinking/technology-driving-innovation/cars-2025/ (accessed on 5 April 2017).

24. Price Waterhouse Coopers. *PWC: Prediction; Electric Cars: A Market Outlook: The Future of Plug-in Hybrid Electric and All-Electric Vehicles in Hungary*; Technical Report; PWC: Prediction: London, UK, 2014.

25. Statista. Available online: https://www.statista.com/statistics/267128/outlook-on-worldwide-passenger-car-sales/ (accessed on 17 March 2017).

26. Information Handling Services Automotive. *Global Light Vehicle Forecast: Readying For The Next Stage*; Technical Report; IHS Automotive: London, UK, 15 October 2015.

27. International Energy Agency. *Electric and Plug-In Hybrid Vehicle Roadmap*; International Environmental Agency: Lisbon, Portugal, 2010; p. 4.

28. Oica. *Oica: Sales Figures*; Technical Report; Oica: Paris, France, 2017.

29. Deng, D. Li-ion batteries: Basics, progress, and challenges. *Energy Sci. Eng.* **2015**, *3*, 385–418.

30. Fuller, T.F.; Doyle, M.; Newman, J. Simulation and Optimization of the Dual Lithium Ion Insertion Cell. *J. Electrochem. Soc.* **1994**, *141*, 1.

31. Xu, G.L.; Wang, Q.; Fang, J.C.; Xu, Y.F.; Li, J.T.; Huang, L.; Sun, S.G. Tuning the structure and property of nanostructured cathode materials of lithium ion and lithium sulfur batteries Gui-Liang. *J. Mater. Chem. A* **2014**, *2*, 19941–19962.

32. Gopalakrishnan, R.; Goutam, S.; Oliveira, L.M.; Timmermans, J.M.; Omar, N.; Messagie, M.; Bossche, P.V.D.; Mierlo, J.V. A comprehensive study on rechargeable energy storage technologies. *J. Electrochem. Energy Convers. Storage* **2017**, *13*, 1–107.

33. Smekens, J.; Gopalakrishnan, R.; Van den Steen, N.; Omar, N.; Hegazy, O.; Hubin, A.; Van Mierlo, J. Influence of electrode density on the performance of Li-ion batteries: Experimental and simulation results. *Energies* **2016**, *9*, 104.

34. Kheirabadi, N.; Shafiekhani, A. Graphene/Li-Ion battery. *J. Appl. Phys.* **2012**, *112*, 1–19.

35. Nitta, N.; Wu, F.; Lee, J.T.; Yushin, G. Li-ion battery materials: Present and future. *Mater. Today* **2015**, *18*, 252–264.

36. Mekonnen, Y.; Sundararajan, A.; Sarwat, A.I. A Review of Cathode and Anode Materials for Lithium-Ion Batteries. In Proceedings of the 2016 SoutheastCon, Norfolk, VA, USA, 30 March–3 April 2016; pp. 2–7.

37. Su, X.; Wu, Q.; Li, J.; Xiao, X.; Lott, A.; Lu, W.; Sheldon, B.W.; Wu, J. Silicon-Based Nanomaterials for Lithium-Ion Batteries: A Review. *Adv. Energy Mater.* **2013**, *4*, 1–23.

38. Blomgren, G.E. The Development and Future of Lithium Ion Batteries. *J. Electrochem. Soc.* **2017**, *164*, A5019–A5025.

39. Fergus, J.W. Recent developments in cathode materials for lithium ion batteries. *J. Power Source* **2010**, *195*, 939–954.

40. Mizushima, K.; Jones, P.C.; Wiseman, P.J.; Goodenough, J.B. LixCoO2: A new cathode material for batteries of high energy density. *Solid State Ion.* **1981**, *3-4*, 171–174.

41. Liu, C.; Neale, Z.G.; Cao, G. Understanding electrochemical potentials of cathode materials in rechargeable batteries. *Mater. Today* **2016**, *19*, 109–123.

42. Julien, C.M.; Mauger, A.; Zaghib, K.; Groult, H. Comparative Issues of Cathode Materials for Li-Ion Batteries. *Inorganics* **2014**, *2*, 132–154.

43. Meyers, R.A.; Doeff, M.M. *Encyclopedia of Sustainability Science and Technology*; Springer: New York, NY, USA, 2012; pp. 529–564.
44. Heyns, M.; Vereecken, P. *Materials for the Next Generation Batteries (Some) Ways of Storing Electricity*; Technical Report; KU Leuven: Leuven, Belgium, 2013.
45. Meeus, M.; Pace, G. *Current and Future Development of Battery Technology and Its Suitability within Smart Grids*; Technical Report; Ghent University: Ghent, Belgium, 2013.
46. Noorden, R.V. The rechargeable revolution: A better battery. *Nature* **2014**, *507*, 26–28.
47. Preefer, M. *Lithium-Sulfur Batteries and Discharge Products from Cycling Why Li-S Batteries?* Technical Report; Materials Research Laboratory at UCSB: Santa Barbara, CA, USA, 2016.
48. Xu, W.; Wang, J.; Ding, F.; Chen, X.; Nasybulin, E.; Zhang, Y.; Zhang, J.G. Lithium metal anodes for rechargeable batteries. *Energy Environ. Sci.* **2014**, *7*, 513–537.
49. Aurbach, D.; Gofer, Y.; Lu, Z.; Schechter, A.; Chusid, O.; Gizbar, H.; Cohen, Y.; Ashkenazi, V.; Moshkovich, M.; Turgeman, R.; et al. A short review on the comparison between Li battery systems and rechargeable magnesium battery technology. *J. Power Source* **2001**, *97–98*, 28–32.
50. Figgemeier, E. Electrode Composition Comprising Carbon Naotubes, Electrochemical Cell and Method of Making Electrochemical Cells. Patent EP 3029759 A1, 6 August 2016.
51. Kurfer, J.; Westermeier, M.; Tammer, C.; Reinhart, G. Production of large-area lithium-ion cells— Preconditioning , cell stacking and quality assurance. *CIRP Ann. Manuf. Technol.* **2012**, *61*, 1–4.
52. Field, F.R. *Fundamentals of Process-Based Cost Modeling Session Goal & Outline Review of Process-Based Cost Model*; Technical Report; MIT: Cambridge, MA, USA, 2017.
53. Field, F.; Kirchain, R.; Roth, R. Process cost modeling: Strategic engineering and economic evaluation of Materials technologies. *JOM J. Miner. Met. Mater. Soc.* **2007**, *59*, 21–32.
54. Patry, G.; Romagny, A.; Martinet, S.; Froelich, D. Cost modeling of lithium-ion battery cells for automotive applications. *Energy Sci. Eng.* **2015**, *3*, 71–82.
55. Isaacs, J.A.; Tanwani, A.; Healy, M.L.; Dahlben, L.J. Economic assessment of single-walled carbon nanotube processes. *J. Nanoparticle Res.* **2010**, *12*, 551–562.
56. Nelson, P.A.; Gallagher, K.G.; Bloom, I.; Dees, D.W. *Modeling the Performance and Cost of Lithium-Ion Batteries for Electric-Drive Vehicles*; Argonne National Lab: Lemont, IL, USA, 2011; pp. 1–102.
57. Wood, D.L.; Li, J.; Daniel, C. Prospects for reducing the processing cost of lithium ion batteries. *J. Power Source* **2015**, *275*, 234–242.
58. Henriksen, G.L.; Amine, K.; Liu, J.; Nelson, P.A. *ANL-03/5 Materials Cost Evaluation Report for High-Power Li-Ion HEV Batteries*; Argonne National Lab: Lemont, IL, USA, 2002; Volume 1.
59. Lieberman, M. The Learning Curve and Pricing in the Chemical Processing industries. *RAND J. Econ.* **1984**, *15*, 213–228.
60. Gaines, L.; Cuenca, R. *Costs of Lithium-Ion Batteries for Vehicles*; Center for Transportation Research, Energy Division, Argonne National Labratorye: Lemont, IL, USA, 2000.
61. Nykvist, B.; Nilsson, M. Rapidly falling costs of battery packs for electric vehicles. *Nat. Clim. Chang.* **2015**, *5*, 329–332.
62. Anderson, D. An evaluation of current and future costs for lithium-ion batteries for use in electrified vehicle powertrains. *arXiv* **2009**, arXiv:1011.1669v3.
63. Alibaba. Available online: https://www.alibaba.com/showroom/li-ion-nmc-battery.html (accessed on 15 April 2017).
64. Aliexpress. Available online: http://www.aliexpress.com/popular/nmc-battery.html (accessed on 15 April 2017).
65. Batteryspace. Available online: http://www.batteryspace.com/LiNiMnCo-Cells/Packs.aspx (accessed on 15 April 2017).
66. Chung, D.; Elgqvist, E.; Santhanagopalan, S. *Automotive Lithium-ion Cell Manufacturing: Regional Cost Structures and Supply Chain Considerations*; Clean Energy Manufacturing Analysis Center (CEMAC): Golden, CO, USA, 2016.
67. Ashley, S. *Battling the High Cost of EV Batteries*; SAE: Washington, DC, USA, 2015.
68. Sweeting, J. *Project Cost Estimating: Principles and Practice*; Institution of Chemical Engineers: Rugby, UK, 1997; pp. 150–170.

69.   Lacy, S.   Stem CTO: Lithium-Ion Battery Prices Fell 70% in the Last 18 Months.   Available online: https://www.greentechmedia.com/articles/read/stem-cto-weve-seen-battery-prices-fall-70-in-the-last-18-months (accessed on 4 June 2017).

70.   Lambert, F. Electric Vehicle Battery Cost Dropped 80% in 6 Years down to $227/kWh—Tesla Claims to Be Below $190/kWh. Available online: https://electrek.co/2017/01/30/electric-vehicle-battery-cost-dropped-80-6-years-227kwh-tesla-190kwh/ (accessed on 6 May 2017).

71.   The Boston Consulting Group. *Batteries for Electric cars: Challanges Opportunities, and the outlook to 2020*; Technical Report; The Boston Consulting Group: Boston, MA, USA, 2010.

72.   Pillot, C. Battery Market Development for Consumer Electronics, Automotive, and Industrial: Materials Requirements and Trends. In Proceedings of the 5th Israeli Power Sources Conference 2015, Herzelia, Israel, 20–21 May 2015; Volume 1.

73.   Roland and Berger.   Technology & Market Drivers for Stationary and Automotive Battery Systems. In Proceedings of the Batteries 2012, Nice, France, 24–26 October 2012.

74.   Roland and Berger. The Lithium-Ion Battery Value Chain. In Proceedings of the Batteries 2012, Nice, France, 24–26 October 2012.

75.   P3 Consulting Group. *Cost Developments of Battery Systems*; Technical Report; P3 Consulting Group, Inc.: Coral Gables, FL, USA, 2016.

76.   Haupt, R.; Kloyer, M.; Lange, M. Patent indicators for the technology life cycle development. *Res. Policy* **2007**, *36*, 387–398.

Article

New Electro-Thermal Battery Pack Model of an Electric Vehicle

Muhammed Alhanouti [1,*], Martin Gießler [1], Thomas Blank [2] and Frank Gauterin [1]

[1] Institute of Vehicle System Technology, Karlsruhe Institute of Technology, Karlsruhe 76131, Germany; martin.giessler@kit.edu (M.G.); frank.gauterin@kit.edu (F.G.)
[2] Institute of Data Processing and Electronics, Eggenstein-Leopoldshafen 76344, Germany; thomas.blank@kit.edu
* Correspondence: muhammed.alhanouti@partner.kit.edu; Tel.: +49-721-608-45328

Academic Editor: Michael Gerard Pecht
Received: 30 May 2016; Accepted: 12 July 2016; Published: 20 July 2016

Abstract: Since the evolution of the electric and hybrid vehicle, the analysis of batteries' characteristics and influence on driving range has become essential. This fact advocates the necessity of accurate simulation modeling for batteries. Different models for the Li-ion battery cell are reviewed in this paper and a group of the highly dynamic models is selected for comparison. A new open circuit voltage (OCV) model is proposed. The new model can simulate the OCV curves of lithium iron magnesium phosphate (LiFeMgPO$_4$) battery type at different temperatures. It also considers both charging and discharging cases. The most remarkable features from different models, in addition to the proposed OCV model, are integrated in a single hybrid electrical model. A lumped thermal model is implemented to simulate the temperature development in the battery cell. The synthesized electro-thermal battery cell model is extended to model a battery pack of an actual electric vehicle. Experimental tests on the battery, as well as drive tests on the vehicle are performed. The proposed model demonstrates a higher modeling accuracy, for the battery pack voltage, than the constituent models under extreme maneuver drive tests.

Keywords: temperature influence; new OCV model; battery circuit model; synthesized battery model; thermal model; electric vehicle

1. Introduction

The global climate change, escalation in fuel cost, and the energy consumption, urged the necessity to replace the fossil fuel with renewable and environment friendly energy sources. Battery electric vehicles (BEV) are one major application, demonstrating the replacement of fossil fuel by renewable energy. Li-ion batteries have become the preferable energy storage for the future electric vehicles [1]. They receive greater attention than other battery types, such as lead-acid and nickel-cadmium batteries, due to their practical physical characteristics. They have a high specific energy, specific power, power density and a long life cycle. Moreover, their self-discharge rate is lower compared to other types of batteries [1–3].

Figure 1 demonstrates a typical discharge characteristic curve of a lithium-ion battery. The battery voltage extends between an upper and lower voltage limits V_{full} and $V_{cut-off}$, respectively. $V_{cut-off}$ represents the empty state of the battery where the minimum allowable voltage is reached. This restriction is meant to protect the battery from deep depletion. The section formed between V_{full}, V_{exp} and the correspondences capacity rates (C-rate) 0 and Q_{exp} is identified as the exponential region of the discharge characteristic curve, at which the discharged voltage changes exponentially regarding to the battery capacity. The voltage holds an approximately steady value for C-rates beyond Q_{exp} up to the nominal C-rate Q_{nom}, where the nominal V_{nom} voltage is reached. Not only is the battery voltage influenced by the discharge rate, but the battery capacity is also diminished at high discharge rates. This occurs as a sharp voltage drop at the end of the discharge process, as indicated in Figure 1. At that rate the discharge terminates at $V_{cut\text{-}off}$ [4].

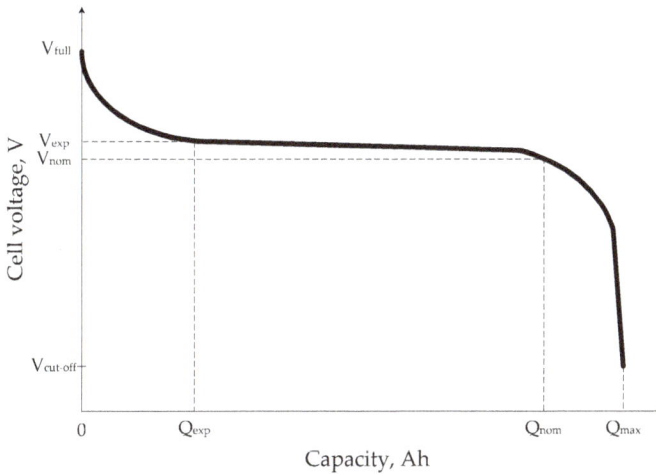

Figure 1. Typical discharge characteristic curve of Li-ion battery [4,5].

The vehicle under test (VUT) is equipped with battery cells of a cathode type LiFeMgPO$_4$. Lithium iron phosphate cells are characterized by their flat open circuit voltage curve (OCV). Hence, the cell voltage stays almost constant over the complete state of charge (SOC) range [2,3,6]. The battery pack of the VUT consists of 19 modules. Each module comprises six cell blocks connected in series; a single cell block is constructed out of 50 LiFeMgPO$_4$–graphite cells, connected in parallel. In total, there are 300 cells within the single battery module. Each cell block has a nominal voltage of 3.2 V, amounting to a total voltage of 19.2 V. The battery specifications are given in Table 1. A Controller Area Network (CAN) communication environment is used for the control and management of the battery modules. More technical information about the VUT is available in the Appendix.

Table 1. Technical data of LiFeMgPO$_4$ Battery [7].

Parameter (Unit)	Value
Nominal Module Voltage (V)	19.2
Nominal Module Capacity (Ah)	69
Max Continuous Load Current (A)	120
Peak Current for 30 s (A)	200

In this paper, we will investigate the different battery modeling methods to decide which approach will be the most appropriate for modeling the battery pack of VUT. Then, we will select a group of the most thorough models for Li-ion batteries. These models will contribute to the development of the proposed battery model, which will be considered for modeling actual battery pack voltage response when the VUT undergoes severe driving maneuvers.

The paper is organized as follows. Section 2 presents the different modeling techniques of Li-ion batteries. Three models are elected from the reviewed models and are discussed in more details in Section 3. In Section 4 the thermal behavior of the battery is elaborated. The experimental tests are described in Section 5. The models are evaluated and a new, improved model is developed and proposed in Section 6. Finally, the conclusions are stated at the end of the paper.

2. Existing Battery Models

Different modeling approaches are found in the literature. The most prominent battery modeling techniques are: Electrochemical, analytical, and circuit-based models [8]. Electrochemical models employ non-linear differential equations to model the chemical and electrical behavior of the cell [4,9]. Detailed knowledge of the battery chemistry, material structure and other physical characteristics are essential to achieve high accuracy and cover a large number of different operating points. However, the producers of batteries will rarely reveal the full parameters set of their products. Another shortcoming of electrochemical models is the high computational effort required to solve the non-linear partial differential equations [8]. Electrochemical models are better suited for research in battery's components fabrication, like electrodes and electrolyte [4,10]. The analytical modeling, on the other hand, reduces the computational complexity for the battery. However, that would be on the expense of capturing the circuit physical features of the battery, such as open circuit voltage, output voltage, internal resistance, and transient response [8].

Lumped electrical circuit models offer low complexity combined with high accuracy and robustness in simulating batteries dynamics [11–13]. Models with single or double resistor-capacitor (RC) networks are the best candidates for simulating the battery module [12–14]. RC parameters employed to model the battery characteristic show a dependency on temperature, charge/discharge rates and the SOC. Several techniques had been discussed in literature [1,15–19] for SOC estimation. Lam and Bauer [20] proposed a circuit model for the Li-ion battery with variable open circuit voltages, resistances and capacitances. The equivalent circuit components were represented as empirical functions of the current direction, the SOC, the battery temperature and the C-rate. Tremblay et al. [5,21] proposed an improved version Shepherd's model [22]. This model considers the influence of SOC on the OCV by considering the polarization voltage in the discharge-charge model. Different dynamic models for Li-ion, lead-acid, NiMH and NiCd battery typed were presented in Reference [5]. However, neither the temperature effect nor the variation of the internal resistance were considered. Saw et al. [23] investigated the thermal behavior for a $LiFePO_4$–graphite battery by coupling the empirical equations of the modified Shephard's battery model with a lumped thermal model for the battery cell. The temperature development of a complete vehicle battery pack under different driving cycles was simulated in [23]. Tan et al. [24] have incorporated the thermal losses to Shephard's model for Li-ion battery cells by adding temperature dependent correction terms to the model. Wijewardana et al. [1] proposed a generic electro-thermal model for Li-ion batteries. The model considers potential correction terms accounting for electrode film formation and electrolyte electron transfer chemistry. In addition, the constant values in the empirical equations that represent the equivalent circuit components of the battery were adjusted. These equations were employed to model the electrical components in dependence of SOC and temperature. Wijewardana et al. consider the C-rate effect in the estimation of SOC by employing an extended Kalman filter technique. Computational thermal models and temperature distribution estimations were proposed in References [25–28]. Additionally, finite element analysis models to estimate the temperature distribution in the battery were presented in References [25,27–29]. This kind of simulation requires knowledge of thermal properties of the battery cell materials, such as thermal capacity, density, mechanical construction and cooling of the battery. For an accurate parameterization intensive and precise measurements are necessary.

3. Overview of Selected Dynamic Battery Models

The equivalent circuit battery model provides a generic, dynamic way of modeling Li-ion batteries with moderate complexity. The moderate model complexity supports the integration of the model in a multiphysical simulation, allowing to analyze dynamic effects of the electric drive train. Three models are selected from the literature as the best candidates for Li-ion battery modeling, since they are the most thorough among the reviewed models. These models are:

- Tremblay et al. [5] (battery model 1)
- Lam and Bauer [20] (battery model 2)
- Wijewardana et al. [1] (battery model 3)

3.1. Tremblay Battery Model (Battery Model 1)

Tremblay and Dessaint [5] improved their own model in Reference [21]. The new model is shown in Figure 2. Only three points on the steady state manufacturer's discharge curve are required to parametrize the model, which are the full voltage, the end point of exponential voltage region, and the nominal voltage. The variation of the OCV from a reference constant voltage (E_0) is related to SOC changes by incorporating a polarization constant (K).

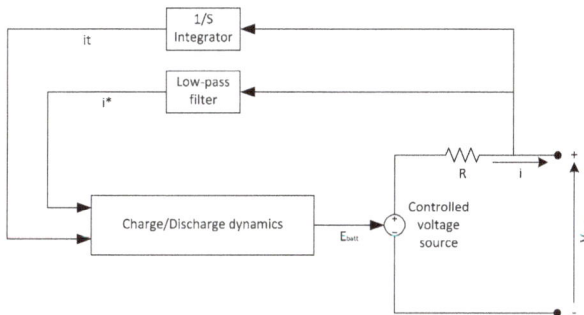

Figure 2. Tremblay and Dessaint battery discharge model.

In case of discharging the output voltage reads:

$$V_{batt} = E_0 - R \cdot i - K \frac{Q}{Q - it} \cdot (it + i^*) + A e^{-B \cdot it} \tag{1}$$

and for charging, the equation becomes:

$$V_{batt} = E_0 - R \cdot i - K \frac{Q}{it - 0.1Q} \cdot i^* - K \frac{Q}{Q - it} \cdot it + A e^{-B \cdot it} \tag{2}$$

the variables and constants in Equations (1) and (2) are defined in Table A1 in the Appendix A.1.

Although, the charging and discharging characteristics are extensively modeled, some other influential factors like the variations in internal resistance (R) and temperature influence are not considered. The capacity fading effect is not taken into account in this model as well.

3.2. Lam and Bauer Battery Model (Battery Model 2)

The Lam and Bauer battery equivalent circuit model is shown in Figure 3. The model demonstrates the V_{OC} as a function of SOC, a variable ohmic resistance R_o, and two variable RC-networks: $R_S C_S$ and $R_l C_l$ for the short and the long time transient responses, respectively. Lam and Bauer also showed the relation between capacity fading due to aging and different stress influences, which are the cell's temperature, C-rate, SOC and intensity of discharge. Lam and Bauer parametrized their equations through curve fitting of experimental measurements. They employed in their tests the LiFePO$_4$ battery cell. The equivalent circuit resistors and capacitors equations for temperatures from 20 °C and above are detailed in Equations (7)–(28) in Reference [20]. We refer also to the V_{OC} equation with the corrected constants as proposed in Reference [20]:

$$V_{OC} (SOC) = -0.5863 \, e^{-21.9 \, SOC} + 3.414 + 0.1102 \, SOC - 0.1718 \, e^{-\frac{8 \times 10^{-3}}{1 - SOC}} \tag{3}$$

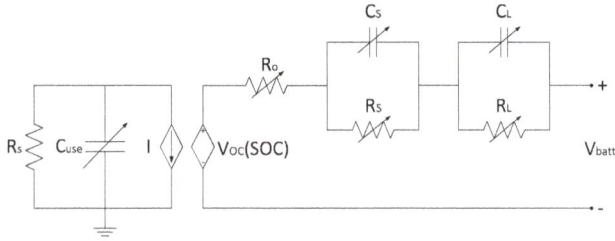

Figure 3. Lam and Bauer battery circuit model.

3.3. Wijewardana Battery Model

A different perception, than the two previous models, is adopted in this model. The electrical components, R_S, C_S, R_L, C_L are functions of SOC and independent of the temperature. Only the series internal resistance resistor is a function of SOC and temperature ($R_{intS}(SOC,T)$). The temperature influence is considered by adding potential correction terms, which are voltage due to electrode film formation (ΔE) and voltage due to electrolyte electrons transfer formation (ΔV_{Che}). The capacity fading effect is modeled as an additional series resistance R_{cyc}. The battery output voltage is computed by subtracting the voltage drop of each circuit element from the V_{OC} value. Figure 4 demonstrates Wijewardana battery model.

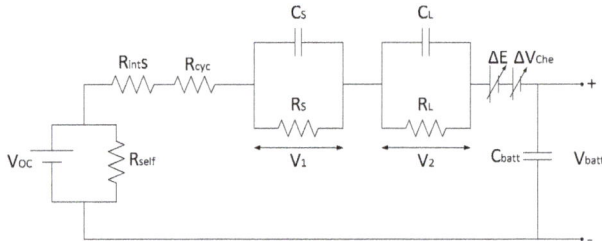

Figure 4. Wijewardana battery circuit model.

The battery output voltage of this model reads [1]:

$$V_{batt} = V_{OC} - i\left(R_{intS} + R_{cyc}\right) - V_1 - V_2 - \Delta E\left(T\right) - \Delta V_{Che}\left(T\right) \tag{4}$$

$$\Delta E\left(T\right) = \left(1 + C_{E1}\Delta T\right)\left.\frac{dV_r}{dT}\right|_{T=T_{cell}}\Delta T \tag{5}$$

$$\Delta V_{Che}\left(T\right) = \beta w \, exp^{(-1/t)}\Delta T + \left.\frac{dV_{Che}}{dT}\right|_{T=T_{cell}}\left(C_{Che} + C_{Che1}\Delta T\right)\left(1 + \beta\right)\Delta T \tag{6}$$

The values of the model parameters are listed in the Appendix.

3.4. Assessment of Battery Models Qualities

Three different highly dynamic to model Li-ion batteries have been proposed. Each model has its dominant features. Table 2 summaries the qualities of each model. The (+) sign implies that the corresponded feature is considered in the model. Whereas, a (++) sign denotes an extensive consideration to the related feature. In contrast, the (-) sign means that the related feature was assigned as a constant or it is not considered at all in the model. Battery model 2 considers more factors than the other models. However, each model must be evaluated with experimental data to investigate its accuracy.

Table 2. Comparison between the battery models.

Feature	Battery Model 1		Battery Model 2		Battery Model 3	
Charge-Discharge hysteresis	++	Considered in the output voltage	+	Considered in the internal resistance (R) equations	-	Charge-discharge
Open circuit voltage	-	Constant value for E_0	+	V_{OC}(SOC)	+	V_{OC}(SOC)
Internal resistance (R)	-	Constant value	++	R(SOC,T,C-rate)	+	R(SOC,T)
Temperature influence	-	Not considered	+	Considered in the internal resistance model	+	Considered as potential correction terms
Capacity fading	-	Not considered	+	Considered in the battery's used capacity estimation	+	Considered in the battery's internal resistance (R) estimation
Total Assessment		2		6		4

4. Battery Thermal Model

A general energy balance is applied to estimate the battery cell temperature. It is assumed that the thermal distribution inside the cell is uniform and that the conduction resistance inside the battery cell is negligible compared with the convection and radiation heat transfer [1,23,27,28]. The change in temperature depends significantly on the battery thermal capacity (C_p) and the difference between the generated heat and the dissipated heat. The dissipation of the heat to the battery surrounding is performed by convection and radiation. Generated heat comprises two sources, irreversible heat generation by means of the effective ohmic resistance of the cell's material, and reversible generated heat due to the entropy change in both cathode and anode. The total entropy changes in the battery cell can be considered as zero according to References [1,29,30]. The temperature development inside the battery cell is described as:

$$mC_p\frac{dT_{cell}}{dt} = i\left(V_{OC} - V_{batt}\right) - hA_{cell}\Delta T - \epsilon\sigma A_{cell}\left(T_{cell}^4 - T_{amp}^4\right) \tag{7a}$$

where ΔT is the difference between the battery cell and the ambient temperatures ($T_{cell} - T_{amp}$), h is the natural convection coefficient, m is the cell mass, i is the cell current, A_{cell} is the surface area of the single battery cell, σ is Stefan-Boltzmann constant, and ϵ is the emissivity of heat. Assuming that the temperature differences between the cells in the single battery module are small, Equation (7a) can be generalized for the whole battery module as:

$$MC_p\frac{dT_{cell}}{dt} = I\left(V_{OC} - V_{batt}\right) - hA\Delta T - \epsilon\sigma A\left(T_{cell}^4 - T_{amp}^4\right) \tag{7b}$$

where M is the total cells mass, I is the battery current, A is the surface area of the cells blocks in the single battery module.

Saw et al. [23] has pointed out to the contribution of the contact resistance in heat generation. The contact resistance can be neglected in the investigated batteries, since the cells are realizing low ohmic over wide cell connectors by means of welded connections. Fifty individual cells are connected in parallel and we have a nominal current about 1.4 A per cell. The resistance of the single contact is 0.2 mΩ. With four welding points per cell connectors, the contact resistance for the single battery cell became 50 $\mu\Omega$. The power loss in the single cell due to contact resistance is determined as: $P_{loss} = I_{cell}^2 R_{contact}$ = 0.098 mW/cell, which is a negligible amount, thermally, as well as electrically.

5. Experimental Characterization of the Battery and the Vehicle under the Test

5.1. Battery Measurements

In order to characterize the LiFeMgPO$_4$-battery, several experimental tests were implemented on the battery at different conditions. OCV vs. SOC measurements were performed at 10, 20 and 40 °C. The battery was discharged until the cut-off voltage of 2 V was reached and then recharged up to the nominal capacity. A Low C-rate of C/10 was used to minimize the dynamic effects and to achieve a good approximation to an open circuit. Figure 5 demonstrates the charge and discharge OCV curves over the SOC for various temperatures. It is noticeable that the lower the operating temperatures, the higher the difference between charging and discharging.

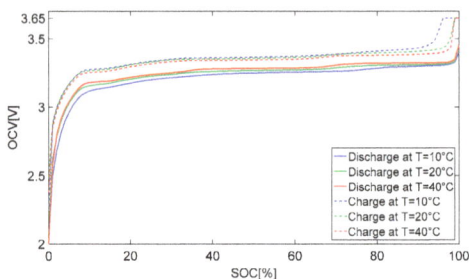

Figure 5. Charge and discharge OCV curves over SOC.

5.2. Driving Tests on the Real Vehicle

The objective of a real test is to find a real driving maneuver reference signal to validate the battery model performance. Nevertheless, the battery model should be able to simulate the actual system in real operating conditions, not merely charging-discharging cycles. Two driving tests were performed with the test vehicle for the purpose of investigation the system and collecting the experimental data. The tests were executed on the testing ground at Karlsruhe Institute of Technology (KIT). The experimental data attained from the CAN bus are displayed in Figures A1 and A2 in the Appendix A.2. Then, the data were processed in MATLAB. In the first test, the vehicle was driven in a counterclockwise circular direction for about 230 s. Then the direction for driving was reversed and the test was resumed. A variable pedal input was implemented in order to cover a broader range of data. This test represents an aggressive driving scenario, leading to discharge rate of up to 3.4 C. The second test was implemented by subjecting the vehicle to a sequence of sudden accelerations and decelerations. These tests were performed this way to create highly fluctuating signals in the battery system, which are shown in Figure 6. The dynamic responses of battery voltage can be used to evaluate the battery models, discussed above.

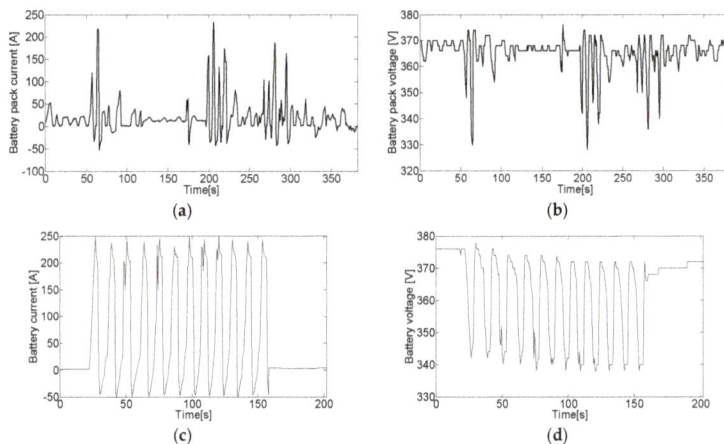

Figure 6. CAN bus measurements: (**a**) Battery pack current for the circular driving test; (**b**) Battery pack voltage for the circular driving test; (**c**) Battery pack current for the rapid acceleration and deceleration driving test; (**d**) Battery pack voltage for the rapid acceleration and deceleration driving test.

6. Battery Models Validation

6.1. Evaluating the Open Circuit Voltage Models (V_{OC})

Wijewardana et al. [1] have employed a common model for VOC that is widely used and found in literature. Lam and Bauer [20] have redefined the model equation to suit the $LiFeMgPO_4$ cathode type batteries. They concluded that the OCV is temperature independent. They justified this conclusion based on small changes of the OCV measurements due to temperature variations, which were in the range 2–8 mV [20]. An absolute error of 30 mV for $LiFeMgPO_4$ battery cell will lead to an uncertainty of 13% in the SOC estimation at 1 C discharge and 25 °C, according to Blank et al. [31]. The battery of our vehicle is $LiFeMgPO_4$-cathode type. Its OCV curves are presented earlier in Figure 6. According to our measurements, the OCV temperature alteration is from 15 to 90 mV, which is about 10 times higher than the result presented in Reference [20]. In our study, we use a battery module that contains 6 cellblocks in series, with 50 cells in parallel for each. Moreover, the vehicle's battery pack has 19 modules. With this combination of battery cells, the range of voltage alteration becomes 1.71–10.26 V, which is a considerable change in the battery pack output voltage.

We validated both V_{OC} models in References [1,20] by comparing the simulation results with our own measurements, as shown in Figure 7. We selected the charge-discharge curves at $T = 20$ °C to be the references for validation. The V_{OC} model utilized in battery model 3 does not fit our measurements. The V_{OC} of battery model 2 better fits the experimental data. The deviation in an SOC range spanning from 10% to 90% is about 0.03 V. This deviation increases at low temperature.

The influence of the temperature variation on the OCV curves is defined as dV_{OC}/dT. From the measurements shown in Figure 6, the value of this term was found to be 1.25 mV in case of discharge and 0.69 mV for charging. The V_{OC} model 2 model is modified for better fitting of the OCV curve along the SOC range and the temperature influence is considered. The new V_{OC} is modeled by Equations (8) and (9) and the constants values of the new V_{OC} are presented in Table 3. The validation results are shown in Figure 8 and in Table 4.

$$V_{OC,discharge} (SOC, T) = a_1 \, e^{-a_2 SOC} + a_3 + a_4 \, SOC + a_5 \, e^{-\frac{a_6}{1-SOC}} + T \, dV_{OC,d}/dT \tag{8}$$

$$V_{OC,charge} (SOC, T) = b_1 \, e^{-b_2 SOC} + b_3 + b_4 \, SOC + b_5 \, e^{-\frac{b_6}{1-SOC}} + T \, dV_{OC,c}/dT \tag{9}$$

Table 3. V_{OC} parameter values.

Constant	Value	Constant	Value
a_1	−1.166	b_1	−0.9135
a_2	−35	b_2	−35
a_3	3.344	b_3	3.484
a_4	0.1102	b_4	0.1102
a_5	−0.1718	b_5	−0.1718
a_6	-2×10^{-3}	b_6	-8×10^{-3}
$dV_{OC,d}/dT$	0.00125	$dV_{OC,c}/dT$	0.00069

6.2. Evaluating the Battery Models Output Voltage

The accuracy of each model is yet to be proved. For objective comparison, the thermal model elaborated in Section 4 is employed for all models. The battery currents in Figure 6a,c are designated as the inputs for all models and the output voltage of each model is investigated against the voltage response signal, shown in Figure 6b,d. The V_{OC} of model 3 [1] showed a large deviation from the actual curve, as shown in Figure 7. Therefore, the V_{OC} derived from model 2 [20] will be also utilized in model 3. Figures 9 and 10 demonstrate the responses of the three models for both driving test. The simulation results gained from model 1 reveal the highest accuracy for the first test. The mean squared error between the measured voltage and the simulated voltage by model 1 is less than 1%. However, it performed the worst in the second test. The good performance of model 1 in the first test ascribed to the fact that the test conditions were nearly matching the standard condition for defining the constant voltage (E_0). E_0 is equal to the nominal voltage at 20 °C, which is equal to 3.21 V and the initial voltage of the single battery cell is estimated as 3.25 V. When the test second driving test was performed at different circumstances, the outcome was not as good as it in the first case. Figure 10a reveals a relatively large offset error in the response of battery model 1 with a mean square error (MSE) of about 2.24%. Model 2 performed moderately with percentage errors between 1% and 2%. The offset errors between the reference signal and initial voltage value of both battery models 2 and 3 were minor, whereas Equation (3) is employed in both models for the estimation of V_{OC}. The simulation results of model 3 indicate less dynamic response than the other two models. It could not conduct the drastic changes in the battery current input signal.

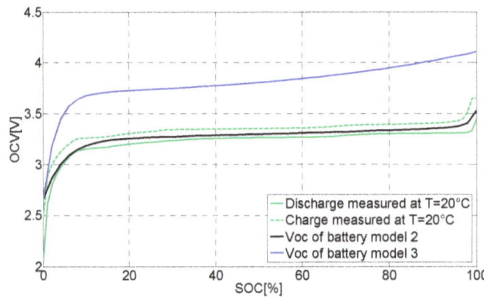

Figure 7. Comparing the V_{oc} model with measured experimental results.

Table 4. Accuracy of the proposed V_{OC} model.

Temperature °C	MSE in Discharge Model %	MSE in Charge Model %
10	0.5232	0.8914
20	0.5320	0.5719
40	0.5751	0.4522

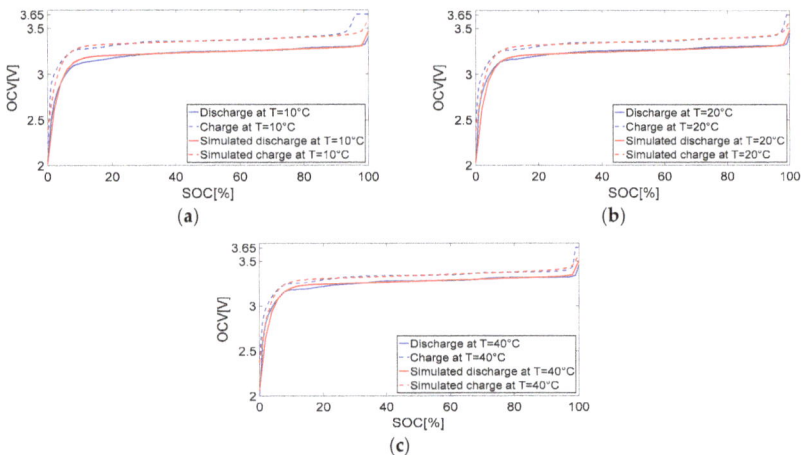

Figure 8. The new V_{oc} model and measured experimental data: (**a**) $T = 10$ °C; (**b**) $T = 20$ °C; (**c**) $T = 40$ °C.

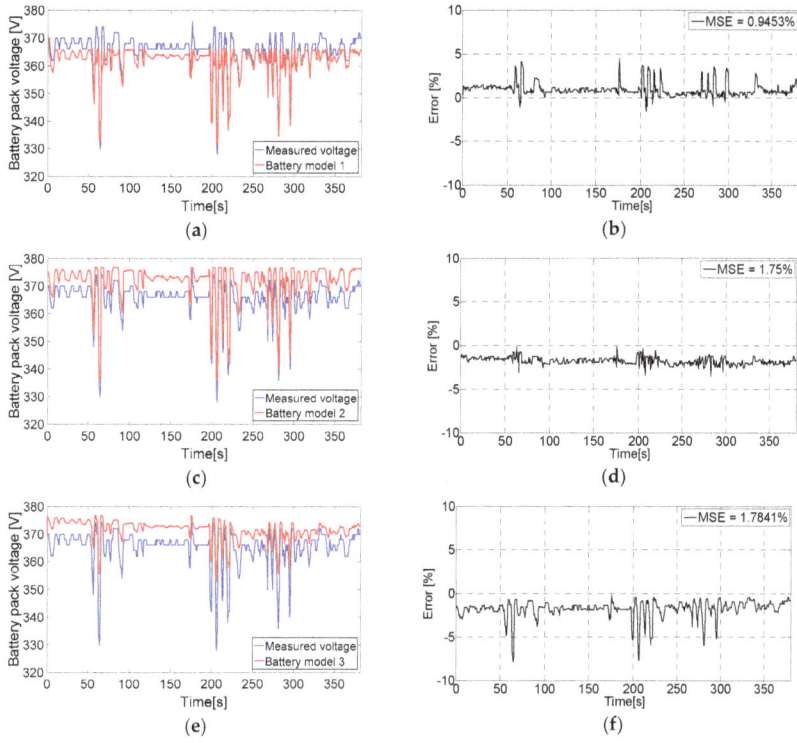

Figure 9. Battery simulation models and reference voltage signal for circular driving test: (**a**) Battery model 1 response; (**b**) Mean square error of battery model 1; (**c**) Battery model 2 response; (**d**) Mean square error of battery model 2; (**e**) Battery model 3 response (**f**) Mean square error in battery model 3.

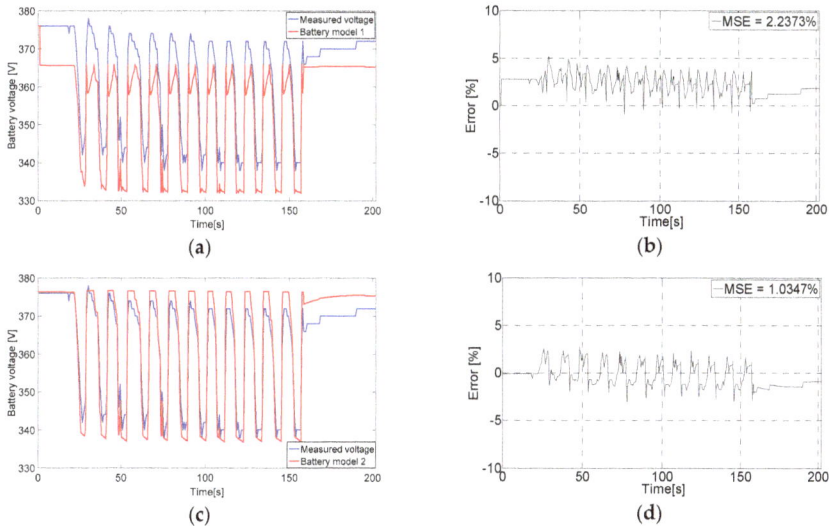

Figure 10. *Cont.*

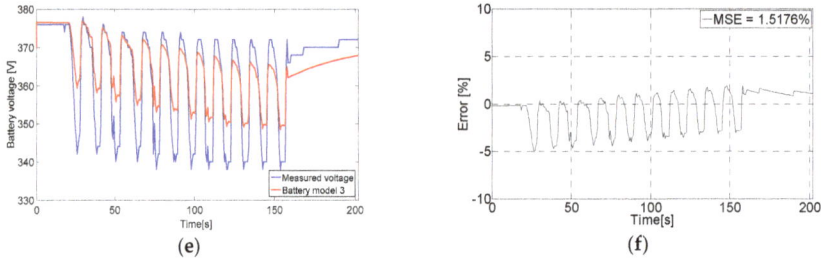

(e)   (f)

Figure 10. Battery simulation models and reference voltage signal for rapid acceleration and deceleration driving test: (**a**) Battery model 1 response; (**b**) Mean square error of battery model 1; (**c**) Battery model 2 response; (**d**) Mean square error of battery model 2; (**e**) Battery model 3 response (**f**) Mean square error in battery model 3.

6.3. The Proposed Synthesized Battery Model

It is explicable that each model has some flaws, as determined in Table 2. The simulation results in Figures 9 and 10 prove that even the best performing models need to be further improved. Accordingly, a synthesized model that holds the best qualities of each model has been developed. First, the charge-discharge characteristics of model 1 are considered. Additionally, the discharging part our proposed V_{OC}(SOC,T) is employed instead of constant (E_0) value. Then, the highly detailed internal resistance model of battery model 2 in case of discharging, which is represented by Equations (7)–(11), (17)–(19), (21), (27) in Referance [20], is taking the place of the constant internal resistance (R). The capacity fading effect is considered by adding R_{cyc} from battery model 3 to the internal resistance. The empirical equations in case of discharging are implemented because the charging-discharging hysteresis is properly modeled by model 1. The synthesized model is shown in Figure 11.

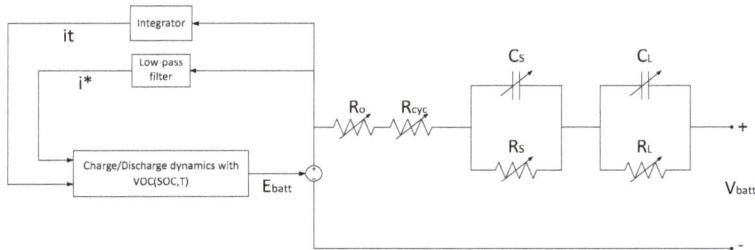

Figure 11. The proposed synthesized battery model.

In case of discharging, the output voltage reads:

$$V_{batt} = V_{OC,discharge}\,(SOC, T) - (R_O + R_{cyc} + V_S + V_L) \cdot i - K\frac{Q}{Q - it} \cdot (it + i^*) + Ae^{-B \cdot it} \quad (10)$$

and for charging, the equation expressed as:

$$V_{batt} = V_{OC,discharge}\,(SOC, T) - (R_O + R_{cyc} + V_S + V_L) \cdot i - K\frac{Q}{it - 0.1Q} \cdot i^* - K\frac{Q}{Q-it} \cdot it + Ae^{-B \cdot it} \quad (11)$$

where

$$\frac{dV_S}{dt} = \frac{i}{C_S} - \frac{V_S}{R_S C_S} \quad (12)$$

$$\frac{dV_L}{dt} = \frac{i}{C_L} - \frac{V_L}{R_L C_L} \quad (13)$$

The resistances and capacitors values are determined through Equations (14)–(18):

$$R_s\left(\text{SOC}, \vartheta\right) = \left(c_1 e^{\left(c_2 \text{SOC}\right)} + c_3 + c_4 \text{SOC}\right) + c_5 \Delta\vartheta + c_6 \text{SOC}\Delta\vartheta \tag{14}$$

$$C_s\left(\text{SOC}, \vartheta\right) = \left(c_7 \text{SOC}^3 + c_8 \text{SOC}^2 + c_9 \text{SOC} + c_{10}\right) + c_{11}\text{SOC}\Delta\vartheta + c_{12}\Delta\vartheta \tag{15}$$

$$R_l\left(\text{SOC}, \vartheta, I_{C-rate}\right) = \left(\left(c_{13}e^{\left(c_{14}\text{SOC}\right)} + c_{15} + c_{16}\text{SOC}\right) + c_{17}\Delta\vartheta e^{\left(c_{18}\text{SOC}\right)} + c_{19}\Delta\vartheta\right) \times \left(c_{20}\left(I_{C-rate}\right)^{c_{21}} + c_{22}\right) \tag{16}$$

$$C_l\left(\text{SOC}, \vartheta\right) = \left(c_{23}\text{SOC}^6 + c_{24}\text{SOC}^5 + c_{25}\text{SOC}^4 + c_{26}\text{SOC}^3 + c_{27}\text{SOC}^2 + c_{28}\text{SOC} + c_{29}\right) + c_{30}\, e^{c_{31}/\vartheta} \tag{17}$$

$$R_o\left(\text{SOC}, \vartheta\right) = \left(c_{32}\text{SOC}^4 + c_{33}\text{SOC}^3 + c_{34}\text{SOC}^2 + c_{35}\text{SOC} + c_{36}\right) c_{37}e^{c_{38}/(\vartheta - c_{39})} \tag{18}$$

where ϑ is the battery cell temperature in Kelvin (°K) and c_1–c_{39} are constants, which their values are listed in Table 5.

Table 5. Constants values of Equations (14)–(18).

Constant	Value	Constant	Value	Constant	Value	Constant	Value
c_1	1.080×10^{-2}	c_{11}	-6.580	c_{21}	-6.919×10^{-1}	c_{31}	-2.398×10^3
c_2	-11.03	c_{12}	12.11	c_{22}	2.902×10^{-1}	c_{32}	1.298×10^{-1}
c_3	1.827×10^{-2}	c_{13}	2.950×10^{-1}	c_{23}	2.130×10^6	c_{33}	-2.892×10^{-1}
c_4	-6.462×10^{-3}	c_{14}	-20.00	c_{24}	-6.007×10^6	c_{34}	2.273×10^{-1}
c_5	-3.697×10^{-4}	c_{15}	4.722×10^{-2}	c_{25}	6.271×10^6	c_{35}	-7.216×10^{-2}
c_6	2.225×10^{-4}	c_{16}	-2.420×10^{-2}	c_{26}	-2.958×10^6	c_{36}	8.980×10^{-2}
c_7	1.697×10^2	c_{17}	6.718×10^{-3}	c_{27}	5.998×10^5	c_{37}	7.613×10^{-1}
c_8	-1.007×10^3	c_{18}	-20.00	c_{28}	-3.102×10^4	c_{38}	10.14
c_9	1.408×10^3	c_{19}	-5.967×10^{-4}	c_{29}	2.232×10^3	c_{39}	2.608×10^2
c_{10}	3.897×10^2	c_{20}	6.993×10^{-1}	c_{30}	3.128×10^3		

The new model shows a significant improvement in the simulation response, as shown in Figure 12. The mean square error was reduced to only 0.256%. The new model shows also the best result when it employed for simulating the rapid acceleration and deceleration driving test, though it had a small offset error at the beginning of 3.2 V. This error yielded from an absolute voltage error of 0.028 V in proposed VOC model, represented by Equation (8), at the specified SOC and temperature values. The battery models are simulated in MATLAB/Simulink environment. Figure 13 shows the Simulink model for the proposed battery model.

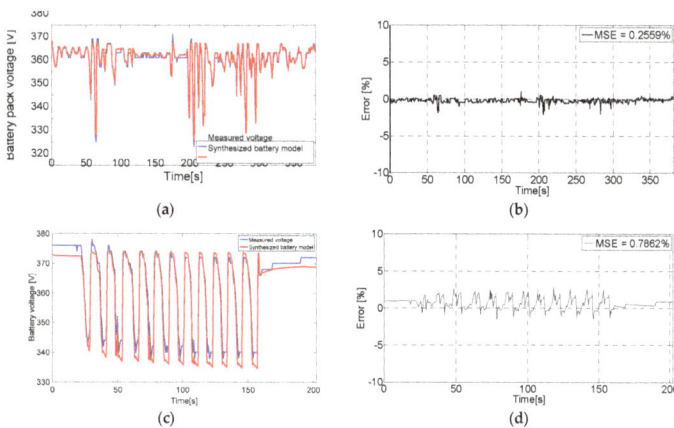

Figure 12. The proposed synthesized battery model and reference voltage signal: (**a**) Proposed model response for the circular driving test; (**b**) Error of proposed model voltage for the circular driving test; (**c**) Proposed model response for the rapid acceleration and deceleration driving test; (**d**) Error of proposed model voltage for the rapid acceleration and deceleration driving test.

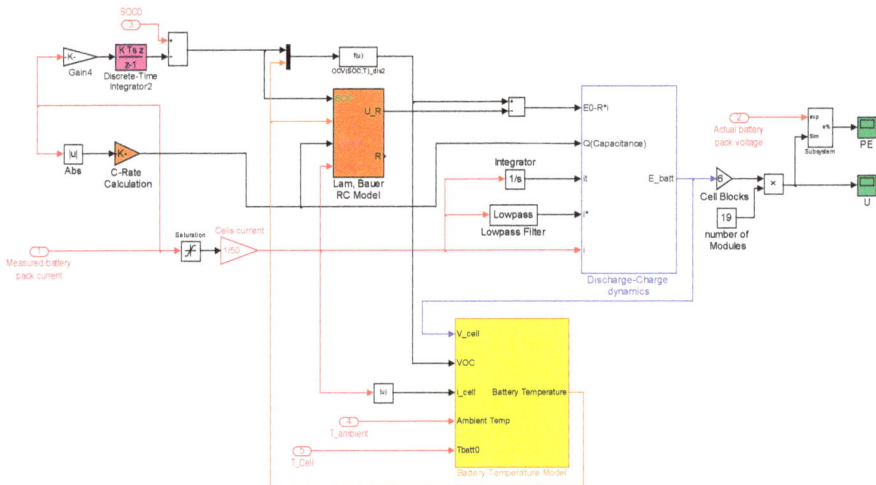

Figure 13. The proposed synthesized battery simulation model.

7. Conclusions

A battery is a sophisticated system, which necessitates a detailed model for accurate simulation. Many factors must be considered in the battery model for accurate simulation results. Charging-discharging dynamics, battery internal resistance, and open circuit voltage are the most significant aspects for battery modeling. Temperature is an influential factor for all of these aspects. The difference between the charging and discharging in the OCV curves increases at low temperatures. This phenomenon occurs due to the decrement in battery capacity, which in turn appears as a result of a rise of the internal resistance. Neglecting the effect of temperature will lead to inaccuracy in simulation. Even the slightest errors in the simulation results of the battery cell model, would grow significantly when the model is extended to the complete battery pack. Battery model 1 has two flaws. Firstly, it assumes the initial voltage value to be the nominal battery cell voltage. This assumption led to large offset from the actual value, when the vehicle was tested at about 30 °C. Secondly, it considers a constant internal resistance of the battery, which is in fact a very fluctuating quantity that affects the battery cell current and the voltage response as well. The proposed battery model has compensated for these shortages and it has accurately simulated the battery pack voltage response on the real vehicle.

Author Contributions: Muhammed Alhanouti made the literature review on the current battery models, proposed the synthesized battery model, developing the simulation models, and wrote the paper. Martin Gießler and Thomas Blank designed and performed the battery measurements, and they helped in editing the paper content. Frank Gauterin supervised the work of this paper and approved the results

Conflicts of Interest: The authors declare no conflict of interest.

Appendix A

Appendix A.1. Nomenclature

Table A1. Battery models parameters.

Parameter (Unit)	Symbol	Value
Constant voltage (V)	E_0	3.21 [23]
Constant internal resistance (Ω)	R	0.0833
Polarization constant (V/(Ah)) or polarization resistance (Ω)	K	0.0119 [23]
Battery capacity (Ah)	Q	Variable

Table A1. *Cont.*

Parameter (Unit)	Symbol	Value
Actual battery charge (Ah)	it	Variable
Exponential zone amplitude (V)	A	0.2711 [23]
Exponential zone time constant inverse $(Ah)^{-1}$	B	152.130 [23]
Battery current (A)	i	Variable
Filtered current (A)	$i*$	Variable
Voltage change due to electrolyte electrons transfer formation	ΔV_{Che}	Variable
the effective voltage gradient	dV_{Che}/dT	0.0016 [1]
Constant property of electrolyte	C_{Che}	0.07 [1]
Constant property of electrolyte	C_{Che1}	0.001 [1]
Constant property of electrolyte	b	0.0012 [1]
Constant property of electrolyte	w	0.012 [1]
Voltage change due to electrode film formation	ΔE	Variable
voltage gradient	dV_r/dT	0.00003 [1]
Constant property	C_{E1}	0.00011 [1]
Battery module surface area (m^2)	A	0.283954
Battery cell mass (kg)	m	0.04 [23]
Battery module mass (kg)	M	12
Specific heat capacity $(J \cdot kg^{-1} \cdot K^{-1})$	C_p	1360 [27]
Stefane-Boltzmann constant $(W \cdot m^{-2} \cdot K^4)$	σ	5.67×10^{-8}
Emissivity of heat	ε	0.95
Natural heat convection constant $(W \cdot m^{-2} \cdot K^{-1})$	h	4

Appendix A.2. Driving Tests

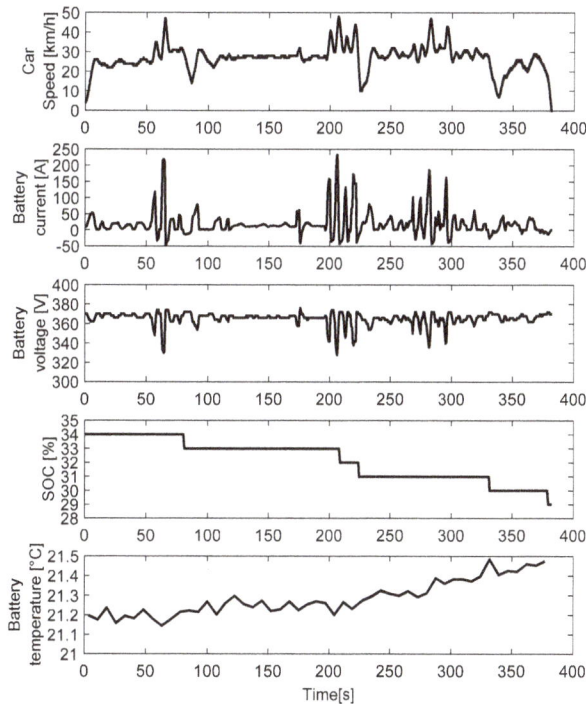

Figure A1. The measured data of the circular drive test.

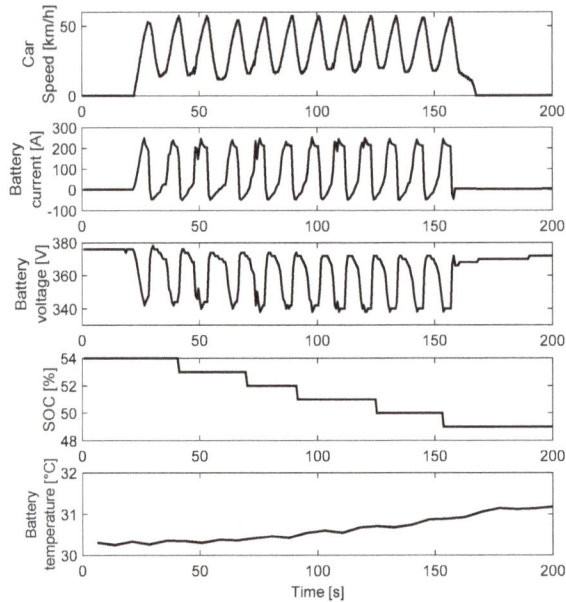

Figure A2. The measured data of the rapid acceleration and deceleration drive test.

Appendix A.3. The Vehicle under the Test

An electric hydrogen front wheel passenger car (Figure A3) was modified by the institute for "FAhrzeug-SystemTechnik" (FAST) of the Karlsruhe Institute of Technology to a battery electric vehicle (BEV), [30]. A battery pack was installed in the vehicle. This vehicle is used for research in different automotive engineering and e-mobility related projects. The propulsion systems comprises an induction motor with a single gear transmission. The motor specifications are listed in Table A2.

Figure A3. Vehicle under test.

Table A2. Technical data of electric motor.

Parameter	Value
Rated Power, PN	45 kW
Peak Power, Pmax	68 kW
Peak Torque, Tmax	210 N·m
Rated Speed, nN	3000 rpm

The motor shown in Figure A4 is powered by the AC current, delivered from the power electronics that converts the DC current supplied by the battery. As the driver presses the accelerator pedal, a corresponding "torque demand" signal is converted by the vehicle control unit (VCU) to an appropriate signal for the motor control unit (power electronics), which in turn transforms it into a current frequency signal. The motor control unit (MCU) is incorporated with a thermal derating system in order to limit the torque demand received by the power electronics and to prevent any critical operating conditions for the motor. The assigned powertrain can accelerate the vehicle to a maximum speed of 120 km/h.

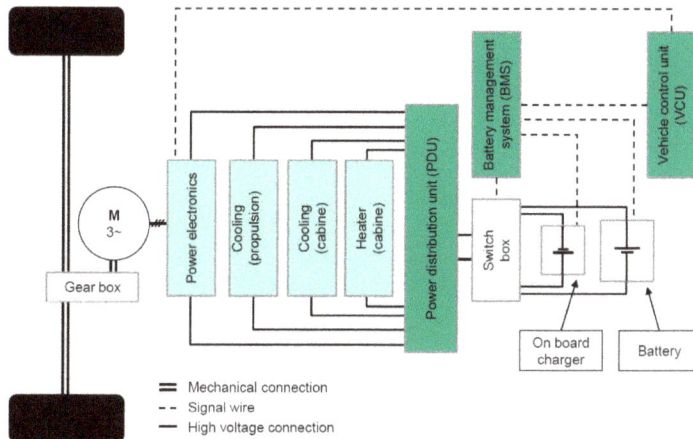

Figure A4. The basic drive train topology of the Mercedes A-Class research vehicle [30].

References

1. Wijewardana, S.; Vepa, R.; Shaheed, M.H. Dynamic battery cell model and state of charge estimation. *J. Power Sources* **2016**, *308*, 109–120. [CrossRef]
2. *Doppebattery Model 2auer, M. Hybrid and Electric Vehicles—Lecture Notes*; ETI-HEV, Karlsruhe Institute of Technology: Karlsruhe, Germany, 2014.
3. Ivers-Tiffee, E. *Batteries and Fuel Cells—Lecture Notes*; IWE, Karlsruhe Institute of Technology: Karlsruhe, Germany, 2012.
4. Illig, J. Physically Based Impedance Modelling of Lithium-Ion Cells. Ph.D. Thesis, Karlsruhe Institute of Technology, Karlsruhe, Germany, 2014.
5. Tremblay, O.; Dessaint, L.-A. Experimental validation of a battery dynamic model for EV applications. *World Electr. Veh. J.* **2009**, *3*, 1–10.
6. Padhi, A.K.; Nanjundaswamy, K.S.; Goodenough, J.B.D. Phospho-olivines as positive-electrode materials for rechargeable lithium batteries. *J. Electrochem. Soc.* **1997**, *144*, 1188–1194. [CrossRef]
7. Valence Technology. *U-Charge®XP Rev 2 User Manual*; Valence Technology, Inc.: Austin, TX, USA, 2011.
8. Lin, N.; Ci, S.; Li, H. An enhanced circuit-based battery model with considerations of temperature effect. In Proceedings of the 2014 IEEE Energy Conversion Congress and Exposition (ECCE), Pittsburgh, PA, USA, 14–18 September 2014; pp. 3985–3989.
9. Doyle, M.; Fuller, T.F.; Newman, J. Modeling of galvanostatic charge and discharge of the lithium/polymer/insertion cell. *J. Electrochem. Soc.* **1993**, *140*, 1526–1533. [CrossRef]
10. Dees, D.W.; Battaglia, V.S.; Bélanger, A. Electrochemical modeling of lithium polymer batteries. *J. Power Sources* **2002**, *110*, 310–320. [CrossRef]
11. He, H.; Xiong, R.; Guo, H.; Li, S. Comparison study on the battery models used for the energy management of batteries in electric vehicles. *Energy Convers. Manag.* **2012**, *64*, 113–121. [CrossRef]

12. Hu, X.; Li, S.; Peng, H. A comparative study of equivalent circuit models for Li-ion batteries. *J. Power Sources* **2012**, *198*, 359–367. [CrossRef]
13. He, H.; Xiong, R.; Fan, J. Evaluation of lithium-ion battery equivalent circuit models for state of charge estimation by an experimental approach. *Energies* **2011**, *4*, 582–598. [CrossRef]
14. Hussein, A.A.; Batarseh, I. An overview of generic battery models. In Proceedings of the 2011 IEEE Power and Energy Society General Meeting, San Diego, CA, USA, 24–29 July 2011; pp. 1–6.
15. Kroeze, R.C.; Krein, P.T. Electrical battery model for use in dynamic electric vehicle simulations. In Proceedings of the 2008 IEEE Power Electronics Specialists Conference, Rhodes, Greece, 15–19 June 2008; pp. 1336–1342.
16. Liu, X.; Ma, Y.; Ying, Z. Research of SOC estimation for lithium-ion battery of electric vehicle based on AMEsim-simulink co-simulation. In Proceedings of the 32nd Chinese Control Conference (CCC), Xi'an, China, 26–28 July 2013; pp. 7680–7685.
17. Szumanowski, A.; Chang, Y. Battery management system based on battery nonlinear dynamics modeling. *IEEE Trans. Veh. Technol.* **2008**, *57*, 1425–1432. [CrossRef]
18. Zhang, C.; Jiang, J.; Zhang, W.; Sharkh, S.M. Estimation of state of charge of lithium-ion batteries used in HEV using robust extended Kalman filtering. *Energies* **2012**, *5*, 1098–1115. [CrossRef]
19. Watrin, N.; Roche, R.; Ostermann, H.; Blunier, B.; Miraoui, A. Multiphysical lithium-based battery model for use in state-of-charge determination. *IEEE Trans. Veh. Technol.* **2012**, *61*, 3420–3429. [CrossRef]
20. Lam, L.; Bauer, P.; Kelder, E. A practical circuit-based model for Li-ion battery cells in electric vehicle applications. In Proceedings of the 2011 IEEE 33rd International Telecommunications Energy Conference (INTELEC), Amsterdam, The Netherlands, 9–13 October 2011; pp. 1–9.
21. Tremblay, O.; Dessaint, L.A.; Dekkiche, A.I. A generic battery model for the dynamic simulation of hybrid electric vehicles. In Proceedings of the 2007 IEEE Vehicle Power and Propulsion Conference, Arlington, TX, USA, 9–12 September 2007; pp. 284–289.
22. Shepherd, C.M. Design of primary and secondary cells II. An equation describing battery discharge. *J. Electrochem. Soc.* **1965**, *112*, 657–664. [CrossRef]
23. Saw, L.H.; Somasundaram, K.; Ye, Y.; Tay, A.A.O. Electro-thermal analysis of Lithium Iron Phosphate battery for electric vehicles. *J. Power Sources* **2014**, *249*, 231–238. [CrossRef]
24. Tan, Y.K.; Mao, J.C.; Tseng, K.G. Modelling of battery temperature effect on electrical characteristics of Li-ion battery in hybrid electric vehicle. In Proceedings of the 2011 IEEE Ninth International Conference on Power Electronics and Drive Systems (PEDS), Singapore, Singapore, 5–8 December 2011.
25. Pesaran, A.A. Battery thermal models for hybrid vehicle simulations. *J. Power Sources* **2002**, *110*, 377–382. [CrossRef]
26. Kim, Y.; Siegel, J.B.; Stefanopoulou, A.G. A computationally efficient thermal model of cylindrical battery cells for the estimation of radially distributed temperatures. In Proceedings of the 2013 American Control Conference (ACC), Washington, DC, USA, 17–19 June 2013; pp. 698–703.
27. Rad, M.S.; Danilov, D.L.; Baghalha, M.; Kazemeini, M.; Notten, P.H. Thermal modeling of cylindrical LiFeMgPO$_4$ Batteries. *J. Mod. Phys.* **2013**, *4*, 1–7. [CrossRef]
28. Fan, L.; Khodadadi, J.M.; Pesaran, A.A. A parametric study on thermal management of an air-cooled lithium-ion battery module for plug-in hybrid electric vehicles. *J. Power Sources* **2013**, *238*, 301–312. [CrossRef]
29. Sun, Y. Construction and Validation of a Thermal FEM-Model of an Automobile Battery. Master's Thesis, Karlsruhe Institute of Technology, Karlsruhe, Germany, 2011.
30. Gießler, M.; Fritz, A.; Paul, J.; Sander, O.; Gauterin, F.; Müller-Glaser, K.D. Converted vehicle for battery electric drive: Aspects on the design of the software-driven vehicle control unit. In Proceedings of the 2nd International Energy Efficient Vehicle Conference (EEVC), Dresden, Germany, 18–19 June 2012.
31. Blank, T.; Lipps, C.; Ott, W.; Hoffmann, P.; Weber, M. Influence of environmental conditions on the sensing accuracy of Li-Ion battery management systems with passive charge balancing. In Proceedings of the 17th European Conference on Power Electronics and Applications, Geneva, Switzerland, 8–10 September 2015; pp. 1–9.

energies

MDPI

Article

Design and Implementation of a Smart Lithium-Ion Battery System with Real-Time Fault Diagnosis Capability for Electric Vehicles

Zuchang Gao [1], Cheng Siong Chin [2,*], Joel Hay King Chiew [1], Junbo Jia [1] and Caizhi Zhang [3]

[1] School of Engineering, Temasek Polytechnic, Singapore 529757, Singapore; zuchang@tp.edu.sg (Z.G.); joelchk@tp.edu.sg (J.H.K.C.); jiajunbo@tp.edu.sg (J.J.)
[2] Faculty of Science, Agriculture and Engineering, Newcastle University, Newcastle upon Tyne NE1 7RU, UK
[3] School of Automotive Engineering, Chongqing University, Chongqing 400044, China; czzhang@cqu.edu.cn
*   Correspondence: cheng.chin@newcastle.ac.uk; Tel.: +44-65-6908-6013

Received: 13 September 2017; Accepted: 26 September 2017; Published: 27 September 2017

Abstract: Lithium-ion battery (LIB) power systems have been commonly used for energy storage in electric vehicles. However, it is quite challenging to implement a robust real-time fault diagnosis and protection scheme to ensure battery safety and performance. This paper presents a resilient framework for real-time fault diagnosis and protection in a battery-power system. Based on the proposed system structure, the self-initialization scheme for state-of-charge (SOC) estimation and the fault-diagnosis scheme were tested and implemented in an actual 12-cell series battery-pack prototype. The experimental results validated that the proposed system can estimate the SOC, diagnose the fault and provide necessary protection and self-recovery actions under the load profile for an electric vehicle.

Keywords: lithium-ion; energy-storage system; fault diagnosis; protection; electric vehicle

1. Introduction

As one of many energy storage solutions, lithium-ion batteries (LIBs) are attracting more and more attention from researchers and users due to their high energy density, high power density, long lifespan and environmental friendliness [1,2]. The LIBs have been used in energy-storage applications in solar panel systems from those that use a few kilowatt-hours in residential systems to multi-megawatt batteries in grid power systems. There are also broad applications in some high-power applications such as electric vehicles using large numbers of serial or parallel battery cells [3–8]. However, despite being a promising candidate for energy storage solutions, these batteries are facing some challenges, such as ensuring safe operation of the battery-power system that depends on the accurate state-of-charge (SOC) estimation [9–11]. The safety of the LIB power system is crucial, especially when the battery-power system is grouped by a considerable number of battery cells in serial or parallel topology, or in a battery stack, to give a higher power density. The LIBs can deteriorate if they are to operate beyond the battery specifications [9,10].

The estimations of SOC in the battery management system (BMS) can improve the system performance and reliability. However, battery discharge and charge involve complex chemical and physical processes while in operation. It is therefore not easy to estimate the SOC accurately under various operational conditions [12–14]. There are several kinds of LIBs in the market, such as those containing $LiFeO_4$, lithium polymers and $LiCoO_2$. With the different dynamic behavior of the batteries and their topology, specific SOC algorithms are sometimes required. There have been many development and research works in recent years to improve SOC estimation accuracy [13,15–18]. Firstly, the standard measurement-based estimation approaches, such as the coulomb-counting or ampere-hour (Ah) methods, as well as the open-circuit voltage (OCV) and impedance measurement methods, give a more intuitive and reliable estimation [19,20]. However, this approach is prone to errors: errors related to initial SOC determination and the accumulative errors from sensors during the measurement of current and time. Secondly, the machine learning-based estimation methods (also called data-driven approaches), such as the artificial neural network–fuzzy logic (FL) [16] and support vector machine [17] methods. However, machine learning-based methods require a high computational effort due to large training datasets for training the model, although they consider the nonlinearities of the battery model. In addition, most machine learning-based SOC estimation models were established offline. They are not suitable for a practical and low-cost embedded battery-power system application. Lastly, the state-space model-based estimation methods (such as using the extended Kalman filter (EKF)) reduce the convergent time but increase the computational load of the BMS [21,22]. The aforementioned literature has its own disadvantages and advantages. However, what is unusual is that most SOC implementation is only meant for a single battery cell. It is not useful for researchers who would like to implement it on multi-cell batteries for actual applications.

Due to the characteristics of LIBs, the faults of the battery-power system may lead to serious safety issues, such as catching fire and explosion. For instance, a lithium–cobalt oxide battery backup power system caught fire in a Boeing 787 of Japan Airlines in 2013 [23]. Hence, the capabilities of fault diagnosis and protection are important and necessary in a battery-power system. Fault-diagnosis technology is an interdisciplinary field that combines control theory, computer network, database, artificial intelligence and other technologies. In the past few years, there were many researchers focused on battery-power systems. Bohlen et al. [24] investigated the internal-resistance fault diagnosis of batteries by a model-based identification method. D.P. Abraham et al. [25] proved that the changes of battery electrodes are the cause of the sudden increase in the battery internal resistance and the power degradation mechanism in the power battery pack. X.J. Liu [26] tried to diagnose the battery faults by using the fuzzy-logic method. Although the research involved different methods or models that produced good results, most of the authors were focusing on overall theoretical aspects of fault diagnosis using nonlinear model-based or intelligent approaches, such as fuzzy-logic and neural network methods, to determine the battery faults. However, these required a higher computational time and further resources to perform the fault diagnosis. Therefore, in this study, one of the research objectives is to use a computational, inexpensive and intuitive approach to detect and diagnose the faults in the battery and provide corresponding remedy actions for the faults. In addition, the self-recovery scheme is also proposed. Firstly, four types of critical faults [9,10,26–32], being the over-charged fault, over-discharged fault, over-current fault and external short-circuit fault, are considered. Secondly, different fault diagnosis algorithms are studied, and the corresponding solutions are proposed. Lastly, the proposed fault diagnosis and self-recovery schemes are applied to the 12-cell battery pack prototype and validated experimentally.

In summary, a structure of a smart multi-cell battery-power system is proposed to improve the safety and its operational intelligence. Hence, the smart battery-power system has the following features: (1) compatibility and flexibility with different kinds of LIBs and battery pack configurations; (2) capability for SOC self-initialization and self-adjustment; (3) capability for fault diagnosis and self-recovery; and (4) ability to provide a human–machine interface for status report and system configuration, locally or in the cloud. A few designed modules, such as battery data acquisition, battery pack SOC estimation, fault diagnosis, data communication, user-interface module for data display and system configuration, and dual-path switching for charging and discharging, are proposed. In the proposed smart battery-power system, the SOC self-initialization scheme coupled with fault diagnosis and the self-recovery algorithm were investigated and implemented in a 12-cell series (12S) battery-pack prototype. The experimental results show that the system can diagnose the faults and carry out the corresponding protection and recovery actions. In addition, the proposed battery pack has been shown to estimate the SOC successfully under the actual load profile from an electric vehicle.

The rest of this paper is organized as follows: Section 2 presents the system design by different modules. It is followed by Section 3 that deals with the implementation and demonstration of the proposed system in a 12S battery-pack prototype. Finally, Section 4 concludes the work.

2. Proposed Battery-Power System Design

This section encompasses the design and development of a smart LIB battery-power system for SOC estimation, intelligent fault diagnosis and protection for a typical energy-storage module consisting of a 36 V battery pack module with 12-cell series LIBs (ANR26650M1-B) that can be scaled up to 120 cells in series.

2.1. System Structure Design

The proposed smart LIB system has three main parts: controller hardware that includes a microcontroller (MCU) with necessary interfaces and peripherals, embedded software for SOC and fault diagnosis implementation, and a 3.5-inch touchscreen thin-film-transistor liquid-crystal display (TFT LCD) as a user interface for data display and system configuration. The overall system structure is shown in Figure 1.

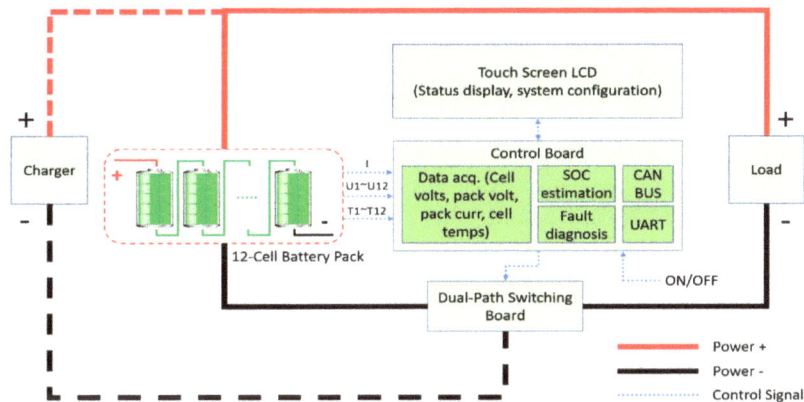

Figure 1. Overall system architecture of proposed smart battery-power system (LCD: liquid-crystal display, SOC: state-of-charge, CAN BUS: controller area network bus, UART: universal asynchronous receiver-transmitter).

The power system periodically measures the voltage value of each cell and the battery pack's current and voltage using suitable analog-to-digital converters (ADCs) and sensors. The controller can perform the SOC estimation and fault-diagnosis algorithms in real-time using measured voltages, current values, temperature values ands the parameters obtained from the touch screen LCD (such as the battery-cell material, battery-cell capacity, battery-cell maximum discharged current and battery topology). The SOC estimation and fault diagnosis results will be displayed on the LCD and sent to the host PC for further processing via a universal asynchronous receiver–transmitter (UART). If the battery pack is grouped into more than 12 cells in series, the controller area network (CAN bus) will communicate with other peer systems or the master system. However, the heat generated from the charging or discharging switches affects the performance of the power system during the high-current application. A dual-path switching board is designed specially to separate the charging and discharging paths to decrease the heat generated from the switches. In addition, a phase-change material (PCM) capable of storing the heat generated will be used. The heat generated is estimated to be reduced by 50% on the discharging path.

2.2. Data Acquisition and SOC Estimation with Self-Initialization Capability

During the system running process, many parameters, such as the cell voltage, temperature, pack voltage and current, must be collected in real-time. The status of the battery pack, such as SOC, is required for continuous fault diagnosis. Therefore, the accuracy of the SOC estimation is critical to the system performance. A high-accuracy data-acquisition method using the LTC6804 is selected as an analog-to-digital converter to monitor the battery pack. It can measure the battery cells with a total measurement error of less than 1.2 mV. The measurement range from 0 to 5 V makes the LTC6804 suitable for the LIB application, as it does not consume much power. The speed of the data acquisition is fast enough with the 12-cell voltages that are sampled at 290 µs. However, lower data-acquisition rates can be used for higher noise reduction. The configuration of this hardware module can be seen in Figure 2. The precision of the LTC6804 was around 0.3%. However, the overall system accuracy can be affected by the print circuit board (PCB) design. Nevertheless, the precision target of 6% can be obtained by continuous calibration for the case of a lower-current application.

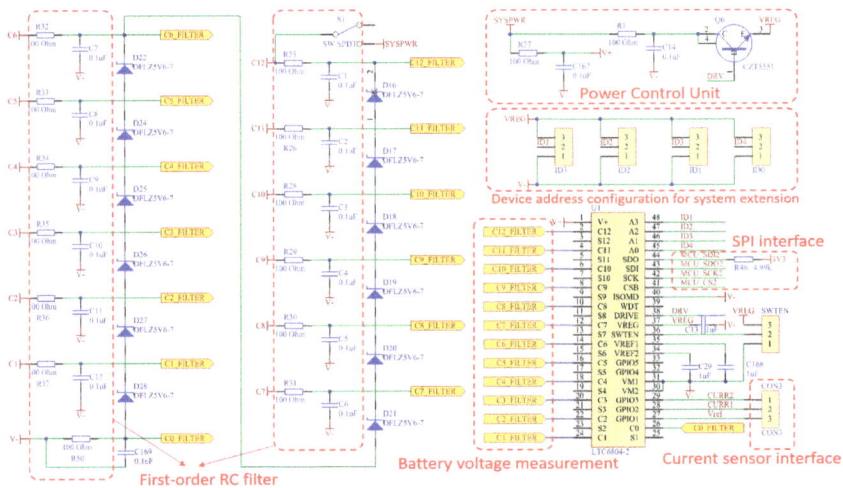

Figure 2. Hardware configuration of data-acquisition module of battery cells (RC: resistor capacitor, SPI: Serial Peripheral Interface).

The releasable capacity ($C_{releasable}$) of an operating battery is the released capacity when it is completely discharged. The SOC is defined as the percentage of the releasable capacity relative to the battery-rated capacity (C_{rated}), given as follows:

$$SOC = C_{releaseable}/C_{rated} \times 100\%. \tag{1}$$

A fully charged battery has the maximal releasable capacity (C_{max}), which can be different from the rated capacity. In general, C_{max} declines as time increases. Hence, the C_{max} can be used for evaluating the state-of-health (SOH) of a battery:

$$SOH = C_{max}/C_{rated} \times 100\%. \tag{2}$$

When a battery is discharging, the depth of discharge (DOD) can be defined as the percentage of the capacity that has been discharged ($C_{released}$) relative to C_{rated}:

$$DOD = C_{released}/C_{rated} \times 100\%. \tag{3}$$

With a measured battery current (I_t), the difference of the DOD in an operating period (T) can be computed by:

$$\Delta\text{DOD} = \frac{\int_{t_0}^{t_0+T} I_t dt}{C_{\text{rated}}} \times 100\%. \tag{4}$$

By considering the SOH, the SOC is estimated as:

$$\text{SOC}(t) = \text{SOH}(t) - \text{DOD}(t). \tag{5}$$

In order to improve the convergence performance of SOC estimation, an accurate method to determine the initial status of the batteries will be used. However, unlike the lab environment, it is hard to know the exact initial SOC and DOD of the battery. Therefore, a self-initialization method for the battery system is proposed and implemented to provide prior configuration during the initialization stage.

At the charging stage, the variations of the battery voltage and battery current when the battery is charged by the constant current–constant voltage (CC–CV) mode are usually specified by the manufacturer. With the constant charging current, the battery voltage increases gradually and reaches the threshold. Once the battery has been charged by the constant voltage mode, the charging current reduces rapidly before it more gradually decreases. Eventually, the current will reach almost zero when it is fully charged. This charging curve can be converted into the relationship between the charging voltage and the SOC, during the constant current stage. By using the relationship between the charging current and the SOC during the constant-voltage stage, the initial SOC value can be obtained.

At the discharging stage, the typical voltage curves at different discharging currents are given by the manufacturer. A higher current will cause a faster decline in the terminal voltage, and lead to a shorter operation time. The relationship between the SOC and the discharging voltage at different currents can be obtained. At the open-circuit stage, the relationship between open-circuit voltage (OCV) and SOC is needed. The battery is discharged by different currents before disconnecting from the load. The OCV can be used to estimate SOC if a long period of relaxation is given for the transient to settle to its steady state. Figure 3 is the flowchart to indicate the self-initialization procedures for SOC estimation when the system restarts after a long period of relaxation.

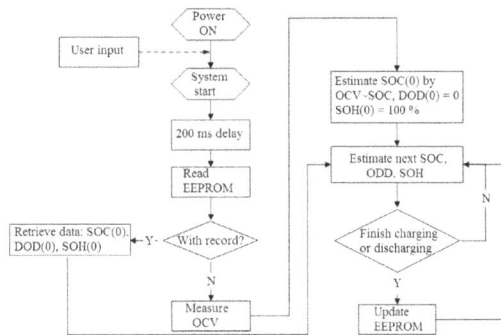

Figure 3. Flowchart of SOC self-initialization and updating (DOD: depth of discharge, SOH: state-of-charge, OCV: open circuit voltage, EEPROM: electrically erasable programmable read-only memory).

2.3. Smart Fault-Diagnosis Strategies

As the LIBs provide a high-energy-density power, the system must be able to detect the abnormality in real-time to ensure the safety of the users and efficient power supply under unexpected conditions, such as an external short circuit. The safety of Li-ion batteries depends on the attributes of system design, such as electronics, estimation algorithms, and thermal and mechanical characteristics, regardless of electrochemistry. They should be equipped with the capability to diagnose the faults, and perform a corresponding corrective action. In addition, they need to carry out a self-recovery action after the fault conditions are eliminated.

2.3.1. External Short-Circuit Fault Diagnosis of a Battery Pack

The external short-circuit to the power system is a common fault that happens during installation or uninstallation processes of the battery-power system. Dual diagnosis and protection schemes are designed and implemented into the system to ensure that the fault can be isolated reliably and timely without affecting other components. The hardware design for the external short-circuit can be seen in Figure 4. The dual diagnosis and protection schemes consist of the following analog and digital diagnosis and protection systems:

(a) Analog diagnosis and protection scheme: this scheme detects the fault by a specially designed analog circuit. Once the fault is detected, the circuit will inform the actuator (i.e., the switching board within the control board in Figure 1) to carry out the protection action immediately. This avoids polling time of the software algorithm. The output signal will reset, and the system will recover once the fault condition has been eliminated.

(b) Digital diagnosis and protection scheme: this method periodically detects the faults by software polling. The strategy is as follows: if the discharge current is more than twice the power-pack maximum current, I_{max}, the current will vary across the batteries to enable the external short-circuit fault to be detected. The MCU will give a signal to perform the protection action. This ensures the fault is isolated effectively when the analog diagnosis and protection scheme does not work properly. The output signal will reset to its normal state once the fault condition has been eliminated. A flowchart of the digital diagnosis and protection scheme of the external fault diagnosis is depicted in Figure 5.

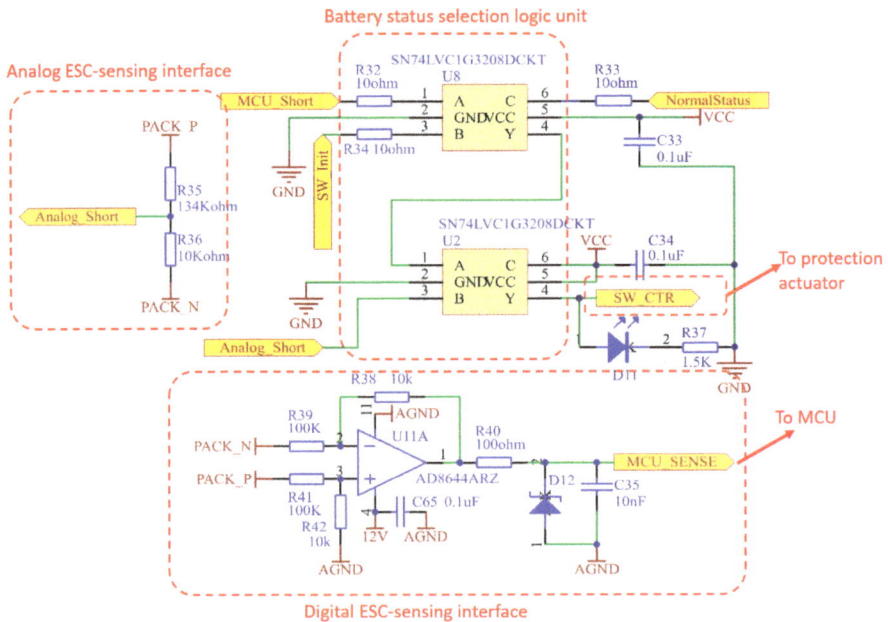

Figure 4. Dual-protection hardware design for external short-circuit (ESC) to microcontroller (MCU).

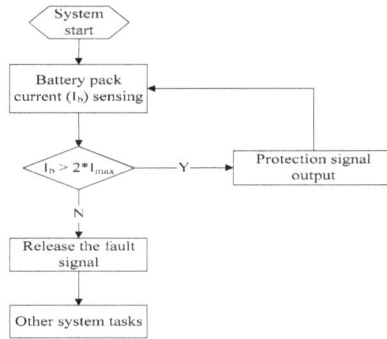

Figure 5. Diagram of the digital diagnosis and protection scheme for external-fault diagnosis.

2.3.2. Fault Diagnosis of Battery Cells

The over-charged, over-discharged, over-current and over-temperature conditions are the four different fault conditions of the battery cells. They can cause permanent damage to the battery cells. As these fault conditions are related to the parameters of the battery, they are included in the fault diagnosis. Battery charge and discharge involve complex chemical and physical processes. However, the over-charged, over-discharged and over-current states can also lead to fire due to the explosion of the cells. To circumvent these problems, the battery-charging or discharging voltage needs to be monitored continuously in real-time. If the upper or lower cut-off voltage or the current reaches the maximum value, the system controller will use the over-charged/over-discharged/over-current protection to stop the charging or discharging of the battery cells.

Similarly, over-temperature is another fault to be avoided. In the 12S system prototype, six thermistors are attached to the gaps between every two battery cells to monitor the inner-cell temperatures. A two-level reporting and protection scheme, as well as the self-recovery scheme with a temperature window, are proposed. If the temperature reaches 50 °C, the controller will trigger the first-level protection action to alert the users without interrupting the charging or discharging processes. However, if the cell's temperature reaches 60 °C (i.e., quite high for a cell), the controller will trigger the second-level protection to stop the charging/discharging process immediately. The controller will trigger the self-recovery process automatically when all the cell's temperatures drop to 40 °C. The flowchart of the fault diagnosis and protection scheme of over-temperature is shown in Figure 6. The system will detect the faults mentioned above to allow protection to take place immediately.

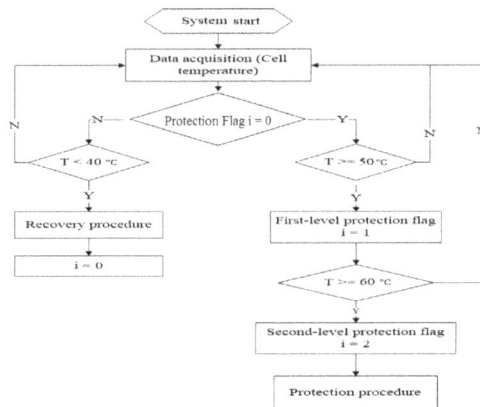

Figure 6. Flowchart of the fault diagnosis and protection scheme for high temperature.

2.3.3. Open-Wire Fault Diagnosis

In the multi-cell LIB pack, the voltage of each cell needs to be monitored in real-time. In addition, battery-cell balancing is required for the cells that are unbalanced, that is, when the SOC of one or more cells are unequal. Therefore, there are $n + 1$ (n is the cell number in the battery pack) wires from the cells to the balancing board. It is often difficult to ensure that all the wires are always connected properly during the actual operation. Some of the wires may be loose or disconnected due to frequency impact or vibration from the mechanical structure that encloses the cells. The floating voltage of the LTC6804 input pins will be around 0, or higher than 5 V when the open-wire condition occurs. Based on the floating-voltage sensor, the fault-diagnosis strategy is proposed to diagnose the open-wire fault. The failure will be displayed via the graphical-user-interface (GUI) to alert the user of the fault that requires attention. In this case, it is not possible to automatically connect the wire, as it will require human involvement. The flowchart of the diagnosis strategy can be seen in Figure 7.

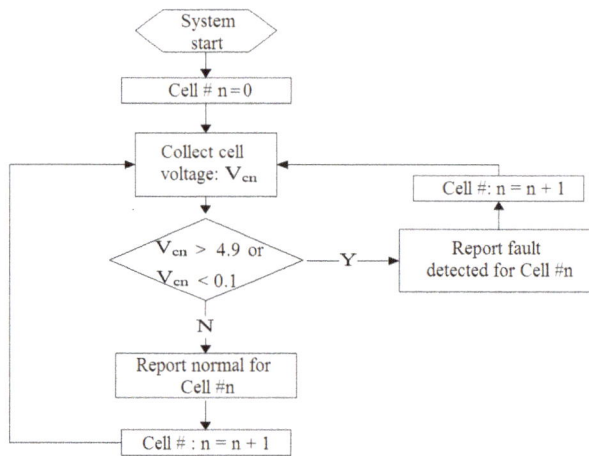

Figure 7. Flowchart of open-wire fault diagnosis strategy.

A dual-entrance interface is designed for different users to configure the system for the initial use. Several default parameters for various batteries have been stored in the system memory for ease of operation. The ordinary users do not have to worry about the exact values. Instead, they need to select the types of batteries, the battery-pack grouping information, the battery-cell rated capacity and the maximum discharging current. The system will automatically initialize the corresponding parameters according to the information provided. Currently, four lithium battery types, such as LiFePO$_4$, Li-ion, LiCoO$_2$ and lithium polymers, are supported in the proposed system. In addition, the system has an option to customize the parameter values such as protection-voltage level and charging- or discharging-current limitation for further analysis. The entrance design for system parameter configuration is described in Figure 8.

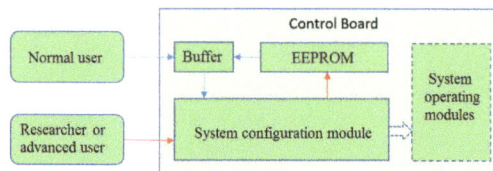

Figure 8. Entrance design for system parameter configuration.

3. Experimental Results

The proposed smart LIB system is implemented as a 12-cell series LIB-pack prototype to test the fault-detection and diagnosis algorithms, as shown earlier. In addition, an actual load profile for an electric vehicle is used to validate the proposed system after the fault-detection and diagnosis algorithms. The overall structure and the physical testing system setups are shown in Figure 9.

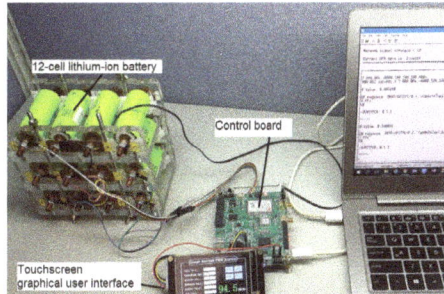

Figure 9. Test setup for 12-cell series Lithium-ion battery (LIB) and graphical-user-interface (GUI).

In order to increase the flexibility of the power system, the proposed battery-power system provides a touchscreen LCD interface for users to configure the battery-power system. It consists of parameters such as battery-cell type, battery-cell rated capacity, maximum discharging current and battery-pack grouping topology for the battery pack during the first initialization. This open feature enables the controller to adapt to various types of the battery cell and different topologies. The front page of the GUI is shown in Figure 10, while the user's configuration page can be seen in Figure 11.

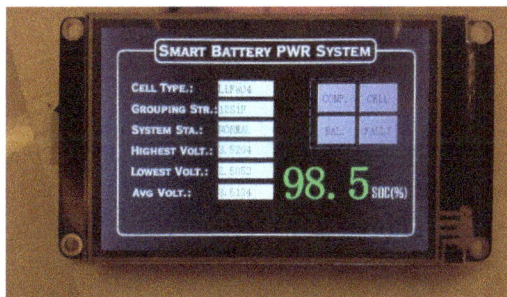

Figure 10. Front panel of GUI for proposed battery-power system.

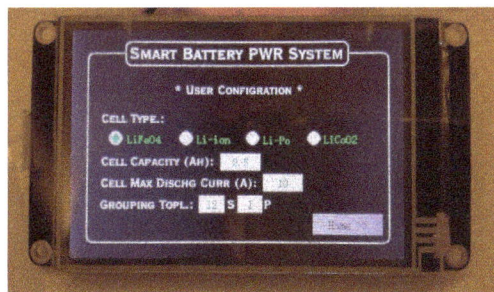

Figure 11. Configuration page of GUI for proposed battery-power system.

The front panel in Figure 10 contains information such as the current configuration of the battery pack and the respective minimum, maximum and mean voltages and SOC value of the battery pack. The other functional modules that contain the cells' information are shown in Figure 10. In the configuration page, as shown in Figure 11, four options of the battery-cell types can be selected. The users can choose the material of the physical cells connected to the control board, the battery-cell capacity and battery-cell maximum discharging current. The last parameter is the grouping topology of the battery pack that depends on the physical battery-cell layout, that is, series or parallel. The system can be initialized to proceed with the proper function and fault diagnosis operation. However, if the series-cell layout is defined by the actual parallel-cell configuration, the system will not function for safety purposes.

The data-acquisition module of the power system is essential for the SOC estimation and fault diagnosis. It will affect the overall performance of the system. In this paper, an accurate data-acquisition module for the battery cell is implemented with a total measurement error of less than 1.2 mV. The precise data ensure that the accuracy of the SOC estimation and the safety of the system can be monitored. The cell-monitoring page of the GUI can be seen in Figure 12. For example, the system can estimate the SOC of each cell. The SOC values are quite close to the values obtained by the extended Kalman filter (EKF). The proposed SOC estimation exhibits a close match with the EKF under the regular pulse discharge test (PDT), as seen in Figure 13. Here, the 2 Resistor-Capacitor (RC) equivalent circuit model (ECM) is applied for the EKF-based SOC estimation. The function of the SOC with respect to U_{OC} is described as follows [33]:

$$U_{OC} = 3.397 - 0.19SOC + 0.087 \log SOC - 0.054 \log(1 - SOC), \tag{6}$$

where $SOC \in (0, 1)$.

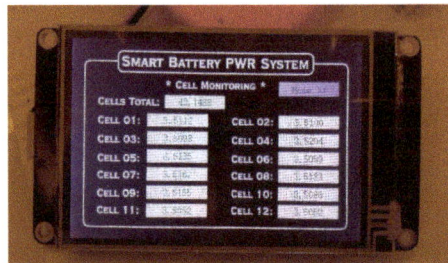

Figure 12. Cell-monitoring panel of GUI for proposed battery-power system.

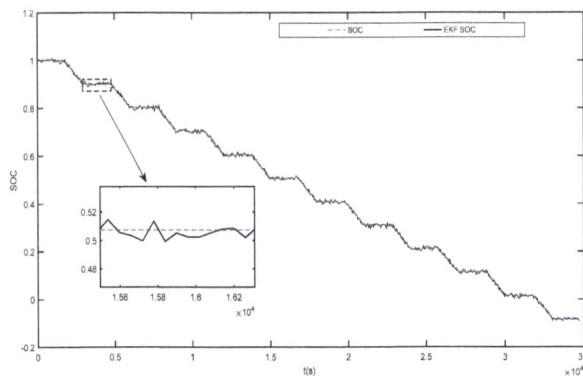

Figure 13. Comparison of Extended Kalman filter (EKF) with proposed SOC estimation under regular pulse discharge test.

A detailed battery modeling and EKF-based SOC estimation can be found in the reference [33]. Note that the proposed battery system can also allow different SOC-estimation algorithms to be programmed and compared.

To further validate the proposed SOC-estimation algorithm under realistic and dynamic situations, the New European Driving Cycle (NEDC) load profile for a typical electric vehicle was applied to the 12-cell series battery-pack prototype to simulate the electric vehicle applications under the ambient temperature effect and compared with the standard EKF-estimation approach. The NEDC load profile is shown in Figure 14a, subjected to the ambient temperature. It is worthy to note that the load profile had been scaled down to fit the battery pack. The programmable DC electronic load was used to run the pre-programmed NEDC load profile. Before the experiment, the battery pack was fully charged to obtain an initial SOC of 100%. The calibrated Ampere-hour (Ah) data reading from the equipment was used as the reference. As shown in Figure 14b, the SOC estimation of the proposed SOC model can follow the reference SOC robustly. Despite the load profile changing drastically, the SOC error can rebound after a short period of drift. On the other hand, the SOC-estimation result from the EKF-based SOC model drifted away after several test cycles. In summary, under the NEDC test profile, which is used to simulate the load in reality, the proposed SOC-estimation method performs better than the EKF-based method.

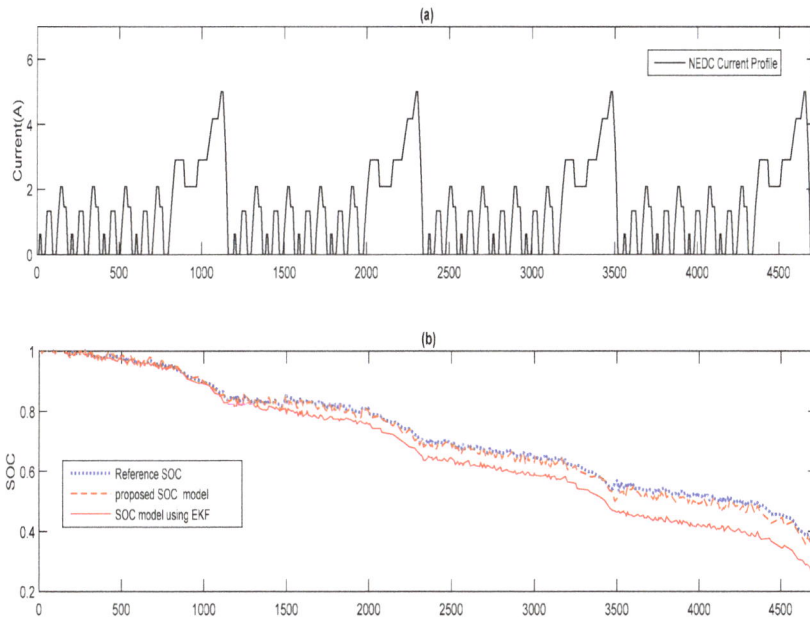

Figure 14. (**a**) New European Driving Cycle (NEDC) current profile; (**b**) SOC comparison with EKF-based model.

After the SOC estimation was validated, the dual diagnosis and protection schemes were tested under different fault conditions. In the 12-cell series battery-pack system, the maximum current of the battery cell was set to 10 A with a maximum delay of 10 s after the fault is detected. This provides limitations on the current and the time delay used in the fault diagnosis. Note that these values can be adjusted. Figure 15 shows the fault-detection and self-recovery curve. Once the external short-circuit occurred, the analog scheme detected the fault within a very short time and generated a protection signal to turn off the discharging switch, as shown in Figure 15. It took around 100 ms before the fault was detected. Thus, the result shows that the external short-circuit fault can be detected and the system can later recover from the fault.

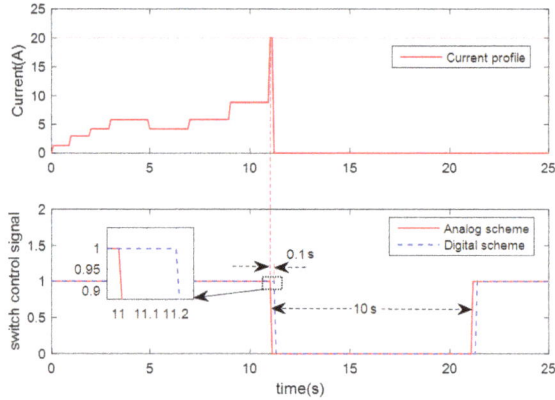

Figure 15. External short-circuit fault detection and self-recovery curve.

The over-charged and over-discharged states are another two faults that can cause permanent damage to the internal structure of batteries. In the prototype system, over-charged and over-discharged fault-diagnosis schemes are both integrated. The protection and self-recovery processes can be seen in Figures 16 and 17, respectively. As observed in Figure 16, after the batteries were fully charged, the over-charged fault was reported, and the charging switch was turned off by the controller to isolate the charging module from the cells. To trigger the charging function, the charger needs to be removed to allow the charging module to recover automatically. Similarly, the over-discharged fault was detected when any cell of the battery pack reached the cut-off voltage, as shown in Figure 17. The discharging switch was turned off to isolate the battery from the output module or load. Using the self-locking and self-recovery function, the over-charged and over-discharged faults can be managed effectively to prevent damage to the batteries and possible injury to the user. Similarly, over-current, over-temperature and broken-wire fault diagnoses of the battery system were implemented using the proposed schemes in the actual battery-power system.

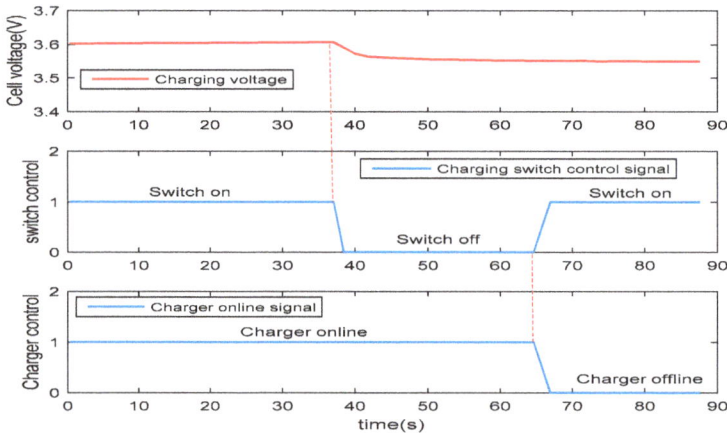

Figure 16. Over-charged fault detection and self-recovery curve.

154

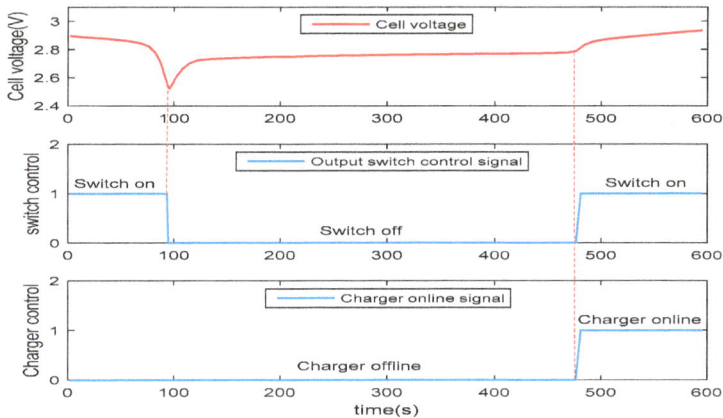

Figure 17. Over-discharged fault detection and self-recovery curve.

4. Conclusions

In this paper, an overall structure of a 12-cell series LIB power system integrated with smart real-time fault-diagnosis and self-recovery schemes was proposed and implemented. Several common faults, such as the external short-circuit, over-charged, over-discharged, over-current, over-temperature and open-wire faults, were investigated and validated in the actual implementation. The experimental results of LIBs under the real load profile confirmed the efficacy of the state-of-charge estimation and fault diagnosis capabilities for the electric vehicle application. It ensured the safety of the users and, in particular, prolonged the life of the battery cells in an electric vehicle with human presence. Future work will include the study of the aging and recycling of cells to improve the battery performance at a reasonable implementation and operating cost.

Acknowledgments: This work was supported by equipment and manpower from the Temasek Polytechnic and Newcastle University. The authors would like to thank all partners for providing research support.

Author Contributions: Zuchang Gao is credited with the theoretical formulation, simulation and experimental works performed in this paper. Cheng Siong Chin from defined the flow of the paper, ideas of smart battery power system with fault diagnosis capabilities and participated in the model simulation and verification that drove this research with Zuchang Gao, Junbo Jia and Joel Hay King Chiew to test its relevance for industrial applications. Caizhi Zhang provided the academic support and advice during the project. All authors discussed and provided comments at all stages.

Conflicts of Interest: The authors declare no conflict of interest.

References

1. Zou, C.; Manzie, C.; Nesic, D. A framework for simplification of pde-based lithium-ion battery models. *IEEE Trans. Control Syst. Technol.* **2016**, *24*, 1594–1609. [CrossRef]
2. Gao, Z.; Jia, J.; Xie, J.; Toh, W.D.; Lin, P.; Lyu, H.; Julyanto, D.; Chin, C.S.; Woo, W.L. Modelling and simulation of a 12-cell battery power system with fault control for underwater robot. In Proceedings of the 2015 IEEE 7th International Conference on Cybernetics and Intelligent Systems (CIS) and IEEE Conference on Robotics, Automation and Mechatronics (RAM), Siem Reap, Cambodia, 15–17 July 2015; pp. 261–267.
3. Xiong, R.; He, H.; Sun, F.; Zhao, K. Evaluation on state of charge estimation of batteries with adaptive extended kalman filter by experiment approach. *IEEE Trans. Veh. Technol.* **2013**, *62*, 108–117. [CrossRef]
4. Kim, T.; Wang, Y.; Fang, H.; Sahinoglu, Z.; Wada, T.; Hara, S.; Qiao, W. Model-based condition monitoring for lithium-ion batteries. *J. Power Sources* **2015**, *295*, 16–27. [CrossRef]
5. Ablay, G. Online condition monitoring of battery systems with a nonlinear estimator. *IEEE Trans. Energy Convers.* **2014**, *29*, 232–239. [CrossRef]

6. Petzl, M.; Danzer, M.A. Advancements in OCV measurement and analysis for lithium-ion batteries. *IEEE Trans. Energy Convers.* **2013**, *28*, 675–681. [CrossRef]

7. Alves, J.; Baptista, P.C.; Gonçalves, G.A.; Duarte, G.O. Indirect methodologies to estimate energy use in vehicles: Application to battery electric vehicles. *Energy Convers. Manag.* **2016**, *124*, 116–129. [CrossRef]

8. Tsang, K.M.; Sun, L.; Chan, W.L. Identification and modelling of lithium ion battery. *Energy Convers. Manag.* **2010**, *51*, 2857–2862. [CrossRef]

9. Lu, L.; Han, X.; Li, J.; Hua, J.; Ouyang, M. A review on the key issues for lithium-ion battery management in electric vehicles. *J. Power Sources* **2013**, *226*, 272–288. [CrossRef]

10. Offer, G.J.; Yufit, V.; Howey, D.A.; Wu, B.; Brandon, N.P. Module design and fault diagnosis in electric vehicle batteries. *J. Power Sources* **2012**, *206*, 383–392. [CrossRef]

11. Iraola, U.; Aizpuru, I.; Gorrotxategi, L.; Segade, J.M.C.; Larrazabal, A.E.; Gil, I. Influence of voltage balancing on the temperature distribution of a Li-ion battery module. *IEEE Trans. Energy Convers.* **2015**, *30*, 507–514. [CrossRef]

12. Zhang, F.; Liu, G.; Fang, L.; Wang, H. Estimation of battery state of charge with H_observer: Applied to a robot for inspecting power transmission lines. *IEEE Trans. Ind. Electron.* **2012**, *59*, 1086–1095. [CrossRef]

13. Chaoui, H.; Golbon, N.; Hmouz, I.; Souissi, R.; Tahar, S. Lyapunov-based adaptive state of charge and state of health estimation for lithium-ion batteries. *IEEE Trans. Ind. Electron.* **2015**, *62*, 1610–1618. [CrossRef]

14. Zou, C.; Manzie, C.; Nešić, D.; Kallapur, A.G. Multi-time-scale observer design for state-of-charge and state-of-health of a lithium-ion battery. *J. Power Sources* **2016**, *335*, 121–130. [CrossRef]

15. Zheng, F.; Xing, Y.; Jiang, J.; Sun, B.; Kim, J.; Pecht, M. Influence of different open circuit voltage tests on state of charge online estimation for lithium-ion batteries. *Appl. Energy* **2016**, *183*, 513–525. [CrossRef]

16. Li, I.H.; Wang, W.Y.; Su, S.F.; Lee, Y.S. A merged fuzzy neural network and its applications in battery state-of-charge estimation. *IEEE Trans. Energy Convers.* **2007**, *22*, 697–708. [CrossRef]

17. Hansen, T.; Wang, C.-J. Support vector based battery state of charge estimator. *J. Power Sources* **2005**, *141*, 351–358. [CrossRef]

18. Gandolfo, D.; Brandão, A.; Patiño, D.; Molina, M. Dynamic model of lithium polymer battery—Load resistor method for electric parameters identification. *J. Energy Inst.* **2015**, *88*, 470–479. [CrossRef]

19. Jeong, Y.M.; Cho, Y.K.; Ahn, J.H.; Ryu, S.H.; Lee, B.K. Enhanced coulomb counting method with adaptive soc reset time for estimating OCV. In Proceedings of the 2014 IEEE Energy Conversion Congress and Exposition (ECCE), Pittsburgh, PA, USA, 14–18 September 2014; pp. 1313–1318.

20. Stockley, T.; Thanapalan, K.; Bowkett, M.; Williams, J. Design and implementation of an open circuit voltage prediction mechanism for lithium-ion battery systems. *Syst. Sci. Control Eng.* **2014**, *2*, 707–717. [CrossRef]

21. Hussein, A.A.H.; Batarseh, I. State-of-charge estimation for a single lithium battery cell using extended kalman filter. In Proceedings of the 2011 IEEE Power and Energy Society General Meeting, Detroit, MI, USA, 24–29 July 2011; pp. 1–5.

22. Yu, Z.; Huai, R.; Xiao, L. State-of-charge estimation for lithium-ion batteries using a kalman filter based on local linearization. *Energies* **2015**, *8*, 7854–7873. [CrossRef]

23. Boeing 787 Aircraft Grounded after Battery Problem in Japan. BBC News. Available online: http://www.bbc.com/news/av/business-25740181/boeing-787-aircraft-grounded-after-battery-problem-in-japan (accessed on 25 September 2017).

24. Bohlen, O.; Buller, S.; Doncker, R.W.D.; Gelbke, M.; Naumann, R. Impedance based battery diagnosis for automotive applications. In Proceedings of the 2004 IEEE 35th Annual Power Electronics Specialists Conference (IEEE Cat. No. 04CH37551), Aachen, Germany, 20–25 June 2004; Volume 2794, pp. 2792–2797.

25. Swiatowska, J.; Barboux, P. Lithium battery technologies: From the electrodes to the batteries. In *Lithium Process Chemistry: Resources, Extraction, Batteries, and Recycling*; Elsevier: Amsterdam, The Netherlands, 2015; Volume 125.

26. Liu, X.J. Research and application of intelligent battery fault diagnosis system. Master's Thesis, Beijing University of Posts and Telecommunications, Beijing, China, 2011.

27. Hannan, M.A.; Lipu, M.S.H.; Hussain, A.; Mohamed, A. A review of lithium-ion battery state of charge estimation and management system in electric vehicle applications: Challenges and recommendations. *Renew. Sustain. Energy Rev.* **2017**, *78*, 834–854. [CrossRef]

28. Cheng, K.W.E.; Divakar, B.; Wu, H.; Ding, K.; Ho, H.F. Battery-management system (BMS) and SOC development for electrical vehicles. *IEEE Trans. Veh. Technol.* **2011**, *60*, 76–88. [CrossRef]

29. Yong, L.; Lifang, W.; Chenglin, L.; Liye, W.; Dongping, X. State-of-charge estimation of lithium-ion battery using multi-state estimate technique for electric vehicle applications. In Proceedings of the Vehicle Power and Propulsion Conference (VPPC), Beijing, China, 15–18 October 2013; pp. 1–5.
30. Chen, Z.; Lin, F.; Wang, C.; Wang, Y.L.; Xu, M. Active diagnosability of discrete event systems and its application to battery fault diagnosis. *IEEE Trans. Control Syst. Technol.* **2014**, *22*, 1892–1898. [CrossRef]
31. Lagorse, J.; Simões, M.G.; Miraoui, A. A multiagent fuzzy-logic-based energy management of hybrid systems. *IEEE Trans. Ind. Appl.* **2009**, *45*, 2123–2129. [CrossRef]
32. Sidhu, A.; Izadian, A.; Anwar, S. Adaptive nonlinear model-based fault diagnosis of li-ion batteries. *IEEE Trans. Ind. Electron.* **2015**, *62*, 1002–1011. [CrossRef]
33. Jia, J.; Lin, P.; Chin, C.S.; Toh, W.D.; Gao, Z.; Lyu, H.; Cham, Y.T.; Mesbahi, E. Multirate strong tracking extended kalman filter and its implementation on lithium iron phosphate (LiFePO$_4$) battery system. In Proceedings of the 2015 IEEE 11th International Conference on Power Electronics and Drive Systems, Sydney, Australia, 9–12 June 2015; pp. 640–645.

energies

MDPI

Article

Improved Battery Parameter Estimation Method Considering Operating Scenarios for HEV/EV Applications

Jufeng Yang [1,2], Bing Xia [2,3], Yunlong Shang [2,4], Wenxin Huang [1,*] and Chris Mi [2,*]

[1]   Department of Electrical Engineering, Nanjing University of Aeronautics and Astronautics,
    Nanjing 211106, China; jufeng.yang@mail.sdsu.edu
[2]   Department of Electrical and Computer Engineering, San Diego State University,
    San Diego, CA 92182, USA; bixia@eng.ucsd.edu (B.X.); shangyunlong@mail.sdu.edu.cn (Y.S.)
[3]   Department of Electrical and Computer Engineering, University of California San Diego,
    San Diego, CA 92093, USA
[4]   School of Control Science and Engineering, Shandong University, Jinan 250061, China
*   Correspondence: huangwx@nuaa.edu.cn (W.H.); cmi@sdsu.edu (C.M.);
    Tel.: +86-138-5149-7182 (W.H.); +1-619-594-3741 (C.M.)

Academic Editor: Rui Xiong
Received: 3 October 2016; Accepted: 13 December 2016; Published: 22 December 2016

Abstract: This paper presents an improved battery parameter estimation method based on typical operating scenarios in hybrid electric vehicles and pure electric vehicles. Compared with the conventional estimation methods, the proposed method takes both the constant-current charging and the dynamic driving scenarios into account, and two separate sets of model parameters are estimated through different parts of the pulse-rest test. The model parameters for the constant-charging scenario are estimated from the data in the pulse-charging periods, while the model parameters for the dynamic driving scenario are estimated from the data in the rest periods, and the length of the fitted dataset is determined by the spectrum analysis of the load current. In addition, the unsaturated phenomenon caused by the long-term resistor-capacitor (RC) network is analyzed, and the initial voltage expressions of the RC networks in the fitting functions are improved to ensure a higher model fidelity. Simulation and experiment results validated the feasibility of the developed estimation method.

Keywords: lithium-ion battery; operating scenario; equivalent circuit modeling; parameter estimation

1. Introduction

Lithium-ion batteries have been widely used in the energy storage systems of hybrid electric vehicles (HEVs) and pure electric vehicles (EVs) because of their low self-discharge rate, high energy and power densities. To ensure the safe and reliable operation of lithium-ion batteries, the battery management system (BMS) is of significant importance. The main task of a BMS includes monitoring of critical states, fault diagnosis and thermal management [1–7].

1.1. Review of the Literature

The performance of a BMS is highly dependent on the accurate description of battery characteristics. Hence, a proper battery model, which can not only correctly characterize the electrochemical reaction processes, but also be easily implemented in embedded microcontrollers, is necessary for a high-performance BMS. There are two common forms of battery models available in the literature: the electrochemical model and the equivalent circuit model (ECM). The electrochemical model expresses the fundamental electrochemical reactions by complex nonlinear partial differential algebraic equations (PDAEs) [8]. It can accurately capture the characteristics of the battery, but requires extensive computational power to obtain the solutions of the equations. Hence, such models are suitable for the battery design rather than the system level simulation. In contrast, the ECM abstracts away the detailed internal electrochemical reactions and characterizes them solely by simple electrical components; thus, it is ideal for circuit simulation software and implementation in embedded microcontrollers. The accuracy of the ECM is highly dependent on the model structure and model parameters. Theoretically, a higher order ECM can represent a wider bandwidth of the battery application and can generate more accurate voltage estimation results. However, the high order ECM can not only increase the computational burden, but also reduce the numerical stability for the further battery states' estimation [9,10]. Hence, considering a tradeoff among the model fidelity, the computational burden and the numerical stability, the second order ECM is employed in this paper [11–18]. The common structure of the second order ECM is illustrated in the top subfigure of Figure 1, where the open circuit voltage (*OCV*), which is a function of state of charge (*SoC*), stands for the open circuit voltage, R_{in} is the internal resistance, which represents the conduction and charge transfer processes [19–21], and two resistor-capacitor (RC) networks approximately describe the diffusion process. Among them, the short-term RC network models the fast dynamics diffusion process (Part A in the bottom subfigure of Figure 1), and the long-term RC network represents the slow dynamics diffusion process (Part B in the bottom subfigure of Figure 1). The above model parameters can be identified either through the time-domain or the frequency-domain parameter extraction experiments. For the time-domain parameter estimation methods, model parameters are usually identified through fitting the voltage response from the parameter extraction experiment with the exponential-based functions. The electrochemical impedance spectroscopy (EIS) test is the commonly-used frequency-domain parameter extraction experiment. Compared to the time-domain test process, one limitation of the EIS test is that the amplitude of the current excitation is so low that the battery can be considered as equalized during the whole test process, which seldom happens in HEV/EV applications. In order to overcome the above drawback, references [22–24] propose superimposing the direct current (DC) offset over the EIS signals to determine the current dependency of impedance parameters. However, since significant time is required for the EIS test, the battery *SoC* changes significantly during the test procedure if the amplitude of the superimposed current is improper. This can reduce the parameter estimation accuracy and make this method practically not applicable at moderate and high current rates [25,26]. Based on the aforementioned analysis, the second order ECM with parameters estimated by the time-domain analysis is discussed in this paper.

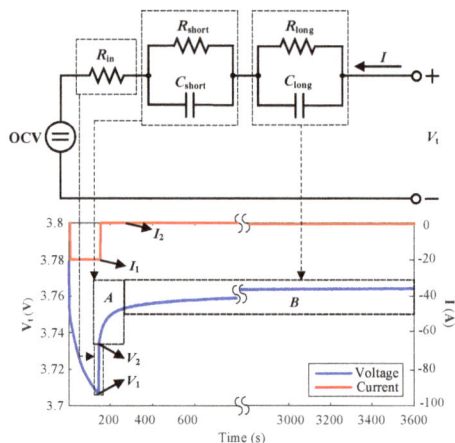

Figure 1. The second order equivalent circuit model (ECM). *OCV*, open circuit voltage.

Generally speaking, batteries usually operate in two scenarios in automotive applications: The constant-current (CC) charging scenario and the dynamic driving scenario [27]. Usually, the motions of lithium ions under the continuous external excitation (representing the CC charging scenario) and the discontinuous external excitation (representing the dynamic driving scenario) show different characteristics, and this difference is related to the diffusivity of ions. In other words, the model parameters, especially the RC network parameters, show diverse values under different operating scenarios [21,28]. Therefore, battery parameters should be identified separately according to the actual operating scenarios. Abundant research work has been conducted to seek the accurate ECM for the specific operating scenario. For the charging scenario, a universal model based on a simple mathematical equation with constant parameters is proposed [29–31]. The mathematical equations include one polynomial component and one or two exponential functions, and relevant parameters can be obtained by fitting collected charging profiles. Verification results in related literature show that the overall model output profiles match well with the experimental data, but there still exists obvious estimation errors during certain periods (at the beginning of the plateau region and the last charging region). This is mainly caused by the constant parameters during the whole charging process since the actual model parameters, such as time constants, may vary greatly at different *SoC* regions [32]. The works in [32–34] estimate the model parameters through the data in the rest periods of the pulse-rest test at different *SoC* points, and the estimated model parameters can be shown as functions of *SoC*. However, the charging concentration process under continuous excitation is different from the charging recovery process under the rest period [19,35]; thus, the estimated model parameters may not accurately represent the charging characteristics of the battery. For the dynamic driving scenario, many modeling approaches have been reported on the basis of the pulse discharge analysis. In [36–38], model parameters are obtained by simple algebraic operations. This is straightforward, but large estimation errors exist. A more accurate method is to fit the voltage response of the whole rest period with an exponential function [39–41]. The limitation of this method is its poor dynamic performance. In order to improve the battery model accuracy, Hu and Wang in [42] propose a two time-scale identification algorithm to separate the identifications of slow and fast battery dynamics. This method shows better frequency response matching without increasing computational complexity. Xiong in [17] uses the bias correction method to ensure the battery model prediction performance. This approach shows excellent performance and high accuracy against uncertain operating scenarios and battery packs. Instead of the conventional pulse-rest test, [43,44] propose two types of application-oriented parameter extraction tests, leading to a fast dynamics battery model with high fidelity. One major limitation of this kind of method is that the parameter extraction test corresponds to a specific operating scenario. If the actual load profiles show obviously different bandwidths under different working conditions, the parameter extraction test should be re-implemented. One solution to overcome this drawback is to conduct as many parameter extraction tests as possible to cover the typical load characteristics, but this requires an extensive amount of time and effort.

1.2. Contributions of This Paper

Based on the battery parameter estimation methods discussed above, it can be concluded that seldom does work in the previous literature discuss a battery model considering both the CC charging and dynamic driving scenarios. Hence, the focus of this paper is to propose a battery parameter estimation method, which is applicable to common operating scenarios in HEV/EV applications. The main contributions are: (1) both the constant-current charging and the dynamic driving scenarios are taken into consideration, and two separate sets of model parameters are estimated through different parts of the pulse-rest test; (2) the model parameters for the constant-current charging scenario are estimated from the data in the pulse-charging periods; (3) the model parameters for the dynamic driving scenario are estimated from the data in the rest periods, and the length of the fitted dataset is determined by the spectrum analysis of the load current; (4) the unsaturated phenomenon caused by the long-term RC network is analyzed, and the initial voltage expressions of the RC networks in the fitting functions are improved to ensure a higher model fidelity; (5) both the simulation and experiment results agree with the analysis and demonstrate the improvement of the proposed battery parameter estimation method over the existing ones.

2. Parameter Extraction Procedure

2.1. Parameter Extraction Test Design

It can be seen from Figure 1 that the second order ECM contains one *OCV-SoC* relationship and five impedance parameters (R_{in}, R_{short}, C_{short}, R_{long} and C_{long}), which need to be estimated. Theoretically, all of the impedance parameters mentioned above should be multivariable functions of *SoC*, the C-rate of the load current (C is the amplitude of the current with which the battery can be fully discharged in 1 h), temperature and cycle numbers [39,45]. These functions not only make the parameter extraction process complex and time consuming, but also increase the computational burden of the BMS. Hence, within certain error tolerance, some relationships can be simplified or ignored. Usually, aging periods are generally in the range of months to years. While for the system-level simulations of automotive applications, the time periods of interest are typically in the range of seconds to hours or days in special cases [43,45]. Hence, the long-term aging effect is usually ignored in the parameter estimation process and handled separately in most cases [39,46].

In this paper, all of the model parameters are estimated through the discharging/charging pulse-rest test at room temperature (22 °C–25 °C). A lithium-ion polymer battery with nickel-manganese-cobalt-based cathode and graphite-based anode is under test. Its specifications are given in Table 1, and the detailed experimental steps are described as follows.

Table 1. Specification of the tested battery.

Charge Capacity	40.99 Ah
Discharge capacity	40.89 Ah
Nominal voltage	3.7 V
Charge cutoff voltage	4.2 V
Discharge cutoff voltage	2.7 V

The discharging pulse-rest test starts with a fully-charged battery. In each cycle of the test, the battery is discharged at a 2% *SoC* step with C/2 constant current, then followed by a rest period.

This cycle is repeated until the battery is fully discharged. Data points (including current, voltage, charging capacity and discharging capacity) are collected with the sampling frequency of 1 Hz. The relevant voltage and current profiles of the discharging pulse-rest test during the 66%–64% *SoC* interval are plotted in the bottom subfigure of Figure 1. The charging pulse-rest test is conducted similarly, that is it begins with a fully-discharged battery, then charged at a 2% *SoC* step with C/2 constant current and followed by a rest period. In order to eliminate the polarization voltage, the *OCV* values are extracted at the end of each rest period. Too short a rest time leads to a large *OCV* estimation error, whereas too long a rest time makes the whole test time consuming. It has been shown previously that for the lithium-ion polymer batteries, electrochemical reactions are negligible after a 2-h rest period [47,48]. Therefore, the rest time in this paper is predetermined as 2 h.

2.2. Parameter Estimation Algorithm

The electrical behavior of the ECM is expressed as the following state space formalism:

$$\begin{bmatrix} dV_{RC,short}/dt \\ dV_{RC,long}/dt \end{bmatrix} = \begin{bmatrix} -1/R_{short}C_{short} & 0 \\ 0 & -1/R_{long}C_{long} \end{bmatrix}\begin{bmatrix} V_{RC,short} \\ V_{RC,long} \end{bmatrix} + \begin{bmatrix} 1/C_{short} \\ 1/C_{long} \end{bmatrix}I \quad (1)$$

$$V_t = OCV(SoC) + IR_{in} + V_{RC,short} + V_{RC,long} \quad (2)$$

where Equation (1) is the state equation and Equation (2) is the output equation, $V_{RC,short}$ and $V_{RC,long}$ represent the voltages across the short-term and the long-term RC networks, respectively, $OCV(SoC)$ is an eighth-order polynomial equation as a function of *SoC*, V_t is the battery terminal voltage and the positive current I represents charging. R_{in} represents the internal resistance; R_{short} and R_{long} denote the diffusion resistances; and C_{short} and C_{long} represent the diffusion capacitances. Among them, R_{in} can be directly obtained from each pulse-rest cycle through Equation (3); the corresponding four variables (V_1, V_2, I_1 and I_2) are marked in the bottom subfigure of Figure 1, and the variation of identified R_{in} with *SoC* is shown in Figure 2. *SoC* can be calculated through Equation (4), in which C_{ap} denotes the capacity of the battery in Ah.

$$R_{in} = \frac{V_2 - V_1}{I_2 - I_1} \quad (3)$$

$$SoC = SoC(0) + \frac{1}{3600C_{ap}}\int_0^t I(\tau)d\tau \quad (4)$$

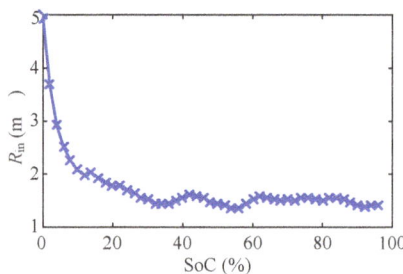

Figure 2. R_{in} variation with different state of charge (*SoC*).

For the CC operating scenario ($I \neq 0$), the analytical solutions of Equation (1) are derived as:

$$\begin{cases} V_{RC,short}(t) = V_{RC,short}(0)e^{-\frac{t}{\tau_{short}}} + IR_{short}(1 - e^{-\frac{t}{\tau_{short}}}) \\ V_{RC,long}(t) = V_{RC,long}(0)e^{-\frac{t}{\tau_{long}}} + IR_{long}(1 - e^{-\frac{t}{\tau_{long}}}) \end{cases} \quad (5)$$

where $V_{RC,short}(0)$ and $V_{RC,long}(0)$ are the initial voltages of corresponding RC networks and $\tau_{short} = R_{short}C_{short}$, $\tau_{long} = R_{long}C_{long}$, which represent the short-term and the long-term time constants, respectively.

Substituting Equation (5) into Equation (2), the output equation is rewritten as:

$$V_t(t) = OCV(SoC) + IR_{in} + V_{RC,short}(0)e^{-\frac{t}{\tau_{short}}} + V_{RC,short}(0)e^{-\frac{t}{\tau_{long}}} + IR_{short}(1 - e^{-\frac{t}{\tau_{short}}}) + IR_{long}(1 - e^{-\frac{t}{\tau_{long}}}) \quad (6)$$

During the rest period, where there is no current excitation ($I = 0$), Equation (6) can be simplified to:

$$V_t(t) = OCV(SoC) + V_{RC,short}(0)e^{-\frac{t}{\tau_{short}}} + V_{RC,long}(0)e^{-\frac{t}{\tau_{long}}} \quad (7)$$

With the knowledge of R_{in} and charging/discharging OCV-SoC relationships, RC network parameters (R_{short}, C_{short}, R_{long} and C_{long}) can be obtained through fitting the experimental data with relevant exponential functions, as

$$\begin{cases} y = IR_{short}(1 - e^{-\frac{t}{\tau_{short}}}) + IR_{long}(1 - e^{-\frac{t}{\tau_{long}}}) & I \neq 0 \\ y = V_{RC,short}(0)e^{-\frac{t}{\tau_{short}}} + V_{RC,long}(0)e^{-\frac{t}{\tau_{long}}} & I = 0 \end{cases} \quad (8)$$

where $y = V_t - OCV(SoC) - IR_{in}$. Since there only exists 2% SoC variation during each pulse-charging/discharging period, it is reasonable to make an assumption that the RC network parameters keep constant during this period. In addition, considering that the battery has converged to the steady state after a 2-h rest, $V_{RC,short}(0)$ and $V_{RC,long}(0)$ are set as zero at the beginning of the pulse-charging/discharging period.

Based on the above analysis, the RC network parameters can be estimated through fitting the experimental dataset with Equation (8). The cost function of the curve fitting method J is to minimize the sum of squared errors between the estimation results and the measured data, subjected to the following constraints:

$$\begin{cases} J = \min_{r,\tau} \sum_{k=1}^{n} [V_t^m(t_k) - V_t^e(r, \tau, t_k)]^2 \\ s.t.\ R_{short}, \tau_{short}, R_{long}, \tau_{long} > 0 \end{cases} \quad (9)$$

where t_k is the input time sequence, n is the length of the fitted experimental dataset, $r = [R_{short}, R_{long}]$, $\tau = [\tau_{short}, \tau_{long}]$, V_t^e is the model estimated voltage and V_t^m is the voltage measurements from the pulse-rest test.

3. RC Network Parameters Estimation

Based on the Introduction in Section 1, the RC network parameters show diverse values under different operating scenarios. In HEV/EV applications, batteries usually work in two typical scenarios: the CC charging scenario and the dynamic driving scenario. In the CC charging scenario, continuous external charging currents are applied to the batteries, and the transport of ions is mainly driven by the electric field. While for the dynamical driving scenario, especially for the urban driving condition, the load current has the characteristics of discontinuous amplitude values and a wide-spread frequency spectrum. In this case, besides the electric field, the gradient in concentration is also largely responsible for the transport of ions within batteries [45]. Therefore, the RC network parameters employed in different operating scenarios should be identified through different identification approaches.

3.1. RC Network Parameters for the CC Charging Scenario

The polarization voltage (V_P) is adopted to illustrate the variation of RC network parameters under the CC excitation. According to the aforementioned battery output equation, V_P can be obtained as:

$$V_P = V_{RC,short} + V_{RC,long} = V_t - OCV(SoC) - IR_{in} \quad (10)$$

The V_P-*SoC* profile during the C/2 rate CC charging process is shown in Figure 3. Since in the HEV/EV application, batteries seldom work in the extremely low or high *SoCs*, the voltage profile from 10%–90% *SoC* is covered. It can be observed from Figure 3 that the polarization voltage increases dramatically in Stage I (10%–18% *SoC*), then it declines slowly and shows a concave shape curve in Stage II, with the local minimum value at around 30% *SoC*. During Stage III (40%–70% *SoC*), the polarization voltage becomes relatively stable. After that (70%–90% *SoC*), the polarization voltage rises sharply.

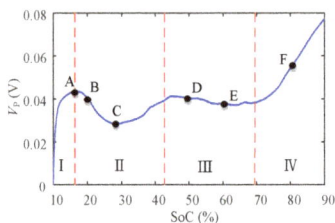

Figure 3. V_P versus *SoC* under constant-current (CC) charging.

The variation of the polarization voltage during the above *SoC* range is closely related to the internal electrochemical reaction process during charging. In the initial *SoC* region, a relatively large amount of energy is needed to form the nucleation on the surfaces of the electrodes; thus, the polarization voltage increases quickly. Once the nuclei are formed, the following lithium ions' removal process needs less energy. This explains the concave shape voltage curve occurring from 18% *SoC* to 40% *SoC*. While in the last charging stage, the lithium-ion concentration increases in the negative materials. Hence, a large amount of energy is needed to insert the lithium ions, which leads to the obvious growth of the polarization voltage in the high *SoC* region. The detailed explanation for the electrochemical reaction mechanism occurring during the CC charging process can be found in [28,32].

As mentioned in Section 2, the model parameters are estimated through fitting the measured data either from the pulse-charging period or the rest period. In order to select the proper experimental datasets that can better describe the charging characteristic of the battery, the profiles of the polarization voltage during the pulse-charging and the following rest periods, which are also calculated from Equation (10), are compared in Figure 4. Figure 4a shows the polarization voltage under the pulse-charging excitation, and Figure 4b plots the absolute values of the polarization voltage during the following rest. It can be seen from both figures that the shape of the polarization voltage curve strongly depends on the *SoC*. In Figure 4a, it is obvious that the final value of the polarization voltage obtained from 26%–28% *SoC* is the lowest, which is similar to point C in Figure 3. In addition, the final values of the voltage curves obtained from 18%–20% *SoC* and 50%–52% *SoC* are almost coincident with each other, which approximately matches the corresponding parts (point B and point D) in Figure 3. Meanwhile, the relations among the final voltage values collected from 14%–16% *SoC*, 60%–62% *SoC* and 80%–82% *SoC* are also identical to the relations among point A, point E and point F in Figure 3, respectively. Hence, it can be summarized from Figure 4a that the final values of the polarization voltage obtained from different pulse-charging periods are approximately consistent with the corresponding points in Figure 3. While in Figure 4b, the variation trend of the predicted stable voltage values differs greatly compared to the results in Figure 4a. This is because in the pulse-charging period, the ion migration is driven by external electric potential. While in the rest period, the transport of ions is mainly dominated by diffusion, owing to the concentration gradient. The detailed explanation of the electrochemical reactions occurring under different load current has been discussed in [21,45].

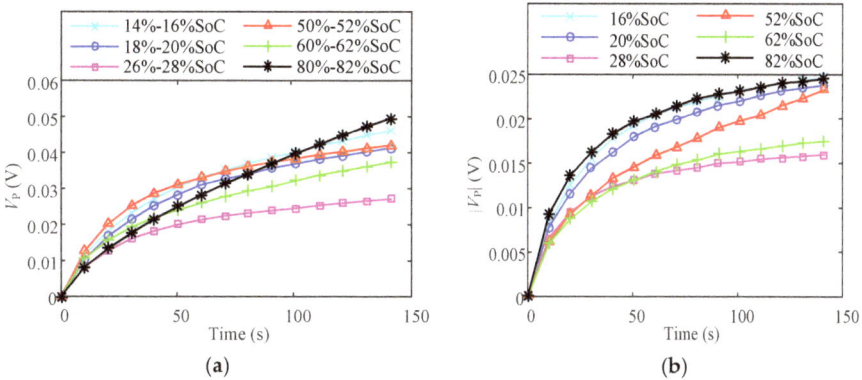

Figure 4. (a) The profiles of V_P at different *SoC* intervals during the pulse-charging period; (b) the profiles of $|V_P|$ at different *SoC* points during the rest period.

Consequently, it can be concluded that the voltage response during the pulse-charging period can better describe the characteristic of the CC charging process because of the similar current excitation.

3.2. RC Network Parameters for the Dynamic Driving Scenario

3.2.1. Typical Dynamic Driving Scenarios

For the dynamic driving scenario, especially for the urban driving scenario, vehicles accelerate and brake frequently, which cause the long lasting load current to seldom exist. There are two typical kinds of standard urban driving cycles, namely the urban dynamometer driving schedule (UDDS) and the worldwide harmonized light vehicles test procedure (WLTP), which are the American and European certification cycles, respectively. The load current profiles and the load current amplitude distributions of the two driving cycles are plotted in Figure 5. It can be observed from Figure 5a,b that both of the dynamic current profiles vary frequently over the test span. Meanwhile, from Figure 5c,d, it can be concluded that: (1) the discharging current accounts for a much larger portion, compared to the charging current during the regenerative process; (2) among the load currents, the low C-rate discharging current, particularly around zero-value amplitudes, accounts for a larger portion in both tests. Hence, the voltage response during the rest period can be employed to estimate the RC network parameters for the dynamic driving scenario.

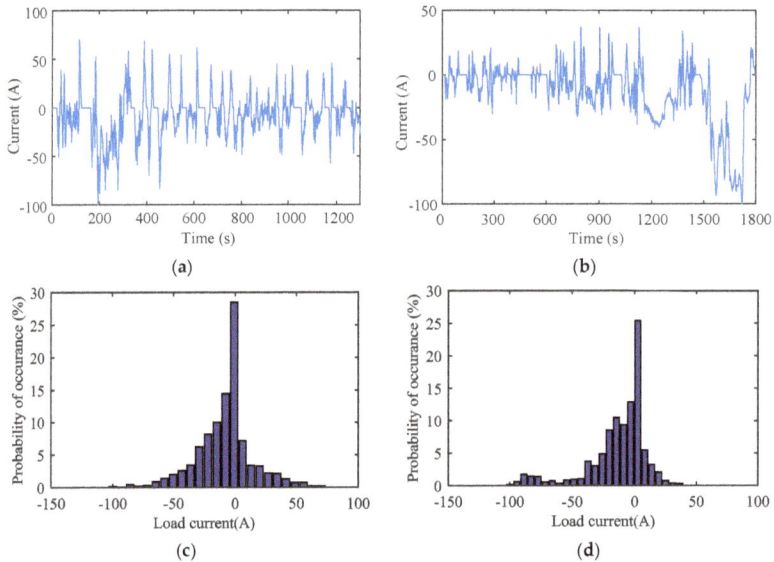

Figure 5. (**a**) The load current profile of the urban dynamometer driving schedule (UDDS) test; (**b**) the load current profile of the worldwide harmonized light vehicles test procedure (WLTP) test; (**c**) the load current amplitude distribution of the UDDS test; (**d**) the load current amplitude distribution of the WLTP test.

3.2.2. Determination of the Length of the Fitted Experimental Dataset

The diffusion process, which is caused by the gradient in concentration, plays a major role in the low C-rate load current and rest cases. Since the electrochemical reactions occurring during the diffusion process are very complex, these reactions can be accurately modeled as infinite series-connected RC networks with a wide range of time constants ($\tau_1, \tau_2, \ldots, \tau_j$). Usually, the values of time constants depend on the electrode thickness and the structure of the battery to a great extent, and typical time constants are in the range of seconds to minutes [45]. The second order RC network can only approximate the diffusion process by two parts: the fast dynamics part (the short-term RC network with τ_{short}) and the slow dynamics part (the long-term RC network with τ_{long}).

In general, the values of the two time constants are closely related to the length of the fitted experimental data Δt. When only the initial segment of the voltage response is employed in parameter estimation, such as Part A in the bottom subfigure of Figure 1, the voltages across the shorter-term RC networks have a larger degree of variability, which means that the shorter-term RC networks have a greater impact on the initial segment of the voltage response. This in turn leads to the smaller estimated time constants and subsequently ignores the slower dynamics diffusion process. On the contrary, after the initial phase of the rest period, such as Part B in the bottom subfigure of Figure 1, the voltages across the shorter-term RC networks have converged to zero; thus, the voltage variation caused by the shorter-term RC networks is negligible. Instead, the voltages across the longer term RC networks make a remarkable contribution to the total voltage response. Subsequently, it can be inferred that the measured data show a slower varying characteristic, which represent the slower dynamics diffusion process and can be modeled by the RC networks with larger time constants. Hence, if the whole voltage response of the long time rest period is adopted, data with slower varying values will account for a large portion, which will lead to the relatively larger estimated time constants. However, too large time constants will make the model output voltage severely lag behind the actual response and result in a poor dynamic performance.

In order to further illustrate the above analysis, a third order RC network circuit is simulated in MATLAB; two equivalent time constants (τ'_{short} and τ'_{long}) are estimated from the different value of Δt. In the simulation, the resistances of the three RC networks are all set as 1 mΩ, and the time constants are predetermined as $\tau_1 = 40$ s, $\tau_2 = 200$ s and $\tau_3 = 2000$ s ($\tau_3 \gg \tau_2 > \tau_1$). The applied excitation consists of a 400-s pulse-discharging current and a 2-h rest period, and the amplitude of the current is 20 A. Time constants estimated by different lengths of the voltage response are given in Table 2. It can be clearly seen from Table 2 that both τ'_{short} and τ'_{long} decrease simultaneously with the reduced value of Δt, which is consistent with the previous analysis. Hence, to obtain the appropriate values of the time constants, Δt should be predetermined properly, which is illustrated in detail as follows.

Table 2. Equivalent time constant estimation results with different values of Δt.

Δt (s)	7200	3600	1800	1400	1200	1000	900	850	800
τ'_{short} (s)	88.67	67.18	48.53	45.10	43.74	42.59	42.08	41.83	41.63
τ'_{long} (s)	971.0	484.3	284.4	256.7	245.3	235.3	230.9	228.8	226.8
k [1]	4.049×10^{-12}	4.395×10^{-5}	0.1448	0.8759	2.154	5.299	8.311	10.41	13.03

[1] k represents the degree of resistor-capacitor (RC) voltage variability; the detailed expression can referred to in Equation (13).

During Δt, the derivative of Equation (13) with respect to τ_i during the rest period is expressed as:

$$\left| \frac{dV_{RC,i}}{d\tau_i} \right| = \frac{\Delta t |V_{RC,i}(0)|}{\tau_i^2} e^{-\frac{\Delta t}{\tau_i}} \tag{11}$$

where $V_{RC,i}$ is the voltage across the i-th RC network, $i \in \{1,2,3,\ldots,j\}$, $V_{RC,i}(0)$ is the corresponding initial voltage, R_i is the resistance of the i-th RC network and τ_i is the time constant of the i-th RC network, which is subject to $\tau_1 < \tau_2 < \ldots < \tau_j$.

After the pulse-discharging period, $|V_{RC,i}(0)|$ can be expressed as:

$$|V_{RC,i}(0)| = |I|R_i(1 - e^{-\frac{D}{\tau_i}}) \tag{12}$$

where D denotes the length of the pulse-discharging period.

For the two well-separated time constants τ_i and τ_{i+m} ($\tau_{i+m} \geq 10\tau_i$ and $0 < m < j - i$), the voltage across the shorter term RC network $V_{RC,i}$ has a larger degree of variability when satisfying the following requirement:

$$\frac{|dV_{RC,i}/d\tau_i|}{|dV_{RC,i+m}/d\tau_{i+m}|} = k \tag{13}$$

where the constant k denotes the degree of variability, and it is subject to $k > 1$.

Substituting Equations (11) and (12) into Equation (13), the value of Δt can be derived as:

$$\Delta t = \ln \left[\frac{R_i(1 - e^{-\frac{D}{\tau_i}})\tau_{i+m}^2}{kR_{i+m}(1 - e^{-\frac{D}{\tau_{i+m}}})\tau_i^2} \right] \frac{\tau_i \tau_{i+m}}{\tau_{i+m} - \tau_i} \tag{14}$$

In Equation (14), since the values of R_i and R_{i+m} are nearly of the same order of magnitude [39,43,46], the value of R_i/R_{i+m} can be neglected when compared to the value of τ_{i+m}^2/τ_i^2; thus, Δt can be simplified as:

$$\Delta t = \ln \left[\frac{(1 - e^{-\frac{D}{\tau_i}})\tau_{i+m}^2}{k(1 - e^{-\frac{D}{\tau_{i+m}}})\tau_i^2} \right] \frac{\tau_i \tau_{i+m}}{\tau_{i+m} - \tau_i} \tag{15}$$

Equation (15) shows that k and τ_i should be determined before calculating Δt. In the aforementioned simulation, the value of k for τ_2 and τ_3 can be obtained directly from Equation (13), as shown in Table 2. This indicates that when k is larger than one, the estimated τ'_{short} and τ'_{long} are closer to τ_1 and τ_2. This is because the voltage across the RC network with τ_3 has a lower degree of variability, compared to those with τ_1 and τ_2. It can be observed from Table 2 that τ'_{short} and τ'_{long} are nearly stable when k is larger than 10. Hence, k is selected as 10 throughout the paper.

In order to set a proper τ_i in Equation (15), the discrete Fourier analysis of the load current is employed to determine the lower bandwidth limitation of the ECM. The current spectrums of UDDS and WLTP tests are shown in Figure 6. It can be observed in Figure 6a,b that there exists a large DC component (Points A and C) due to the nonzero mean value of the two current profiles. Since the characteristics of the DC component cannot be modeled by the RC circuit, they are neglected when determining the length of the fitted dataset. The major low frequency components for the two profiles are around 0.00146 Hz (point B) and 0.00138 Hz (the mean value from point D to point E), respectively. Hence, the mean value of the long-term time constant is selected as 704 s. In order to exclude the voltage variation caused by the larger time constants (larger than $10\tau_i$), the prior 1-h measured battery voltage dataset is employed to estimate the RC parameters.

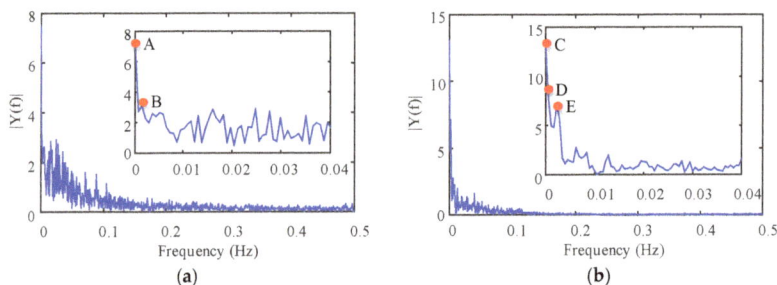

Figure 6. The spectral analysis of the load current: (**a**) the urban dynamometer driving schedule (UDDS) test; (**b**) the worldwide harmonized light vehicles test procedure (WLTP) test.

3.2.3. Improved Fitting Function

From Equations (6) and (7), it can be observed that only the initial values $V_{RC,short}(0)$, $V_{RC,long}(0)$ and time constants τ_{short}, τ_{long} can be obtained directly from the fitting results; thus, we should do the further computations to obtain the resistances and capacitances of RC networks.

In [37,39–41], two initial voltages across the RC networks are predetermined as IR_{short} and IR_{long} respectively, from which the resistances of the RC networks can be derived under the knowledge of the current value. In [49], the capacitances of the RC networks are firstly obtained from the initial voltage values. Both of the above two methods have an assumption that the capacitors of the RC networks have already converged to the steady state at the end of the pulse-discharging period.

Usually, in the parameter extraction test, in order to obtain as much data as possible at different *SoC* intervals, the length of the pulse-charging/discharging period is usually set as several minutes (resulting in 2% *SoC* variation in this paper), while the rest time is usually set as one or more hours (such as 2 h in this paper) to get an accurate *OCV* value. For the short-term RC network, the voltage can easily converge to the equilibrium state during the pulse-discharging process, which is shown in Figure 7. In other words, there is no current flowing through the capacitor branch of the short-term RC network during the last stage of the pulse-discharging period; thus, $V_{RC,short}(0)$ at the beginning of the rest period can be expressed as:

$$V_{RC,short}(0) = IR_{short} \tag{16}$$

However, for the long-term RC network, the voltage varies continuously due to a relatively large time constant, as illustrated in Figure 7. The voltage across the long-term RC network has not reached the equilibrium state at the end of the pulse-discharging period; thus, there always exists a significant proportion of the load current $I(1 - e^{-D/\tau_{long}})$ flowing through the corresponding capacitor. Consequently, $V_{RC,long}(0)$ at the beginning of the rest period should be written as:

$$V_{RC,long}(0) = IR_{long}\left(1 - e^{-\frac{D}{\tau_{long}}}\right) \qquad (17)$$

where I is the value of the pulse-discharging current. Since the *SoC* variation in each test cycle is set as 2% in this paper, it can be assumed that the model parameters keep constant during the pulse-discharging period.

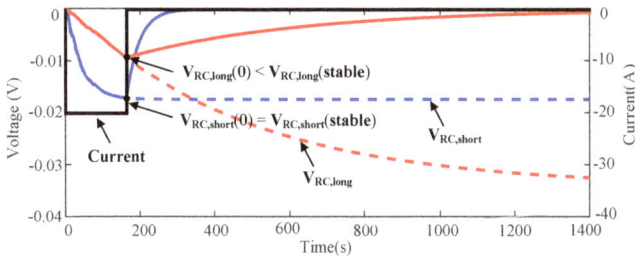

Figure 7. The voltage curve of RC networks during one cycle of the discharging pulse-rest test.

4. Experimental Results and Discussions

4.1. RC Network Parameter Estimation Results

Based on the aforementioned analysis in Section 3.1, for the case of the CC charging scenario, the charging pulse-rest test is implemented firstly. The parameters are estimated from the voltage response of the pulse-charging period, and the estimation results are shown in Figure 8. Figure 8a plots two estimated time constants; it can be seen that the general order of the magnitude of the short-term time constant is 10 s; it fluctuates greatly when the *SoC* changes, especially in the middle *SoC* region, while the order of the magnitude of the long-term time constant is 100 s; it is relatively flat during the whole *SoC* region. Figure 8b plots two estimated resistances; it can be observed that in the middle *SoC* range, the short-term resistance has a larger value, which means that the voltage across the short-term RC network accounts for more weight during this period. Hence, it can be observed from Figures 3 and 8b that the variation tendencies of the polarization voltage and the short-term resistance are similar during the middle *SoC* range. At the end of the charging process, the short-term resistance decreases and stabilizes around a very small value, while the long-term resistance increases almost linearly after 60% *SoC*, leading to a similar variation tendency of the polarization voltage, compared to the corresponding part in Figure 3. Hence, it can be concluded that the long-term diffusion process plays a major role in this stage.

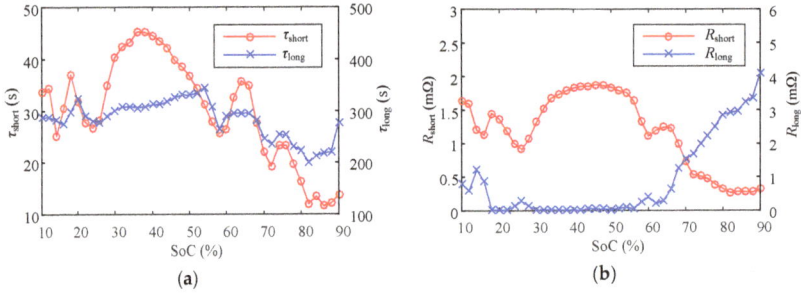

Figure 8. Parameter estimation results for the constant-current (CC) charging scenario: (**a**) time constant; (**b**) resistance.

For the case of the dynamic driving scenario, the discharging pulse-rest test is implemented, and the data from the rest periods are adopted in the parameter estimation. According to the analysis in Section 3.2.2, different time constants will be obtained from the fitted experimental datasets in different lengths. Firstly, in order to compare the best fit performances for the measured datasets in different lengths, the measured battery terminal voltage response at 60% *SoC* during a 2-h rest period is adopted, and the curve fitting results are shown in Figure 9. It can be observed from Figure 9a that the fitting result of the whole measured voltage response shows a better performance during most of the rest period, especially in the equilibrium state. Whereas for the performance of the first 200 s, the fitting result through the prior 0.5-h measured voltage response yields less errors, which is illustrated in Figure 9b. Parameter estimation results in Figure 10 show the time constants estimated from the measured voltage dataset in different lengths, ranging from 30 min–2 h with a 30-min interval. It can be observed that the time constants, both for the long term and the short term, increase simultaneously when the length of the fitted dataset increases. In addition, by comparing Figure 10 with Figure 8a, it can be concluded that the time constants applied in the CC charging scenario and the dynamic driving scenario show different variation tendencies. Hence, it is essential to adopt different sets of model parameters for different operating scenarios.

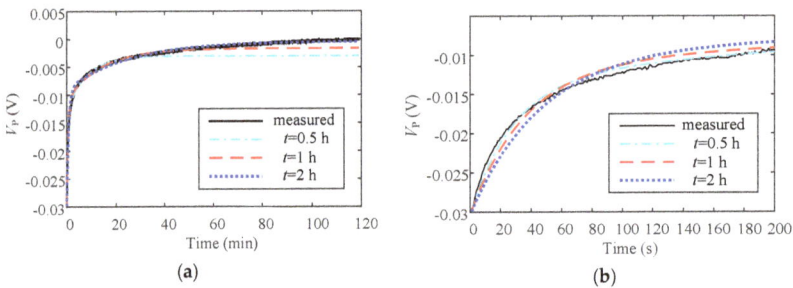

Figure 9. Curve fitting results of V_P during the rest period of the discharging pulse-rest test at 60% *SoC*: (**a**) the overall result; (**b**) a close look at the transient part at the beginning.

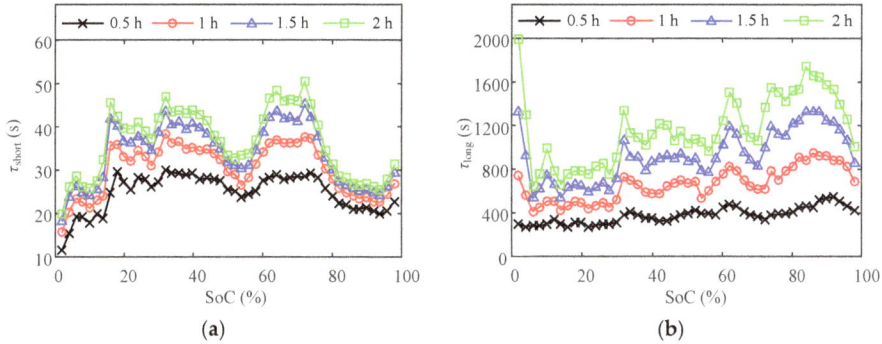

Figure 10. Time constant estimation results with different lengths of the experimental dataset: (**a**) τ_{short}; (**b**) τ_{long}.

After determining the length of the fitted experimental dataset, we can subsequently obtain the resistances. Figure 11 shows the R_{long} estimation results by the conventional fitting function and the improved fitting function. It can be concluded from Figure 11 that the R_{long} estimated by the conventional fitting function is generally less than the one estimated by the improved fitting function, because it neglects the $(1 - e^{-D/\tau_{long}})$ part. In order to demonstrate the advantage of the improved fitting function, data from the 20th cycle of the discharging pulse-rest test are adopted. In this cycle, *SoC* changes from 62% to 60% during the pulse-discharging period, then keeps the value of 60% during the following rest period. The current profile of the 20th discharging pulse-rest test is applied on the ECM MATLAB/SIMULINK model as an excitation. Figure 12a,b shows the model output voltage responses with two sets of estimated model parameters. It can be seen that the model with parameters estimated by the proposed fitting function outputs better estimation results. The lower voltage error is mainly contributed by the higher voltage drop across the long-term RC network, as plotted in Figure 12c. In addition, the root mean square errors (RMSEs) between the measured voltage and the model output voltage at different *SoCs* are given in Table 3. It can also be seen that the model parameters estimated by the proposed fitting function show a better performance for a wide range of *SoC*.

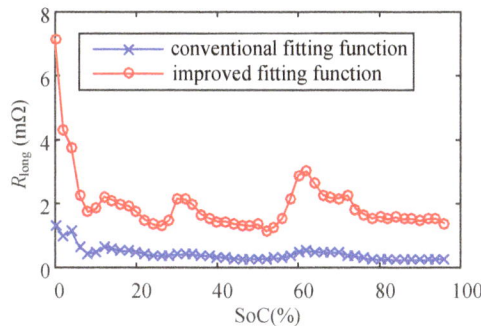

Figure 11. R_{long} estimation results.

Table 3. Comparison of RMSE at different *SoC*.

SoC (%)		10	20	30	40	50	60	70	80	90
RMSE (mV)	Conventional fitting function	1.802	1.714	2.167	1.540	1.268	2.803	2.416	1.558	1.444
	Improved fitting function	0.7658	0.7582	0.9707	0.7643	0.5000	1.202	1.242	0.7104	0.6482

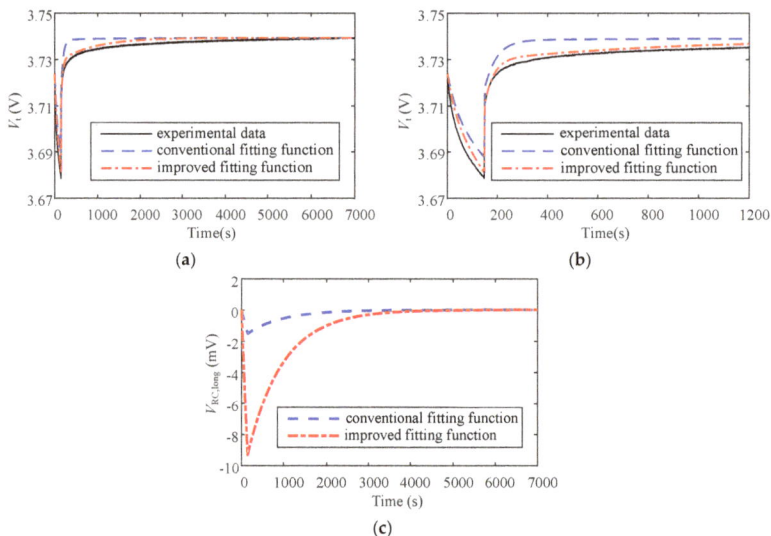

Figure 12. Voltage curves of one cycle of the discharging pulse-rest test (62%–60%): (**a**) the overview; (**b**) a close look; (**c**) the voltage across the long-term RC network.

4.2. Model Verification

In this paper, the CC charging test and the consecutive UDDS test, which respectively represent two typical operating scenarios in HEV/EV applications, are conducted separately to verify the effectiveness of the model. For the charging condition, the battery is charged from 10%–90% SoC. The typical charging current in practice varies from C/8 to 2C [50], and a C/2 rate current is employed in the charging test. The consecutive UDDS test starts from 90% SoC to 20% SoC, with a 10-min rest period in between to simulate a short parking time. In the real application, a specific set of parameters can be selected by the characteristics of the measured load current. For example, if the values of the current are approximately constant over a certain time interval, parameters estimated from the data in the pulse-charging periods are employed. On the other hand, parameters estimated from the data in the rest periods are employed when the load current shows the characteristics of high dynamics over a certain time interval.

Firstly, for the CC charging scenario, three model outputs and measured battery terminal voltage curves are plotted in Figure 13, and the corresponding RMSEs are given in Table 4. It can be observed that during the whole charging process, the model with parameters estimated from the data in pulse-charging periods outputs a voltage curve matching the measured curve better because of considering the continuous external electric driving forces. However, parameters estimated from the data in the rest periods result in relatively larger errors, especially in the high SoC region. In addition, during most part of the charging period, the model with parameters used in the dynamic driving scenarios outputs a voltage higher than the experimental voltage. Comparing the corresponding curves in Figures 8b and 11, it can be deduced that the higher estimated voltage is mainly caused by the larger value of estimated R_{long}, especially during the middle range of the SoC region.

Table 4. RMSE of model voltage estimation under the CC charging test.

Modeling Methods	Dynamic Condition	Rest-Period	Pulse-Period
RMSE (mV)	18.41	19.76	5.448

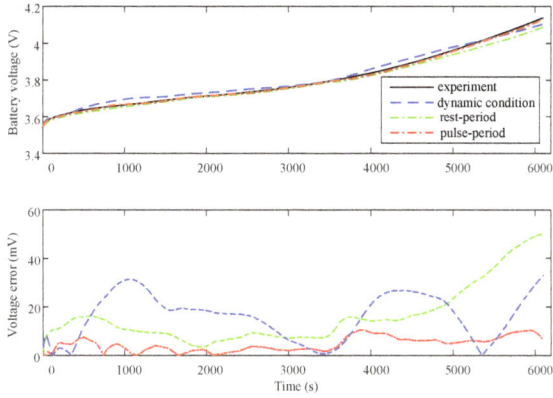

Figure 13. Verification results of different parameter estimation methods under the CC charging test.

In order to verify the robustness of the proposed parameter estimation method, the CC charging voltage profiles at different initial *SoC* are plotted in Figure 14. This shows that the estimated voltage curves match well with the measurement voltage curves, despite the different initial *SoC*.

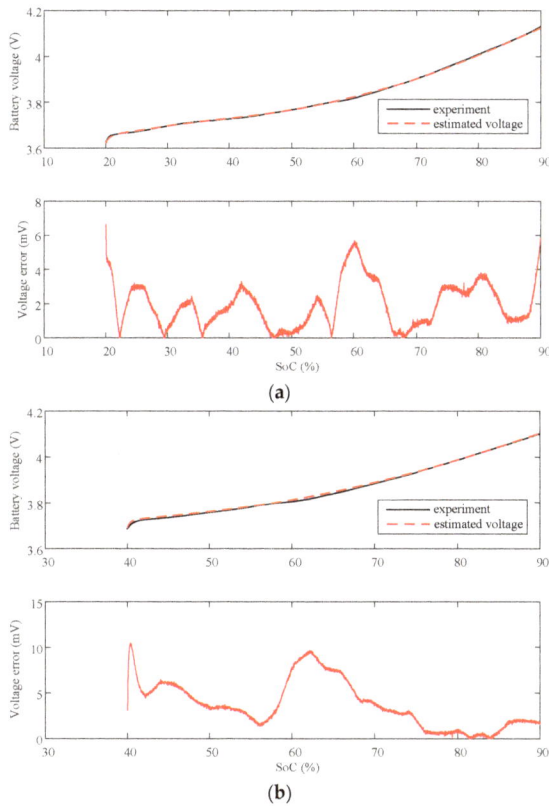

(a)

(b)

Figure 14. CC charging voltage profiles at different initial *SoC*: (**a**) initial *SoC* = 20%; (**b**) initial *SoC* = 40%.

Secondly, in order to demonstrate the improvement of the proposed battery modelling approach during the dynamic driving scenario, the model and experimental voltage outputs in the consecutive UDDS validation are plotted in Figure 15a, the corresponding calculated *SoC* profile is shown in Figure 15b, and the detailed figure from 10,000 s to 12,000 s is plotted in Figure 15c. The RMSE of the aforementioned estimation methods during the whole consecutive UDDS test are also shown in Table 5. Figure 15b shows that the consecutive UDDS test is started from 90% *SoC*, and terminated when the value of *SoC* drops below 20%. It can be observed from Figure 15c that parameters estimated by the improved fitting function generally demonstrate a better performance, especially during the dynamic period (ranging from 10,000 s to 11,400 s), because considering the unsaturated phenomenon of the long-term RC network. It can also be concluded that the model containing parameters estimated by the prior 1-h experimental data from the rest period gives voltage output with the least error, especially during the short-time rest period. In addition, it can be seen from Figure 5a that there exists a relatively long-time and high C-rate discharging current in the UDDS cycle approximately ranging from 150 s to 300 s. Since larger time constants are obtained from the data of the whole rest period, this causes the corresponding voltage output not to recover fast after a relatively long-time discharging current, which leads to an offset of voltage errors in comparison to the voltage error caused by the proposed approach.

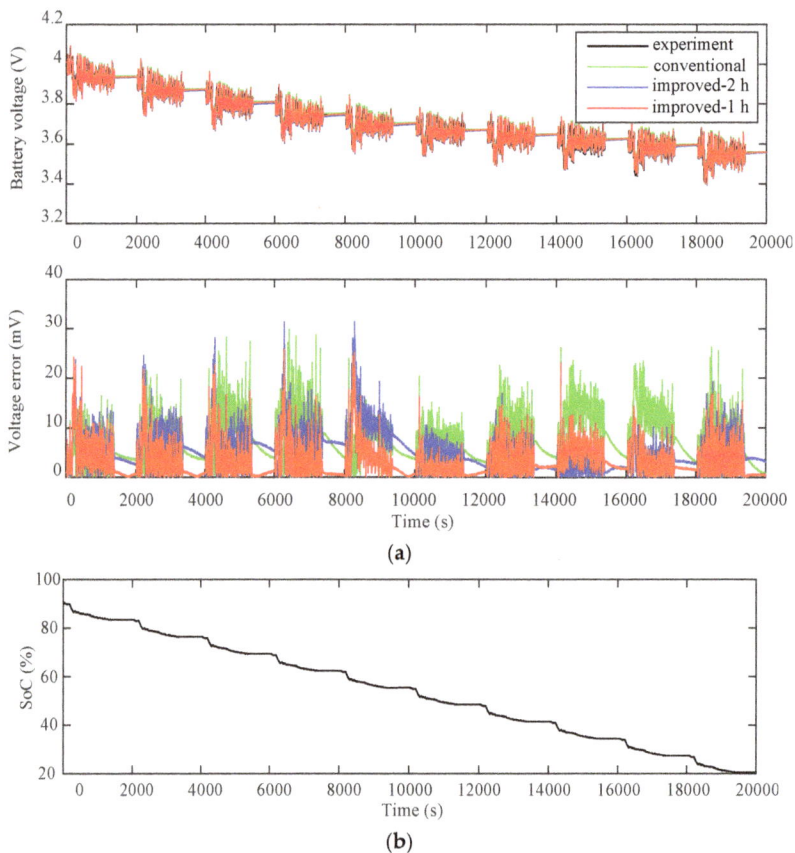

Figure 15. *Cont.*

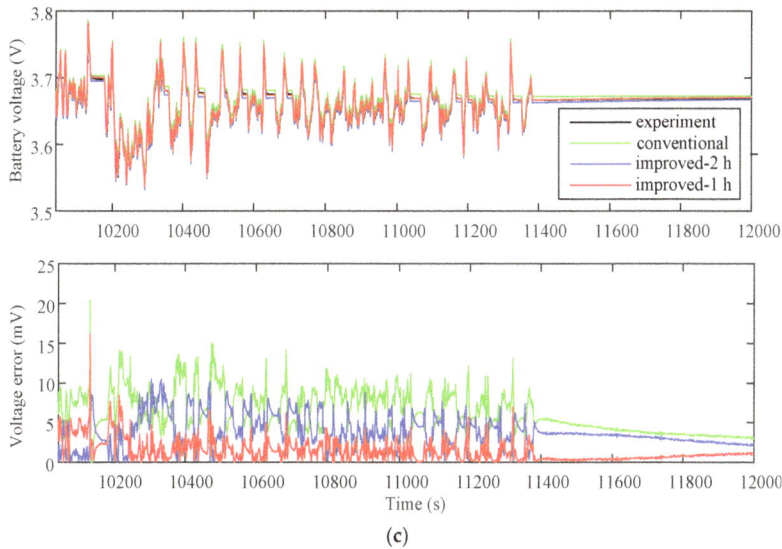

Figure 15. Verification results of different parameter estimation methods under the UDDS tests: (**a**) The overall look; (**b**) The calculated *SoC* profile (**c**) The close look.

Table 5. RMSE of the model voltage estimation under the urban dynamometer driving schedule (UDDS) test.

Modeling Methods	Conventional	Improved-2 h	Improved-1 h
RMSE (mV)	8.504	6.329	4.244

5. Conclusions

In this paper, an advanced battery parameter estimation method based on two general operating scenarios in HEV/EV applications is proposed. Firstly, the second order ECM is employed, and the model parameter extraction process is described in detail. Considering the typical operating scenarios in HEV/EV applications, namely the CC charging scenario and the dynamic driving scenario, two sets of model parameters are extracted from the charging/discharging pulse-rest tests. Specifically, voltage responses of the pulse-charging phases are selected to estimate model parameters applied in the CC charging scenario. For the dynamic driving scenario, the model parameters are identified through the measured data from the rest period. Instead of employing the data from the whole rest period, only the prior portion of the collected data is selected, and the length of the fitted data is determined by the frequency spectrum analysis of the load current under two typical urban driving conditions. In addition, an unsaturated phenomenon caused by the long-term RC network is analyzed in detail, and subsequently, an improved fitting equation with more accurate initial voltage expression of the RC network is adopted. Finally, verification tests simulating the CC charging scenario and the dynamic driving scenario are conducted, respectively, and comparisons between the conventional and the proposed battery parameter estimation methods are given. Experimental results show that in both cases, the voltage profiles predicted from the proposed model show a better conformity to the experimental data.

It is important to note that the proposed battery parameter estimation method for the dynamic driving scenario only considers the typical urban driving conditions at room temperature. However, the characteristics of the load current under the other special conditions (such as the highway driving condition and the extremely cold condition) will be obviously different. For the future work, the influence caused by different C-rates of the current profiles, bandwidths of the current profiles and temperature effects will be considered, and the parameter extraction test will be modified accordingly.

Acknowledgments: The authors would like to acknowledge the funding support from the China Scholarship Council (CSC); the U.S. DOE Graduate Automotive Technology Education (GATE) Center of Excellence; and Nanjing Golden Dragon Bus Co., Ltd.

Author Contributions: Jufeng Yang handled the technical modeling, drafted and revised the manuscript. Bing Xia revised the manuscript. Jufeng Yang and Bing Xia designed the experiments and analyzed the data. Yunlong Shang participated in the experiment. Wenxin Huang revised the manuscript. Chris Mi contributed the experiment platform, gave great suggestions and polished the manuscript.

Conflicts of Interest: The authors declare no conflict of interest.

References

1. Xiong, R.; Sun, F.; Chen, Z.; He, H. A data-driven multi-scale extended kalman filtering based parameter and state estimation approach of lithium-ion olymer battery in electric vehicles. *Appl. Energy* **2014**, *113*, 463–476. [CrossRef]
2. Zou, Z.; Xu, J.; Mi, C.; Cao, B.; Chen, Z. Evaluation of model based state of charge estimation methods for lithium-ion batteries. *Energies* **2014**, *7*, 5065–5082. [CrossRef]
3. Shang, Y.; Zhang, C.; Cui, N.; Guerrero, J.M. A cell-to-cell battery equalizer with zero-current switching and zero-voltage gap based on quasi-resonant lc converter and boost converter. *IEEE Trans. Power Electron.* **2015**, *30*, 3731–3747. [CrossRef]
4. Xia, B.; Mi, C. A fault-tolerant voltage measurement method for series connected battery packs. *J. Power Sources* **2016**, *308*, 83–96. [CrossRef]
5. Sun, F.; Xiong, R.; He, H. A systematic state-of-charge estimation framework for multi-cell battery pack in electric vehicles using bias correction technique. *Appl. Energy* **2016**, *162*, 1399–1409. [CrossRef]
6. Xia, B.; Shang, Y.; Nguyen, T.; Mi, C. A correlation based fault detection method for short circuits in battery packs. *J. Power Sources* **2017**, *337*, 1–10. [CrossRef]
7. Salameh, M.; Schweitzer, B.; Sveum, P.; Al-Hallaj, S.; Krishnamurthy, M. Online temperature estimation for phase change composite-18650 lithium ion cells based battery pack. In Proceedings of the 2016 IEEE Applied Power Electronics Conference and Exposition (APEC), Long Beach, CA, USA, 20–24 March 2016; pp. 3128–3133.
8. Seaman, A.; Dao, T.-S.; McPhee, J. A survey of mathematics-based equivalent-circuit and electrochemical battery models for hybrid and electric vehicle simulation. *J. Power Sources* **2014**, *256*, 410–423. [CrossRef]
9. Zou, Y.; Hu, X.; Ma, H.; Li, S.E. Combined state of charge and state of health estimation over lithium-ion battery cell cycle lifespan for electric vehicles. *J. Power Sources* **2015**, *273*, 793–803. [CrossRef]
10. Wei, Z.; Tseng, K.J.; Wai, N.; Lim, T.M.; Skyllas-Kazacos, M. Adaptive estimation of state of charge and capacity with online identified battery model for vanadium redox flow battery. *J. Power Sources* **2016**, *332*, 389–398. [CrossRef]
11. He, H.; Xiong, R.; Guo, H.; Li, S. Comparison study on the battery models used for the energy management of batteries in electric vehicles. *Energy Convers. Manag.* **2012**, *64*, 113–121. [CrossRef]
12. He, H.; Zhang, X.; Xiong, R.; Xu, Y.; Guo, H. Online model-based estimation of state-of-charge and open-circuit voltage of lithium-ion batteries in electric vehicles. *Energy* **2012**, *39*, 310–318. [CrossRef]
13. Nejad, S.; Gladwin, D.; Stone, D. A systematic review of lumped-parameter equivalent circuit models for real-time estimation of lithium-ion battery states. *J. Power Sources* **2016**, *316*, 183–196. [CrossRef]
14. Xia, B.; Zhao, X.; De Callafon, R.; Garnier, H.; Nguyen, T.; Mi, C. Accurate lithium-ion battery parameter estimation with continuous-time system identification methods. *Appl. Energy* **2016**, *179*, 426–436. [CrossRef]
15. Pérez, G.; Garmendia, M.; Reynaud, J.F.; Crego, J.; Viscarret, U. Enhanced closed loop state of charge estimator for lithium-ion batteries based on extended kalman filter. *Appl. Energy* **2015**, *155*, 834–845. [CrossRef]
16. Chen, Z.; Fu, Y.; Mi, C.C. State of charge estimation of lithium-ion batteries in electric drive vehicles using extended kalman filtering. *IEEE Trans. Veh. Technol.* **2013**, *62*, 1020–1030. [CrossRef]
17. Sun, F.; Xiong, R. A novel dual-scale cell state-of-charge estimation approach for series-connected battery pack used in electric vehicles. *J. Power Sources* **2015**, *274*, 582–594. [CrossRef]

18. Li, K.; Tseng, K.J. An equivalent circuit model for state of energy estimation of lithium-ion battery. In Proceedings of the 2016 IEEE Applied Power Electronics Conference and Exposition (APEC), Long Beach, CA, USA, 20–24 March 2016; pp. 3422–3430.
19. Fuller, T.F.; Doyle, M.; Newman, J. Relaxation phenomena in lithium-ion-insertion cells. *J. Electrochem. Soc.* **1994**, *141*, 982–990. [CrossRef]
20. Smith, K.A. Electrochemical Modeling, Estimation and Control of Lithium Ion Batteries. Ph.D. Thesis, The Pennsylvania State University, State College, PA, USA, 2006.
21. Park, M.; Zhang, X.; Chung, M.; Less, G.B.; Sastry, A.M. A review of conduction phenomena in Li-ion batteries. *J. Power Sources* **2010**, *195*, 7904–7929. [CrossRef]
22. Karden, E.; Buller, S.; De Doncker, R.W. A method for measurement and interpretation of impedance spectra for industrial batteries. *J. Power Sources* **2000**, *85*, 72–78. [CrossRef]
23. Thele, M.; Bohlen, O.; Sauer, D.U.; Karden, E. Development of a voltage-behavior model for nimh batteries using an impedance-based modeling concept. *J. Power Sources* **2008**, *175*, 635–643. [CrossRef]
24. Buller, S.; Thele, M.; De Doncker, R.; Karden, E. Impedance-based simulation models of supercapacitors and Li-ion batteries for power electronic applications. *IEEE Trans. Ind. Appl.* **2005**, *41*, 742–747. [CrossRef]
25. Waag, W.; Käbitz, S.; Sauer, D.U. Experimental investigation of the lithium-ion battery impedance characteristic at various conditions and aging states and its influence on the application. *Appl. Energy* **2013**, *102*, 885–897. [CrossRef]
26. Howey, D.A.; Mitcheson, P.D.; Yufit, V.; Offer, G.J.; Brandon, N.P. Online measurement of battery impedance using motor controller excitation. *IEEE Trans. Veh. Technol.* **2014**, *63*, 2557–2566. [CrossRef]
27. Zheng, Y.; Lu, L.; Han, X.; Li, J.; Ouyang, M. Lifepo 4 battery pack capacity estimation for electric vehicles based on charging cell voltage curve transformation. *J. Power Sources* **2013**, *226*, 33–41. [CrossRef]
28. Nakayama, M.; Iizuka, K.; Shiiba, H.; Baba, S.; Nogami, M. Asymmetry in anodic and cathodic polarization profile for LiFePO4 positive electrode in rechargeable Li ion battery. *J. Ceram. Soc. Jpn.* **2011**, *119*, 692–696. [CrossRef]
29. Musio, M.; Damiano, A. A simplified charging battery model for smart electric vehicles applications. In Proceedings of the 2014 IEEE International Energy Conference (ENERGYCON), Dubrovnik, Croatia, 13–16 May 2014; pp. 1357–1364.
30. Tsang, K.; Sun, L.; Chan, W. Identification and modelling of lithium ion battery. *Energy Convers. Manag.* **2010**, *51*, 2857–2862. [CrossRef]
31. Yao, L.W.; Aziz, J.; Kong, P.Y.; Idris, N.; Alsofyani, I. Modeling of lithium titanate battery for charger design. In Proceedings of the 2014 IEEE Australasian Universities Power Engineering Conference (AUPEC), Perth, Australia, 28 September–1 October 2014; pp. 1–5.
32. Jiang, J.; Liu, Q.; Zhang, C.; Zhang, W. Evaluation of acceptable charging current of power Li-ion batteries based on polarization characteristics. *IEEE Trans. Ind. Electron.* **2014**, *61*, 6844–6851. [CrossRef]
33. Kim, N.; Ahn, J.-H.; Kim, D.-H.; Lee, B.-K. Adaptive loss reduction charging strategy considering variation of internal impedance of lithium-ion polymer batteries in electric vehicle charging systems. In Proceedings of the 2016 IEEE Applied Power Electronics Conference and Exposition (APEC), Long Beach, CA, USA, 20–24 March 2016; pp. 1273–1279.
34. Chen, Z.; Xia, B.; Mi, C.C.; Xiong, R. Loss-minimization-based charging strategy for lithium-ion battery. *IEEE Trans. Ind. Appl.* **2015**, *51*, 4121–4129. [CrossRef]
35. Rao, R.; Vrudhula, S.; Rakhmatov, D.N. Battery modeling for energy aware system design. *Computer* **2003**, *36*, 77–87.
36. Fleischer, C.; Waag, W.; Heyn, H.-M.; Sauer, D.U. On-line adaptive battery impedance parameter and state estimation considering physical principles in reduced order equivalent circuit battery models: Part 1. Requirements, critical review of methods and modeling. *J. Power Sources* **2014**, *260*, 276–291. [CrossRef]
37. Schweighofer, B.; Raab, K.M.; Brasseur, G. Modeling of high power automotive batteries by the use of an automated test system. *IEEE Trans. Instrum. Meas.* **2003**, *52*, 1087–1091. [CrossRef]
38. Castano, S.; Gauchia, L.; Voncila, E.; Sanz, J. Dynamical modeling procedure of a Li-ion battery pack suitable for real-time applications. *Energy Convers. Manag.* **2015**, *92*, 396–405. [CrossRef]
39. Chen, M.; Rincon-Mora, G.A. Accurate electrical battery model capable of predicting runtime and iv performance. *IEEE Trans. Energy Convers.* **2006**, *21*, 504–511. [CrossRef]

40. Baronti, F.; Fantechi, G.; Leonardi, E.; Roncella, R.; Saletti, R. Enhanced model for lithium-polymer cells including temperature effects. In Proceedings of the IECON 2010—36th Annual Conference on IEEE Industrial Electronics Society, Glendale, AZ, USA, 7–10 November 2010; pp. 2329–2333.

41. Lam, L.; Bauer, P.; Kelder, E. A practical circuit-based model for li-ion battery cells in electric vehicle applications. In Proceedings of the 2011 IEEE 33rd International Telecommunications Energy Conference (INTELEC), Amsterdam, The Netherlands, 9–13 October 2011; pp. 1–9.

42. Hu, Y.; Wang, Y.-Y. Two time-scaled battery model identification with application to battery state estimation. *IEEE Trans. Control Syst. Technol.* **2015**, *23*, 1180–1188. [CrossRef]

43. Li, J.; Mazzola, M.S. Accurate battery pack modeling for automotive applications. *J. Power Sources* **2013**, *237*, 215–228. [CrossRef]

44. Widanage, W.; Barai, A.; Chouchelamane, G.; Uddin, K.; McGordon, A.; Marco, J.; Jennings, P. Design and use of multisine signals for Li-ion battery equivalent circuit modelling. Part 1: Signal design. *J. Power Sources* **2016**, *324*, 70–78. [CrossRef]

45. Jossen, A. Fundamentals of battery dynamics. *J. Power Sources* **2006**, *154*, 530–538. [CrossRef]

46. Hentunen, A.; Lehmuspelto, T.; Suomela, J. Time-domain parameter extraction method for thévenin-equivalent circuit battery models. *IEEE Trans. Energy Convers.* **2014**, *29*, 558–566. [CrossRef]

47. Petzl, M.; Danzer, M.A. Advancements in *OCV* measurement and analysis for lithium-ion batteries. *IEEE Trans. Energy Convers.* **2013**, *28*, 675–681. [CrossRef]

48. Barai, A.; Widanage, W.D.; Marco, J.; McGordon, A.; Jennings, P. A study of the open circuit voltage characterization technique and hysteresis assessment of lithium-ion cells. *J. Power Sources* **2015**, *295*, 99–107. [CrossRef]

49. Hariharan, K.S.; Kumar, V.S. A nonlinear equivalent circuit model for lithium ion cells. *J. Power Sources* **2013**, *222*, 210–217. [CrossRef]

50. Gong, X.; Xiong, R.; Mi, C.C. A data-driven bias-correction-method-based lithium-ion battery modeling approach for electric vehicle applications. *IEEE Trans. Ind. Appl.* **2016**, *52*, 1759–1765.

![energies logo] *energies*

MDPI

Article

Optimal Siting of Charging Stations for Electric Vehicles Based on Fuzzy Delphi and Hybrid Multi-Criteria Decision Making Approaches from an Extended Sustainability Perspective

Huiru Zhao [1] and Nana Li [1,2,]

[1] School of Economics and Management, North China Electric Power University, Beijing 102206, China; huiruzhao@163.com

[2] School of Natural Resources and Environment, University of Michigan, Ann Arbor, MI 48108-1041, USA

* Correspondence: nancyli1007@163.com or lnana@umich.edu; Tel.: +86-178-8884-0018 or +1-734-709-2069

Academic Editor: Michael Gerard Pecht
Received: 25 January 2016; Accepted: 22 March 2016; Published: 6 April 2016

Abstract: Optimal siting of electric vehicle charging stations (EVCSs) is crucial to the sustainable development of electric vehicle systems. Considering the defects of previous heuristic optimization models in tackling subjective factors, this paper employs a multi-criteria decision-making (MCDM) framework to address the issue of EVCS siting. The initial criteria for optimal EVCS siting are selected from extended sustainability theory, and the vital sub-criteria are further determined by using a fuzzy Delphi method (FDM), which consists of four pillars: economy, society, environment and technology perspectives. To tolerate vagueness and ambiguity of subjective factors and human judgment, a fuzzy Grey relation analysis (GRA)-VIKOR method is employed to determine the optimal EVCS site, which also improves the conventional aggregating function of fuzzy Vlsekriterijumska Optimizacija I Kompromisno Resenje (VIKOR). Moreover, to integrate the subjective opinions as well as objective information, experts' ratings and Shannon entropy method are employed to determine combination weights. Then, the applicability of proposed framework is demonstrated by an empirical study of five EVCS site alternatives in Tianjin. The results show that A3 is selected as the optimal site for EVCS, and sub-criteria affiliated with environment obtain much more attentions than that of other sub-criteria. Moreover, sensitivity analysis indicates the selection results remains stable no matter how sub-criteria weights are changed, which verifies the robustness and effectiveness of proposed model and evaluation results. This study provides a comprehensive and effective method for optimal siting of EVCS and also innovates the weights determination and distance calculation for conventional fuzzy VIKOR.

Keywords: electric vehicle charging station; optimal siting; fuzzy Delphi method; combination weights; fuzzy Grey relation analysis-Vlsekriterijumska Optimizacijia I Kompromisno Resenje (fuzzy GRA-VIKOR); sustainability; sensitivity analysis

---

1. Introduction

With the rapid economic development and depletion of natural resources, energy shortages and climate change have become severe issues for the sustainable development of the present world. China, as the largest greenhouse gas (GHG) emitter and energy consumer, has proposed the corresponding strategies for energy utilization. In past a few years, urbanization development and an explosive demand for automobiles have stimulated an increase in energy consumption and carbon emissions in the transportation sector. The Chinese transportation sector accounted for about 21% of the total national energy consumption, as well as 7% of China's gross carbon emissions [1]. Electric vehicles

(EVs), as a kind of new environmentally-friendly means of transportation, are an effective way to tackle the problems related to environment pollution and fossil resource utilization [2]. Therefore, the Chinese government has devoted considerable resources to promote the adoption of electric vehicles, and has set up a target of putting five million EVs on the road by 2020 [3]. Meanwhile, a significant amount of investment has been made to subsidize EV manufacturers and buyers, build charging stations and posts, and offer tax breaks and other non-monetary incentives.

Charging infrastructure, as the energy provider of electric vehicles, is critical to the development of an electric vehicle system. The availability of efficient, convenient and economic EVCSs could enhance the willingness to buy of consumers and promote the development of the sector. Low availability of charging infrastructure could hinder EV adoption, which could then in turn reduce incentives to invest in charging infrastructure development [4]. EVCS siting is the preliminary stage of EVCS construction, and has a significant impact on the service quality and operation efficiency of EVCSs during their whole life cycle. Therefore, it is essential to establish a proper framework to determine the optimal sites for EVCSs.

Sustainability in the scope of energy management aims to meet present demand without compromising the energy utilization by future generations. Sustainable development can be realized by renewable resources, cleaner production and more efficient technologies. The "sustainability" in energy management is described as a long-term development integrating three pillars: economic growth, social development and environment protection [5]. To promote the sustainable development of the EV industry, optimal EVCS siting should be performed from a sustainability perspective. Moreover, concerning the diversity of advanced science and technical constraints, technology is another important perspective to determine the optimal site of EVCSs. Therefore, integrating the issues of technology, an extended concept of sustainability is proposed to determine the optimal EVCS site, which has not been addressed in previous studies. In this study, extended sustainability is employed to determine the initial evaluation criteria for optimal siting of EVCSs, which covers four perspectives, such as "economy", "society", "environment" and "technology". On this basis, 13 final sub-criteria are determined by a fuzzy Delphi method (FDM) through a series of intensive questionnaires.

Considering that optimal siting of EVCSs includes multiple factors, a Multiple Criteria Decision Making model is employed to evaluate the performance of all alternatives under conflicting criteria in this study. Vlsekriterijumska Optimizacijia I Kompromisno Resenje (VIKOR) is a compensatory aggregation MCDM method, which has been used to appraise performance in many fields [6–8]. VIKOR has a simple and logical computation procedure that simultaneously considers the closeness to positive ideal as well as negative ideal solutions [9]. Due to the increasing complexity of decision-making, more and more qualitative, uncertain and imprecise factors are involved in MCDM problems [10,11], and thus a fuzzy VIKOR method is constructed to determine the fuzzy compromise solutions for multiple criteria, which can efficiently grasp ambiguous information as well as the essential fuzziness of human judgment [12–14]. Moreover, Grey relation analysis (GRA) is used to modify the conventional aggregating function of fuzzy VIKOR, which can better measure the distance between fuzzy numbers as well as give a ranking order of alternatives with precise numbers [15–18]. On the other hand, in the application of VIKOR for optimal siting of EVCSs, weighting determination turns out to be crucial for the final ranking of alternatives. To obtain a better weights determining system for fuzzy VIKOR, a combination weights system based on subjective judgment and objective information are introduced in this study. The subjective weights are determined by experts' opinions, and the objective weights are obtained by the Shannon entropy method. Therefore, in our research, a hybrid framework on the basis of FDM, combination weights and fuzzy GRA-VIKOR methods will be employed to determine the optimal sites for EVCSs.

The remainder of this paper is organized as follows: a review of the literature related to the EV industry, optimal EVCS site determination, and the main contributions of this research can be found in Section 2. In Section 3, the basic theories of related methods are elaborated. Section 4 presents the proposed framework for optimal siting of electric vehicle station. The evaluation index system for optimal siting of EVCSs is established by FDM in Section 5. Section 6 performs the EVCS siting by employing combination weighting and a fuzzy GRA-VIKOR model. Results discussion and sensitivity analysis are performed to check the rationality and robustness of the proposed model and results in Section 7. Conclusions are drawn in Section 8.

2. Literature Review

The construction of electric vehicle charging stations is important in the whole life cycle of the electric vehicle industry. Meanwhile, an appropriate site and capacity for EVCS can not only benefit the related stakeholders, but also promote the sustainable development of the EV industry. Over the last decade, many studies related to the economic and environmental benefit, influence and technology in the EV industry have been conducted. Simpson [19] presented a comparison of the costs (vehicle purchase costs and energy costs) and benefits (reduced petroleum consumption) of PHEVs related to hybrid-electric and conventional vehicles. By 2011 little was known about the economic rationale for public fast chargers for electric vehicles, Schroeder *et al.* [20] aimed to provide an insight into the business case for this technology in a case study for Germany. Hawkins *et al.* [21] developed and provided a transparent life cycle inventory of conventional vehicles and electric vehicles, which verified that EVs have decreased global warming potential (GWP) relative to conventional diesel or gasoline vehicles. Matsuhashi *et al.* [22] developed a process-relational model to estimate lifecycle CO_2 emissions from electric vehicles (EVs) and gasoline vehicles (GVs), which indicated that the manufacture and driving of EVs produces less CO_2 emissions than that of GVs. Putrus *et al.* [23] analyzed the impact of electric vehicles on existing power distribution networks, including supply/demand matching and potential violations of statutory voltage limits, power quality and imbalance. Clement-Nyns *et al.* [24] pointed out that uncoordinated power consumption on a local scale would lead to grid problems, and computed the optimal charging profile of plug-in hybrid electric vehicles by minimizing the power losses and maximizing the main grid load. Mets *et al.* [25] presented smart energy control strategies based on quadratic programming for charging PHEVs, aiming to minimize the peak load and flatten the overall load profile. Rivera *et al.* [26] proposes a novel architecture for PEV DC charging stations by using a grid-tied neutral point clamped converter.

Research focused on siting and sizing of EVCSs has received much more attention in recent years. Liu *et al.* [27] presented a modified primal-dual interior point algorithm to solve the optimal sizing of EV charging stations, in which environmental factors and the service radius of EV charging stations were considered. Wirges *et al.* [28] presented a dynamic spatial model of a charging infrastructure development for electric vehicles in the German metropolitan region of Stuttgart, and generated several scenarios of a charging infrastructure development until 2020. Jia *et al.* [29] introduced an optimization process for the sizing and siting of electric vehicle charging stations with minimized integrated cost of charging stations and consumers' costs, in which the charging demand and road network structure were variables. Aiming at minimizing users' losses on the way to the charging station, Ge *et al.* [30] determines the best location by using a Genetic Algorithm (GA) considering the traffic density and the charging station's capacity constraints. Xi *et al.* [31] developed a simulation–optimization model to determine the location of electric vehicle chargers, and explored the interactions between the optimization criterion and the available budget. Sathaye *et al.* [32] utilized a continuous facility location model for optimally siting electric vehicle infrastructure in highway corridors, and carefully dealt with the influence of demand uncertainty. Pashajavid *et al.* [33] proposed a scenario optimization based on a particle swarm optimization (PSO) algorithm to allocate charging stations for plug-in electric vehicles (PEVs), and a multivariate stochastic modeling methodology based on the notion of copula is provided in order to develop a probabilistic model of the load demand due to PEVs. Zi *et al.* [34] presented an adaptive particle swarm optimization (APSO) algorithm to optimize the siting and sizing of electric vehicle charging stations, which considered geographic information, construction costs and running costs. In order to install alternative fuel charging stations at suitable locations for alternative-fuel vehicles (AFVs), You *et al.* [35] developed a mixed-integer programming model to address budget limitations and to maximize the number of people who can complete round-trip itineraries. Yao *et al.* [36] developed a multi-objective collaborative planning strategy to address the optimal planning issue in integrated power distribution and EV charging systems, in which the overall annual cost of investment and energy losses are minimized simultaneously with maximization of the annual traffic flow captured by fast charging stations (FCSs). An equilibrium-based traffic assignment model and decomposition-based multi-objective evolutionary algorithm were developed for obtaining the optimal solution. Sadeghi *et al.* [37] presented a Mixed-Integer Non-Linear (MINLP) optimization approach for the optimal placement and sizing of fast charging stations, which considered the station development cost, EV energy loss, and electric gird loss as well as the location of electric substations and urban roads. Chung *et al.* [38] formulated a multi-period optimization model based on a flow-refueling location model for strategic charging station location planning, and then developed a case study based on the real traffic flow data of the Korean Expressway network in 2011.

After analyzing the literature, it can be concluded that the majority of existing studies related to the optimal siting of EVCSs are concentrated on Multi-Objective Decision Making (MODM) methods, such as linear/nonlinear programming, stochastic programming, mixed-integer programming and multilayer programming. In most cases, heuristic algorithms such as GA and PSO were applied to tackle the optimal solution. However, there are two major critiques with such MODM approaches. First, although the aforementioned optimization models are remarkable it is less likely they can be implemented in practice due to the complexity of modeling real-world problems. Second, optimization models can only account for quantitative variables such as construction cost and running cost, electric grid loss, EV energy loss and so on, but are not capable of modeling important qualitative variables such as ecological environment (e.g., deterioration on soil and vegetation), *etc.*

In view of this, herein another kind of decision-making method, *i.e.,* the Multiple-Criteria Decision-Making method will be employed to determine the optimal site of electric vehicle charging stations from an extended sustainability perspective. The MCDM method can comprehensively capture the quantitative and qualitative criteria that both play important roles in EVCS site selection. The main contributions of this paper are as follows:

(1)   This is the first study that involves both quantitative and qualitative criteria for EVCS siting from an extended sustainability perspective, which overcomes the defects of traditional mathematical programming in addressing qualitative but nevertheless important factors.

(2)   The conventional concept of sustainability is improved through integrating the issues of technology, namely economy, society, environment and technology perspectives, which have not been considered in previous studies. In this study, the initial criteria are established based on extended sustainability. Furthermore, to obtain the most reliable consensus among a group of experts in a shorter time, FDM is employed to determine the final sub-criteria for EVCS site selection.

(3)   The fuzzy VIKOR method, which shows good performance in the decision-making of alternatives selection, has been applied in many fields. To the best of our knowledge, this is a novel hybrid MCDM technique based on combination weights and fuzzy GRA-VIKOR for the optimal siting of EVCSs, which also extends the application domains of the fuzzy VIKOR method. The proposed model addresses the fuzziness and uncertainty of subjective factors and human judgment, and additionally it considers subjective and objective information within the weights calculation process. Moreover, GRA are used to measure the distances of fuzzy numbers between alternatives to ideal solutions in this study, which can better measure the distance between fuzzy numbers as well as provide a ranking order of alternatives with precise numbers.

(4)   Since experts with various knowledge backgrounds may have different priorities as their main objective, it is essential to probe the impacts of sub-criteria weights on the final results. This study is the first paper to research the economy, society, environment and technology perspectives for optimal siting of EVCSs by changing the sub-criteria weights.

3. Research Method

3.1. Fuzzy Logic

Fuzzy theory, proposed by Zadeh in 1965, is used to map linguistic terms to numerical terms within human decisions. The fuzzy set is often defined to solve the uncertainty and vagueness in criteria weighting and alternatives ratings of multi-criteria decision making problems [39]. A fuzzy set, featured by a membership function, assigns each criterion a membership rating among (0, 1), reflects criteria grades belonging to a set. In addition, linguistic terms such as "good", "fair" and "bad" are put forward to define numerical intervals [40].

A triangular fuzzy number $\widetilde{M}$, denoted by (a,b,c), is the most popular fuzzy number in fuzzy applications [41]. The membership function is defined as follows:

$$\mu_M(x) = \begin{cases} \dfrac{x-a}{b-a}, & a \leqslant x \leqslant b \\ \dfrac{c-x}{c-b}, & b \leqslant x \leqslant c \\ 0, & otherwise \end{cases} \tag{1}$$

and $-\infty < a \leqslant x \leqslant b < \infty$.

In concrete terms, the membership function $\mu_M(x) = 1$ indicates that variable x fully belongs to the fuzzy set $\widetilde{M}$. Conversely, if the variable x does not belong to the fuzzy set $\widetilde{M}$, then $\mu_M(x) = 0$ [42].

Let $\widetilde{M}_1 = (l_1, m_1, r_1)$ and $\widetilde{M}_2 = (l_2, m_2, r_2)$ be two triangular fuzzy numbers, the operation laws are shown as below:

$$\widetilde{M}_1 \oplus \widetilde{M}_2 = (l_1 + l_2, m_1 + m_2, r_1 + r_2) \tag{2}$$

$$\widetilde{M}_1 \odot \widetilde{M}_2 \approx (l_1 l_2, m_1 m_2, r_1 r_2) \tag{3}$$

$$\lambda \widetilde{M}_1 = (\lambda l_1, \lambda m_1, \lambda r_1), \lambda > 0 \tag{4}$$

$$\widetilde{M}_1^{-1} \approx (1/l_1, 1/m_1, 1/r_1) \tag{5}$$

And the distance of $\widetilde{M}_1 = (l_1, m_1, r_1)$ and $\widetilde{M}_2 = (l_2, m_2, r_2)$ can be defined as follows [43]:

$$d\left(\widetilde{M}_1, \widetilde{M}_2\right) = \frac{1}{2} \int_0^1 [l_1 + (m_1 - l_1)\alpha + r_1 - (r_1 - m_1)\alpha - l_2 - (m_2 - l_2)\alpha - r_2 + (r_2 - m_2)\alpha] d\alpha \tag{6}$$

In most MCDM processes, decision makers often provide uncertain answers rather than precise values. Linguistic values and fuzzy set theory are recommended to rate preference instead of traditional numerical method. Therefore, the fuzzy set theory has been integrated into various MCDM methods, such as fuzzy AHP, fuzzy TOPSIS, fuzzy VIKOR, and so on, which should be more appropriate and effective than conventional ones in real problems involving uncertainty and vagueness [44–46].

3.2. Fuzzy Delphi Method

The Delphi method (DM) is a technique used to obtain the most reliable consensus among a group of experts. It was proposed by Dalky and Helmer in 1963 and has been widely used in decision and prediction making. This technique offers experts opportunities to receive feedback and modify previous opinions through several rounds of consulting. Furthermore, owning to its deficiency in handling ambiguity and uncertainty within expert surveys, fuzzy Delphi method (FDM) was proposed to solve these defects combing DM with fuzzy logic theory. Experts can provide their opinions through triangular fuzzy numbers (TFNs), and are not required to modify them again and again. Moreover, no useful information would be lost, because all opinions can be effectively taken into account by the membership degrees. Due to its advantages in evoking group decisions, FDM is embraced in various studies to construct evaluation. To recognize the vital criteria for the optimal siting of EVCS, the FDM is introduced in this paper. Essential steps of the FDM are listed as follows:

Step 1: Administer questionnaires and determine the most conservative value and the most optimistic value ranging from 0 to 10 for each criterion among a group of experts.

Step 2: Gather the minimum and maximum values and calculate the geometric mean for each criterion. Then, compute the conservative TFN (C_L^i, C_M^i, C_U^i) and optimistic TFN (O_L^i, O_M^i, O_U^i) in terms of each criterion. C_L^i and O_L^i represent the minimum remaining conservative value and minimum remaining optimistic value, respectively; C_U^i and O_U^i represent the maximum remaining conservative value and maximum remaining optimistic value, respectively; and C_M^i and O_M^i represent the geometric mean of the remaining conservative value and the geometric mean of the remaining optimistic value, respectively.

Step 3: Check that the consistency of expert opinions, and compute the consensus significance G_i for each criterion.

(1) If $C_U^i \leqslant O_L^i$, the criterion i holds consensus, and the value of the consensus significance G_i is computed by Equation (7):

$$G_i = \frac{G_M^i + O_M^i}{2} \tag{7}$$

(2) If $C_U^i > O_L^i$, and the gray zone interval value $(Z^i = C_U^i - O_L^i)$ is smaller than the interval value $M^i = O_U^i - C_M^i$, correspondingly, the value of the consensus significance is computed by Equation (2):

$$G_i = \frac{[(C_U^i \times O_M^i) - (O_L^i \times C_M^i)]}{[[(C_U^i - C_M^i) + (O_M^i - O_L^i)]]} \tag{8}$$

When $C_U^i > O_L^i$, however, the gray zone interval value $(Z^i = C_U^i - O_L^i)$ is greater than the interval value $(M^i = O_U^i - C_M^i)$, which means that the expert opinions are inconsistent. Thus, Steps 1–3 should be repeated until each criterion converges and the value of the consensus significance can be calculated.

3.3. Fuzzy GRA-VIKOR Method

The Vlsekriterijumska Optimizacijia I Kompromisno Resenje (VIKOR) method, put forward by Opricovic in 1998, was developed for multi-criteria optimization of complex systems. This model focuses on ranking different alternatives against various even conflicting decision criteria. It relies on an aggregating function that can reflect distance to both positive and negative ideal points [47]. In order to handle the imprecision and subjectivity of decision makers, linguistic values are introduced into the conventional VIKOR. The fuzzy VIKOR combines the advantages of the conventional VIKOR and fuzzy set theory, which is much more sufficient to model problems in the real world than precise values [12–14,48,49].

In fuzzy VIKOR, the multi-criteria measurement process for ranking alternatives is developed from an aggregating function, which represents the distance of each alternative from positive ideal point and negative ideal point. As mentioned in the introduction, in order to better examine the extent of the connection between alternative and ideal points, grey relation analysis is used to modify the conventional aggregating function, which can better identify relationships among fuzzy numbers in a system [15,16]. Moreover, the fuzzy VIKOR based on GRA method can efficiently overcome the deficiencies of fuzzy results and the inconsistent ranking of alternatives [17,18]. More details of this approach are shown as below:

Here, the ratings of criteria are expressed in linguistic terms (the triangular fuzzy numbers), as in Table 1.

Table 1. Fuzzy evaluation scores for the alternatives.

Linguistic Terms	Fuzzy Score
Very poor	$(0, 0, 1)$
Poor	$(0, 1, 3)$
Medium poor	$(1, 3, 5)$
Fair	$(3, 5, 7)$
Medium good	$(5, 7, 9)$
Good	$(7, 9, 10)$
Very good	$(9, 10, 10)$

Step 1. Calculate the aggregated fuzzy linguistic ratings for criteria performance of alternatives.

Suppose that there are m alternatives $A = \{A_1, A_2, \dots A_m\}$ to be evaluated. The performance of n criteria are by linguistic terms which are obtained from r decision makers.

Let $\tilde{x}_{ijk} = \left(x_{ijk}^L, x_{ijk}^M, x_{ijk}^U \right), 0 \leqslant x_{ijk}^L \leqslant x_{ijk}^M \leqslant x_{ijk}^U \leqslant 1, i = 1, 2, \cdots m, j = 1, 2, \cdots n, k = 1, 2, \cdots r$ be the linguistic rating on the performance criteria C_j respect to alternative A_i by expert E_k. Then the aggregated fuzzy linguistic rating $\tilde{x}_{ij} = \left(x_{ij}^L, x_{ij}^M, x_{ij}^U \right)$ can be obtained as follows:

$$\tilde{x}_{ij} = \left(x_{ij}^L, x_{ij}^M, x_{ij}^U \right) = \left(\sum_{k=1}^{r} \frac{x_{ijk}^L}{r}, \sum_{k=1}^{r} \frac{x_{ijk}^M}{r}, \sum_{k=1}^{r} \frac{x_{ijk}^U}{r} \right) \tag{9}$$

Step 2. Assemble the initial fuzzy decision matrix.

According to Equation (9), the initial fuzzy decision matrix $\tilde{D}$ can be obtained, as shown in Equation (10). A MCDM problem can be expressed concisely in the form of triangular fuzzy number, as follows:

$$\tilde{D} = (\tilde{x}_{ij})_{m \times n} = \begin{bmatrix} \tilde{x}_{11} & \tilde{x}_{12} & \cdots & \tilde{x}_{1n} \\ \tilde{x}_{21} & \tilde{x}_{22} & \cdots & \tilde{x}_{2n} \\ \vdots & \vdots & \vdots & \vdots \\ \tilde{x}_{m1} & \tilde{x}_{m2} & \cdots & \tilde{x}_{mn} \end{bmatrix} = \begin{bmatrix} (x_{11}^L, x_{11}^M, x_{11}^U) & (x_{12}^L, x_{12}^M, x_{12}^U) & \cdots & (x_{1n}^L, x_{1n}^M, x_{1n}^U) \\ (x_{21}^L, x_{21}^M, x_{21}^U) & (x_{22}^L, x_{22}^M, x_{22}^U) & \cdots & (x_{2n}^L, x_{2n}^M, x_{2n}^U) \\ \vdots & \vdots & \cdots & \vdots \\ (x_{m1}^L, x_{m1}^M, x_{m1}^U) & (x_{m2}^L, x_{m2}^M, x_{m2}^U) & \cdots & (x_{mn}^L, x_{mn}^M, x_{mn}^U) \end{bmatrix} \tag{10}$$

Step 3. Normalize the initial fuzzy decision matrix using linear scale transformation.

To ensure the compatibility among evaluation criteria, the initial fuzzy decision matrix should be transformed into a comparable scale. The normalized fuzzy decision matrix is denoted by $\tilde{R}$ [50]:

$$\tilde{R} = [\tilde{r}_{ij}]_{m \times n}$$

For the benefit criteria:

$$\tilde{r}_{ij} = \left(\frac{x_{ij}^L}{u_{ij}^+}, \frac{x_{12}^M}{u_{ij}^+}, \frac{x_{12}^U}{u_{ij}^+} \right) \text{ and } u_{ij}^+ = \max_i x_{ij}^U \tag{11}$$

For the cost criteria:

$$\tilde{r}_{ij} = \left(\frac{u_{ij}^-}{x_{ij}^L}, \frac{u_{ij}^-}{x_{ij}^M}, \frac{u_{ij}^-}{x_{ij}^U} \right) \text{ and } u_{ij}^- = \min_i x_{ij}^L \tag{12}$$

Step 4. Define the referential sequences of fuzzy positive ideal solution and negative ideal solution.

The referential sequences of positive ideal solution A^+ and negative ideal solution A^- can be determined as follows:

$$A^+ = [\tilde{r}_{01}^+, \tilde{r}_{02}^+, \cdots \tilde{r}_{0n}^+], A^- = [\tilde{r}_{01}^-, \tilde{r}_{02}^-, \cdots \tilde{r}_{0n}^-] \tag{13}$$

where $\tilde{r}_{0j}^+ = \max_i (\tilde{r}_{ij}), \tilde{r}_{0j}^- = \min_i (\tilde{r}_{ij}), j = 1, 2, \dots n$

Step 5: Compute the distances of each alternative from the positive ideal solution and negative ideal solution [51].

$$\tilde{S}_i = \sum_{j=1}^{n} w_j \left(\frac{\tilde{r}_{0j}^+ - \tilde{r}_{ij}}{\tilde{r}_{0j}^+ - \tilde{r}_{0j}^-} \right) \tag{14}$$

$$\tilde{R}_i = \max_j \left[w_j \left(\frac{\tilde{r}_{0j}^+ - \tilde{r}_{ij}}{\tilde{r}_{0j}^+ - \tilde{r}_{0j}^-} \right) \right] \tag{15}$$

$$i = 1, 2 \cdots m, j = 1, 2 \cdots n$$

where w_j represents the weight of criteria C_j, $\tilde{S}_i$ denotes the distance rate of A_i to the positive ideal solution A^+, and $\tilde{R}_i$ denotes the distance rate of A_i to the negative ideal solution A^-.

In order to better reflect distance of each alternative to the positive and negative ideal points, the fuzzy grey relation coefficient (FGRC) is introduced to modify the conventional formula of fuzzy VIKOR, which are shown as follows [16]:

$$\gamma \left(\tilde{r}_{0j}^u, \tilde{r}_{ij} \right), u = +, -$$

$$\gamma \left(\tilde{r}_{0j}^u, \tilde{r}_{ij} \right) = \frac{\min_i \min_j \tilde{d}_{ij}^u + \xi \max_i \max_j \tilde{d}_{ij}^u}{\tilde{d}_{ij}^u + \xi \max_i \max_j \tilde{d}_{ij}^u} = \frac{\min_i \min_j \tilde{d} \left(\tilde{r}_{0j}^u, \tilde{r}_{ij} \right) + \xi \max_i \max_j \tilde{d} \left(\tilde{r}_{0j}^u, \tilde{r}_{ij} \right)}{\tilde{d} \left(\tilde{r}_{0j}^u, \tilde{r}_{ij} \right) + \xi \max_i \max_j \tilde{d} \left(\tilde{r}_{0j}^u, \tilde{r}_{ij} \right)} \tag{16}$$

$$\tilde{S}_i = \sum_{j=1}^{n} w_j \gamma \left(\tilde{r}_{0j}^+, \tilde{r}_{ij} \right) \tag{17}$$

$$\tilde{R}_i = \max_j w_j \gamma \left(\tilde{r}_{0j}^-, \tilde{r}_{ij} \right) \tag{18}$$

Step 6: Compute the value of Q_i for each alternative as below:

$$Q_i = v \frac{S_i - S^+}{S^- - S^+} + (1 - v) \frac{R_i - R^+}{R^- - R^+} \tag{19}$$

where $S^+ = \max_i S_i$, $S^- = \min_i S_i$, $R^+ = \max_i R_i$, $R^- = \min_i R_i$, and v is the weight of the strategy of "the maximum group utility", whereas $(1-v)$ represents the weight of the individual regret.

Step 7: Rank the alternatives according to the value of Q_i in Step 5.

On the basis of the concepts of GRA and fuzzy VIKOR method, all alternatives can be ranked by the value of Q_i. Obviously, for the alternative A_i which is closer to the positive ideal point and farther from the negative ideal point, the value of Q_i is zero.

In addition, only when the alternative which is the best ranked by the value of Q_i satisfies the following conditions, it can be selected as the optimal solution.

(I) Acceptable advantage:

$$Q \left(A^{(2)} - A^{(1)} \right) \geqslant DQ$$

where $A^{(2)}$ is the second in the list of priorities by Q_i; $DQ = 1/(N-1)$, N is the number of alternatives [51].

(II) Acceptable stability in decision-making:

The alternative $A^{(1)}$ must also be the best ranked by $\tilde{S}_i$ or R_i. This compromise solution is stable within the decision-making process, which could be the strategy of maximum group utility (when $v > 0.5$ is needed), or "by consensus" ($v \approx 0.5$), or "with veto" ($v < 0.5$). If one of the conditions is not satisfied, then the set of solutions is proposed [16], which consists of:

$A^{(1)}$ and $A^{(2)}$, if only the condition (II) is not satisfied, or

$A^{(1)}, A^{(2)}, A^{(3)}, \ldots A^{(M)}$, if the condition (II) is not satisfied; $A^{(M)}$ is determined by the relation $Q(A^{(M)} - A^{(1)}) < DQ$, for maximum M (the positions of these alternatives are "in closeness").

3.4. The Combination Weights

The weighted sum of the "distances" from an alternative to corresponding ideal points over all criteria is essential for performance comparison among all designated alternatives. From the previous literature on MCDM, the weights of criteria are usually subjective weights determined by decision makers. However, critiques of human errors and inconsistency are often associated with subjective weights for such weighting processes in MCDM. With this regard, to improve the weighting accuracy, some objective weighting models are applied by mathematical techniques. To obtain a better weight determining system for fuzzy VIKOR, combination weights based on subjective methods and objective methods are introduced in this study, which can composite subjective judgment and objective information. On the one hand, the subjective weights are determined by experts' opinions. On the other hand, the objective weights are obtained by the Shannon entropy method.

3.4.1. The Subjective Weights

On the one hand, the subjective weights could be obtained from experts' opinions. Here, the ratings of criteria are expressed in linguistic terms (the triangular fuzzy numbers), as in Table 2.

Table 2. Fuzzy evaluation scores for criteria weights.

Linguistic Terms	Membership Function
Of little importance	(0, 0, 0.3)
Moderately important	(0, 0.3, 0.5)
Important	(0.2, 0.5, 0.8)
Very important	(0.5, 0.7, 1)
Absolutely important	(0.7, 1, 1)

Let $\tilde{s}_{jk} = \left(s_{jk}^L, s_{jk}^M, s_{jk}^U\right)$, $0 \leqslant s_{jk}^L \leqslant s_{jk}^M \leqslant s_{jk}^U \leqslant 1, j = 1, 2, \cdots n, k = 1, 2, \cdots r$ be the superiority linguistic rating on criteria weight assigned to criteria C_j by expert D_k can be calculated by:

$$\tilde{s}_{jk} = \left(s_{jk}^L, s_{jk}^M, s_{jk}^U\right) = \left(\sum_{k=1}^r \frac{s_{jk}^L}{r}, \sum_{k=1}^r \frac{s_{jk}^M}{r}, \sum_{k=1}^r \frac{s_{jk}^U}{r}\right) \tag{20}$$

In order to maintain the consistency between objective weights and subjective weights, the criteria weights based on triangular fuzzy numbers should be also defuzzied based on Equation (21). In this paper, the graded mean integration approach is employed to transform a triangular fuzzy number $M = (l,m,u)$ into a precise number:

$$P\left(\tilde{M}\right) = M = \frac{l + 4m + u}{6} \tag{21}$$

3.4.2. Shannon Entropy and Objective Weights

The entropy concept proposed by Shannon in 1948 is a measure of uncertainty in formulated information, which has been widely used in many fields such as management, engineering and so on. According to the ideal of entropy theory, the number or quality of information from decision-making process is determined by the accuracy and reliability of the decision-making problem [52]. Therefore, entropy can be applied to the assessment problem in different decision-making processes. Moreover, entropy can also be used to analyze the quantity of information provided by data [53]. The basic theory and specific steps of Shannon entropy weighting method are shown as below:

Shannon entropy is capable of evaluating the decision making units and being employed as a weighting decision method. Assume that a MCDM problem contains m alternatives and n criteria, thus the decision making matrix is defined as below:

$$\begin{bmatrix} x_{11} & x_{12} & \cdots & x_{1n} \\ x_{21} & x_{22} & \cdots & x_{2n} \\ \vdots & \vdots & \vdots & \vdots \\ x_{m1} & x_{m2} & \cdots & x_{mn} \end{bmatrix} \tag{22}$$

Then, the criteria weights can be determined based on the entropy concept through the following steps:

Step 1: Normalize the evaluation criteria as below:

$$P_{ij} = \frac{x_{ij}}{\sum_j x_{ij}} \tag{23}$$

Specially, for the fuzzy MCDM problem, the fuzzy decision making matrix should be defuzzyed firstly, according to Equation (21).

Step 2: Calculate the entropy measure of each criterion as [51]:

$$e_j = -k \sum_{j=1}^{n} P_{ij} \ln(P_{ij}) \tag{24}$$

where $k = (\ln(m))^{-1}$.

Step 3: Define the divergence of each criterion through:

$$d_i v_j = 1 - e_j \tag{25}$$

The more the $d_i v_j$ is, the more important the jth criterion is.

Step 4: Determined the normalized weights of all criteria through [54]:

$$w_j = \frac{d_i v_j}{\sum_j d_i v_j} \tag{26}$$

Finally, the combination weights of all criteria are equal to the average of subjective and objective weights.

4. The Framework of the Integrated MCDM Model

The proposed framework for optimal siting of electric vehicle station based on FDM, combination weighting and fuzzy GRA-VIKOR methods involves the following three phases (Figure 1).

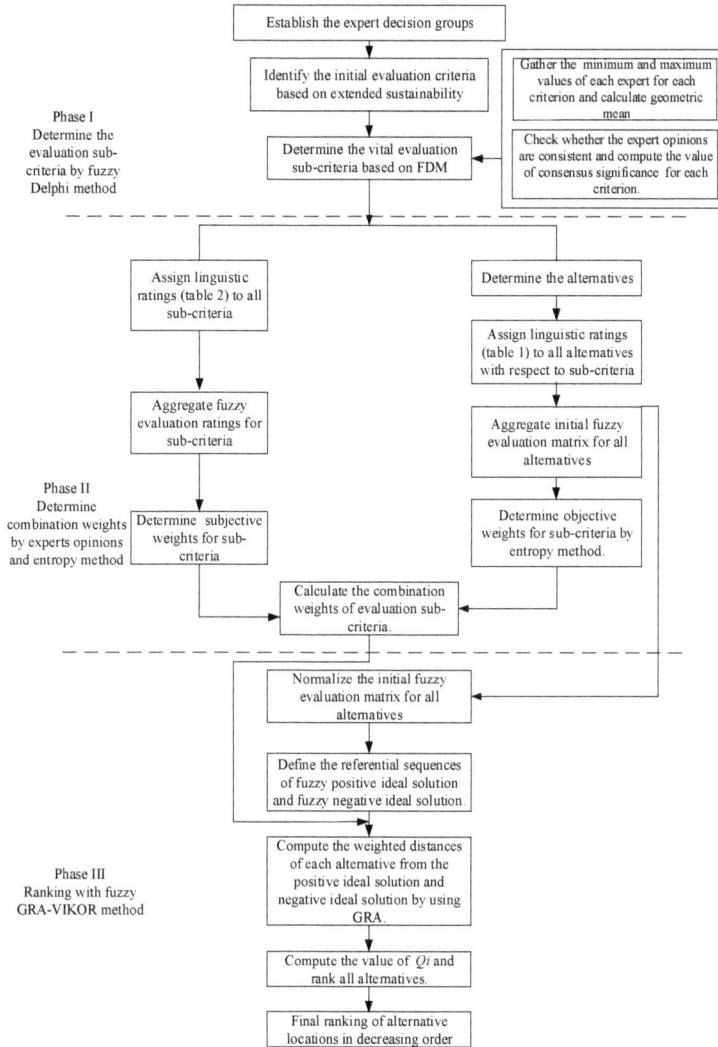

Figure 1. The framework of the proposed model for optimal siting of charging stations for electric vehicles.

Phase 1: Identify the vital evaluation sub-criteria based on extended sustainability and FDM

In the first phase, professors, scholars, residents, governors, EV users and producers, as well as the project management personnel in the field of electric power system, electric vehicle industry, transportation system and sustainability are selected to establish three expert decision groups. According to the extended sustainability concept and industry background, the initial evaluation criteria are determined, which are associated with economy, society, environment and technology perspectives. Further, the vital (final) sub-criteria for optimal siting of EVCS are determined based on FDM technique.

Phase 2: Determine the combination weights of the evaluation sub-criteria based on the fuzzy experts' ratings and entropy approach

In this step, the vital (final) evaluation sub-criteria are weighted by integrating the subjective weights and objective weights. For the subjective weights, three groups of experts firstly assign linguistic ratings to all sub-criteria by using the rating scales given in Table 2. Then, the fuzzy evaluations for sub-criteria are aggregated and the subjective weights for sub-criteria can be computed. On the other side, for the objective weights based on entropy method, linguistic ratings to all alternatives with respect to sub-criteria are firstly allocated by using rating scales in Table 1, and then are transformed to triangular fuzzy numbers. After aggregating the initial fuzzy evaluation matrix for all alternatives, the objective weights for all sub-criteria are determined by using entropy method. Based on above results, the combination weights for all sub-criteria are eventually aggregated by combining subjective weights and objective weights simultaneously.

Phase 3: Rank all alternatives for EVCS and determine the optimal site using the fuzzy GRA-VIKOR

In this step, the normalized fuzzy decision matrix is assembled based on the aggregated initial fuzzy evaluation matrix in phase 2. Next, define the referential sequences of fuzzy positive ideal solution and negative solution. Then, the GRA method is employed to compute the weighted distances of each alternative from ideal solutions. Finally, all EVCS site alternatives are ranked in a descending order of performance based on the values of Q_i.

5. Evaluation Index System for Optimal Siting of Vehicle Charging Station

Evaluation criteria are very important to the optimal EVCS siting. It is important to establish an evaluation index system to comprehensively reflect the inherent characteristics of EVCS siting. However, the electric-vehicle industry is still in the early stages of management and technological exploration, so there is no consistent list of criteria for EVCS site selection in China. Since electric vehicles are a sustainable way of energy development, the evaluation index system for optimal EVCS siting is built from the perspective of extended sustainability. The conventional sustainability theory put forward a new development way which can achieve economic growth and social development without environmental damage, and sustainability is designed as economy sustainability, society sustainability and environment sustainability. Moreover, since EVCS construction involves large numbers of technical conditions, the technology sustainability is introduced to improve the classical idea of sustainability. Therefore, the evaluation index system for optimal EVCS siting includes economy criteria, society criteria, environment criteria and technology criteria. Further, the sub-criteria that are affiliated with above four criteria are determined by fuzzy Delphi method as follows.

First of all, based on the extended sustainability theory, academic literatures and feasibility reports of EVCS, 37 initial sub-criteria are collected according to relative industry standards and expert consultation, in which economy, society, environment and technology are covered. Furthermore, the vital sub-criteria are selected as the final evaluation sub-criteria based on the FDM.

Experts firstly express their opinions on the sub-criteria importance through conservative and optimistic values. And the scores of sub-criteria lies on the scale from 0 to 10. Subsequently, according to Equations (1) and (2), the conservative TFN (C_L^i, C_M^i, C_U^i) and optimistic TFN (O_L^i, O_M^i, O_U^i) of each expert respect to each criterion are calculated (Table 3). Then, the consistency of the experts' opinions are verified by calculating the values of Z^i and M^i. Finally, the vital sub-criteria are determined based on the consensus value G^i. Particularly, the threshold value of G^i in is set to 6.0, which has been accepted by more than 92% of experts [15]. Therefore, 13 evaluation sub-criteria are selected to realize the optimal site selection of vehicle charging station (Table 3). The evaluation index system is summarized in the flowchart shown in Figure 2.

Table 3. Calculation results of evaluation sub-criteria based on FDM.

Perspectives	Initial Sub-Criteria	Pessimistic Value		Optimistic Value		Geometric Mean		$M^i - Z^i$	Consensus Value
		C_L^i	C_U^i	O_L^i	O_U^i	C_M^i	O_M^i		G_i
	Investment pay-back period	1	3	5	8	3.54	6.98	3.02	5.26 < 6.0
	Total construction cost	1	4	7	9	4.97	7.54	4.46	6.26 > 6.0
	Annual economic benefit	2	7	5	9	5.11	6.16	0.84	5.98 < 6.0
	Internal rate of return	2	6	7	10	5.31	7.65	3.35	6.48 > 6.0
Economy	Land acquisition costs	2	8	6	9	3.36	6.07	0.93	3.86 < 6.0
	Annual operation and maintenance cost	1	6	8	9	4.36	8.69	2.31	6.53 > 6.0
	Removal cost	2	6	7	10	3.55	5.99	5.01	4.77 < 6.0
	Causeway construction costs	3	7	6	9	2.67	6.54	1.46	3.63 > 6.0
	EV ownership in the service area	2	8	6	10	5.84	7.34	0.66	5.47 < 6.0
	Service area population	2	5	7	9	3.75	5.68	5.32	4.72 < 6.0
	Service radius	1	6	5	9	2.59	7.65	0.35	3.89 < 6.0
	Service capacity	1	5	7	10	4.59	8.49	3.51	6.54 > 6.0
Society	Residents professional habit	1	6	7	8	4.05	6.27	2.73	5.16 < 6.0
	Residents consumption habits	3	4	7	10	3.56	5.24	7.59	4.40 < 6.0
	Traffic convenience	1	6	7	9	4.35	7.84	2.16	6.10 > 6.0
	Impact on living level of resident	1	6	5	10	4.58	7.65	1.35	5.12 < 6.0
	Coordinate level of EVCS with urban development planning	3	6	7	9	5.06	7.64	2.36	6.35 > 6.0
	Level of public facilities	2	7	6	9	4.52	7.68	0.32	4.72 < 6.0
	Deterioration on water resource	1	6	5	9	3.54	7.24	0.76	4.18 < 6.0
	Deterioration on soil and vegetation	2	7	8	10	5.24	7.35	3.65	6.30 > 6.0
	Waste discharge	2	6	5	10	3.75	8.26	0.74	4.56 < 6.0
	Noise pollution	2	6	7	9	3.64	6.84	3.16	5.24 < 6.0
Environment	Atmospheric particulates emission reduction	1	6	7	9	4.59	8.06	1.94	6.33 > 6.0
	Industrial electromagnetic field	2	5	7	10	3.68	5.64	6.36	4.66 < 6.0
	Radio interference	3	8	7	10	5.16	8.59	0.41	5.20 < 6.0
	GHG emission reduction	4	6	8	9	4.96	8.85	2.15	6.91 > 6.0
	Ecological influence	1	5	7	9	4.36	6.84	4.16	5.60 < 6.0
	Substation capacity permits	1	5	7	10	4.16	8.64	3.36	6.40 > 6.0
	Distance from the substation	1	5	7	10	4.35	6.89	5.11	5.62 < 6.0
	Power quality influence	3	7	6	10	5.89	7.68	1.32	6.35 > 6.0
	Power balance level	3	7	6	10	3.64	8.04	0.96	4.44 < 6.0
Technology	Power grid security implications	4	7	8	10	5.68	6.54	4.46	6.11 > 6.0
	Transformer capacity-load ratio	2	5	6	9	4.64	6.87	3.13	5.76 < 6.0
	Interface flow margin	3	8	7	10	3.74	8.94	0.06	4.79 < 6.0
	Voltage fluctuation	1	8	7	9	5.66	5.98	2.02	3.09 < 6.0
	Power grid frequency deviation	2	7	5	9	4.21	7.64	−0.64	5.21 < 6.0
	Harmonic pollution	2	6	7	7	3.95	4.68	3.32	4.32 < 6.0

Energies 2016, 9, 270

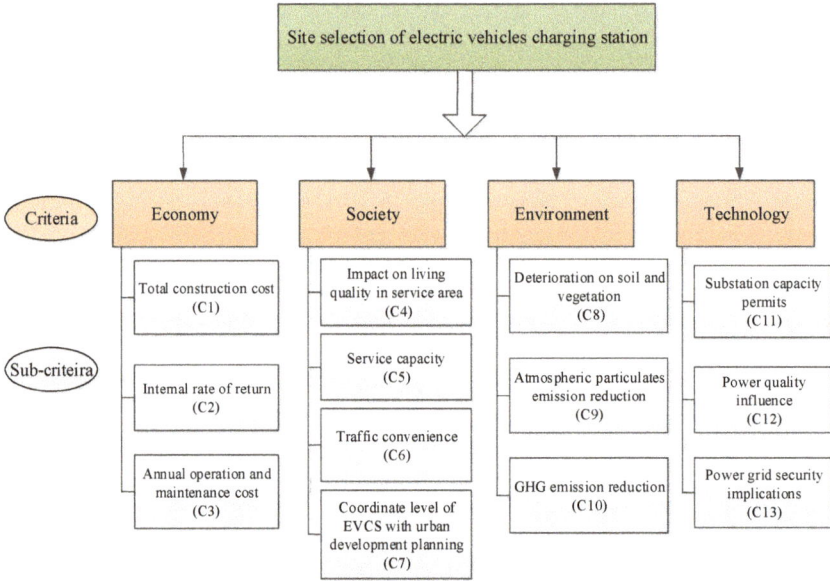

Figure 2. Evaluation index system for optimal siting of charging stations for electric vehicles.

6. Empirical Analysis

Tianjin is one of the most famous modern cities in China, which has been devoted to developing the electric-vehicle industry. In order to promote the sustainable development and management of the EV industry in Tianjin city, it is necessary to select the optimal sites for EVCSs. After reviewing the project feasibility research reports, the expert groups finally determine five EVCS site alternatives. The geographical locations of these five alternatives are shown in Figure 3. Five alternatives $A_i(1,2, \ldots 5)$ are located in the Beichen district, Dongli district, Nankai district, Jinnan district and Tanggu district in Tianjin, respectively.

The MCDM problem related to optimal siting of EVCS includes four criteria (economy, society, environment and technology) and thirteen sub-criteria. After reviewing the literatures and research reports related to all alternatives, each experts group give the linguistic ratings judgments for sub-criteria weights and sub-criteria performance of all alternatives. The rating results are listed in Tables 4 and 5.

Table 4. Linguistic ratings for sub-criteria weights.

	C1	C2	C3	C4	C5	C6	C7	C8	C9	C10	C11	C12	C13
E1	I	I	MI	MI	I	I	MI	MI	VI	VI	I	MI	MI
E2	I	VI	I	I	VI	VI	LI	I	AI	VI	LI	I	I
E3	I	VI	I	I	LI	MI	MI	I	I	AI	MI	LI	I
E4	VI	AI	MI	I	VI	MI	I	VI	VI	AI	I	MI	VI

192

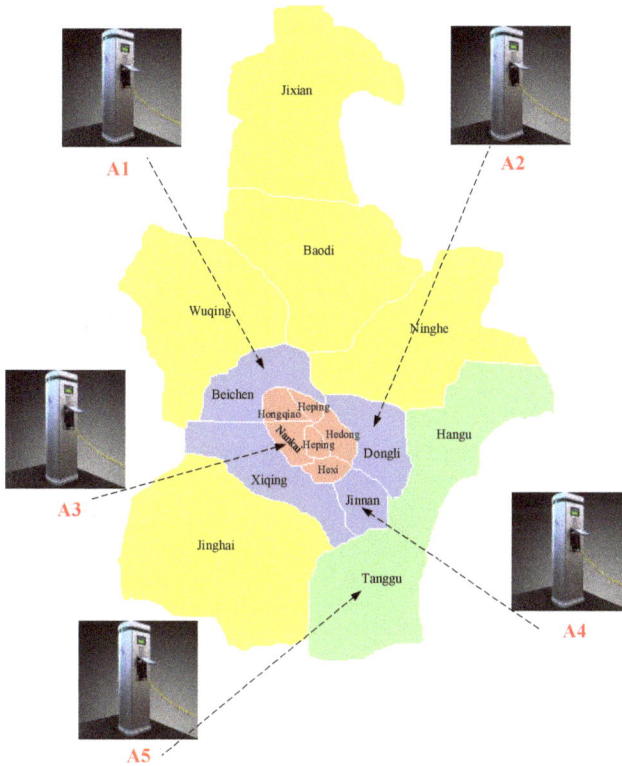

Figure 3. The geographical locations of five EVCS site alternatives

Table 5. Linguistic ratings for sub-criteria performances of five EVCS site alternatives.

		C1	C2	C3	C4	C5	C6	C7	C8	C9	C10	C11	C12	C13
E1	A1	MG	F	MG	MP	F	MG	F	F	MP	G	MP	F	MG
	A2	F	MG	MP	MP	MP	F	MP	MP	F	F	MP	F	F
	A3	MP	F	MG	F	F	MG	MP	F	VP	MP	F	MG	MG
	A4	VG	F	MG	F	F	MG	F	MG	VP	MP	MP	MG	F
	A5	P	F	F	MG	MP	MG	P	MP	F	G	F	P	P
E2	A1	F	MG	G	F	MG	MP	F	MG	F	F	P	F	G
	A2	MG	MP	F	F	F	F	MG	MP	F	MP	F	F	MG
	A3	F	F	F	MG	MP	MP	MP	MG	P	F	MP	MG	F
	A4	F	F	F	MG	MP	MP	F	F	MP	F	F	G	MG
	A5	MP	MP	MG	F	F	F	MP	MG	F	MG	MG	P	MP
E3	A1	MP	F	MP	MP	F	F	G	F	F	F	MP	VG	F
	A2	F	MP	F	MG	F	MP	F	MG	MG	MG	F	MG	MP
	A3	MG	F	G	MP	F	F	P	MP	F	VG	VP	G	VG
	A4	MG	F	G	MP	F	F	MP	MP	MG	MG	P	MG	G
	A5	MG	F	F	MG	G	G	P	F	MG	G	G	MP	F
E4	A1	F	F	F	F	P	P	F	MP	MP	MG	P	MG	G
	A2	F	F	MP	F	F	F	F	F	MG	MG	MG	F	MP
	A3	VG	P	MG	MP	MP	MP	P	F	MP	MP	MP	VG	MG
	A4	G	MP	MG	MP	MP	MP	MP	MG	MP	MP	MP	VG	MG
	A5	P	G	MG	MG	MG	F	G	MG	G	MG	MG	MP	MP

Then, according to Table 4 and Equations (9) and (10), the initial fuzzy decision matrix $\tilde{D}$ can be obtained, as below:

$$
\tilde{D} =
\begin{bmatrix}
\begin{array}{ccc}
 & C1 & \\
0.50 & 0.30 & 0.21 \\
0.43 & 0.27 & 0.20 \\
0.75 & 0.38 & 0.25 \\
0.25 & 0.19 & 0.17 \\
1.00 & 0.50 & 0.30
\end{array}
&
\begin{array}{ccc}
 & C2 & \\
0.38 & 0.59 & 0.81 \\
0.27 & 0.49 & 0.70 \\
0.70 & 0.89 & 1.00 \\
0.24 & 0.43 & 0.65 \\
0.38 & 0.59 & 0.78
\end{array}
&
\begin{array}{ccc}
 & C3 & \\
0.29 & 0.18 & 0.14 \\
0.50 & 0.25 & 0.17 \\
1.00 & 0.40 & 0.22 \\
0.20 & 0.14 & 0.11 \\
0.25 & 0.17 & 0.13
\end{array}
&
\begin{array}{ccc}
 & C4 & \\
0.23 & 0.46 & 0.69 \\
0.34 & 0.57 & 0.80 \\
0.57 & 0.80 & 1.00 \\
0.29 & 0.51 & 0.74 \\
0.51 & 0.74 & 0.97
\end{array}
&
\begin{array}{ccc}
 & C5 & \\
0.33 & 0.55 & 0.79 \\
0.30 & 0.55 & 0.79 \\
0.55 & 0.79 & 1.00 \\
0.24 & 0.48 & 0.73 \\
0.48 & 0.73 & 0.94
\end{array}
\\[3em]
\begin{array}{ccc}
 & C6 & \\
0.26 & 0.46 & 0.69 \\
0.29 & 0.51 & 0.74 \\
0.63 & 0.83 & 1.00 \\
0.29 & 0.51 & 0.74 \\
0.51 & 0.74 & 0.94
\end{array}
&
\begin{array}{ccc}
 & C7 & \\
0.44 & 0.67 & 0.86 \\
0.33 & 0.56 & 0.78 \\
0.61 & 0.83 & 1.00 \\
0.11 & 0.28 & 0.50 \\
0.36 & 0.56 & 0.75
\end{array}
&
\begin{array}{ccc}
 & C8 & \\
0.17 & 0.10 & 0.07 \\
0.20 & 0.11 & 0.08 \\
1.00 & 0.29 & 0.14 \\
0.14 & 0.09 & 0.07 \\
0.20 & 0.11 & 0.08
\end{array}
&
\begin{array}{ccc}
 & C9 & \\
0.21 & 0.42 & 0.63 \\
0.42 & 0.63 & 0.84 \\
0.68 & 0.87 & 1.00 \\
0.11 & 0.24 & 0.42 \\
0.47 & 0.68 & 0.87
\end{array}
&
\begin{array}{ccc}
 & C10 & \\
0.51 & 0.72 & 0.90 \\
0.36 & 0.56 & 0.77 \\
0.67 & 0.87 & 1.00 \\
0.26 & 0.46 & 0.67 \\
0.62 & 0.82 & 0.97
\end{array}
\\[3em]
\begin{array}{ccc}
 & C11 & \\
0.06 & 0.23 & 0.46 \\
0.34 & 0.57 & 0.80 \\
0.63 & 0.83 & 0.97 \\
0.14 & 0.34 & 0.57 \\
0.57 & 0.80 & 1.00
\end{array}
&
\begin{array}{ccc}
 & C12 & \\
0.11 & 0.08 & 0.06 \\
0.14 & 0.09 & 0.07 \\
1.00 & 0.25 & 0.13 \\
0.08 & 0.06 & 0.05 \\
1.00 & 0.25 & 0.13
\end{array}
&
\begin{array}{ccc}
 & C13 & \\
0.23 & 0.17 & 0.14 \\
0.50 & 0.28 & 0.19 \\
1.00 & 0.42 & 0.25 \\
0.25 & 0.18 & 0.14 \\
1.00 & 0.42 & 0.25
\end{array}
\end{bmatrix}
$$

According to Table 4, Equations (20) and (21), the subjective weights of sub-criteria can be obtained. On the other side, the objective weights of sub-criteria can also be obtained by fuzzy decision matrix $\tilde{D}$ and Equations (21)–(26). Finally, the combination weights of all sub-criteria equal to the average of subjective and objective sub-criteria weights, which are shown in Table 6, can be obtained that C2, C4, C5, C6, C7, C10, C11 are benefit sub-criteria and C1, C3, C8, C9, C12, C13 are cost sub-criteria.

Table 6. Combination weights of evaluation criteria.

	C1	C2	C3	C4	C5	C6	C7	C8	C9	C10	C11	C12	C13
$w_{subjective}$	0.0853	0.1109	0.0603	0.0686	0.0763	0.0686	0.0429	0.0769	0.1109	0.1282	0.0513	0.0429	0.0769
$w_{objective}$	0.0832	0.0845	0.0764	0.0769	0.0787	0.0795	0.0798	0.0681	0.0787	0.0728	0.0689	0.0781	0.0743
w_j	0.0842	0.0977	0.0683	0.0727	0.0775	0.0741	0.0614	0.0725	0.0948	0.1005	0.0601	0.0605	0.0756

The normalized fuzzy decision matrix can be obtained based on Equations (11) and (12). Then the distances of alternatives from the positive ideal solutions and negative ideal solutions can be calculated according to Equations (6) and (16)–(18). Finally, compute the values of Q_i for five EVCS site alternatives according to Equation (13). And thus rank and determine the optimal site for EVCS based on the principle of VIKOR. The results are shown in Table 7.

Table 7. The values of S_i, R_i and Q_i for each alternative.

	A1	A2	A3	A4	A5
S_i	0.512	0.532	0.972	0.443	0.759
R_i	0.070	0.066	0.049	0.084	0.064
Q_i	0.733	0.655	0.000	1.000	0.408
Rank	4	3	1	5	2

Obviously, EVCS site alternative A3 outranks other four alternatives. Therefore, A3, namely the EVCS site in Nankai district of Tianjin should be selected as the optimal EVCS site.

7. Discussion

The EVCS site alternatives are ranked by using FDM, combination weights and fuzzy GRA-VIKOR methods. Based on the Q_i, the ranking of all EVCS selections in descending order are A3, A5, A2, A1 and A4. The best alternative is found to be A3, and the second best alternative is A5. Based on above results, this proposed model can easily evaluate and select a best alternative. In this section, to examine the rationality and stability of the proposed framework and analysis results, the sensitivity analysis of v value and sub-criteria weights are presented.

Table 6 shows that the sub-criteria C9 and C10 affiliated with the environmental aspect obtain much more attention from the expert group, which reflects the strategy and energy saving and environment protection goals of the Chinese government. Meanwhile, the sub-criteria affiliated with economic development are not so important as before, which is consistent with the development goals of China. As we all know, in recent years, transportation and electricity industry has suffered pressures and challenges from the "twelfth five-year" plan and the environmental protection law of China, which indicates the responsibility and target of these industries for environment protection. Moreover, the severe environment and resource issues have posed undesirable conditions to humans for living. Therefore, the environmental aspect has been given more consideration by experts for the optimal siting for EVCSs in China.

As mentioned above, this study uses the variation of v values to demonstrate that all of them do not affect the analysis results (Figure 4). The v values are postulated to change from 0.1 to 0.9, while the ranking orders of five EVCSs are same, namely A3 > A5 > A2 > A1 > A4. And thus, this study can confirm that the results obtained by using the proposed model are reliable and effective.

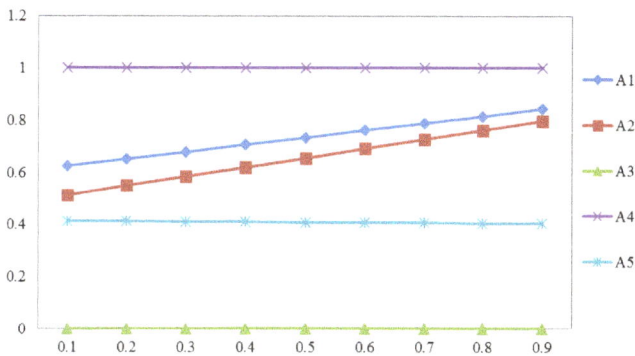

Figure 4. Sensitivity analysis of v value for each alternative.

Next, a sensitivity analysis on the impacts of sub-criteria weights for optimal EVCS siting is presented, so as to obtain better insight of evaluation results and verify the robustness of evaluation results. According to the criteria, thirteen sub-criteria are divided into four analysis aspects, namely economy, society, environment and technology. All sub-criteria have 10%, 20% and 30% less weight than the base weight and 10%, 20% and 30% more weight than the base weight (all base weights are shown in Table 6).

It can be seen that in Figure 5, the Q_i of A5 and A1 decrease when the sub-criterion C1 becomes less important. The Q_i of A2 increases when the weight of C1 becomes more important, and it ranks fourth, surpassed by A1. However, no matter how the C1 weight changes, the Q_i of A3 always has the lowest score, indicating the best alternative. As C2 is given more importance, only the Q_i of A2 shows a small rising tendency, while the scores of other alternatives remain relatively stable although C2 carries large weight in the optimal EVCS siting. In the case of C3, the Q_i of A1 and A5 dramatically rise along with weight increase, which gets closer to that of A4 and A2, respectively. A3 and A4 are still the optimal and worst sites the same as in the base case. Apparently, C1 and C3 are sensible sub-criteria which dramatically affect the optimal EVCS siting results. However, no matter how the weights in the economy group change, A3 is always the best choice in the optimal siting of EVCS in Tianjin.

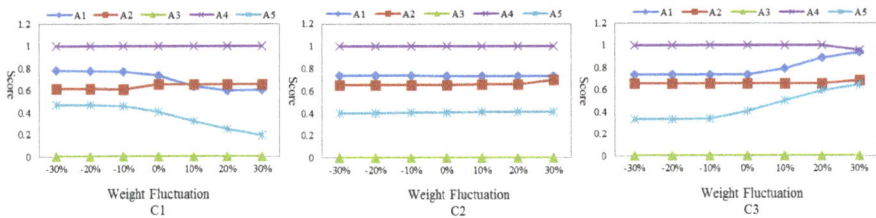

Figure 5. Sensitivity analysis results of sub-criteria in the economy group.

The case where the society sub-criteria have 10%, 20% and 30% more or less weight than the base weight are shown in Figure 6. The Q_i scores of the five EVCSs have tiny variations, no matter how the sub-criteria C4, C5, C6 and C7 change. Therefore, the sub-criteria in the society group are not sensitive factors, and A3 and A4 are the optimal and worst site in the optimal siting for EVCSs, no matter how the sub-criteria weights in the society group change.

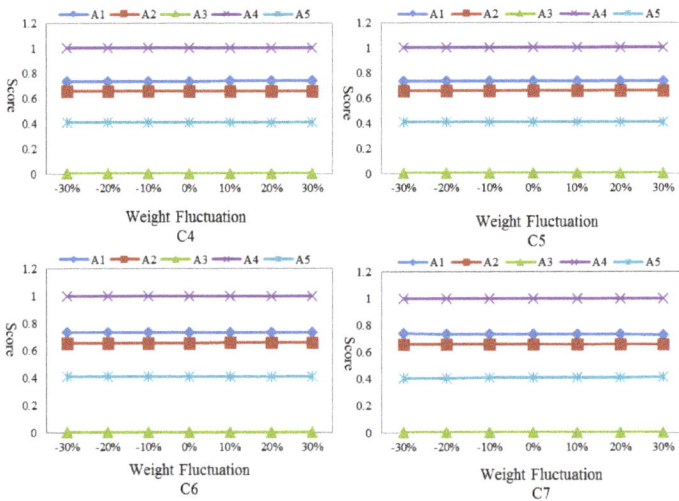

Figure 6. Sensitivity analysis results of the sub-criteria in the society group.

For the sub-criteria in the environment group, the Q_i of A1, A2 and A5 increase when the sub-criteria C8 becomes more important, and A2 ranks third, surpassed by A1(Figure 7). Meanwhile, the score of A3 and A4 remain stable with the weight variation of C8. For the weight changes of C9, scores of A1, A2 and A5 show a decreasing tendency along with increase of weights, while the rank of all alternatives keep consistent with the base situation. In the case of C10, the Q_i of the five alternatives remain stable with increasing weight. Therefore, C8 in the sensitive sub-criterion which obviously affects the EVCS site selection results. No matter how the weights in the economy group change, A3 is always the best choice in the optimal EVCS siting.

For the sub-criteria in the technology group, the Q_i of the five alternatives remain stable when the sub-criteria C11 and C12 become more important (Figure 8). Moreover, when the weight of C13 becomes more important, the Q_i of A1 and A2 present a rising tendency, while the Q_i of A5 shows a decreasing tendency. However, no matter how the sub-criteria weights in the technology group change, the ranking order of the five EVCSs remains relatively stable, and A3 is always the top choice in the EVCS site selection.

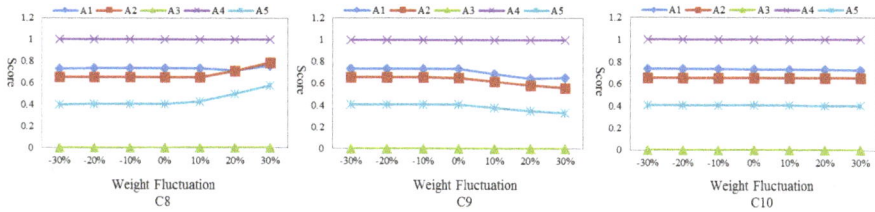

Figure 7. Sensitivity analysis results of the sub-criteria in the environment group.

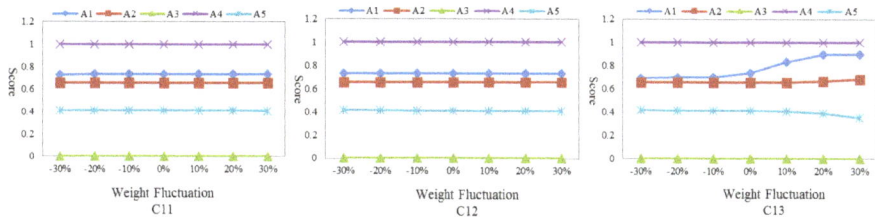

Figure 8. Sensitivity analysis results of the sub-criteria in the technology group.

Above all, electric vehicle charging station A3 always secures its best ranking, no matter how the sub-criteria weights change. It can be verified that the optimal EVCS siting results using GRA-VIKOR and combination weighting techniques is robust and effective. This study can distinguish the priorities of alternatives more easily, and thus help decision makers evaluate and identify the best alternative and more improvement items.

8. Conclusions

A comprehensive framework for selecting the optimal site for EVCSs is studied in this paper. Considering the concept of extended sustainability, experts' opinions, and industry background, the final evaluation sub-criteria for optimal EVCS siting are determined based on FDM, which consists of four pillars: economy, society, environment and technology. To address the fuzziness and uncertainty of subjective factor and human judgment, a fuzzy GRA-VIKOR method is employed to determine the optimal EVCS site. It is worth mentioning that GRA is used to measure the distances of fuzzy numbers between alternatives to ideal solutions in this study, which can efficiently avoid the priority result of fuzzy numbers, as well as ensure a consistent ranking list for all alternatives. Moreover, in order to provide a scientific weighting system, the sub-criteria weights are determined combining the subjective weights of experts' opinions as well as the objective weights of the entropy method, which updates the weighting process of conventional fuzzy VIKOR. The evaluation results shows that the sub-criteria C9 and C10 affiliated with the environment obtain much more attention from the experts group, and the alternative A3 in Tianjin Nankai district is determined as the optimal EVCS site. Last but not least, to test the robustness and effectiveness of decision results, a sensitivity analysis is presented, which showed that the siting results remain stable no matter how the v value and sub-criteria weights change. Moreover, it can also be found that C1, C3 and C8 are the sensitive sub-criteria which dramatically affect the optimal EVCS siting result.

Although this study realized the optimal siting of EVCSs by using FDM, combination weighting and fuzzy GRA-VIKOR techniques, limitations may still exist due to the fact the evaluation criteria will change along with objective conditions. Moreover, from a methodological perspective, it would be helpful to test the proposed framework with other approaches. The results from these approaches could be compared with the results in this paper, which is an outline for the future research.

Acknowledgments: This study is supported by China Scholarship Council, the National Natural Science Foundation of China (Project number: 71373076), the Fundamental Research Funds for the Central Universities (2015XS31), and the Science and Technology Project of State Grid Corporation of China (Contract number: SGHB0000DKJS1400116).

Author Contributions: Huiru Zhao and Nana Li conceived and designed the research method used in this paper; Nana Li performed the empirical analysis and wrote the paper.

Conflicts of Interest: The authors declare no conflict of interest.

References

1. Mao, X.; Yang, S.; Liu, Q.; Tu, J.; Jaccard, M. Achieving CO_2 emission reduction and the co-benefits of local air pollution abatement in the transportation sector of China. *Environ. Sci. Policy* **2012**, *21*, 1–13. [CrossRef]
2. Tie, S.F.; Tan, C.W. A review of energy sources and energy management system in electric vehicles. *Renew. Sustain. Energy Rev.* **2013**, *20*, 82–102. [CrossRef]
3. Wu, Y.; Yang, Z.; Lin, B.; Liu, H.; Wang, R.; Zhou, B.; Hao, J. Energy consumption and CO_2 emission impacts of vehicle electrification in three developed regions of China. *Energy Policy* **2012**, *48*, 537–550. [CrossRef]
4. Cai, H.; Jia, X.; Chiu, A.S.; Hu, X.; Xu, M. Siting public electric vehicle charging stations in Beijing using big-data informed travel patterns of the taxi fleet. *Transp. Res. Part D Transp. Environ.* **2014**, *33*, 39–46. [CrossRef]
5. Subhadra, B.G. Sustainability of algal biofuel production using integrated renewable energy park (IREP) and algal biorefinery approach. *Energy Policy* **2010**, *38*, 5892–5901. [CrossRef]
6. Opricovic, S.; Tzeng, G.H. Compromise solution by MCDM methods: A comparative analysis of VIKOR and TOPSIS. *Eur. J. Oper. Res.* **2004**, *156*, 445–455. [CrossRef]
7. Rojas-Zerpa, J.C.; Yusta, J.M. Application of multicriteria decision methods for electric supply planning in rural and remote areas. *Renew. Sustain. Energy Rev.* **2015**, *52*, 557–571. [CrossRef]
8. Tansel İç, Y.; Yıldırım, S. Moora-based taguchi optimisation for improving product or process quality. *Int. J. Prod. Res.* **2013**, *51*, 3321–3341. [CrossRef]
9. San Cristóbal, J. Multi-criteria decision-making in the selection of a renewable energy project in Spain: The vikor method. *Renew. Energy* **2011**, *36*, 498–502. [CrossRef]
10. Mardani, A.; Jusoh, A.; Zavadskas, E.K. Fuzzy multiple criteria decision-making techniques and applications—Two decades review from 1994 to 2014. *Expert Syst. Appl.* **2015**, *42*, 4126–4148. [CrossRef]
11. Tzeng, G.-H.; Huang, C.-Y. Combined DEMATEL technique with hybrid MCDM methods for creating the aspired intelligent global manufacturing & logistics systems. *Ann. Oper. Res.* **2012**, *197*, 159–190.
12. Kaya, T.; Kahraman, C. Multicriteria renewable energy planning using an integrated fuzzy VIKOR & AHP methodology: The case of Istanbul. *Energy* **2010**, *35*, 2517–2527.
13. Mardani, A.; Zavadskas, E.K.; Govindan, K.; Amat Senin, A.; Jusoh, A. VIKOR technique: A systematic review of the state of the art literature on methodologies and applications. *Sustainability* **2016**, *8*, 37. [CrossRef]
14. Mardani, A.; Jusoh, A.; Zavadskas, E.K.; Cavallaro, F.; Khalifah, Z. Sustainable and renewable energy: An overview of the application of multiple criteria decision making techniques and approaches. *Sustainability* **2015**, *7*, 13947–13984. [CrossRef]
15. Chang, S.-C.; Tsai, P.-H.; Chang, S.-C. A hybrid fuzzy model for selecting and evaluating the e-book business model: A case study on Taiwan e-book firms. *Appl. Soft Comput.* **2015**, *34*, 194–204. [CrossRef]
16. Kuo, M.-S.; Liang, G.-S. Combining VIKOR with GRA techniques to evaluate service quality of airports under fuzzy environment. *Expert Syst. Appl.* **2011**, *38*, 1304–1312. [CrossRef]
17. Zhang, S.-F.; Liu, S.-Y. A GRA-based intuitionistic fuzzy multi-criteria group decision making method for personnel selection. *Expert Syst. Appl.* **2011**, *38*, 11401–11405. [CrossRef]
18. Hou, J. Grey relational analysis method for multiple attribute decision making in intuitionistic fuzzy setting. *J. Converg. Inf. Technol.* **2010**, *5*, 194–199.
19. Simpson, A. *Cost-Benefit Analysis of Plug-in Hybrid Electric Vehicle Technology*; National Renewable Energy Laboratory: Golden, CA, USA, 2006.
20. Schroeder, A.; Traber, T. The economics of fast charging infrastructure for electric vehicles. *Energy Policy* **2012**, *43*, 136–144. [CrossRef]

21. Hawkins, T.R.; Singh, B.; Majeau-Bettez, G.; Strømman, A.H. Comparative environmental life cycle assessment of conventional and electric vehicles. *J. Ind. Ecol.* **2013**, *17*, 53–64. [CrossRef]
22. Matsuhashi, R.; Kudoh, Y.; Yoshida, Y.; Ishitani, H.; Yoshioka, M.; Yoshioka, K. Life cycle of CO_2-emissions from electric vehicles and gasoline vehicles utilizing a process-relational model. *Int. J. Life Cycle Assess.* **2000**, *5*, 306–312. [CrossRef]
23. Putrus, G.; Suwanapingkarl, P.; Johnston, D.; Bentley, E.; Narayana, M. Impact of electric vehicles on power distribution networks. In Proceedings of the Vehicle Power and Propulsion Conference, Dearborn, MI, USA, 7–11 September 2009; pp. 827–831.
24. Clement-Nyns, K.; Haesen, E.; Driesen, J. The impact of charging plug-in hybrid electric vehicles on a residential distribution grid. *IEEE Trans. Power Syst.* **2010**, *25*, 371–380. [CrossRef]
25. Mets, K.; Verschueren, T.; Haerick, W.; Develder, C.; De Turck, F. Optimizing smart energy control strategies for plug-in hybrid electric vehicle charging. In Proceedings of the Network Operations and Management Symposium Workshops (NOMS Wksps), Osaka, Japan, 19–23 April 2010; pp. 293–299.
26. Rivera, S.; Wu, B.; Kouro, S.; Yaramasu, V.; Wang, J. Electric vehicle charging station using a neutral point clamped converter with bipolar DC bus. *IEEE Trans. Ind. Electron.* **2015**, *62*, 1999–2009. [CrossRef]
27. Liu, Z.; Wen, F.; Ledwich, G. Optimal planning of electric-vehicle charging stations in distribution systems. *IEEE Trans. Power Deliv.* **2013**, *28*, 102–110. [CrossRef]
28. Wirges, J.; Linder, S.; Kessler, A. Modelling the development of a regional charging infrastructure for electric vehicles in time and space. *Eur. J. Transp. Infrastruct. Res.* **2012**, *12*, 391–416.
29. Jia, L.; Hu, Z.; Song, Y.; Luo, Z. Optimal siting and sizing of electric vehicle charging stations. In Proceedings of the 2012 IEEE International Electric Vehicle Conference (IEVC), Greenville, NC, USA, 4–8 March 2012; pp. 1–6.
30. Ge, S.; Feng, L.; Liu, H. The planning of electric vehicle charging station based on grid partition method. In Proceedings of the 2011 International Conference on Electrical and Control Engineering (ICECE), Yichang, China, 16–18 September 2011; pp. 2726–2730.
31. Xi, X.; Sioshansi, R.; Marano, V. Simulation-optimization model for location of a public electric vehicle charging infrastructure. *Transp. Res. Part D Transp. Environ.* **2013**, *22*, 60–69. [CrossRef]
32. Sathaye, N.; Kelley, S. An approach for the optimal planning of electric vehicle infrastructure for highway corridors. *Transp. Res. Part E Logist. Transp. Rev.* **2013**, *59*, 15–33. [CrossRef]
33. Pashajavid, E.; Golkar, M. Optimal placement and sizing of plug in electric vehicles charging stations within distribution networks with high penetration of photovoltaic panels. *J. Renew. Sustain. Energy* **2013**, *5*, 053126. [CrossRef]
34. Zi-fa, L.; Wei, Z.; Xing, J.; Ke, L. Optimal planning of charging station for electric vehicle based on particle swarm optimization. In Proceedings of the Innovative Smart Grid Technologies-Asia (ISGT Asia), Tianjing, China, 21–24 May 2012; pp. 1–5.
35. You, P.-S.; Hsieh, Y.-C. A hybrid heuristic approach to the problem of the location of vehicle charging stations. *Comput. Ind. Eng.* **2014**, *70*, 195–204. [CrossRef]
36. Yao, W.; Zhao, J.; Wen, F.; Dong, Z.; Xue, Y.; Xu, Y.; Meng, K. A multi-objective collaborative planning strategy for integrated power distribution and electric vehicle charging systems. *IEEE Trans. Power Syst.* **2014**, *29*, 1811–1821. [CrossRef]
37. Sadeghi-Barzani, P.; Rajabi-Ghahnavieh, A.; Kazemi-Karegar, H. Optimal fast charging station placing and sizing. *Appl. Energy* **2014**, *125*, 289–299. [CrossRef]
38. Chung, S.H.; Kwon, C. Multi-period planning for electric car charging station locations: A case of Korean expressways. *Eur. J. Oper. Res.* **2015**, *242*, 677–687. [CrossRef]
39. Yager, R.R.; Zadeh, L.A. *An Introduction to Fuzzy Logic Applications in Intelligent Systems*; Springer Science & Business Media: New York, NY, USA, 2012; Volume 165.
40. Novák, V.; Perfilieva, I.; Mockor, J. *Mathematical Principles of Fuzzy Logic*; Springer Science & Business Media: New York, NY, USA, 2012; Volume 517.
41. Kumar, A.; Kaur, J.; Singh, P. A new method for solving fully fuzzy linear programming problems. *Appl. Math. Model.* **2011**, *35*, 817–823. [CrossRef]
42. Li, D.-F. A ratio ranking method of triangular intuitionistic fuzzy numbers and its application to MADM problems. *Comput. Math. Appl.* **2010**, *60*, 1557–1570. [CrossRef]

43. Yao, J.-S.; Wu, K. Ranking fuzzy numbers based on decomposition principle and signed distance. *Fuzzy Sets Syst.* **2000**, *116*, 275–288. [CrossRef]
44. Lee, A.H.; Chen, W.-C.; Chang, C.-J. A fuzzy AHP and BSC approach for evaluating performance of it department in the manufacturing industry in Taiwan. *Expert Syst. Appl.* **2008**, *34*, 96–107. [CrossRef]
45. Chen, T.-Y.; Tsao, C.-Y. The interval-valued fuzzy TOPSIS method and experimental analysis. *Fuzzy Sets Syst.* **2008**, *159*, 1410–1428. [CrossRef]
46. Chen, L.Y.; Wang, T.-C. Optimizing partners' choice in IS/IT outsourcing projects: The strategic decision of fuzzy VIKOR. *Int. J. Prod. Econ.* **2009**, *120*, 233–242. [CrossRef]
47. Hsu, C.-H.; Wang, F.-K.; Tzeng, G.-H. The best vendor selection for conducting the recycled material based on a hybrid MCDM model combining danp with VIKOR. *Resour. Conserv. Recycl.* **2012**, *66*, 95–111. [CrossRef]
48. Zhang, N.; Wei, G. Extension of VIKOR method for decision making problem based on hesitant fuzzy set. *Appl. Math. Model.* **2013**, *37*, 4938–4947. [CrossRef]
49. Shemshadi, A.; Shirazi, H.; Toreihi, M.; Tarokh, M.J. A fuzzy VIKOR method for supplier selection based on entropy measure for objective weighting. *Expert Syst. Appl.* **2011**, *38*, 12160–12167. [CrossRef]
50. Sanayei, A.; Mousavi, S.F.; Yazdankhah, A. Group decision making process for supplier selection with VIKOR under fuzzy environment. *Expert Syst. Appl.* **2010**, *37*, 24–30. [CrossRef]
51. Lihong, M.; Yanping, Z.; Zhiwei, Z. Improved vikor algorithm based on AHP and Shannon entropy in the selection of thermal power enterprise's coal suppliers. In Proceedings of the International Conference on Information Management, Innovation Management and Industrial Engineering, Taipei, Taiwan, 19–21 December 2008; pp. 129–133.
52. Stahura, F.L.; Godden, J.W.; Xue, L.; Bajorath, J. Distinguishing between natural products and synthetic molecules by descriptor Shannon entropy analysis and binary QSAR calculations. *J. Chem. Inf. Comput. Sci.* **2000**, *40*, 1245–1252. [CrossRef] [PubMed]
53. Godden, J.W.; Stahura, F.L.; Bajorath, J. Variability of molecular descriptors in compound databases revealed by Shannon entropy calculations. *J. Chem. Inf. Comput. Sci.* **2000**, *40*, 796–800. [CrossRef] [PubMed]
54. Ghorbani, M.; Arabzad, S.M.; Bahrami, M. Implementing Shannon entropy, SWOT and mathematical programming for supplier selection and order allocation. *Int. J. Suppl. Chain Manag.* **2012**, *1*, 43–47.

Article

Robust Peak-Shaving for a Neighborhood with Electric Vehicles

Marco E. T. Gerards * and Johann L. Hurink

Faculty of Electrical Engineering, Mathematics and Computer Science, 7500 AE Enschede, The Netherlands; j.l.hurink@utwente.nl
* Correspondence: m.e.t.gerards@utwente.nl; Tel.: +31-53-489-4896

Academic Editor: Chunhua Liu
Received: 4 May 2016; Accepted: 21 July 2016; Published: 28 July 2016

Abstract: Demand Side Management (DSM) is a popular approach for grid-aware peak-shaving. The most commonly used DSM methods either have no look ahead feature and risk deploying flexibility too early, or they plan ahead using predictions, which are in general not very reliable. To counter this, a DSM approach is presented that does not rely on detailed power predictions, but only uses a few easy to predict characteristics. By using these characteristics alone, near optimal results can be achieved for electric vehicle (EV) charging, and a bound on the maximal relative deviation is given. This result is extended to an algorithm that controls a group of EVs such that a transformer peak is avoided, while simultaneously keeping the individual house profiles as flat as possible to avoid cable overloading and for improved power quality. This approach is evaluated using different data sets to compare the results with the state-of-the-art research. The evaluation shows that the presented approach is capable of peak-shaving at the transformer level, while keeping the voltages well within legal bounds, keeping the cable load low and obtaining low losses. Further advantages of the methodology are a low communication overhead, low computational requirements and ease of implementation.

Keywords: adaptive scheduling; demand side management; electric vehicles; optimal scheduling; smart grids

---

1. Introduction

In the future, we expect an increasing penetration of electric vehicles, photovoltaic panels, heat pumps and wind turbines. As especially heat pumps and electric vehicles create relatively large and synchronized peaks, often at times when there is little renewable energy available, the balance between production and consumption of electricity becomes more and more an urgent issue.

To counter the problems that arise due to this trend (for a survey, see [1]), Demand Side Management (DSM) techniques can be deployed to prevent peaks. Here, DSM is the collective term for a set of techniques that control the production or consumption within the customers' premises. A central entity (e.g., the network operator) requests customers, via a steering signal, to adapt their production or consumption in order to shape the load profile of a certain subgroup of the grid (e.g., a group of houses). These customers can either adapt their behavior manually, or install a device that is referred to as a Home Energy Management System (HEMS), which makes such decisions on their behalf. In the latter situation, the HEMS should start appliances (referred to as smart appliances) when the consumption should be increased, or should advance or delay the use of such an appliance when the consumption should be decreased. In addition, energy storage (e.g., batteries) can be used to mitigate a surplus of renewable energy by charging, and a peak in the consumption by discharging. In this article, we focus on electric vehicles as they introduce both a significant load, and provide a lot of flexibility.

There are several types of DSM approaches described in the literature. In this article, we consider two important classes of such approaches. The first class, called *control-based DSM*, makes control decisions for appliances for the next time period based on available data at the given time such as consumption, flexibility, priorities, etc. The second class, called *planning-based DSM*, plans control decisions for a longer period in the future (e.g., one day ahead). Because control-based DSM does not plan ahead, it risks using the available flexibility too early, such that no flexibility is left when later on, e.g., a large peak occurs. Planning-based DSM does take this into account, but requires predictions of future production/consumption. For example, the planning-based DSM approaches in [2,3] require such predictions at a house level. In Section 2, we discuss control and planning-based DSM in more detail, and argue that both approaches have serious disadvantages.

In Section 3, we describe how a HEMS can be used to control the charging of a single electric vehicle (EV) at the house level to follow some desired load profile. In contrast to the related work (e.g., [4]), no prediction of a load profile is required. We show that a prediction of a single parameter characterizing the optimal solution and a prediction of the load for the upcoming interval is sufficient to make a near-optimal planning of an electric vehicle. We provide an analysis that studies how the results of our online approach approximates the optimal solution, and give a bound on the maximal relative deviation. This bound shows that, under reasonable circumstances, the costs for the achieved solution are only a few percent more than that of the optimal solution.

Section 4 generalizes these results to the case of a neighborhood with multiple EVs. The presented solution is a peak-shaving approach that keeps the overall load of the neighborhood below a certain level, and simultaneously keeps the load profiles of each individual house as flat as possible. By this, also the risk that the voltage exceeds the legal bounds (e.g., 207 V–253 V in The Netherlands [5]) is minimized, the cable load is kept low and thereby unnecessary losses are avoided. Note, that a DSM approach that does not take these aspects into account may cause more problems than it solves, as was shown by Hoogsteen et al. [6].

Since Sections 3 and 4 depend on predictions of a few characterizing values, we study how these predictions are obtained. Section 5 uses measured data and discusses how the required predictions of the characteristics can be easily obtained.

The potential of our approach is evaluated in Section 6 and compared to the state-of-the-art research. We show that the few parameters we need can be predicted accurately and are sufficient for near optimal operation of a group of houses. Section 7 concludes this article.

Summarizing, the contribution of this article is a demand side management methodology with the following properties:

- It only uses a prediction of a single parameter that characterizes the optimal solution, together with a power prediction of the house for the upcoming interval to plan the charging of an electric vehicle in a house.
- A peak at the transformer can be counteracted with low communication overhead using a decision making process that also requires predictions of characteristics that aid at making trade-offs at the neighborhood level.
- A prediction scheme for the required parameters, combined with a sensitivity study that shows that the results do not suffer much from prediction errors.

2. Related Work

There are several classes of DSM approaches in the literature. In Section 2.1, we discuss control-based DSM approaches, and in Section 2.2 planning-based DSM approaches. As we use electric vehicles as a proof of concept, we treat DSM applied for a group of electric vehicles specifically in Section 2.3.

2.1. Control-Based DSM

The first class of approaches is control-based DSM. These approaches use an estimation of the current state of the system, and make online control decisions based on this state. As these approaches do not take future decisions into account, they may deploy flexibility at an early stage, while this flexibility might be of more use at a later time. As a result, it can occur that a large peak cannot be prevented because the system has already used most of its flexibility.

Example 1. *A typical example for this is the control of a battery in combination with photovoltaic (PV) panels in a house. A control-based DSM approach would charge the battery starting at the time more energy is produced by the PV than is consumed inside the house. On a sunny summer day, this generally leads to a full battery before noon. As a consequence, the high PV peak cannot be reduced by the battery and the local distribution grid may get capacity problems. In Germany, this is nowadays already a serious problem (see, e.g., [7]).*

An example of control-based DSM is the PowerMatcher (see, e.g., [8,9]). Recently, some of these control-based approaches have been adapted to incorporate some form of predictions to mitigate to some extent the mentioned disadvantages (see, e.g., [10]).

2.2. Planning-Based DSM

The second class of approaches, planning-based DSM, makes a more detailed prediction of the future power production/consumption (e.g., one day ahead in time) and uses this information to plan the control decisions for smart appliances (a planning) to attain a given goal (e.g., peak-shaving). The strength of this approach, compared to control-based DSM, is that flexibility can be preserved for when it is required the most. A disadvantage may be that to make the needed plannings, specialized algorithms are required at device level (e.g., [4]) and/or for groups of appliances or houses (e.g., [2]).

In the approaches described in [2,3], each house makes a prediction of its average power consumption for each 15 min interval within the upcoming day (i.e., 96 intervals). These predictions are sent to a neighborhood controller, and summed up to obtain the predicted neighborhood profile. This predicted neighborhood profile indicates when peaks occur, and gives also hints on how the house profiles could be adopted to counter these peaks. Based on this, the neighborhood controller requests some or all houses to follow a new desired (or difference) profile [2], or it sends incentives to houses to adapt their profile [3]. Each house on its turn uses both the information sent by the neighborhood controller and its previous predictions and planning to make a new planning for all its controllable appliances. This procedure is repeated iteratively until the neighborhood controller is satisfied with the resulting planned neighborhood profile.

The drawback of these type of approaches is that they are sensitive to poor predictions. When the predictions are not accurate, the derived planning may be of a low quality or even not valid. Although some information is relatively easy to predict at the neighborhood level [11,12], it is often hard to predict the same information at the household level [11]. Furthermore, the error made due to poor predictions at the household level may accumulate at the neighborhood level. This effect can be explained the best by a simple artificial example:

Example 2. *Household power prediction problem. Consider a group of 100 houses, where each house contains a television and where each of the television is used with a 90% probability at 8 p.m. When a house controller has to decide if it incorporates the TV within its planning at 8 p.m., it should do this since the probability is close to 1. Within the planning at the neighborhood level, this implies that all 100 televisions are likely to be switched on at 8 p.m. However, when one would estimate, on a neighborhood level, how many televisions are turned on at 8 p.m., the best estimator is the expected value, which is 90 TVs. This shows that if predictions are based on probabilities, prediction errors may accumulate at the neighborhood level.*

The example indicates that summing up local predictions of events may cause prediction errors that would not have been made at a more global level. Furthermore, predictions based on statistics alone work in general only for large groups, e.g., for neighborhoods instead of houses [11]. When this observation is used as a work-around in the planning-based DSM approach, details about user behavior and measurements are accumulated at the neighborhood controller level such that decisions may be made there based on probabilities over large groups (instead of within the house). However, such an approach would no longer be (decentralized) DSM, but is, e.g., closer to demand response [1] and no longer distributed. Furthermore, it would require sharing of privacy sensitive information. The goal of this article is to avoid the mentioned restrictions. Summarizing, we may state that robust demand side management should only rely on house level predictions if this methodology is robust against prediction errors.

2.3. Groups of Electric Vehicles

For the application of EV charging, the state-of-the-art research on DSM (see, e.g., [13–17]) requires EV owners to share information (e.g., EV arrival) with a centralized controller. The approach presented in this article does not require such sharing of privacy sensitive information and has a low communication footprint.

The approach ORCHARD presented in [14] also flattens the load profile for a group of EVs without requiring knowledge of future EV arrivals. It has a time complexity of $\mathcal{O}(N^5)$ for N EVs and is similar to our work in the sense that we also do not know about EV arrivals beforehand. Contrary to [14], our work requires $\mathcal{O}(M \log M)$ time (for M time intervals) when an EV arrives (within each house). Furthermore, our research flattens the load both at the house level and at the neighborhood level.

3. Online Electric Vehicle Planning

This section describes for the case of charging an Electric Vehicle (EV) how a house can follow some profile, for example to make the load profile as flat as possible, by using proper charging settings. Section 3.1 introduces the corresponding EV planning problem. Section 3.2 presents our approach, which is inspired by online optimization, rolling horizon planning, and model predictive control: at the start of every time interval, we use a prediction of the power for this interval together with a prediction of a single value that characterizes the requirements for the future intervals to determine the charging power for the next time interval.

By delaying the decision for the amount of charging done in an interval until the very start of this interval, we can hope that more accurate predictions are possible. Furthermore, we can compensate for errors made in previous time intervals. Finally, Section 3.3 discusses the influence of prediction errors on this algorithm.

3.1. The EV Charging Problem

We first introduce some notation before we give a formal definition of the EV planning problem. Let a_n be the time interval at which the electric vehicle arrives at house $n \in \{1, \ldots, N\}$ and is ready to be charged, let d_n be the the interval at which the charging has to be completed, and let C_n be the energy that needs to be charged in the intervals $a_n, \ldots, d_n$. The EV charging decisions can be described by a vector $\vec{x}_n = (x_{n,1}, \ldots, x_{n,M})$, where $x_{n,m}$ is the amount of electricity charged for vehicle n during time interval m. The maximum charging power is given by $\bar{x}$ (in watts), and the vector $\vec{x}_n$ is feasible if $x_{n,m} = 0$ for $m < a_n$ and $m > d_n$, $0 \leq x_{n,m} \leq \bar{x}$ in the remaining intervals, and $\sum_{m=a_n}^{d_n} \tau x_{n,m} = C_n$, where τ is the length of a time interval. To ease the notation, we use $\tau = 1$.

Let the desired power profile for the house and EV load together be $\vec{q}_n$. The goal is to determine a charging vector $\vec{x}_n$ such that the uncontrollable house power consumption $(\vec{p}_n)$ together with the charging $\vec{x}_n$ matches this desired power profile $\vec{q}_n$ as well as possible. More precisely, we look for a vector $\vec{x}_n$ that minimizes the Euclidean distance between the vectors $\vec{q}_n$ and $(\vec{p}_n + \vec{x}_n)$. To express the difference between the actual and desired load, we define $z_{n,m} := p_{n,m} + x_{n,m} - q_{n,m}$, $m \in \{1, \ldots, M\}$.

Note that the aim of the objective is to bring the values of $z_{n,m}$ as close as possible to zero, where large values are addressed first in order to minimize the Euclidean distance (i.e., due to quadratic terms).

To ease the discussion and notation further, we omit the index n (we consider a single house), and we may assume w.l.o.g. that $a := a_n = 1$ and $d := d_n = M$. This leads to the following optimization problem:

Problem 1.

$$\min_{\vec{x}} \sqrt{\sum_{m=1}^{M} z_m^2}, \tag{1}$$

$$\text{subject to} \sum_{m=1}^{M} x_m = C, \tag{2}$$

$$z_m = p_m + x_m - q_m, \; for \; m \in \{1,\dots,M\}, \tag{3}$$

$$0 \le x_m \le \bar{x}, \; for \; m \in \{1,\dots,M\}. \tag{4}$$

One commonly considered objective is to create a power profile that is as flat as possible, i.e., to set all elements of the vector $\vec{q}$ to a constant (e.g., the average of the loads). A well-known property of Problem 1 is that all constant vectors $\vec{q} = (q,q,\dots,q)$ lead to the same optimal result $\vec{x}$, hence, we may use $\vec{q} = \vec{0}$ (and thus $\vec{z} = \vec{p} + \vec{x}$, which is the total load) to ease the notation.

Several papers study algorithms that can solve this problem optimally. An intuitive approach is the so-called valley filling approach [16]. This approach is demonstrated graphically in Figure 1, wherein an EV (arriving at 18:00, to be fully charged at 07:00) is charged with the goal to obtain a flat profile (i.e., $\vec{q} = \vec{0}$ as mentioned before). The optimal solution $\vec{x}$ flattens the total house profile by charging in such a way that the "valleys" are filled up to a certain fill level Z. This optimal fill level Z depends on the shapes of the "valleys" and the amount C to be charged. In Figure 1, C corresponds to the gray area. Note that Z can also be interpreted as the optimal minimal mismatch between the obtained and desired profile that we have to accept. If we have a lower mismatch in some other interval ℓ (i.e., $z_m < Z$), this needs to be compensated in some other interval ℓ by a mismatch even greater than Z (i.e., $z_\ell > Z$), which is, in terms of Problem 1, less flat and thus suboptimal. Charging up to the level Z (i.e., $z_m \ge Z$) makes sure that the EV battery is fully charged exactly at the deadline without introducing peaks.

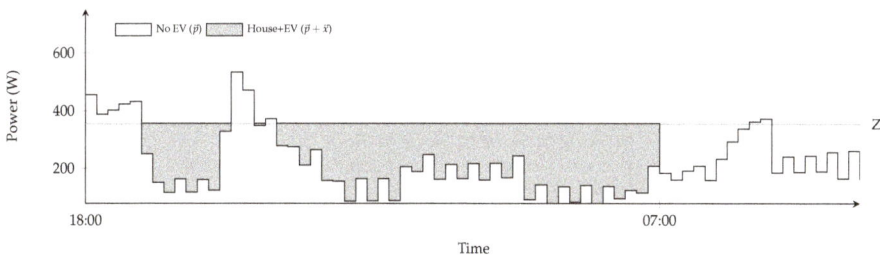

Figure 1. Optimal EV planning.

In [16], a line search to find the value Z is proposed. A different approach is taken by van der Klauw et al. [4]. They determine the intervals in which no charging should take place, after which they can straightforwardly calculate Z. Their algorithm finds the optimal planning in $\mathcal{O}(M \log M)$ time.

The EV charging problem falls in a more general class of problems that are referred to as "resource allocation problems" [18]. Hochbaum and Hong [19] study several resource allocation problems, and they present an algorithm for a generalized version of Problem 1, which also can take into account integer restrictions for the values of $\bar{x}$ and has an asymptotic time complexity of $\mathcal{O}(M)$. However, this algorithm has a large constant since it relies on median finding [20]. Therefore, the algorithm is in practice only fast for a relatively large M.

3.2. Online Optimization

When the power consumption in future intervals is known beforehand, the aforementioned algorithms can be used to calculate the optimal charging power for each future interval. However, in most cases, these future values are not known and must be predicted. Obviously, prediction errors may lead to non-optimal decisions, and the impact of such errors at a house level is rarely considered in the literature. Furthermore, it is in general difficult to predict the power consumption for all future charging intervals at the beginning of the planning. Therefore, we take a different approach that does not need this detailed information.

Central in our approach is the observation that for the aforementioned algorithms, a single value Z uniquely characterizes the optimal solution. For all feasible instances of Problem 1, there exists a Z such that the optimal solution $x_1, \ldots, x_M$ can be constructed by setting

$$x_m := \max\left(0, \min\left(Z - p_m + q_m, \bar{x}\right)\right). \tag{5}$$

Note that this calculation, for time interval m, can be delayed until the beginning of the time interval, and, thus we can delay the prediction of p_m to this time.

The sketched approach is in principle interval based; however, it can easily be adapted to an event based approach that recalculates the charging power whenever the power consumption of the household changes and all results in this section are also valid for the event based approach.

As a consequence of the above, the main challenge of our approach is that we do not know Z and, therefore, this value must be predicted. However, compared to approaches that predicts power for each interval, this approach has two advantages, namely, that only a single value Z is predicted (instead of the complete vector $\vec{p}$), and that the resulting error due to an incorrect prediction can be bounded as we discuss in the next section.

3.3. Predictions

As we, in general, do not know Z, we use a prediction of Z, which we denote by $\hat{Z}$. When $\hat{Z} < Z$, Equation (5) chooses charging powers that are not sufficient to charge the car up to the desired level. This can be resolved by charging at the maximum power $\bar{x}$ (or some other pre-set charging power) starting from the interval where this becomes the minimum charging power that is needed to charge the EV within the remaining intervals. Note that this may result in large peaks and high costs (i.e., high objective values) at the end of the charging period, as we discuss later on. On the contrary, in the situation where $\hat{Z} > Z$, the algorithm charges faster than required resulting in some time intervals at the end of the planning period with low or zero charging power values.

Algorithm 1 takes these two cases into account and guarantees that the SoC target is reached in both situations. Note that Algorithm 1 produces the same (optimal) solution as Equation (5) when $\hat{Z} = Z$. Before the first invocation of the algorithm, the variable T, which expresses the amount of energy already charged, is initialized to zero. Then, iteratively before each interval m, this algorithm is used to calculate the charging power x_m for this interval. Hence, at the beginning of the m-th interval, $T = \sum_{i=1}^{m} x_m$ expresses the amount already charged up to this interval of the total C to be charged.

Figure 2 illustrates an instance with $\hat{Z} = 1.1Z$ and $\vec{q} = \vec{0}$, which shows that the prediction error is spread out evenly over all intervals where charging takes place, but the charging stops before the deadline. The ratio between the objective value of the optimal solution and of the online algorithm using $\hat{Z}$ is 1.008 in this example. On the website [21], an interactive demonstration of Algorithm 1 can be found.

Algorithm 1 Online EV planning for time interval m.

$x_m = \max\left(0, \min\left(\hat{Z} - p_m + q_m, \bar{x}\right)\right)$
$x_m = \min(x_m, C - T)$ {needed for $Z < \hat{Z}$}
if $T + x_m + (M - m)\bar{x} < C$ **then** {needed for $Z > \hat{Z}$}

$\quad x_m := \min\{C - T, \bar{x}\}$
end if
$T = T + x_m$

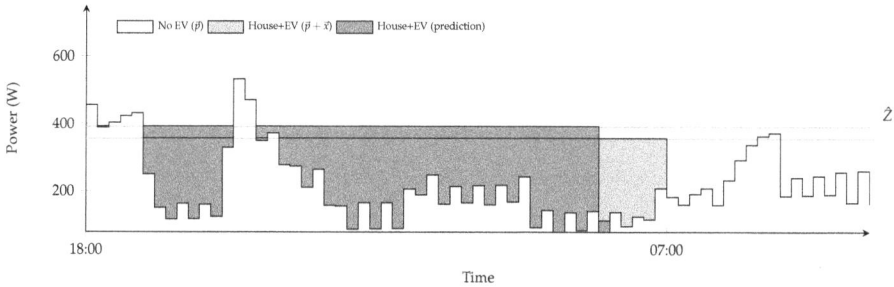

Figure 2. Optimal EV planning and EV planning using the prediction $\hat{Z}$.

In the following, we consider the case that $\hat{Z} \geq Z$. For this case, we derive a bound on the ratio between the objective value of the optimal solution and the objective value of the solution that uses the prediction $\hat{Z}$, which only depends on Z and $\hat{Z}$.

To ease the discussion, we use $C(Z)$ and $C(\hat{Z})$ to denote the objective value of the optimal solution and the objective value of the solution that uses the prediction $\hat{Z}$, respectively. To derive bounds on the ratio $C(\hat{Z})/C(Z)$, we need the following lemma.

Lemma 1. *Let $\bar{x}$ be an optimal solution to an instance where $x_m < \bar{x}$ (for all m), and let $Z_\Sigma = \sum_{m=1}^{M} z_m$. Then, we have:*

$$C(Z) \geq \sqrt{ZZ_\Sigma}.$$

Proof. Assume that in the optimal solution we have a $z_m < Z$ for some m. This implies that $p_m - q_m < Z$ and the algorithm tried to fill this interval up to Z, but did not succeed. The only reason for this is that the maximum amount of charging was not sufficient to reach Z. As a consequence, the algorithm will charge $\bar{x}$ in this interval, which is a contradiction to $x_m < \bar{x}$ for all m.

Hence, we have

$$C(Z) = \sqrt{\sum_{m=1}^{M} z_m^2}$$

$$\geq \sqrt{\sum_{m=1}^{M} z_m Z}$$

$$= \sqrt{ZZ_\Sigma}$$

□

This lemma is used to derive a bound on the ratio between the objective value of the optimal solution, and the objective value of the solution based on the prediction $\hat{Z}$.

Theorem 1. *When $Z \leq \hat{Z}$, the ratio between the objective value of the optimal solution and that based on the prediction $\hat{Z}$ is given by*

$$\frac{\mathcal{C}(\hat{Z})}{\mathcal{C}(Z)} \leq \sqrt{\frac{\hat{Z}}{Z}}.$$

Proof. Note that when $\hat{z}_m > \hat{Z}$ for an interval m, we have $\hat{z}_m = z_m$ because $\hat{Z} \geq Z$, i.e., both solutions are the same for such intervals m. As a consequence, if the instance has an interval m with $\hat{z}_m > \hat{Z}$, replacing this instance by an instance with $p_m - q_m = \hat{Z}$ leads to an increase of the ratio $\frac{\mathcal{C}(\hat{Z})}{\mathcal{C}(Z)}$ as $\mathcal{C}(\hat{Z}) \geq \mathcal{C}(Z)$, and both $\hat{z}_m$ and z_m decrease by the same amount; i.e., we may assume for this proof w.l.o.g. that $\hat{z}_m \leq \hat{Z}$ for all intervals m. Similarly, we may assume w.l.o.g. that $x_m < \bar{x}$, since otherwise $Z \leq z_m \leq \hat{Z}$ and the ratio only improves. Because of this latter assumption, the instance meets the requirements for Lemma 1.

Let $Z_\Sigma = \sum_{m=1}^{M} z_m$. First, we show that $\mathcal{C}(\hat{Z}) \leq \sqrt{\hat{Z} Z_\Sigma}$. For this, consider a solution of Algorithm 1, which is characterized by $\hat{z}_m = \hat{x}_m - q_m + p_m$. Since the obtained result is feasible, we have

$$\sum_{m=1}^{M} \hat{z}_m = Z_\Sigma.$$

Now, we have

$$\mathcal{C}(\hat{Z}) = \sqrt{\sum_{m=1}^{M} \hat{z}_m^2} \leq \sqrt{\sum_{m=1}^{M} \hat{Z} \hat{z}_m}$$

$$= \sqrt{\hat{Z} \sum_{m=1}^{M} \hat{z}_m} = \sqrt{\hat{Z} Z_\Sigma}.$$

Combining this inequality with $\mathcal{C}(Z) \geq \sqrt{Z Z_\Sigma}$ (due to Lemma 1), we get:

$$\frac{\mathcal{C}(\hat{Z})}{\mathcal{C}(Z)} \leq \frac{\sqrt{\hat{Z} Z_\Sigma}}{\sqrt{Z Z_\Sigma}} = \sqrt{\frac{\hat{Z}}{Z}},$$

which proves the theorem. $\square$

This theorem gives a bound on the relative deviation of the objective value of the solution of Algorithm 1 compared to the objective value of the optimal solution. Note that this bound depends only on the relative deviation of the used estimate for the fill level compared to the optimal fill level, and that this dependency on this relative deviation is dampened by the square root function.

4. Fleet Planning

This section extends the results from the previous section to charging multiple EVs. In a domestic situation, cars are typically charged when their owners arrive at home, which commonly is in the evening and coincides with a domestic consumption peak. Especially when multiple EVs are charging simultaneously within a neighborhood, there is a risk of high peaks and therefore of overloading the transformer. As we argued in the previous sections, predicting a load profile is difficult, while controlling an individual EV with only few predictions can be done. We extend this approach to multiple EVs by adding another method that shaves load peaks at the neighborhood level at the moment they are noticed (e.g., in an online setting). For this, we propose the following solution:

1. The charging of EVs is planned locally within the houses such that the total household consumption power profile (including the EV) becomes as flat as possible.

2. When the total power P of a group of houses is above a given threshold of X watts, the EVs are requested to decrease their total charging in the next time interval by $\Delta = P - X$ in a way that keeps the individual local power profiles as flat as possible for their remaining charging window.

3. When the total power P of a group of houses is below a given threshold of Y watts (e.g., PV production peak), the EVs are requested to increase their total charging in the next time interval by $\Delta = Y - P$ in a way that keeps the individual local power profiles as flat as possible for their remaining charging window.

This approach, which is detailed below, has several important advantages. By starting with flat profiles on house level in the first step, the individual peak loads, and with it the probability of overloading the network, decreases. Furthermore, flattening the house load increases self-consumption of locally generated electricity (e.g., PV), has a positive impact on the voltage, avoids overloading the grid, and decreases losses (as is demonstrated in Section 6).

When there is still a peak consumption for the group of houses (e.g., in the evening), the second step of the proposed approach coordinates the charging while avoiding new local peaks. To quantify these local peaks, we use a cost function $\hat{C}_n(\delta)$ that expresses how much changing the power for the upcoming interval from $x_{n,m}$ to $x_{n,m} + \delta$ influences the flatness of the entire power profile for house n (i.e., also considering impact on the future). This function aids us with finding the EVs that can contribute to obtaining a total power difference with minimal impact to the local flatness, and thus preventing problems in the future. More specifically, our objective is to obtain a total power difference Δ for the group of houses in the next time interval, while retaining the flatness of the power profiles of individual houses as much as possible. This is expressed mathematically as:

Problem 2.

$$\min_{\delta_1,\dots,\delta_N} \; f(\vec{\delta}) = \sum_{n=1}^{N} \hat{C}_n(\delta_n),$$

$$\text{subject to } \; g(\vec{\delta}) = \sum_{n=1}^{N} \delta_n = \Delta,$$

where f and g are introduced for reference in the later results. To ease the presentation, maximal charging powers are not considered in this formulation. However, to solve Problem 2 *with* additional maximal charging power constraints, we just can solve Problem 2 (without these constraints) and fix any violations using a Pegging Method; see [22] for a discussion of such methods.

Before we can solve this problem, we first need a formal description of $\hat{C}_n(\delta_n)$. These costs depend on the charging level to be attained in the charging intervals (i.e., the fill level Z_n), and the uncontrollable loads in the other intervals. The subset of intervals where charging up to Z takes place is denoted by $\mathcal{I}_n \subseteq \{1,\dots,M\}$, i.e., $\mathcal{I}_n$ contains the intervals where $p_{n,m} - q_{n,m} < Z_n$. Let $I_n = |\mathcal{I}_n|$ denote the number of such intervals (excluding the first interval). Using this notation, the costs for house n can be determined as function of Z_n (note that $\mathcal{C}(Z)$ now received a subscript to indicate the house):

$$\mathcal{C}_n(Z_n) = \sum_{m \in \mathcal{I}_n} Z_n^2 + \sum_{m \in \{1,\dots,M\}\setminus\mathcal{I}_n} (p_{n,m} - q_{n,m})^2$$

$$= I_n Z_n^2 + \sum_{m \in \{1,\dots,M\}\setminus\mathcal{I}_n} (p_{n,m} - q_{n,m})^2.$$

Intuitively, this means that the incurred costs are the costs of charging up to Z_n in the active intervals $\mathcal{I}_n$ (first term), and the costs of the intervals where no charging takes place (second term). We notice that, for practical data, (see the discussion in the next section) I_n rarely changes with δ and may be estimated to be constant, and therefore the last term of the equation above does not have to be considered for optimization. This means that we can approximate $\hat{C}_n(\delta)$ by:

$$\hat{C}_n(\delta) := (Z_n + \delta)^2 + I_n \left(Z_n - \frac{\delta}{I_n} \right)^2,$$
(6)

where the first term gives the costs of changing the first interval and the second term expresses the costs of the other active charging intervals. To study the influence of δ_n on the costs, we need the derivative of $\hat{C}_n(\delta_n)$, given by

$$\hat{C}_n{}'(\delta) = 2(Z_n + \delta) - 2 \left(Z_n - \frac{\delta}{I_n} \right) = 2\delta \left(1 + \frac{1}{I_n} \right).$$

The following lemma provides a sufficient condition of optimality for Problem 2, which is later used to find an optimal solution.

Lemma 2. *Let $\hat{C}_n(\delta_n)$ be given by (6). If we have a solution $\vec{\delta} = (\delta_1, \ldots, \delta_N)$ to Problem 2 that satisfies*

$$\delta_i \left(1 + \frac{1}{I_i} \right) = \delta_j \left(1 + \frac{1}{I_j} \right), \quad \text{for all } i, j,$$
(7)

and

$$\sum_{n=1}^{N} \delta_n = \Delta,$$
(8)

this solution is optimal.

Proof. Using the method of Lagrange multipliers (i.e., $\nabla f + \lambda \nabla g = 0$), we obtain a set of equations that give sufficient conditions for optimality:

$$\hat{C}_1{}'(\delta_1) + \lambda = 2\delta_1 \left(1 + \frac{1}{I_1} \right) + \lambda = 0$$

$$\vdots$$

$$\hat{C}_N{}'(\delta_N) + \lambda = 2\delta_N \left(1 + \frac{1}{I_N} \right) + \lambda = 0$$

$$\sum_{n=1}^{N} \delta_n = \Delta,$$

with λ being the Lagrange multiplier. It can be readily checked that Equations (7) and (8) solve this set of equations. $\square$

The intuition behind this lemma is that when $C_i'(\delta_i) < C_j'(\delta_j)$, the costs can be decreased by slightly increasing δ_i and decreasing δ_j by the same amount.

Using this lemma, we can derive the following theorem that provides the optimal values δ_n.

Theorem 2. *Let $\hat{C}_n(\delta)$ be given by (6). Then, the optimal solution to Problem 2 is for all n:*

$$\delta_n = \frac{\Delta}{S(2 + 2/I_n)},$$
(9)

where

$$S := \sum_{i=1}^{N} \frac{1}{2 + 2/I_i}.$$

Proof. As the optimal solution is unique (the problem is strictly convex), finding one optimal solution proves this theorem. It can be readily verified that Equation (9) satisfies both Equations (7) and (8). Hence, by Lemma 2, the theorem is proven. □

A consequence of this theorem is that only the number of charging intervals I_n determines the change δ_n in charging power.

5. Predictions

The methods presented in the previous two sections are based on proper predictions of the fill level Z_n and the number of active charging intervals I_n for each house n. To validate that these values are relatively easy to predict, we analyzed these values using 90 consecutive days of measurements from a Dutch house, and used this to calculate the optimal charging profile for each of the days between 18:00 and 24:00, subdivided into 24 intervals of 15 min.

The results for different charging amounts C_n (in kWh) are presented in Table 1. This table shows that the values for Z_n are in a small band for all 90 days. See also Figure 3, where the four graphs correspond to the different C_n values from this table are given. When we take the maximum encountered Z_n as prediction $\hat{Z}_n$ and use this prediction for all 90 days, the last column shows that highest realized costs increase by at most 16%, although the median is significantly lower (7% additional costs). The column $\sqrt{\frac{\hat{Z}}{Z}}$ gives the bound of the costs as estimated by Theorem 1 and shows that this theorem provides a reasonable bound of the costs.

Table 1. Analysis of Z_n and C_n for 90 days for charging between 18:00 and 24:00.

C_n (kWh)	Z_n			I_n			Max. Power			$\sqrt{\frac{\hat{Z}}{Z}}$	$\frac{c(\hat{Z})}{c(Z)}$		
	Min	Med	Max	Min	Med	Max	Min	Med	Max		Min	Med	Max
6	1188	1463	1713	21	22	24	1118	1361	1584	1.20	1.00	1.07	1.16
12	2188	2492	2776	22	24	24	2118	2385	2643	1.13	1.00	1.05	1.11
18	3188	3492	3798	24	24	24	3118	3388	3643	1.09	1.00	1.04	1.09
24	4188	4492	4798	24	24	24	4118	4388	4643	1.07	1.00	1.03	1.07

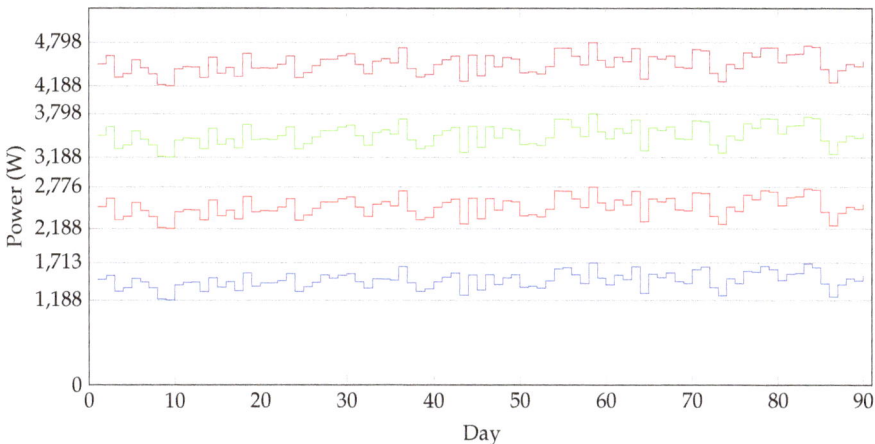

Figure 3. Fill levels Z_n corresponding to Table 1.

These results suggest an efficient way of predicting Z_n: use measurements from the previous days to calculate the Z_n for each of these days, and use the highest encountered value as $\hat{Z}_n$. To calculate

Z_n, the deterministic EV planning algorithm from [4] can be used, and, as indicated in this paper, this algorithm can calculate the result in order of milliseconds and has a time complexity of $\mathcal{O}(M \log M)$. Since we use this approach for each EV that arrives, our approach has a complexity of $\mathcal{O}(KM \log M)$, for K EVs.

Charging the EV within a relatively short interval, such as the interval between 18:00 and 24:00 above, results in a relatively high charging power, especially if the charging amount C_n is also relatively high. This results in a profile wherein most intervals are used for charging (i.e., I_n is close to the total number of intervals) and the profile itself is flat (i.e., it precisely reaches Z_n). When a longer charging interval is taken with multiple higher peaks (e.g., during the day and/or evening), the charging is spread out and done at lower power. To illustrate this, we repeat the experiment with a charging between 14:00 and 24:00 (when consumption is highest). The achieved results are as presented in Table 2. Note that, in this table, the values I_n are more diverse since the house consumption plays a larger role. Note that I_n can be interpreted as the number of intervals with low house consumption compared to the fill level Z_n. Since even with a long charging horizon, the variation in I_n is small, this is an additional indication that our assumption from the last section that I_n is (almost) constant is reasonable. The results also indicate that I_n is easy to predict, and this can be done similar to (and simultaneously with) obtaining $\hat{Z}_n$, as was described above.

To make this overview complete, we considered the highest charging power (Max. power) within each day, and present the minimum, median and maximum of this value over the 90 days in Tables 1 and 2, and the results show that these values grow approximately linearly with C_n.

Table 2. Analysis of Z_n and C_n for 90 days for charging between 14:00 and 24:00.

C_n (kWh)	Z_n			I_n			Max. Power			$\sqrt{\frac{\hat{Z}}{Z}}$	$\frac{c(\hat{Z})}{c(Z)}$		
	Min	Med	Max	Min	Med	Max	Min	Med	Max		Min	Med	Max
6	809	1057	1268	31	35	40	749	969	1184	1.25	1.00	1.06	1.18
12	1409	1721	1962	35	38	40	1349	1627	1851	1.18	1.00	1.06	1.15
18	2009	2340	2603	37	39	40	1949	2253	2468	1.14	1.00	1.05	1.12
24	2609	2943	3221	39	40	40	2549	2859	3079	1.11	1.00	1.04	1.10

6. Evaluation

In this section, we compare our work with the state-of-the-art research and several other charging strategies. The basic variant of our approach aims to make the house load as flat as possible without coordination between the houses and is denoted by NOCOORD. The extension of this technique that shaves the neighborhood peak by adding coordination is referred to by COORD. For both of our approaches, we make predictions upon arrival of the EVs using the approach from the previous section with ten days of historical data as input. We compare our approaches with the state-of-the-art research on Profile Steering (PS) from [2]. The profile steering algorithm is a heuristic that predicts the load of each house, makes a plan that is as flat as possible, and coordinates between houses to further flatten the load peak. In order to compare with the best possible behavior of PS, we assume perfect predictions for profile steering, which gives this approach a significant advantage. For completeness, we also compare to no control (NC), where EVs charge at full power at arrival, and with a grid unaware peak-shaving (PEAKS), which is a simple approach that iteratively selects EVs and decreases their charging as much as possible until the desired load at the transformer is accomplished.

To effectively compare all approaches with the state-of-the-art research, we reproduce the case used in [2]. It considers 121 houses, all equipped with identical electric vehicles that charge 12 kWh between 18:00 and 07:00 and have a maximum charging power of 3.8 kW. We used the same data set and network files as used in [2] to calculate the load flows to obtain the active power at the transformer (incl. losses), the lowest observed voltage in the grid and the highest observed voltage in the grid. For COORD and PEAKS, we need to set a limit for the peak-shaving by the neighborhood controller, and we set this limit to $X = 165\,\text{kW}$. The results are summarized in Table 3 and further detailed in Figure 4 (power) and Figure 5.

Table 3. Comparison between DSM approaches and no control.

	PEAKS	NC	PS	COORD (This Paper)	NOCOORD (This Paper)
Total losses (kWh)	61.31	89.22	33.90	34.36	35.95
Lowest voltage (V)	209.92	199.77	219.15	219.11	218.88
Highest voltage (V)	232.28	232.28	231.51	231.51	231.51
Max. peak (kW)	171.04	575.09	175.70	170.82	195.28
Max. cable load (%)	106.87	143.35	57.41	57.41	57.41

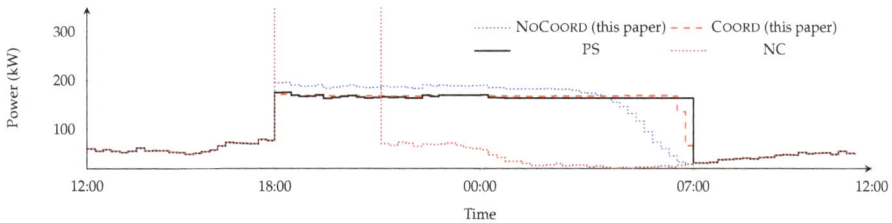

Figure 4. Power at transformer (incl. losses).

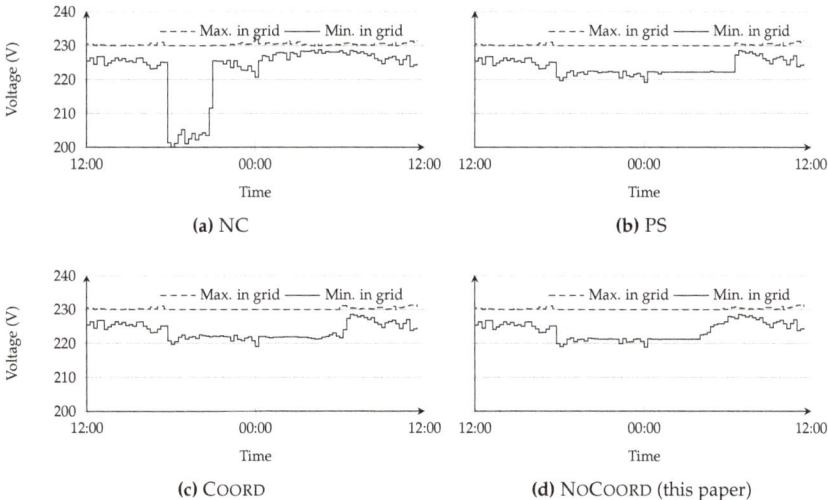

(a) NC

(b) PS

(c) COORD

(d) NOCOORD (this paper)

Figure 5. Lowest and highest observed voltages.

When no control would be applied, a large peak occurs at the transformer (see Figure 4), cables are overloaded and the voltage does not stay within the legal bounds that are required in the Netherlands (i.e., NEN-EN 50160:2010 [5], 207 V–253 V). The transformer load is reduced with PEAKS; however, Table 3 shows that this approach still overloads cables.

In contrast to NC and PEAKS, PS reduces the load peak significantly and keeps the voltage within legal bounds (Figure **??**). Although it performs well, this algorithm depends on hard to make predictions (namely, a power value for each interval, for each house) and requires a lot of communication due to exchanging power profiles.

When no coordination controller is used (NoCoord), the resulting profile is already rather flat (Figure 4) and the voltages are well within legal bounds, and, at the end of the charging intervals, this approach is on par with PS (Figure ??). By using the coordination controller (Coord), the profile is further flattened and the voltages are similar to the no coordination controller for this specific case.

Table 3 gives the exact losses, voltage bounds, maximum peak load and the highest overloading of a cable within the grid. Note that several scenarios share the maximum cable load of 57.41%, and the reason for this is that this cable load occurs before the EVs arrive and cannot be influenced. The table shows that the developed approaches perform similar to profile steering, while, for profile steering, knowledge about the future was used (in the real world, profile steering has to predict these values, whereas, in our experiments, we used the actual values).

7. Conclusions

Existing demand side management approaches either do not plan ahead, and thereby often may not deploy the flexibility of smart appliances when it is needed the most, or they do make a plan but this planning is based on often hard to predict (inaccurate) household power consumption.

To plan the appliances within a house, we propose using online planning. As a proof of concept, we presented an online electric vehicle planning algorithm that only requires a prediction of the fill level characteristic and a prediction of load in the next time interval as an input. The algorithm distributes the prediction error evenly over all charging intervals. This makes the algorithm very robust against incorrect predictions, especially if the predictions are higher than the actual realization. Furthermore, we presented a bound that quantifies the sensitivity of our approach to prediction errors.

We extended the house control mechanism by a neighborhood control mechanism, which initially asks the houses to make a flat profile. When the load exceeds a certain threshold, this coordination controller requests less charging from houses for the next time period in such a way that the house profiles remain as flat as possible. This method only requires a prediction of the number of active charging intervals within a house to make the required trade-off.

Both the house and neighborhood approaches require only few predictions, and we showed that these predictions are easy to obtain. In the evaluation, we studied the combination of predictions with our approach, and demonstrated that it works effectively for peak-shaving. The evaluation furthermore shows that this approach is very robust against prediction errors, and performs adequately even when the predictions are very imprecise. Compared to a naive approach, it leads to lower transportation losses, keeps the voltage within required bounds and keeps cable loads low. Furthermore, it is on par with the state-of-the-art research, which—in contrast to our work—requires predictions of flexibility and predictions of a load profile (24 hours ahead) for each house.

In future work, we aim to extend the online planning within a house to cope with appliances other than electric vehicles.

Acknowledgments: This research was conducted within the DREAM project supported by STW (#11842) and the e-balance project that has received funding from the European Union Seventh Framework Programme (FP7/2007-2013) under grant agreement n° [609132].

Author Contributions: The concepts, ideas and theory described in this article are joint work by both authors. The simulations were conducted by Marco E. T. Gerards.

Conflicts of Interest: The authors declare no conflict of interest.

References

1. Siano, P. Demand response and smart grids–A survey. *Renew. Sustain. Energy Rev.* **2014**, *30*, 416–478.
2. Gerards, M.E.T.; Toersche, H.A.; Hoogsteen, G.; van der Klauw, T.; Hurink, J.L.; Smit, G.J.M. Demand side management using profile steering. In Proceedings of the 2015 IEEE Eindhoven PowerTech, Eindhoven, The Netherlands, 29 June–2 July 2015; pp. 1–6.
3. Molderink, A.; Bakker, V.; Bosman, M.G.C.; Hurink, J.L.; Smit, G.J.M. Management and Control of Domestic Smart Grid Technology. *IEEE Trans. Smart Grid* **2010**, *1*, 109–119.
4. van der Klauw, T.; Gerards, M.E.T.; Smit, G.J.M.; Hurink, J.L. Optimal scheduling of electrical vehicle charging under two types of steering signals. In Proceedings of the Europe IEEE PES Innovative Smart Grid Technologies Conference (ISGT-Europe), Istanbul, Turkey, 12–15 October 2014; IEEE Power & Energy Society: Piscataway, NJ, USA, 2014; pp. 1–6.
5. NEN-EN 50160:2010. Voltage characteristics of electricity supplied by public distribution network, August 2010.
6. Hoogsteen, G.; Molderink, A.; Hurink, J.L.; Smit, G.J.M. Managing energy in time and space in smart grids using TRIANA. In Proceedings of the Europe IEEE PES Innovative Smart Grid Technologies Conference, ISGT-Europe 2014, Istanbul, Turkey, 12–15 October 2014; IEEE Power & Energy Society: Piscataway, NJ, USA, 2014; pp. 1–6.
7. Nykamp, S.; Rott, T.; Dettke, N.; Kueppers, S. The project "ElChe" Wettringen: Storage as an alternative to grid reinforcements—Experiences, benefits and challenges from a DSO point of view. In Proceedings of the International ETG Congress 2015; Die Energiewende - Blueprints for the New Energy Age, Bonn, Germany, 17–18 November 2015.
8. Kok, K. Dynamic pricing as control mechanism. In Proceedings of the 2011 IEEE Power and Energy Society General Meeting, San Diego, CA, USA, 24–29 July 2011; pp. 1–8.
9. Kok, K. The PowerMatcher: Smart Coordination for the Smart Electricity Grid. Ph.D. Thesis, Vrije Universiteit Amsterdam and TNO, Amsterdam, The Netherlands, 2013.
10. Wijbenga, J.; MacDougall, P.; Kamphuis, R.; Sanberg, T.; van den Noort, A.; Klaassen, E. Multi-goal optimization in PowerMatching city: A smart living lab. In Proceedings of the IEEE PES Innovative Smart Grid Technologies Conference Europe, ISGT-Europe 2014, Istanbul, Turkey, 12–15 October 2014; pp. 1–5.
11. Javed, F.; Arshad, N.; Wallin, F.; Vassileva, I.; Dahlquist, E. Forecasting for demand response in smart grids: An analysis on use of anthropologic and structural data and short term multiple loads forecasting. *Appl. Energy* **2012**, *96*, 150–160.
12. Alfares, H.K.; Nazeeruddin, M. Electric load forecasting: Literature survey and classification of methods. *Int. J. Syst. Sci.* **2002**, *33*, 23–34.
13. Ruelens, F.; Vandael, S.; Leterme, W.; Claessens, B.; Hommelberg, M.; Holvoet, T.; Belmans, R. Demand side management of electric vehicles with uncertainty on arrival and departure times. In Proceedings of the 2012 3rd IEEE PES International Conference and Exhibition on Innovative Smart Grid Technologies (ISGT Europe), Berlin, Germany, 14–17 October 2012; pp. 1–8.
14. Tang, W.; Bi, S.; Zhang, Y. Online coordinated charging decision algorithm for electric vehicles without future information. *IEEE Trans. Smart Grid* **2014**, *5*, 2810–2824.
15. Gan, L.; Topcu, U.; Low, S. Optimal decentralized protocol for electric vehicle charging. *IEEE Trans. Power Syst.* **2013**, *28*, 940–951.
16. Mou, Y.; Xing, H.; Lin, Z.; Fu, M. Decentralized optimal demand-side management for PHEV charging in a smart grid. *IEEE Trans. Smart Grid* **2015**, *6*, 726–736.
17. Del Razo, V.; Goebel, C.; Jacobsen, H. Vehicle-originating-signals for real-time charging control of electric vehicle fleets. *IEEE Trans. Transp. Electrification* **2015**, *1*, 150–167.
18. Patriksson, M. A survey on the continuous nonlinear resource allocation problem. *Eur. J. Oper. Res.* **2008**, *185*, 1–46.
19. Hochbaum, D.; Hong, S.P. About strongly polynomial time algorithms for quadratic optimization over submodular constraints. *Math. Progr.* **1995**, *69*, 269–309.
20. Blum, M.; Floyd, R.W.; Pratt, V.; Rivest, R.L.; Tarjan, R.E. Time bounds for selection. *J. Comput. Syst. Sci.* **1973**, *7*, 448–461.

Energies **2016**, *9*, 594

21. Gerards, M.E.T. Online EV Planning. Available online: http://wwwhome.ewi.utwente.nl/ gerardsmet/vis/ oev/ (accessed on 29 April 2016).

22. Patriksson, M.; Strömberg, C.   Algorithms for the continuous nonlinear resource allocation problem—New implementations and numerical studies. *Eur. J. Oper. Res.* **2015**, *243*, 703–722.

energies

MDPI

Article

Development of a Novel Bidirectional DC/DC Converter Topology with High Voltage Conversion Ratio for Electric Vehicles and DC-Microgrids

Ching-Ming Lai

Department of Vehicle Engineering, National Taipei University of Technology, 1, Sec. 3, Chung-Hsiao E. Rd., Taipei 106, Taiwan; pecmlai@gmail.com; Tel.: +886-2-2771-2171 (ext. 3612); Fax: +886-2-2731-4990

Academic Editor: Neville Watson
Received: 3 February 2016; Accepted: 19 May 2016; Published: 26 May 2016

Abstract: The main objective of this paper was to study a bidirectional direct current to direct current converter (BDC) topology with a high voltage conversion ratio for electric vehicle (EV) batteries connected to a dc-microgrid system. In this study, an unregulated level converter (ULC) cascaded with a two-phase interleaved buck-boost charge-pump converter (IBCPC) is introduced to achieve a high conversion ratio with a simpler control circuit. In discharge state, the topology acts as a two-stage voltage-doubler boost converter to achieve high step-up conversion ratio (48 V to 385 V). In charge state, the converter acts as two cascaded voltage-divider buck converters to achieve high voltage step-down conversion ratio (385 V to 48 V). The features, operation principles, steady-state analysis, simulation and experimental results are made to verify the performance of the studied novel BDC. Finally, a 500 W rating prototype system is constructed for verifying the validity of the operation principle. Experimental results show that highest efficiencies of 96% and 95% can be achieved, respectively, in charge and discharge states.

Keywords: bidirectional dc/dc converter (BDC); electric vehicle (EV); dc-microgrid; high voltage conversion ratio

1. Introduction

In recent years, to reduce fossil energy consumption, the development of environmentally friendly dc-microgrid technologies have gradually received attention [1–7]. As shown in Figure 1, a typical dc-microgrid structure includes a lot of power electronics interfaces such as bidirectional grid-connected converters (GCCs), PV/wind distributed generations (DGs), battery energy systems (BES), electric vehicles (EVs), and so on [4]. They connect together with a high-voltage dc-bus, so that dc home appliances can draw power directly from the dc-bus. In this system, the main function of GCCs is to maintain the dc-bus voltage constant, while in order to ensure the reliability of operation for dc-microgrids, a mass of BES can usually be accessed into the system. Electric vehicles (EVs) can also provide auxiliary power services for dc-microgrids, which makes clean and efficient battery-powered conveyance possible by allowing EVs to power and be powered by the electric utility. Usually, in dc-microgrid systems, when the voltage difference between the EV battery, BES and the dc-bus is large, a bidirectional dc/dc converter (BDC) with a high voltage conversion ratio for both buck and boost operations is required [4,7]. In the previous literatures, BDCs circuit topologies of the isolated [8–10] and non-isolated type [11–23] have been described for a variety of system applications. Isolated BDCs use the transformer to implement the galvanic isolation and to comply with the different standards. Personnel safety, noise reduction and correct operation of protection systems are the main reasons behind galvanic isolation. In contrast with isolated BDCs, non-isolated BDCs lack the galvanic isolation between two sides, however, they offer the benefits of smaller volume, high reliability, *etc.*, so they have been widely used for hybrid power system [24,25].

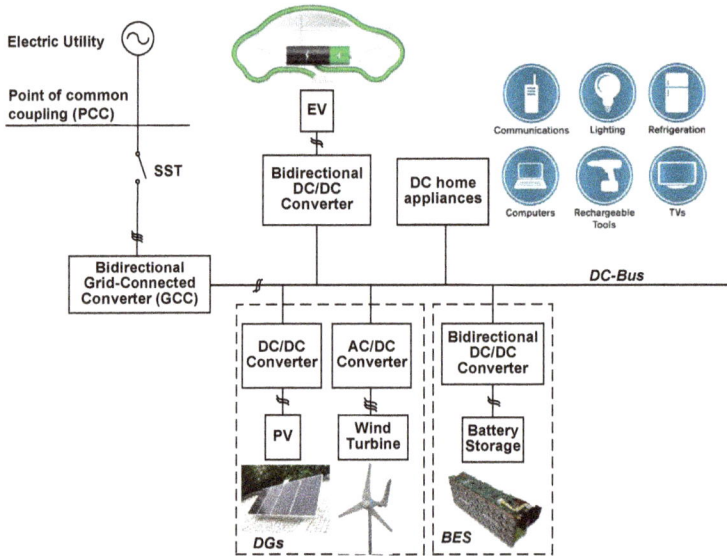

Figure 1. A typical dc-microgrid structure [4].

Compared with isolated types, BDCs with coupled-inductors for non-isolated applications possess simpler winding structures and lower conduction losses [12–17]. Furthermore, the coupled-inductor techniques can achieve easily the high voltage conversion ratio by adjusting the turn ratio of the coupled-inductor. However, the energy stored in the leakage inductor of the coupled inductor causes a high voltage spike in the power devices. Wai *et al.* [12,13] investigated a high-efficiency BDC, which utilizes only three switches to achieve the objective of bidirectional power flow. Also, the voltage-clamped technique was adopted to recycle the leakage energy so that the low-voltage stress on power switches can be ensured. To reduce the switching losses, Hsieh *et al.* proposed a high efficiency BDC with coupled inductor and active-clamping circuit [16]. In this reference, a low-power prototype was built to verify the feasibly.

As shown in Figure 2, Liang *et al.* [17] proposed a bidirectional double-boost cascaded topology for a renewable energy hybrid supply system, in which the energy is transferred from one stage to another stage to obtain a high voltage gain. Hence their conduction losses are high and it requires a large number of components.

Chen *et al.* [18] proposed a reflex-based BDC to achieve the energy recovery function for batteries connected to a low-voltage micro dc-bus system. In [18], a traditional buck-boost BDC was adopted, however, the voltage conversion ratio is limited because of the equivalent series resistance (ESR) of the inductors and capacitors and effect of the active switches [19].

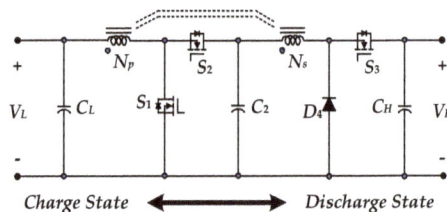

Figure 2. Circuit structure of the bidirectional double-boost cascaded topology [17].

To increase the voltage gain of the converter, the capacitors are switched and it will act as a charge-pump. The main advantage of the switched capacitor-based boost converter is that there is no need of a transformer or inductors. The main drawbacks of this topology are the complexity of the topology, high cost, low power level and high pulsating current in the input side [11,21]. In order to increase the conversion efficiency and voltage conversion ratio, multilevel combined the switched-capacitor techniques have been proposed to achieve lower stress on power devices [20–23]. As shown in Figure 3, in [22,23] two converters regulated the reasonable voltage conversion ratio with a simple pulse-width_modulation (PWM) control. However, if a high voltage conversion ratio must be provided, more power switches and capacitors are indeed required. Furthermore, although the extreme duty cycle can be avoided, the input current ripple is large due to their single-phase operation which renders these BDCs unsuitable for high current and low ripple applications.

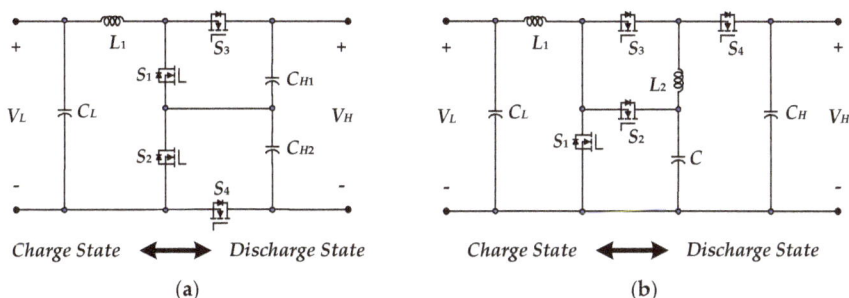

Figure 3. Two multilevel combined the switched-capacitor topologies: (**a**) circuit structure in [22]; (**b**) circuit structure in [23].

The objective of this paper is to study and develop a novel BDC for applications involving EVs connected to dc-microgrids. To meet the high current, low current ripple, and high voltage conversion ratio demands, the studied topology consists of an unregulated level converter (ULC) cascaded with a two-phase interleaved buck-boost charge-pump converter (IBCPC). In discharge state, the topology acts as a two-stage cascaded two-phase boosting converter to achieve a high step-up ratio. In charge state, the topology acts as two-stage cascaded two-phase bucking converter to achieve a high step-down ratio. The extreme duty cycle of power devices will not occur for bidirectional power flow conditions, thus not only can the output voltage regulation range be further extended but also the conduction losses can be reduced. In addition, the two-stage structure benefits reducing the voltage stress of active switches, which enables one to adopt the low-voltage rating and high performance devices, thus the conversion efficiency can be improved. The remainder of this paper is organized as follows: first, the converter topology and the operation principles of the studied BDC are illustrated in Section 2. Then, steady-state characteristic analyzes are presented in Section 3. A 500 W laboratory prototype is also constructed, and the corresponding simulation results, as well as experimental results, are provided to verify the feasibility of the studied BDC in Section 4. Finally, some conclusions are offered in the last section.

2. Proposed BDC Topology and Operation Principles

The system configuration for the studied BDC topology is depicted in Figure 4. The system contains two parts, including a ULC and a two-phase IBCPC. The major symbol representations are summarized as follows: V_H and V_L denote the high-side voltage and low-side voltage, respectively. L_1 and L_2 represent two-phase inductors of IBCPC. C_B denotes the charge-pump capacitor. C_H and C_L are the high-side and low-side capacitors, respectively. The symbols, $Q_1 \sim Q_4$, and $S_1 \sim S_4$, respectively, are the power switches of the IBCPC and ULC.

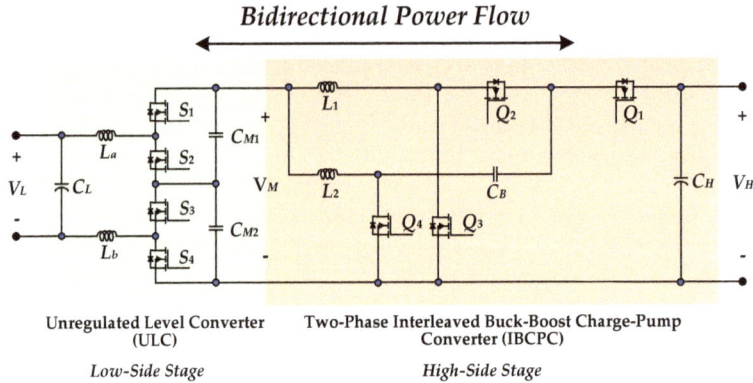

Figure 4. System configuration of the novel BDC topology.

In this study, as the low-side stage, a high efficiency magnetic-less ULC with bidirectional power flow is adopted to output a fixed voltage for a given input voltage. Because only a small sized high frequency line filter (L_a, L_b) is required, it can substantially boost the power density of the low-side stage. Furthermore, by leaving the voltage regulation to another high-side stage, the studied BDC for the low-side stage with fixed 2:1 under charge state operation or 1:2 conversion ratio under discharge state operation, can achieve high efficiency with a relatively low-side voltage in whole load range. As to the high-side stage, the structure of two-phase IBCPC is similar to a conventional buck-boost converter except two active switches in series and a charge-pump capacitor (C_B) employed in the power path. The circuit structure is simple and it can reach the high voltage conversion ratio with a reasonable duty cycle. Therefore, it can reduce the conduction loss of the switch, to further upgrade the efficiency of the whole bidirectional converter.

The studied BDC topology can deliver energy in both directions. When the energy flows from V_H to V_L, it operates in charge state (*i.e.*, buck operation); Q_1 and Q_2 are controlled to regulate the output. Thus, Q_1 and Q_2 are defined as the active switches, while Q_3 and Q_4 are the passive switches. The passive switches work as synchronous rectification (SR). When the energy flows from V_L to V_H, it operates in discharge state (*i.e.*, boost operation); Q_3 and Q_4 are controlled to regulate the output. Thus, Q_3 and Q_4 are defined as the active switches, while Q_1 and Q_2 are the passive switches.

In this study, the following assumptions are made to simplify the converter analyzes as follows: (1) the converter is operated in continuous conduction mode (CCM); (2) capacitors C_H and C_L is large enough to be considered as a voltage source; (3) the middle-link voltage $V_M = V_{M1} + V_{M2}$ is treated as a pure dc and considered as constant; (4) the two inductor L_1 and L_2 have the same inductor L_s; (5) all power semiconductors are ideal; (6) the charge-pump voltage V_{CB} is treated as a pure dc and considered as constant.

2.1. Charge State Operation

Figures 5 and 6 show the circuit configuration and characteristic waveforms of the studied BDC in charge state, respectively. It can be seen that switches Q_1 and Q_2 are driven with the phase shift angle of 180°; Q_3 and Q_4 work as synchronous rectification. In charge state, when S_1, S_3 are turned on and S_2, S_4 are turned off; or else S_2, S_4 are turned on and S_1, S_3 are turned off. The low-side voltage V_L is half the middle-link voltage V_M, *i.e.*, $V_L = 0.5V_M$. In this state, one can see that, when duty ratio of Q_1 and Q_2 are smaller than 50%, there are four operating modes according to the on/off status of the active switches.

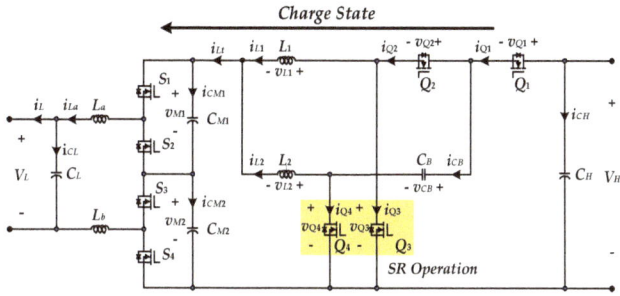

Figure 5. Circuit configuration of the studied BDC in charge state.

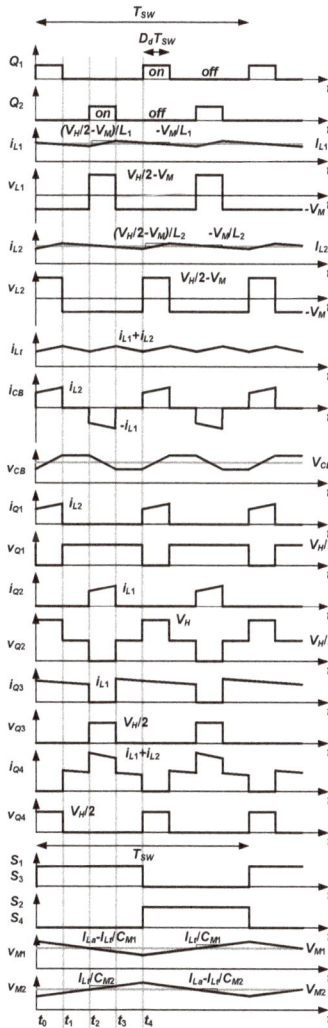

Figure 6. Characteristic waveforms of the studied BDC in charge state.

Referring to the equivalent circuits shown in Figure 7, the operating principle of the studied BDC can be explained briefly as follows.

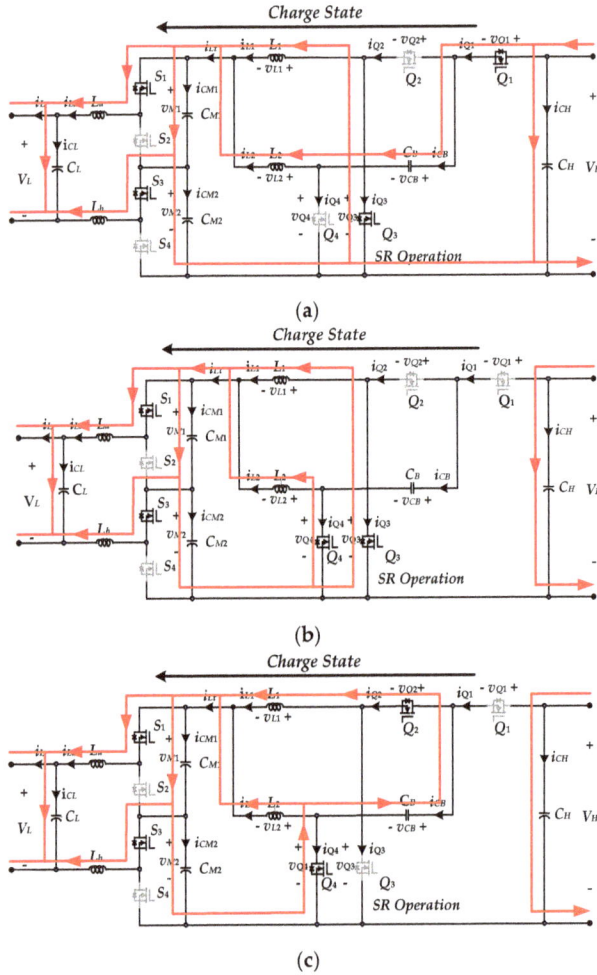

(a)

(b)

(c)

Figure 7. Equivalent circuits of the modes during different intervals in charge state: (**a**) Mode 1; (**b**) Mode 2, Mode 4; (**c**) Mode 3.

2.1.1. Mode 1 [$t_0 < t \leq t_1$]

The interval time is $D_d T_{sw}$, in this mode, switches Q_1, Q_3 turned on and switches Q_2, Q_4 are all off. The voltage across L_1 is the negative middle-link voltage, and hence i_{L1} decreases linearly from the initial value. Also, the voltage across L_2 is the difference of the high-side voltage V_H, the charge-pump voltage V_{CB}, and the middle-link voltage V_M, and its level is positive. The voltages across inductances L_1 and L_2 can be represented as:

$$L_1 \frac{di_{L1}}{dt} = -V_M = -2V_L \tag{1}$$

$$L_2 \frac{di_{L2}}{dt} = V_H - V_{CB} - V_M \tag{2}$$

2.1.2. Mode 2 [$t_1 < t \leqslant t_2$]

For this operation mode, the interval time is $(0.5 - D_d)T_{sw}$, switches Q_3, Q_4 are turned on and switches Q_1, Q_2 are all off. Both voltages across inductors L_1 and L_2 are the negative middle-link voltage V_M, hence i_{L1} and i_{L2} decrease linearly. The voltages across inductances L_1 and L_2 can be represented as:

$$L_1\frac{di_{L1}}{dt} = L_2\frac{di_{L2}}{dt} = -V_M = -2V_L \tag{3}$$

2.1.3. Mode 3 [$t_2 < t \leqslant t_3$]

For this operation mode, the interval time is $D_d T_{sw}$, switches Q_2, Q_4 are turned on and switches Q_1 and Q_3 are all off. The voltage across L_1 is the difference between the charge-pump voltage V_{CB} with the middle-link voltage V_M, and L_2 is keeping the negative middle-link voltage, the voltages across inductances L_1 and L_2 can be represented as follows:

$$L_1\frac{di_{L1}}{dt} = V_{CB} - V_M \tag{4}$$

$$L_2\frac{di_{L2}}{dt} = -V_M \tag{5}$$

2.1.4. Mode 4 [$t_3 < t \leqslant t_4$]

From this operation mode, the interval time is $(0.5 - D_d)T_{sw}$. Switches Q_3, Q_4 are turned on and switches Q_1, Q_2 are all off, and its operation is the same with that of Mode 2.

2.2. Discharge State Operation

Figures 8 and 9 show the circuit configuration and characteristic waveforms of the studied BDC in discharge state, respectively. As can be seen these figures, switches Q_3, Q_4 are driven with the phase shift angle of 180°; Q_1, Q_2 are used for the synchronous rectifier. In discharge state, when S_1, S_3 are turned on and S_2, S_4 are turned off; or else S_2, S_4 are turned on and S_1, S_3 are turned off. The low voltage V_L will charge the C_{M1} and C_{M2} to make V_{M1} and V_{M2} equal to V_L, the middle-link voltage V_M is then twice the low-side voltage V_L, i.e., $V_M = 2V_L$.

Referring to the equivalent circuits shown in Figure 10, the operating principle of the studied BDC can be explained briefly as follows:

2.2.1. Mode 1 [$t_0 < t \leqslant t_1$]

The interval time is $(D_b - 0.5)T_{sw}$, switches Q_3 and Q_4 are turned on; switches Q_1 and Q_2 are all off. For the high-side stage, the middle-link voltage V_M stays between inductance L_1 and L_2, making the inductance current increase linearly, and begins to deposit energy. The voltages across inductances L_1 and L_2 can be represented as:

$$L_1\frac{di_{L1}}{dt} = L_2\frac{di_{L2}}{dt} = V_M = 2V_L \tag{6}$$

2.2.2. Mode 2 [$t_1 < t \leqslant t_2$]

In this operation mode, the interval time is $(1 - D_b)T_{sw}$. Switch Q_1, Q_3 remains conducting and Q_2, Q_4 are turned off. The voltages across inductances L_1 and L_2 can be represented as:

$$L_1\frac{di_{L1}}{dt} = V_M = 2V_L \tag{7}$$

$$L_2\frac{di_{L2}}{dt} = V_M - V_H + V_{CB} = 2V_L - V_H + V_{CB} \tag{8}$$

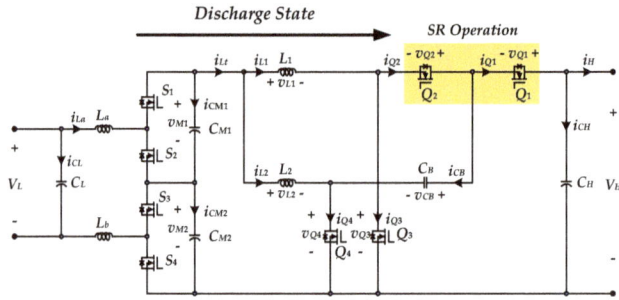

Figure 8. Circuit configuration of the studied BDC in discharge state.

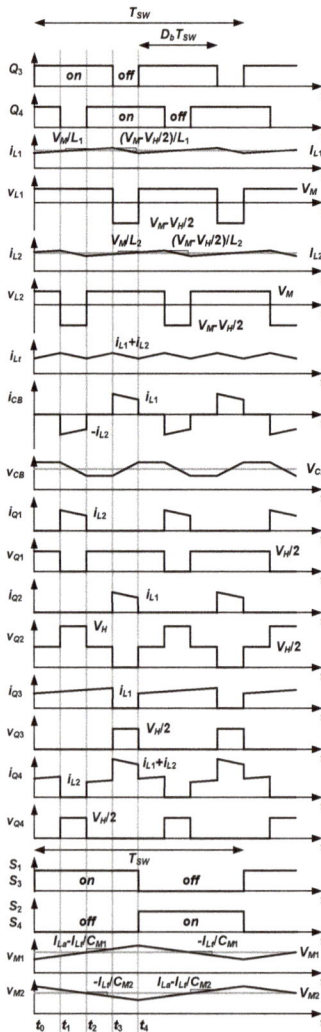

Figure 9. Characteristic waveforms of the studied BDC in discharge state.

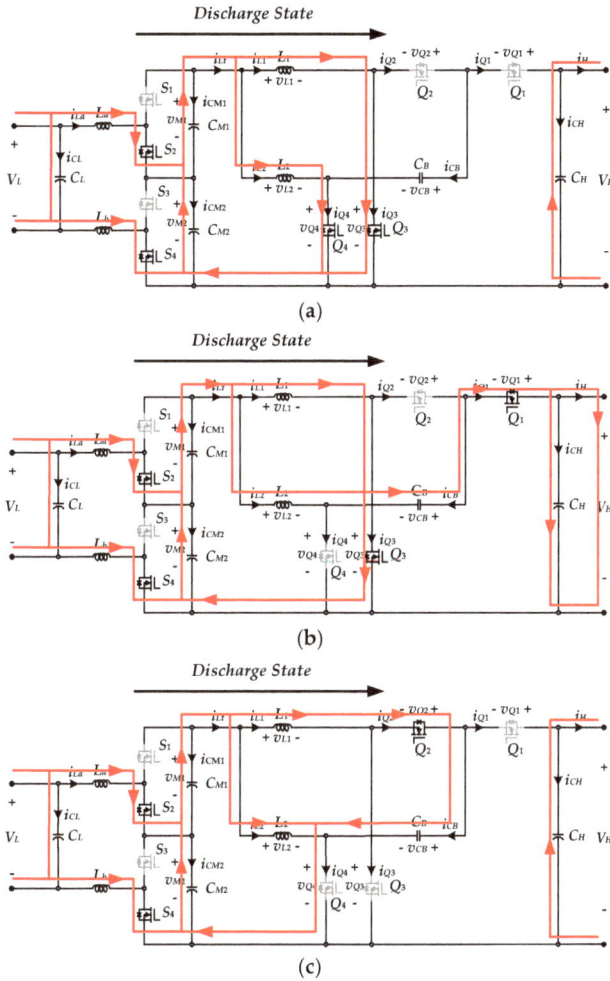

Figure 10. Equivalent circuits of the modes during different intervals in discharge state: (**a**) Mode 1, Mode 3; (**b**) Mode 2; (**c**) Mode 4.

2.2.3. Mode 3 [$t_2 < t \leqslant t_3$]

In this operation mode, the circuit operation is same as Mode 1.

2.2.4. Mode 4 [$t_3 < t \leqslant t_4$]

In this operation mode, the interval time is $(1 - D_b)T_{sw}$. For the low-side stage, switches Q_1, Q_3 are turned off and Q_2, Q_4 are turned on. The energy stored in inductor L_1 is now released energy to charge-pump capacitor C_B for compensating the lost charges in previous modes. The output power is supplied from the capacitor C_H. The voltages across inductances L_1 and L_2 can be represented as:

$$L_1 \frac{di_{L1}}{dt} = V_M - V_{CB} \tag{9}$$

$$L_2 \frac{di_{L2}}{dt} = V_M \tag{10}$$

3. Steady-State Analysis

3.1. Voltage Conversion Ratio

In charge state, V_H is the input and V_L is the output. According to Equations (1)–(5) and based on the voltage-second balance principle in L_1 and L_2, the voltage conversion ratio M_d in charge state can be derived as:

$$M_d = \frac{V_L}{V_H} = \frac{D_d}{4} \tag{11}$$

In Equation (11), D_d is the duty cycle of the active switches Q_1 and Q_2. As can be seen, the voltage conversion ratio in charge state is one-fourth of that of the conventional buck converter. Similarly, in discharge state, V_L is the input and V_H is the output. According to Equations (6)–(10) and based on the voltage-second balance principle in L_1 and L_2, the voltage conversion ratio M_b in discharge state can be derived as:

$$M_b = \frac{V_H}{V_L} = \frac{4}{1 - D_b} \tag{12}$$

where D_b is the duty cycle of the active switches Q_3 and Q_4. As can be seen, the voltage conversion ratio in discharge state is four times of that of the conventional boost converter.

Figure 11 shows that the studied BDC demands a smaller duty cycle for the active switches to produce the same voltage conversion ratio, or can produce a higher voltage conversion ratio at the same duty cycle when compared with the traditional BDC [18] and the previous BDC in [22]. Furthermore, the voltage conversion ratio of studied BDC is higher than that of the BDC proposed in [23], under a reasonable range of 25%~75% duty cycles.

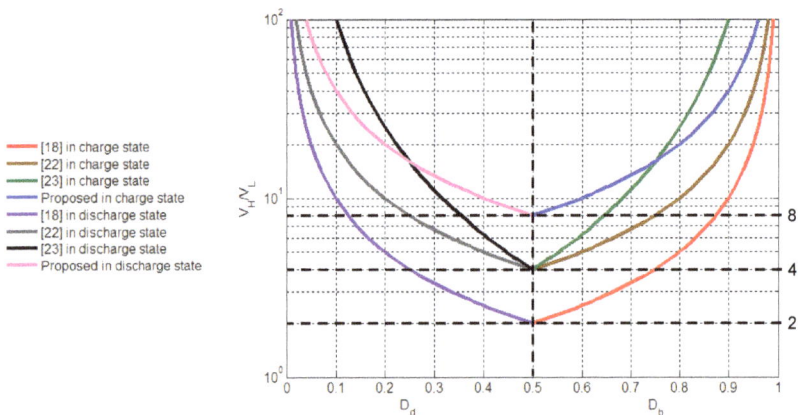

Figure 11. Comparison of voltage conversion ratios produced by the studied BDC, the converters introduced in [18,22,23].

3.2. Voltage Stress of the Switches

Whenever the ULC works as a back or front-end stage, the open circuit voltage stress on the switches S_1~S_4 of ULC is equal to the low-side input voltage V_L, as follows:

$$V_{S1,max} = V_{S2,max} = V_{S3,max} = V_{S4,max} = V_L \tag{13}$$

The particular inherent feature of the ULC benefits the low conduction losses can be achieved by adopting the low-voltage MOSFETs.

As to the high-side stage of the studied BDC, based on the aforementioned operation analyzes in Section 2, the open circuit voltage stress of switches Q_1~Q_4 can be obtained directly as:

$$VQ_{1,max} = V_{Q3,max} = V_{Q4,max} = \frac{V_H}{2} \tag{14}$$

$$V_{Q2,max} = V_H \tag{15}$$

3.3. Inductor Current Ripple

The studied BDC can operate not only in charge state but also in discharge state. Thus, the inductor can be calculated in either charge or discharge state. According to Equations (1)–(5), the total ripple current of the inductor of the studied BDC in charge state can be expressed as:

$$\Delta i_{Lt}|_{charge} = \frac{V_H T_{sw}}{L_s}(0.5 - D_d)D_d \tag{16}$$

Similarly, in discharge state, according to Equations (6)–(10), the total ripple current of the inductor of the studied BDC in discharge state can be expressed as:

$$\Delta i_{Lt}|_{discharge} = \frac{V_H T_{sw}}{L_s}(D_b - 0.5)(1 - D_b) \tag{17}$$

Figure 12 shows the normalized ripple current of the inductor of the studied BDC, the traditional BDC [18], and previous BDCs in [22,23], where the inductor and the switching frequency of these three BDCs are equal, respectively. The ripple current of the traditional BDC at 50% duty cycle is normalized as one.

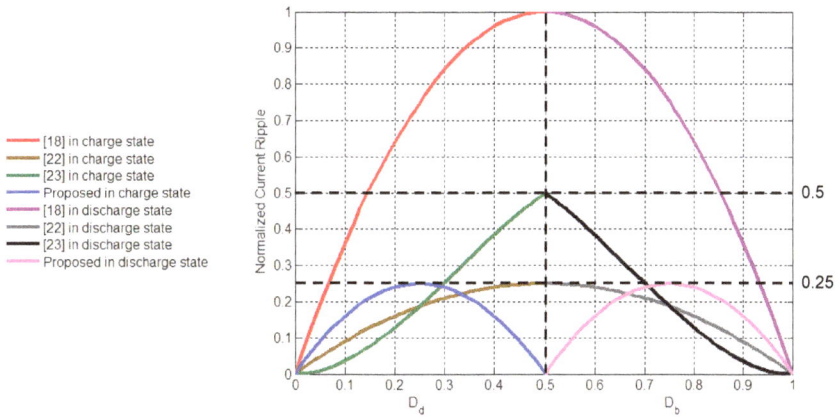

Figure 12. Comparison of the normalized ripple current of the inductor among the studied BDC, the converters introduced in [18,22,23].

It can be seen that from Figure 12, the maximum ripple current of the inductor of studied BDC is only one-fourth of that of a traditional BDC. On the other and, if the ripple currents are equal, the inductor of the studied BDC is only one-fourth of that of traditional BDC [18], which means that the studied BDC has a better dynamic response. From Figure 12, the ripple current of studied BDC is smaller than that of the converter in [22], under a reasonable range of 35%~65% duty cycles. Furthermore, the ripple current of the previous BDC proposed in [23] is higher than that of the one proposed in this study, under a reasonable range of 30%~70% duty cycles.

3.4. Boundary Conduction Mode

The boundary normalized inductor time constant $\tau_{L,B}$ can be defined as:

$$\tau_{L,B} = \frac{L_s f_{sw}}{R} \tag{18}$$

where R is low-side input equivalent resistance.

During boundary conduction mode (BCM), the input current BDC can be derived as:

$$I_L = \frac{4V_L}{L_s f_{sw}}(1 - D_d) \tag{19}$$

Substituting Equation (19) into (18), the boundary normalized time constant in charge state can be expressed as:

$$\tau_{Ld,B} = 4(1 - D_d) \tag{20}$$

Similarly, in discharge state, the input current of the studied BDC can be obtained as:

$$I_L = \frac{4V_L}{L_s f_{sw}} D_b \tag{21}$$

The boundary normalized time constant in discharge state can be expressed as:

$$\tau_{Lb,B} = 4D_b \tag{22}$$

Figure 13 shows the plots of boundary normalized inductor time constant curves $\tau_{Ld,B}$ and $\tau_{Lb,B}$ in charge and discharge states. The BDC in charge state operates in CCM when τ_{Ld} is designed to be higher than the boundary curve of $\tau_{Ld,B}$. The studied BDC in discharge state operates in discontinuous conduction mode (DCM) when τ_{Lb} is selected to be lower than the boundary curve of $\tau_{Lb,B}$.

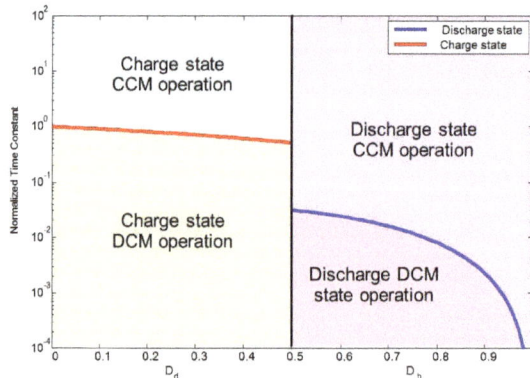

Figure 13. Normalized boundary inductances time constant in charge and discharge states.

Figure 14 shows the boundary inductances curve of the studied BDC in charge and discharge states. If the inductance is selected to be larger than the boundary inductance, the studied BDC will operate in CCM. The studied BDC can operate not only in charge state but also in discharge state, the boundary inductance can be derived as below from Equations (19) and (21), respectively.

$$L_{d,B} = \frac{4(1 - D_d)}{f_{sw}} \frac{V_L^2}{P_{out}} \tag{23}$$

$$L_{b,B} = \frac{4D_b}{f_{sw}} \frac{V_L^2}{P_{out}} \tag{24}$$

where P_{out} is the output power.

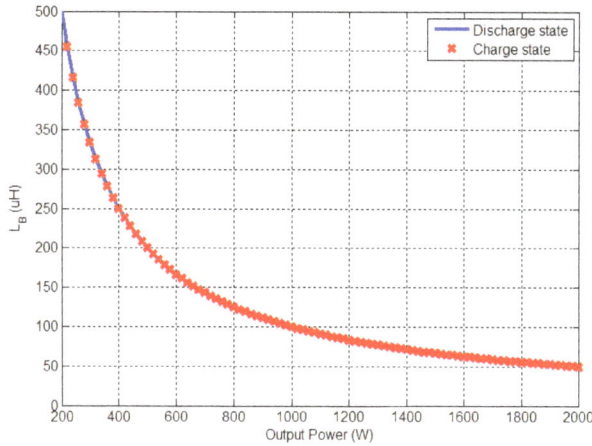

Figure 14. Boundary inductances in various power conditions.

3.5. Selection Considerations of Charge-Pump Capacitor

For the proposed BDC in charge state operation, the ripple voltage of the charge-pump capacitor C_B can be obtained as follows:

$$\Delta V_{CB} = \frac{1}{C_B} \int_{t_0}^{t_1} i_{CB}(t)dt = \frac{I_{Lt}D_d}{2C_B f_{sw}} \cong \frac{I_L D_d}{4C_B f_{sw}} \tag{25}$$

where:

$$i_{CB}(t) = \frac{I_L}{4} - \frac{\Delta i_{ripple}}{2} + \frac{0.5V_H - 2V_L}{L_s f_{sw}}(t - t_0) \tag{26}$$

$$\Delta i_{ripple} = \frac{0.5V_H - 2V_L}{L_s f_{sw}}(t_1 - t_0), t_1 = D_d T_{sw} + t_0 \tag{27}$$

From Equation (25), it is known that although a capacitor with low capacitance is used for charge-pump capacitor C_B, the voltage ripple can be reduced by increasing the switching frequency. The root mean square (RMS) value of the current through the charge-pump capacitor is

$$I_{CB(RMS)} = \sqrt{\frac{2}{f_{sw}} \int_{t_0}^{t_1} i_{CB}^2(t)dt} \cong \frac{I_L}{4}\sqrt{2D_d} \tag{28}$$

3.6. Summaries of Component Stress and Loss

For stress and loss analysis, it is assumed that the studied BDC operates with $D_d < 0.5$ and $D_b > 0.5$ for charge and discharge modes, respectively. The results of component stress can be summarized as in Table 1. Furthermore, equations for loss analysis can be summarized as in Table 2, where Q_g represents the MOSFET total gate charge; t_r is rise time, it's the period after the v_{GS} reaches threshold voltage $v_{GS(th)}$ to complete the transient MOSFET gate charge; t_f is fall time, it's the time where the gate voltage reaches the threshold voltage $v_{GS(th)}$ after MOSFET turn-off delay time [26].

Table 1. Stress analysis results at steady-state.

Items	Charge State	Discharge State
Voltage Stress of Q_1, Q_3, Q_4 (v_{Q1}, v_{Q3}, v_{Q4})	$0.5V_H$	$0.5V_H$
Voltage Stress of Q_2 (v_{Q2})	V_H	V_H
Voltage Stress of $S_1 \sim S_4$ ($v_{S1} \sim v_{S4}$)	V_L	V_L
RMS Current Stress of Q_1 (i_{Q1})	$I_{L2(RMS)} \sqrt{D_d}$	$I_{L2(RMS)} \sqrt{1-D_b}$
RMS Current Stress of Q_2 (i_{Q2})	$I_{L1(RMS)} \sqrt{D_d}$	$I_{L1(RMS)} \sqrt{1-D_b}$
RMS Current Stress of Q_3 (i_{Q3})	$I_{L1(RMS)} \sqrt{1-D_d}$	$I_{L1(RMS)} \sqrt{D_b}$
RMS Current Stress of Q_4 (i_{Q4})	$\sqrt{\dfrac{(I_{Lt(RMS)})^2(D_d)+}{(I_{L2(RMS)})^2(0.5-D_d)}}$	$\sqrt{\dfrac{(I_{Lt(RMS)})^2(1-D_b)+}{(I_{L2(RMS)})^2(D_b-0.5)}}$
RMS Current Stress of $S_1 \sim S_4$ ($i_{S1} \sim i_{S4}$)	$I_{Lt(RMS)}/\sqrt{2}$	$I_{Lt(RMS)}/\sqrt{2}$
RMS Current Stress of L_1 (i_{L1})	$\sqrt{I_{L1}^2 + (\frac{\Delta i_{L1}}{2\sqrt{3}})}$	$\sqrt{I_{L1}^2 + (\frac{\Delta i_{L1}}{2\sqrt{3}})}$
RMS Current Stress of L_2 (i_{L2})	$\sqrt{I_{L2}^2 + (\frac{\Delta i_{L2}}{2\sqrt{3}})}$	$\sqrt{I_{L2}^2 + (\frac{\Delta i_{L2}}{2\sqrt{3}})}$
RMS Current Stress of L_a (i_{La})	$\sqrt{I_{La}^2 + (\frac{\Delta i_{La}}{2\sqrt{3}})}$	$\sqrt{I_{La}^2 + (\frac{\Delta i_{La}}{2\sqrt{3}})}$
RMS Current Stress of L_b (i_{Lb})	$\sqrt{I_{Lb}^2 + (\frac{\Delta i_{Lb}}{2\sqrt{3}})}$	$\sqrt{I_{Lb}^2 + (\frac{\Delta i_{Lb}}{2\sqrt{3}})}$
RMS Current Stress of C_B (i_{CB})	$\left(I_L\sqrt{2D_d}\right)/4$	$\left(I_L\sqrt{2(1-D_b)}\right)/4$
RMS Current Stress of C_H (i_{CH})	$\sqrt{(I_{Q1(RMS)})^2 - I_H}$	$\sqrt{(I_{Q1(RMS)})^2 - I_H}$
RMS Current Stress of C_L (i_{CL})	$\sqrt{I_L^2 - \frac{4\Delta i_{La}I_L}{\pi} + \frac{4\Delta i_{La}^2}{\pi^2} + \frac{\Delta i_{La}^2}{2}}$	$\sqrt{I_L^2 - \frac{4\Delta i_{La}I_L}{\pi} + \frac{4\Delta i_{La}^2}{\pi^2} + \frac{\Delta i_{La}^2}{2}}$
RMS Current Stress of C_{M1}, C_{M2} (i_{CM1}, i_{CM2})	$\sqrt{I_{Lt(RMS)}^2 - I_{S1(RMS)}^2}$	$\sqrt{I_{Lt(RMS)}^2 - I_{S2(RMS)}^2}$

Table 2. Loss equations at steady-state.

Items	Equations
Conduction loss of $Q_1 \sim Q_4$	$R_{DS(Q1)} \times [i_{Q1(RMS)}]^2$; $R_{DS(Q2)} \times [i_{Q2(RMS)}]^2$; $R_{DS(Q3)} \times [i_{Q3(RMS)}]^2$; $R_{DS(Q4)} \times [i_{Q4(RMS)}]^2$
Conduction loss of $S_1 \sim S_4$	$R_{DS(S1)} \times [i_{S1(RMS)}]^2$; $R_{DS(S2)} \times [i_{S2(RMS)}]^2$; $R_{DS(S3)} \times [i_{S3(RMS)}]^2$; $R_{DS(S4)} \times [i_{S4(RMS)}]^2$
Switching loss of Q_1	$(V_{DS(Q1)} \times i_{Q1(ON)} \times T_r)/6T_{sw}$; $(V_{DS(Q1)} \times i_{Q1(OFF)} \times T_f)/6T_{sw}$
Switching loss of Q_2	$(V_{DS(Q2)} \times i_{Q2(ON)} \times T_r)/6T_{sw}$; $(V_{DS(Q2)} \times i_{Q2(OFF)} \times T_f)/6T_{sw}$
Switching loss of Q_3	$(V_{DS(Q3)} \times i_{Q3(ON)} \times T_r)/6T_{sw}$; $(V_{DS(Q3)} \times i_{Q3(OFF)} \times T_f)/6T_{sw}$
Switching loss of Q_4	$(V_{DS(Q4)} \times i_{Q4(ON)} \times T_r)/6T_{sw}$; $(V_{DS(Q4)} \times i_{Q4(OFF)} \times T_f)/6T_{sw}$
Switching loss of S_1	$\left(V_{DS(S1)} \times i_{S1(ON)} \times T_r\right)/6T_{sw}$; $(V_{DS(S1)} \times i_{S1(OFF)} \times T_f)/6T_{sw}$
Switching loss of S_2	$(V_{DS(S2)} \times i_{S2(ON)} \times T_r)/6T$; $\left(V_{DS(S2)} \times i_{S2(OFF)} \times T_f\right)/6T$
Switching loss of S_3	$(V_{DS(S3)} \times i_{S3(ON)} \times T_r)/6T_{sw}$; $(V_{DS(S3)} \times i_{S3(OFF)} \times T_f)/6T_{sw}$
Switching loss of S_4	$(V_{DS(S4)} \times i_{S4(ON)} \times T_r)/6T_{sw}$; $(V_{DS(S4)} \times i_{S4(OFF)} \times T_f)/6T_{sw}$
Conduction loss of $L_1 \sim L_2$	$R_{L1} \times [i_{L1(RMS)}]^2$; $R_{L2} \times [i_{L2(RMS)}]^2$
Conduction loss of $L_a \sim L_b$	$R_{La} \times [i_{La(RMS)}]^2$; $R_{Lb} \times [i_{Lb(RMS)}]^2$
Conduction loss of C_B, C_H, C_L	$R_{CB} \times [i_{CB(RMS)}]^2$; $R_{CH} \times [i_{CH(RMS)}]^2$; $R_{CL} \times [i_{CL(RMS)}]^2$
Conduction loss of $C_{M1} \sim C_{M2}$	$R_{CM1} \times [i_{CM1(RMS)}]^2$; $R_{CM2} \times [i_{CM2(RMS)}]^2$
Gate driving loss of $Q_1 \sim Q_4$	$Q_{g(Q1 \sim Q4)} \times V_{GS(Q1 \sim Q4)} \times f_{sw}$
Gate driving loss of $S_1 \sim S_4$	$Q_{g(S1 \sim S4)} \times V_{GS(S1 \sim S4)} \times f_{sw}$

4. Simulation and Experimental Results

In order to illustrate the performance of the studied BDC, a laboratory prototype circuit is simulated and experimented. To avoid all elements suffer from high-current stress at DCM operation, resulting in high conduction and core losses. The studied BDC operates at CCM, and its parameters and specifications of the constructed hardware prototype are given as below:

(1) high-side voltage V_H: 385 V;
(2) low-side voltage V_L: 48 V;
(3) rated power P_o: 500 W;

(4) switching frequency f_{sw}: 20 kHz;

(5) capacitors $C_H = C_L = 33$ μF, $C_{M1} = C_{M2} = 33$ μF, $C_B = 10$ μF; (ESR of C_H, $R_{CH} = 0.064$ Ω; ESR of C_L, $R_{CL} = 0.062$ Ω, ESR of C_{M1}, $R_{CM1} = 0.16$ Ω; ESR of C_{M2}, $R_{CM2} = 0.16$ Ω; ESR of C_B, $R_{CB} = 0.062$ Ω);

(6) inductors $L_1 = L_2 = L_s = 800$ μH; $L_a = L_b = 1.5$ μH (IHLP-6767GZ-A1); (ESR of L_1, $R_{L1} = 0.18$ Ω, ESR of L_2, $R_{L2} = 0.18$ Ω, ESR of L_a, $R_{La} = 13.6$ mΩ; ESR of L_b, $R_{Lb} = 13.6$ mΩ);

(7) power switches $S_1 \sim S_4$: IXFH160N15T2, 150 V/160 A/$R_{DS(on)} = 9$ mΩ, TO-247AC; Q_1, Q_3, Q_4: FDA59N30, 300 V/59 A/$R_{DS(on)} = 56$ mΩ, TO-247AC; Q_2: W25NM60, 650 V/21 A/$R_{DS(on)} = 160$ mΩ, TO-247AC.

Figure 15 show the simulated low-side filter currents (i_{La}, i_{Lb}), gate signals of active switches (Q_1, Q_2) and two-phase inductor currents (i_{L1}, i_{L2}) in charge state at full load condition. Also the corresponding experimental results are shown in Figure 16. One can observe that both results are in very close agreement as well. From Figures 15a and 16a, as can be seen, the low-side filter (L_a, L_b) can effectively limit the switching current spike and shape the current to a nearly rectified sinusoidal waveform. Also, from the figures it is observed that by interleaved controlling the duty cycles of 0.48 for the switches (Q_1, Q_2), the two-phase currents (i_{L1}, i_{L2}) are in complementary relation and in CCM.

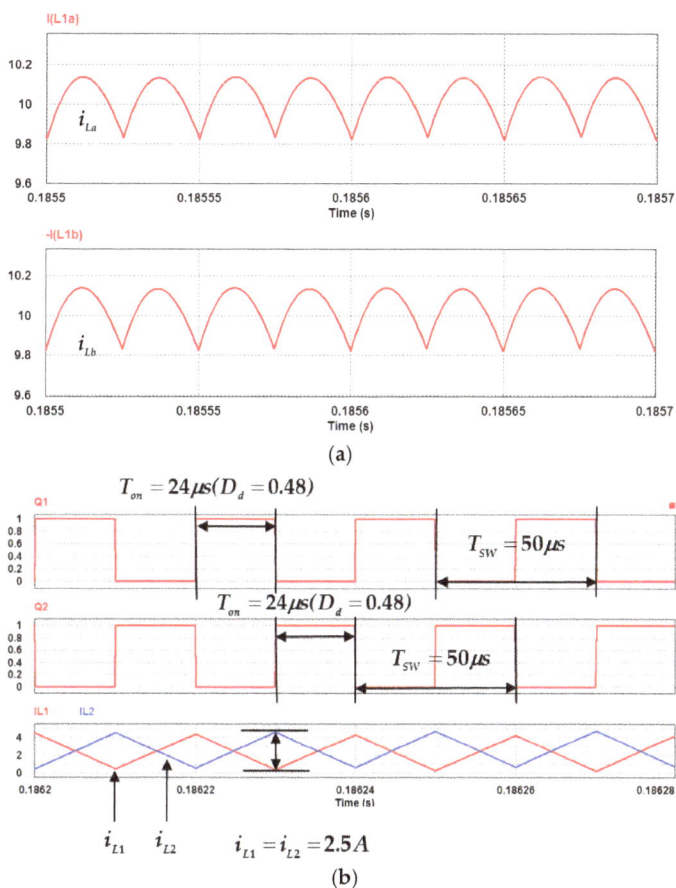

Figure 15. Simulated waveforms of the studied BDC in charge state at full load: (**a**) low-side filter currents i_{La}, i_{Lb}; (**b**) gate signals of Q_1, Q_2 and two-phase inductor currents i_{L1}, i_{L2}.

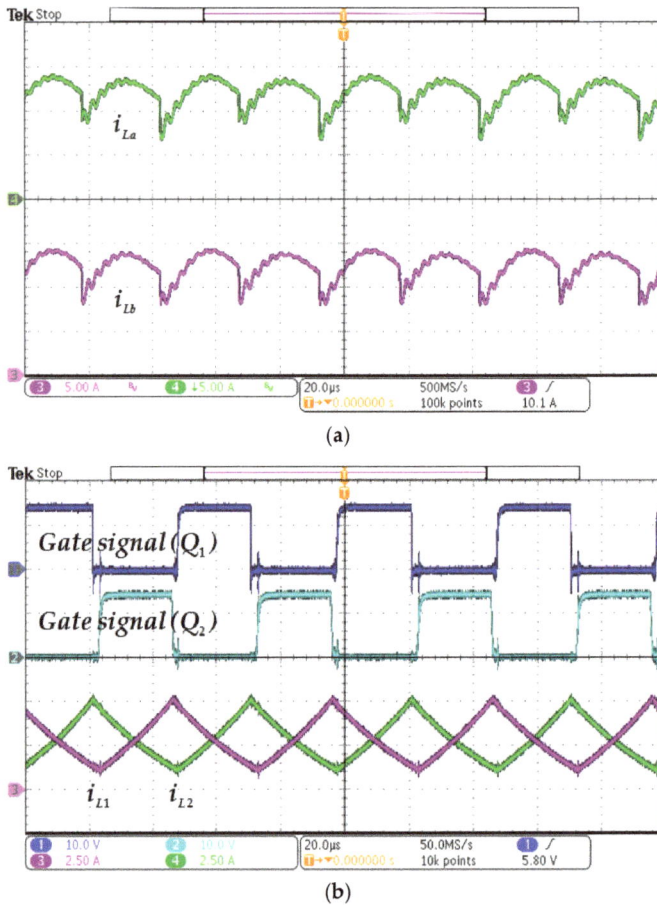

Figure 16. Measured waveforms of the studied BDC in charge state at full load: (**a**) low-side filter currents i_{La}, i_{Lb}; (**b**) gate signals of Q_1, Q_2 and two-phase inductor currents i_{L1}, i_{L2}.

Figures 17 and 18 show the simulated and measured waveforms of charge-pump capacitor voltage (V_{CB}), middle-link voltage (V_M), middle-link capacitor voltages (V_{M1}, V_{M2}), low-side voltage (V_L), and low-side switch voltages (V_{S1}, V_{S2}, V_{S3}, V_{S4}). From Figures 17 and 18 with the ULC of studied BDC, the low-voltage side (V_L) is well regulated at 48 V. The middle-link voltage is 96 V, it does quite reach twice of the regulated low-side voltage (V_L) of 48 V. The charge-pump capacitor voltage (V_{CB}) of 192 V can be achieved easily and indeed can share one-half of the high-side voltage to reduce the voltage stress of active switches. It is observed that the steady-state voltage stresses of low-side active switches (V_{S1}, V_{S2}, V_{S3}, V_{S4}) are only about 48 V, which means that lower on-resistance MOSFETs can be used to achieve the improved conversion efficiency. Also, both the simulated results are in close agreement with the corresponding experimental results.

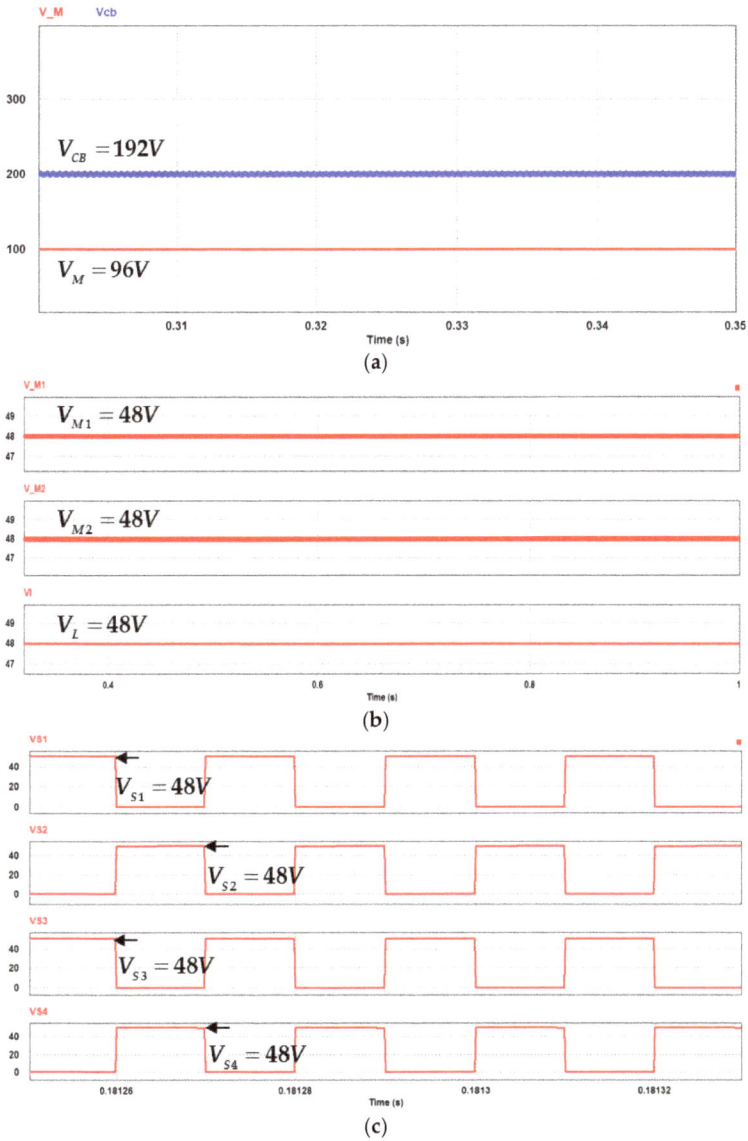

Figure 17. Simulated waveforms of the studied BDC in charge state at full load: (**a**) charge-pump capacitor voltage V_{CB}, middle-link voltage V_M; (**b**) middle-link capacitor voltages V_{M1}, V_{M2}, and low-side voltage V_L; (**c**) switch voltages of S_1, S_2, S_3, S_4.

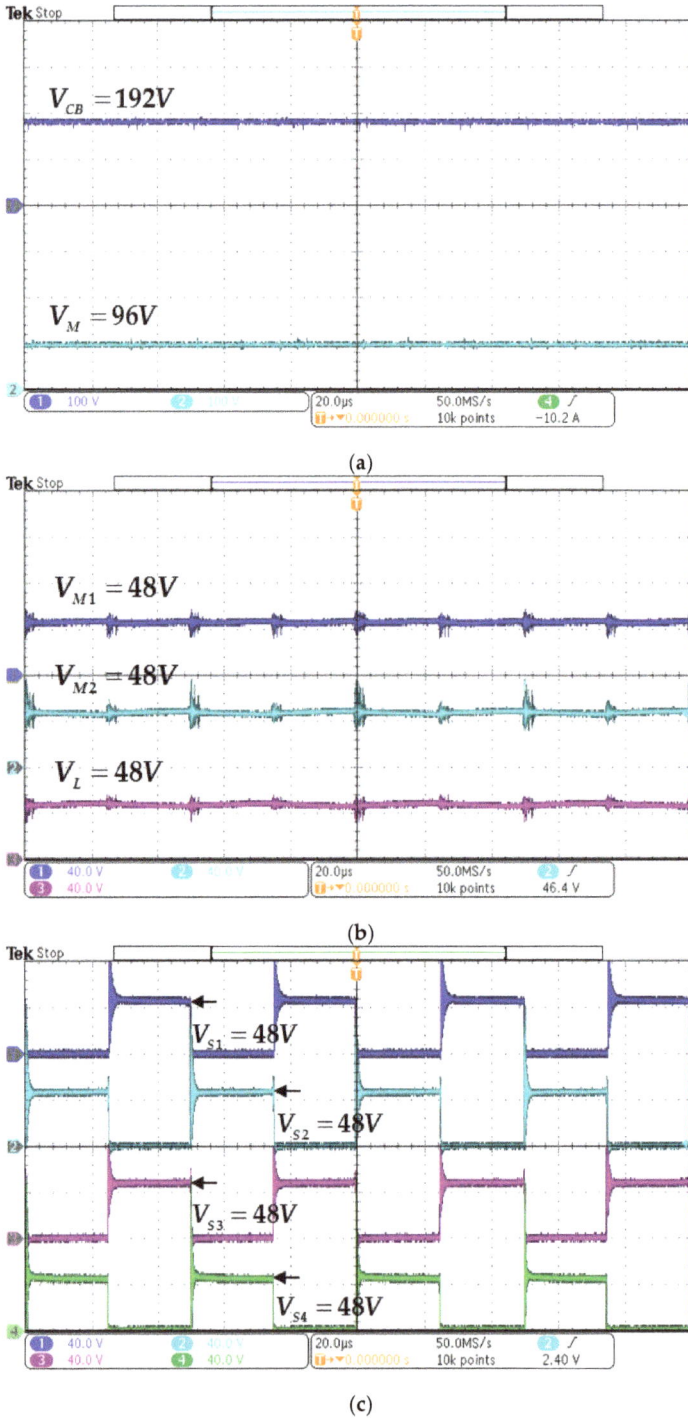

Figure 18. Measured waveforms of the studied BDC in charge state at full load: (**a**) charge-pump capacitor voltage V_{CB} and middle-link voltage V_M; (**b**) middle-link capacitor voltages V_{M1}, V_{M2}, and low-side voltage V_L; (**c**) switch voltages of S_1, S_2, S_3, S_4.

Figure 19 shows the simulated waveforms of gate signals of Q_3, Q_4, the two-phase inductor currents (i_{L1}, i_{L2}) and the switch voltages of (V_{Q3}, V_{Q4}) in charge state at full load condition. The corresponding experimental results are also shown in Figure 20. One can observe that both results are in very close agreement as well. From the figures it is observed that by interleaved controlling the duty cycles of 0.52 for the switches (Q_3, Q_4), the two-phase currents (i_{L1}, i_{L2}) are in complementary relation and in CCM. Also, from Figures 19b and 20b, the charge-pump capacitor voltage (V_{CB}) is about 192.5 V, it can clamp the switch voltages of active switches (Q_3, Q_4) to be nearly one-half of the regulated high-side voltage V_H of 385 V.

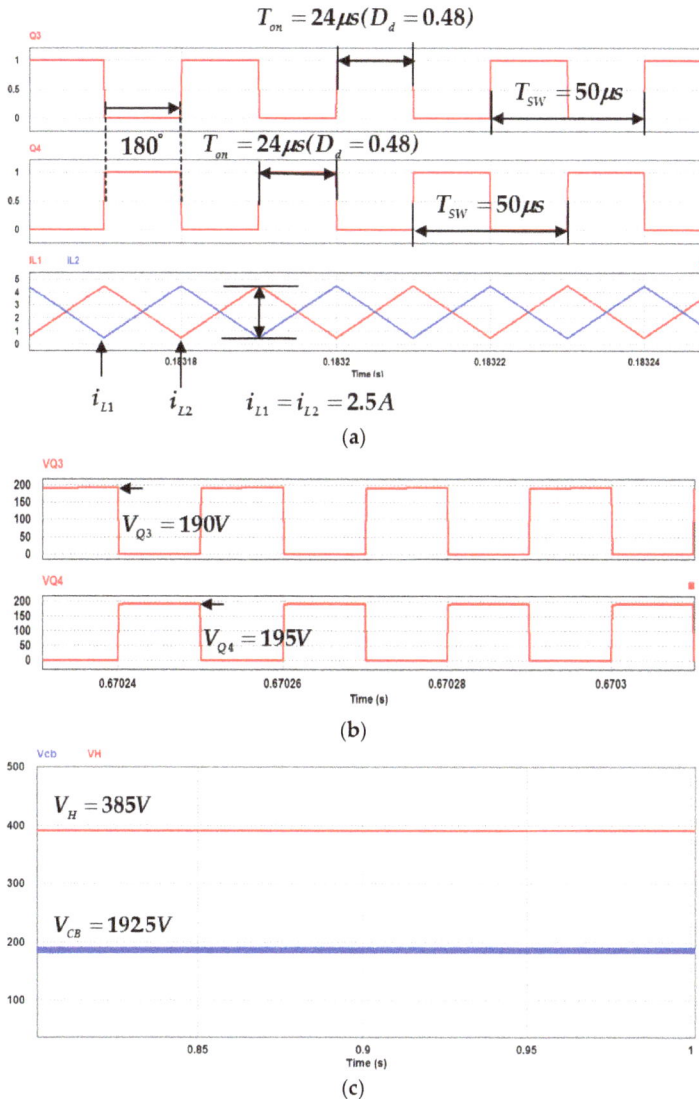

Figure 19. Simulated waveforms of the studied BDC in discharge state at full load: (**a**) gate signals of Q_3, Q_4, two-phase inductor currents i_{L1}, i_{L2}; (**b**) switch voltages of Q_3, Q_4; (**c**) charge-pump capacitor voltage V_{CB} and high-side voltage V_H.

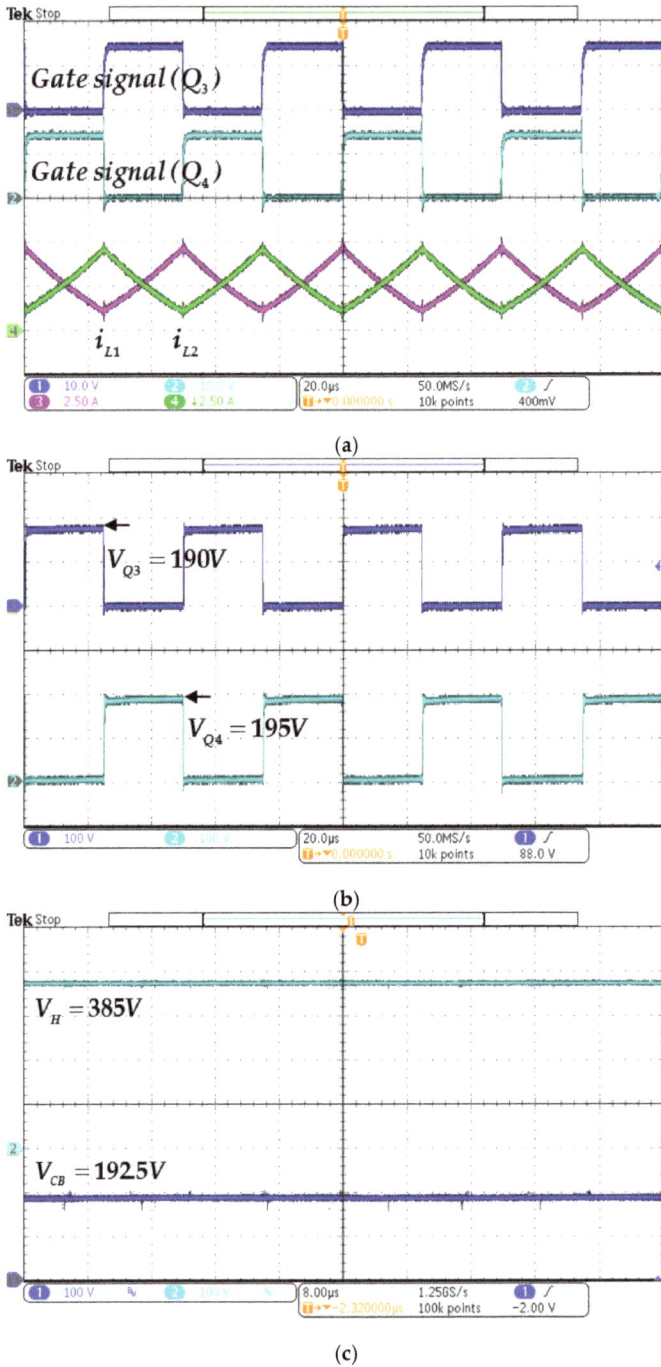

Figure 20. Measured waveforms of the studied BDC in discharge state at full load: (**a**) gate signals of Q_3, Q_4, two-phase inductor currents i_{L1}, i_{L2}; (**b**) switches voltages of Q_3, Q_4; (**c**) charge-pump capacitor voltage V_{CB} and high-side voltage V_H.

Figure 21 summarizes the measured conversion efficiency of the studied BDC in charge and discharge states. On the experimental porotype system, the conversion efficiency is measured via precise digital power meter WT310 equipment, manufactured by the Yokogawa Electric Corporation (Tokyo, Japan). The accuracy of the measured power is within +/−0.1%. It can be seen that from Figure 21, the measured highest conversion efficiency is 95% in discharge state and is around 96% in charge state. In order to clarify the actual measured conversion efficiency further, based on the equations in Table 2, the calculated power loss distribution at the rated load condition is listed in Table 3, and furthermore, the calculated losses breakdown diagrams of the studied BDC are depicted in Figure 22. From Table 3 and Figure 22, one can see that the power losses mainly occur in the copper loss of the inductors, switching loss and conduction loss of the MOSFETs. The total power losses in charge and discharge states are 28.5 W and 28.6 W, accounting for 5.70% and 5.73%, in rated load condition, respectively. These match well the measured conversion efficiency of the studied BDC in charge (94.29%) and discharge (94.25%) states.

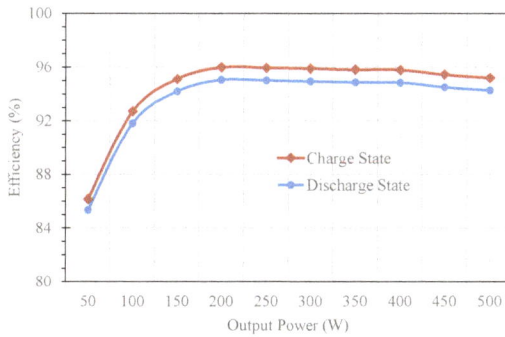

Figure 21. Measured conversion efficiency of the studied BDC for low-side voltage V_L = 48 V and high-side voltage V_H = 385 V under different loads.

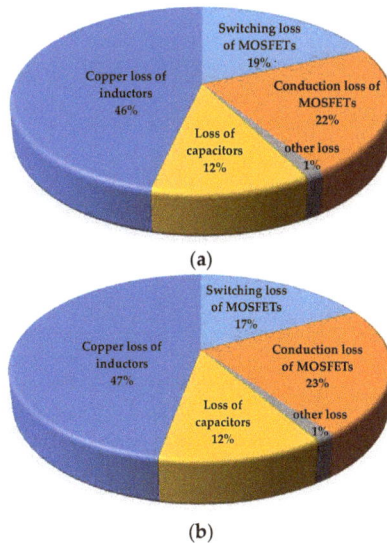

(a)

(b)

Figure 22. Calculated losses breakdown diagrams at rated load condition: (**a**) in charge state; (**b**) in discharge state.

Table 3. Power loss distribution (500 W rated load condition).

Items	Charge State Calculated Results	Discharge State Calculated Results
Conduction loss of Q_1	0.62 W	0.62 W
Conduction loss of Q_2	1.58 W	1.58 W
Conduction loss of Q_3	0.67 W	0.67 W
Conduction loss of Q_4	1.29 W	1.29 W
Conduction loss of S_1	0.58 W	0.58 W
Conduction loss of S_2	0.58 W	0.58 W
Conduction loss of S_3	0.58 W	0.58 W
Conduction loss of S_4	0.58 W	0.58 W
Switching loss of Q_1 (turn on/off transition)	on: 0.09 W; off: 0.52 W	on: 0.10 W; off: 0.72 W
Switching loss of Q_2 (turn on/off transition)	on: 0.19 W; off: 1.01 W	on: 0.17 W; off: 0.87 W
Switching loss of Q_3 (turn on/off transition)	on: 0.09 W; off: 0.62 W	on: 0.09 W; off: 0.52 W
Switching loss of Q_4 (turn on/off transition)	on: 0.10 W; off: 0.69 W	on: 0.09 W; off: 0.54 W
Switching loss of S_1 (turn on/off transition)	on: 0.07 W; off: 0.44 W	on: 0.05 W; off: 0.55 W
Switching loss of S_2 (turn on/off transition)	on: 0.05 W; off: 0.60 W	on: 0.06 W; off: 0.35 W
Switching loss of S_3 (turn on/off transition)	on: 0.05 W; off: 0.47 W	on: 0.05 W; off: 0.29 W
Switching loss of S_4 (turn on/off transition)	on: 0.06 W; off: 0.34 W	on: 0.05 W; off: 0.46 W
Conduction loss of L_1	4.94 W	4.94 W
Conduction loss of L_2	4.94 W	4.94 W
Conduction loss of L_a	1.80 W	1.80 W
Conduction loss of L_b	1.80 W	1.80 W
Conduction loss of C_B	1.61 W	1.61 W
Conduction loss of C_H	1.67 W	1.67 W
Conduction loss of C_L	0.02 W	0.02 W
Conduction loss of C_{M1}	0.01 W	0.01 W
Conduction loss of C_{M2}	0.01 W	0.01 W
Gate driving loss of $Q_1 \sim Q_4$	0.02 W	0.02 W
Gate driving loss of $S_1 \sim S_4$	0.08 W	0.08 W
Total losses	28.5 W	28.64 W
% in rated load condition	5.70%	5.73%
Calculated Efficiency	94.30%	94.27%
Measured Efficiency	94.29%	94.25%

The performance comparisons between the studied BDC and a variety of published research results are summarized in Table 4. As can be seen from the comparative data, though the amounts of components in the proposed converter are more than the requirement in the other previous BDCs. The studied two-phase BDC indeed performs the higher conversion efficiency, bidirectional power flow, lower output ripples under 500 W power rating than other announced works [17,22,23]. Finally, the practical photograph of the realized BDC prototype and the test bench system are depicted in Figure 23.

Table 4. Performance comparisons with other published converters.

Items	Topology			
	This Work	**[17]**	**[22]**	**[23]**
Switching control structure	two-phase	single-phase	single-phase	single-phase
Output ripple	Low	High	Medium	Medium
Step-up conversion ratio	$4/(1 - D_b)$	$n/(1 - D_b)$	$2/(1 - D_b)$	$1/(1 - D_b)^2$
Step-down conversion ratio	$D_d/4$	$D_d/(1 + n - nD_d)$	$D_d/2$	$(D_d)^2$
High-side voltage	385 V	400 V	200 V	62.5 V
Low-side voltage	48 V	48 V	24 V	10 V
Realized prototype power rating	500 W	200 W	200 W	100 W
Number of main switches	8	4	4	4
Number of storage components	7	5	5	5
Maximum efficiency (charge state)	96%	91.6%	94.8%	91.5%
Maximum efficiency (discharge state)	95%	94.3%	94.1%	92.5%

n: the turns ratio of coupled inductor [17].

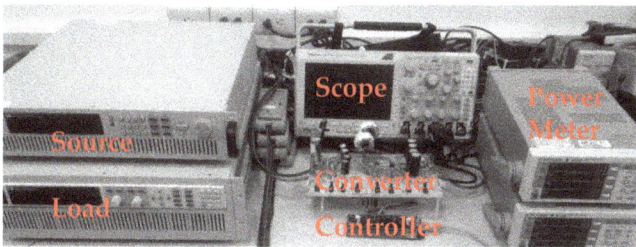

Figure 23. Photograph of the realized BDC prototype and the test bench system.

5. Conclusions

A novel BDC topology with high voltage conversion ratio is developed and a 500 W rating prototype system with 48 V battery input is constructed. Applying the developed BDC topology to the 48 V mini-hybrid powertrain system is also expected in the future [27]. In this study, thanks to the ULC located at the low-side stage, high power density and efficiency in all load range make the studied BDC a promising two-stage power architecture. Furthermore, the IBCPC located at the high-side stage can achieve a much higher voltage conversion ratio under a reasonable duty cycle. In summary, the proposed novel BDC offers the following improvements: (1) high voltage conversion ratio; (2) low ripple current; (3) it is simpler to design, implement and control. Finally, a 500 W rating low-power prototype system is given as an example for verifying the validity of the operation principle. Experimental results show that a highest efficiency of 96% and 95% can be achieved, respectively, in charge and discharge states. Certainly, by making a suitable printed circuit board (PCB) layout, and with good component placement and good heat dissipation transfer process, the novel BDC can be implemented for higher power conversion applications.

Acknowledgments: This research is sponsored by the Ministry of Science and Technology, Taiwan, under contracts 104-2221-E-027-125, 104-2623-E-027-005-ET, and 104-2622-E-027-023-CC3. The author would like to thank the student, Jie-Ting Li for for his help in the experiment and Dr. Yuan-Chih Lin for his suggestions.

Conflicts of Interest: The author declares no conflict of interest.

References

1. Lai, C.M.; Pan, C.T.; Cheng, M.C. High-efficiency modular high step-up interleaved boost converter for dc-microgrid applications. *IEEE Trans. Ind. Appl.* **2012**, *48*, 161–171. [CrossRef]
2. Boroyevich, D.; Cvetkovic, I.; Burgos, R.; Dong, D. Intergrid: A future electronic energy network? *IEEE J. Emerg. Sel. Top. Power Electron.* **2013**, *1*, 127–138. [CrossRef]

3. Yilmaz, M.; Krein, P.T. Review of the impact of vehicle-to-grid technologies on distribution systems and utility interfaces. *IEEE Trans. Power Electron.* **2013**, *28*, 5673–5689. [CrossRef]
4. Lai, C.M.; Lin, Y.C.; Lee, D.S. Study and implementation of a two-phase interleaved bidirectional DC/DC converter for vehicle and dc-microgrid systems. *Energies* **2015**, *8*, 9969–9991. [CrossRef]
5. Takeda, T.; Miyoshi, H.; Yukita, K.; Goto, Y.; Ichiyanagi, K. Power interchange by the DC bus in micro grids. In Proceedings of the IEEE International Conference on DC Microgrids, Atlanta, GA, USA, 7–10 June 2015; pp. 135–137.
6. Wunder, B.; Ott, L.; Kaiser, J.; Han, Y.; Fersterra, F.; Marz, M. Overview of different topologies and control strategies for DC micro grids. In Proceedings of the IEEE International Conference on DC Microgrids, Atlanta, GA, USA, 7–10 June 2015; pp. 349–354.
7. Hu, K.W.; Liaw, C.M. Incorporated operation control of DC microgrid and electric vehicle. *IEEE Trans. Ind. Electron.* **2016**, *63*, 202–215. [CrossRef]
8. Du, Y.; Lukic, S.; Jacobson, B.; Huang, A. Review of high power isolated bi-directional DC-DC converters for PHEV/EV DC charging infrastructure. In Proceedings of the IEEE Energy Conversion Congress and Exposition, Phoenix, AZ, USA, 17–22 September 2011; pp. 553–560.
9. Zhao, B.; Song, Q.; Liu, W.; Sun, Y. Overview of dual-active-bridge isolated bidirectional DC-DC converter for high-frequency-link power-conversion system. *IEEE Trans. Power Electron.* **2014**, *29*, 4091–4106. [CrossRef]
10. Wai, R.J.; Liaw, J.J. High-efficiency-isolated single-input multiple-output bidirectional converter. *IEEE Trans. Power Electron.* **2015**, *30*, 4914–4930. [CrossRef]
11. Du, Y.; Zhou, X.; Bai, S.; Lukic, S.; Huang, A. Review of non-isolated bi-directional DC-DC converters for plug-in hybrid electric vehicle charge station application at municipal parking decks. In Proceedings of the IEEE Applied Power Electronics Conference and Exposition, Palm Springs, CA, USA, 21–25 February 2010; pp. 1145–1151.
12. Wai, R.J.; Duan, R.Y. High-efficiency bidirectional converter for power sources with great voltage diversity. *IEEE Trans. Power Electron.* **1997**, *22*, 1986–1996. [CrossRef]
13. Wai, R.J.; Duan, R.Y.; Jheng, K.H. High-efficiency bidirectional DC-DC converter with high-voltage gain. *IET Power Electron.* **2012**, *5*, 173–184. [CrossRef]
14. Yang, L.S.; Liang, T.J. Analysis and implementation of a novel bidirectional DC-DC converter. *IEEE Trans. Ind. Electron.* **2012**, *59*, 422–434. [CrossRef]
15. Jiang, L.; Mi, C.; Li, S.; Zhang, M.; Zhang, X.; Yin, C. A novel soft-switching bidirectional DC-DC converter with coupled inductors. *IEEE Trans. Ind. Appl.* **2013**, *49*, 2730–2740. [CrossRef]
16. Hsieh, Y.P.; Chen, J.F.; Yang, L.S.; Wu, C.Y.; Liu, W.S. High-conversion-ratio bidirectional DC-DC converter with coupled inductor. *IEEE Trans. Ind. Electron.* **2014**, *61*, 210–222. [CrossRef]
17. Liang, T.J.; Liang, H.H.; Chen, S.M.; Chen, J.F.; Yang, L.S. Analysis, design, and implementation of a bidirectional double-boost DC-DC converter. *IEEE Trans. Ind. Appl.* **2014**, *50*, 3955–3962. [CrossRef]
18. Chen, L.R.; Chu, N.Y.; Wang, C.S.; Liang, R.H. Design of a reflex-based bidirectional converter with the energy recovery function. *IEEE Trans. Ind. Electron.* **2008**, *55*, 3022–3029. [CrossRef]
19. Han, D.; Noppakunkajorn, J.; Sarlioglu, B. Comprehensive efficiency, weight, and volume comparison of SiC- and Si-based bidirectional DC-DC converters for hybrid electric vehicles. *IEEE Trans. Veh. Technol.* **2014**, *63*, 3001–3010. [CrossRef]
20. Jin, K.; Yang, M.; Ruan, X.; Xu, M. Three-level bidirectional converter for fuel-cell/battery hybrid power system. *IEEE Trans. Ind. Electron.* **2010**, *57*, 1976–1986. [CrossRef]
21. Monge, S.B.; Alepuz, S.; Bordonau, J. A bidirectional multilevel boost-buck DC-DC converter. *IEEE Trans. Power Electron.* **2011**, *26*, 2172–2183. [CrossRef]
22. Lin, C.C.; Yang, L.S.; Wu, G.W. Study of a non-isolated bidirectional DC-DC converter. *IET Power Electron.* **2013**, *6*, 30–37. [CrossRef]
23. Ardi, H.; Ahrabi, R.R.; Ravadanegh, S.N. Non-isolated bidirectional DC-DC converter analysis and implementation. *IET Power Electron.* **2014**, *7*, 3033–3044. [CrossRef]
24. Kabalo, M.; Paire, D.; Blunier, B.; Bouquain, D.; Simoes, M.G.; Miraoui, A. Experimental validation of high-voltage-ratio low input-current-ripple converters for hybrid fuel cell supercapacitor systems. *IEEE Trans. Veh. Technol.* **2012**, *61*, 3430–3440. [CrossRef]
25. Lai, C.M.; Yang, M.J. A high-gain three-port power converter with fuel cell, battery sources and stacked output for hybrid electric vehicles and DC-microgrids. *Energies* **2016**, *9*, 180. [CrossRef]

26. MOSFET Basics. Available online: https://www.fairchildsemi.com/application-notes/AN/AN-9010.pdf (accessed on 10 April 2016).

27. Sattler, M.; Smetana, T.; Meyerhofer, T.; Kuhlkamp, L. 48 V minihybrid—A new solution for the minimale hybridization of vehicles. In Proceedings of the 22nd Aachen Colloquium Automobile and Engine Technology, Germany, 8 October 2013; pp. 995–1008.

![energies logo] *energies*

MDPI

Article

Research on Control Strategies of an Open-End Winding Permanent Magnet Synchronous Driving Motor (OW-PMSM)-Equipped Dual Inverter with a Switchable Winding Mode for Electric Vehicles

Liang Chu [1], Yi-fan Jia [1], Dong-sheng Chen [1], Nan Xu [1,*], Yan-wei Wang [1], Xin Tang [1] and Zhe Xu [2]

[1] State Key Laboratory of Automotive Simulation and Control, Jilin University, Changchun 130022, China; chuliang@jlu.edu.cn (L.C.); jiayf16@mails.jlu.edu.cn (Y.-f.J.); chends2313@mails.jlu.edu.cn (D.-s.C.); ywwang15@mails.jlu.edu.cn (Y.-w.W.); tangxin16@mails.jlu.edu.cn (X.T.)

[2] R&D Center, China FAW Group Corporation, Changchun 130011, China; xuzhe1@rdc.faw.com.cn

* Correspondence: nanxu@jlu.edu.cn; Tel.: +86-431-8509-5165

Academic Editor: King Jet Tseng
Received: 1 March 2017; Accepted: 27 April 2017; Published: 2 May 2017

Abstract: An open-end winding permanent magnet synchronous motor (PMSM) has a larger range of speed regulation than normal PMSM with the same DC voltage, and the control method is more flexible. It can also manage energy distribution between two power sources without a DC/DC converter. This paper aims at an electric vehicle equipped with OW-PMSM drive system with dual power sources and dual inverters; based on analyzing the external characteristics of each winding mode, we propose a winding mode switching strategy whose torque saturation judgmental algorithm, which is insensitive to motor's parameters, could automatically realize upswitching of the winding mode. The proposed multi-level current hysteresis modulation algorithm could set the major power source and switch it at any time in independent mode, which accomplishes energy distribution between two power sources; its two control methods, low switching frequency method and high power difference method, could achieve different energy distribution effects. Simulation results confirm the validity and effectiveness of the winding mode switching strategy and current modulation method. They also show that an electric vehicle under the proposed control methods has better efficiency than one equipped with a traditional OW-PMSM drive system under traditional control.

Keywords: energy management; electric vehicle; open-end winding permanent magnet synchronous motor (OW-PMSM); multi-level current hysteresis modulation; winding mode switch

1. Introduction

With more rigorous demands for energy savings and environmental protection in industry and the vigorous development of electric vehicles driven by electric motors nowadays, PMSM has become the typical electric motor in electric vehicles because of its advantages such as high-power density and a simple control scheme. In recent years, a drive system constituting an OW-PMSM (open-end winding permanent magnet synchronous motor) and dual inverters has received extensive application [1–5]. This system is actuated via attaching both ends of OW-PMSM's stator windings with an inverter. Compared to the traditional Y-connection PMSM drive system, it has an expanded speed range and a more flexible control method [6–13]; the speed range is larger with the same DC bus voltage, or the DC bus voltage reduces by half with the same speed range. Moreover, it allows dual inverters to use electricity from different electric sources and, when working in a dual electric sources condition, it could manage energy distribution between the two sources without a DC/DC converter [14–16].

Control strategies of OW-PMSM and dual inverters have always been a research focus. Winding topology and configurations are widely discussed in order to optimize the motor's working range and efficiency [17–19]. Fault-tolerant operation [20–22] and zero-sequence voltage elimination method [23,24] are also under extensive research. Loncarski, J. et al. compared the output current ripple in single and dual inverter motor drives for electric vehicles, concluding that the dual-2L inverter can act as a 3L inverter and offers a significant ripple reduction [2], which can also be observed in this paper. AnQun-tao, Sun Li et al. proposed a dual inverters SVPWM modulation method for current control and in this way voltage vector synthesized in the range of a hexagon and when the bus voltage remains constant, the base speed of the electric motor could be 1.7 times that in Y-connection without generating zero-sequence voltage [7,8]. Zhan H., Heng N. et al. studied a common DC bus-based dual inverters SVPWM modulation strategy. In this way, the amplitude of the voltage vector could be maximized and at the same time a zero-sequence current could be suppressed [23,24]. Welchko studied dual sources-based dual inverters' voltage vectors distribution, and proposed three types of voltage distribution methods to adapt to different working conditions of hybrid vehicles to achieve energy management functions between two power sources [16]. However, the two inverters work independently without coordination, and thus the switching frequency of inverter devices has also doubled, which increases the inverter switching loss. All the studies discussed above have proposed current control methods of dual inverters but division of OW-PMSM winding modes and winding modes switching are not involved. Nguyen N.K. et al. studied dual inverters-five-phase OW-PMSM's winding modes, divided into three types, star mode, pentagon mode, and pentacle mode, and analyzed the external motor characteristics of each mode [18]. However, the winding modes are only decided based on rotational speed. A detailed winding modes switching method has not been proposed, nor has there been an analysis of the energy distribution issue between dual sources. Therefore, it is necessary to propose a control method that could manage energy distributions between power sources, and make full use of each winding mode's working range to reduce inverter loss and increase system efficiency for electric vehicles.

This paper first analyzes the three-phase OW-PMSM winding modes division method and the external motor characteristics of each winding mode. On that basis, we propose a torque saturation winding modes switching strategy, which is insensitive to motor parameters, to accomplish automatic winding modes switching according to different working conditions. Then, a multi-level current hysteresis modulation algorithm used in independent mode is put forward. This algorithm could set the major power source and achieve real-time major power source switching to distribute energy between two power sources. In addition, we design two control methods: a low switching frequency method and a high power difference method for different energy distribution purposes. At the end of this paper, the feasibility and effectiveness of the proposed winding modes switching strategy and multi-level current hysteresis modulation are verified through simulations. Vehicle economy performance simulation also showed that the electric vehicle under the proposed control method has better efficiency.

In particular, coordination transformations between the three-phase stator and two-phase rotator in this paper were equivalent power conversions.

2. Winding Mode Features Analysis and Switching Strategy

The dual sources OW-PMSM drive system structure is shown in Figure 1.

According to different winding connections, three-phase OW-PMSM can be divided into three types: star mode, triangle mode, and independent mode. Star mode's winding connection copies traditional PMSMs. It requires heads or tails of three-phase stator windings connected at one point. If it is a dual inverters configuration, it could be completed by making each bridge's lower arm turned on simultaneously in the idle inverter. Triangle mode is accomplished through connecting the heads and tails of three-phase stator windings in a circle. In dual inverters configuration, it could be achieved by shutting down one power source and turning on a triangle circuit switch. Independent mode, controlled by both sides' inverter bridges of each winding, is a specific one that only OW-PMSM has. It is independent because every winding directly connects with two sides' inverter bridges and there is no direct connection between windings. Independent mode requires both sides' power sources to supply voltage and can accomplish energy flow between power sources through windings.

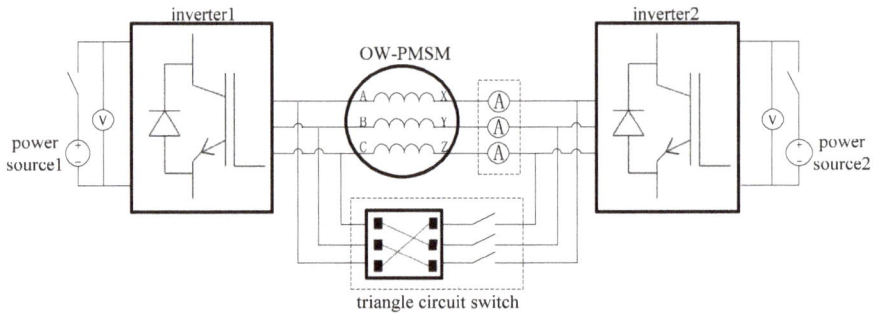

Figure 1. Structure of OW-PMSM drive system.

2.1. Winding Mode Features Analysis

First, we will discuss power source limitations on winding modes. In different winding modes, the maximum amplitude of each winding's phase voltage is different and the connection between each phase is different too, which leads to a difference in basic voltage vectors' amplitude. When the winding mode is star or independent, a certain phase voltage cannot be determined according to the corresponding inverter bridge's switching status and will be influenced by inverter bridges' switching status of other phases due to the load neutral point in star mode and the mid-point potential difference in independent mode floating. For instance, in star mode, when inverter switch statuses are (110) (phase A and B upper bridge arm on, phase C lower bridge arm on), phase voltage relations: $u_A = u_B = V_{dc}/3$, $u_C = -2V_{dc}/3$; when switch statuses are (100), phase voltage relations: $u_A = 2V_{dc}/3$, $u_B = u_C = -V_{dc}/3$. It can be seen that, although phase A's corresponding inverter bridge's switch statuses are the same in both situations, owing to the load neutral point's fluctuation in star mode, phase voltages u_A in both situations are different. Using the schematic diagram of mid-point voltage in Figure 2, we can analyze voltage vectors. We first equally divide power source 1 and power source 2 into two parts according to voltage to determine virtual mid-points m and n. The voltage difference between each inverter bridge's output and the corresponding mid-point is the mid-point voltage and is unaffected by floating neutral potential or mid-point voltage difference. Each inverter bridge's switching status and mid-point voltage has a one-to-one correspondence. The results are identical when using mid-point voltage and phase voltage for voltage vector synthesis.

Star and triangle modes are powered via a single power source and the motor's three-phase stator windings are connected in a specific topology structure. Switching statuses of three-phase inverter bridges can form a voltage vector in the motor space plane. Given that DC bus voltage is V_{dc}, when switching statuses are (110), synthesizing voltage vector in star and triangle modes are shown in Figure 3a,b. Independent mode is powered by dual sources. Supposed that the voltage of power source 1 is V_{dc1} and the voltage of power source 2 is V_{dc2}, when left switch statuses are (110) and right switch statuses are (001), voltage vector synthesis in independent mode is shown in Figure 3c.

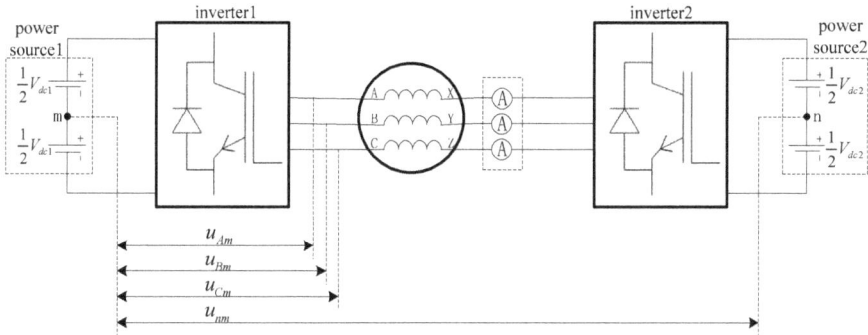

Figure 2. Schematic diagram of mid-point voltage.

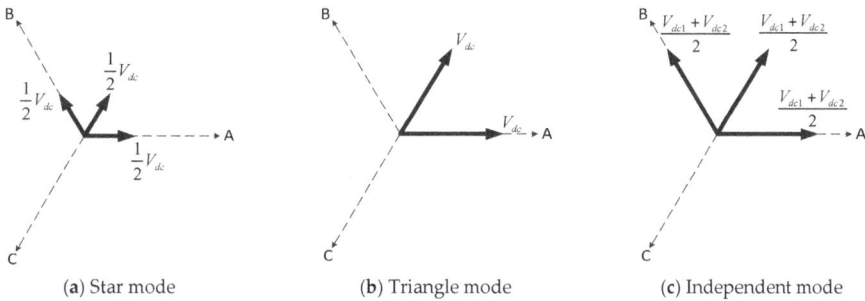

(a) Star mode   (b) Triangle mode   (c) Independent mode

Figure 3. Synthesizing voltage vector of each winding mode.

The amplitudes of basic voltage vector in star, triangle, and independent modes u_{sbY}, $u_{sb\Delta}$ and u_{sbD} can be obtained from Figure 3 as $\sqrt{2/3}V_{dc}$, $\sqrt{2/3} \times \sqrt{3}V_{dc}$, and $\sqrt{2/3}(V_{dc1} + V_{dc2})$, respectively. In these equations, $\sqrt{2/3}$ is the equal-power conversion coefficient. We can see that when the bus voltage is V_{dc}, the amplitude of the basic voltage vector in triangle mode is $\sqrt{3}$ times that in star mode. In independent mode, when bus voltage $V_{dc1} = V_{dc2} = V_{dc}$, the amplitude of its basic voltage vector is twice that in star mode and $2/\sqrt{3}$ times that in triangle mode. It has to be noted that the above basic voltage vectors are distributed with an $\pi/3$ angle interval, and in each winding mode there are six equal-amplitude basic voltage vectors forming a voltage vector hexagon in the motor vector plane. It is likely that voltage vector may not equal the amplitude of a hexagon vertex's voltage vector in any angle. Hence, the amplitude of the voltage vector in the motor's appropriate linear range is supposed to be the radius of the inscribed circle of a hexagon $\sqrt{3}/2$ times the amplitude of basic voltage vectors when the switch status are as above. Then, the maximum amplitude of the voltage vector at any angle in star, triangle, and independent modes u_{smaxY}, $u_{smax\Delta}$, and u_{smaxD} are: $V_{dc}/\sqrt{2}$, $\sqrt{6}V_{dc}/2$, and $(V_{dc1} + V_{dc2})/\sqrt{2}$, respectively.

Now we will discuss the stator current's limitations on winding modes. Because of windings' inductance characteristics, the phase current of the motor has inertia and cannot abruptly change like voltage. Therefore, in steady state, the waveforms of the three-phase current in the time domain are equal-amplitude sinusoid with $2\pi/3$ phase difference. Three-phase current vectors are demonstrated in Equation (1):

$$\begin{cases} i_A = i\cos(\omega_s t + \phi_1)e^{j0} \\ i_B = i\cos(\omega_s t + \phi_1 - \frac{2\pi}{3})e^{j\frac{2\pi}{3}} \\ i_C = i\cos(\omega_s t + \phi_1 - \frac{4\pi}{3})e^{j\frac{4\pi}{3}} \end{cases} . \tag{1}$$

The synthesized current vector is shown in Equation (2):

$$i_s = \sqrt{\frac{2}{3}}(i_A + i_B + i_C) = \sqrt{\frac{2}{3}} \times \frac{3}{2}ie^{j(\omega_s t + \phi_1)}. \tag{2}$$

It can be observed that the amplitude of the synthesized stator current vector is $\sqrt{3/2}$ times that of the phase current. If we take axis A as a referential axis, the space angle of the current vector corresponds with the phase A current in the time domain. In star and independent modes, the line current of the inverter is the phase current and the current capacity i_{max} is the maximum of the phase current. In triangle mode, line currents i_1, i_2, and i_3 and phase currents i_A, i_B, and i_C have the relationship:

$$i_1 = i_A - i_C, i_2 = i_B - i_A, i_3 = i_C - i_B. \tag{3}$$

According to this relationship, when motors are in steady state, the vector diagrams of line and phase currents in the time domain are as shown in Figure 4.

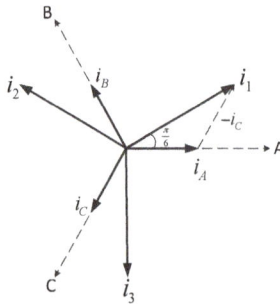

Figure 4. Time domain vector of line current and phase current.

This indicates that, in triangle mode, the amplitude of the line current is $\sqrt{3}$ times that of the phase current in steady state. Consequently, when the current capacity of the inverter is i_{max}, the maximum of the phase current in triangle mode and the maximum amplitude of the synthesized stator current vector is $1/\sqrt{3}$ times that in star and independent modes. In star, independent, and triangle modes, the relationships between maximum amplitude of current vector i_{smaxY}, i_{smaxD}, and $i_{smax\Delta}$ are as shown in Equations (4) and (5):

$$i_{smaxY} = i_{smaxD} = \frac{\sqrt{6}}{2}i_{max} \tag{4}$$

$$i_{smax\Delta} = \frac{\sqrt{6}}{2} \times \frac{i_{max}}{\sqrt{3}} = \frac{\sqrt{2}}{2}i_{max}. \tag{5}$$

Limitations on voltage and current vectors determine the working ranges of each winding mode. In star and triangle modes, flux-weakening regions are not set. In independent mode, there are non-flux-weakening and flux-weakening regions. In non-flux-weakening regions, if the stator voltage increases to the saturation threshold, the motor starts working in the flux-weakening region to expand the speed regulation range. In the non-flux-weakening region of each winding mode, the motor is controlled by the MTPA (maximum torque per ampere) method [25]. In flux-weakening regions, the motor is controlled by the direct flux control method. Based on the present motor's angular velocity ω_s, the algorithm calculates the specific stator's maximum flux, ψ_s^*. Then, on the basis of requested torque T_e^*, axis d and q's expected current i_d^*, i_q^* are worked out [26].

The OW-PMSM drive system's parameters are shown in Table 1; the external characteristic curves controlled by the above current strategy are demonstrated in Figure 5.

Table 1. Parameters of OW-PMSM drive system.

Items	Parameters
Motor type	Interior open-end winding PMSM
Number of pole pairs p_0	4
Stator resistance R_s/Ω	0.3
Equivalent iron loss resistance R_c/Ω	90
Fundamental amplitude and third harmonic amplitude of permanent magnet flux linkage $[\psi_f, \psi_{f3}]/\mathrm{Wb}$	[0.2, 0.01]
d-axis inductance L_d/F	0.0012
q-axis inductance L_q/F	0.0015
Zero sequence inductance L_0/F	0.0003
Rotational inertia of rotor J_m/kgm^{-2}	0.011
Cullen resistance coefficient and viscous resistance coefficient	[0.001, 0.0005]
DC bus voltage of power source 1 V_{dc1}/V	240
DC bus voltage of power source 2 V_{dc2}/V	230
Current capacity of inverter device i_{max}/A	160

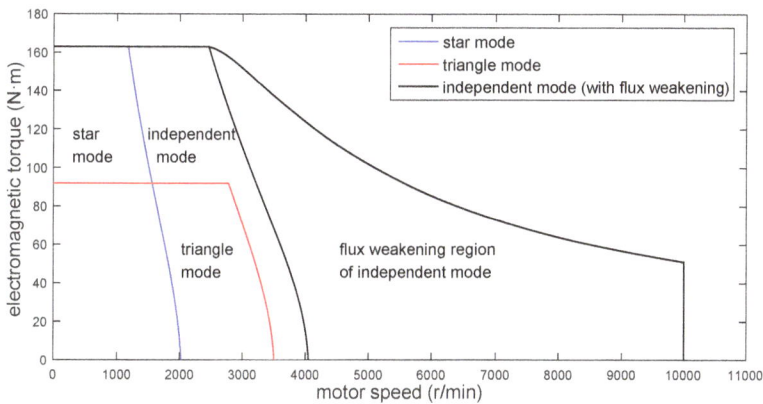

Figure 5. External characteristic curves of each winding mode.

We can tell from Figure 5 that the speed regulation range of triangle mode is about $\sqrt{3}$ times that in star mode but the maximum torque is $1/\sqrt{3}$ times that in star mode; the speed regulation range in the non- flux-weakening region in independent mode is approximately twice that in star mode and its maximum torque is equal to that in star mode.

2.2. Winding Modes Switching Strategy

We first determine general winding modes switching strategies. When the motor is in star or triangle mode, only one inverter is working with on-state losses and switching losses. When another inverter has lower bridge arms on, there is only on-state loss. In triangle mode, it is different than when the triangle circuit switch is turned on; there are on-state losses on triangle circuit. What is more, the topology of stator winding in triangle mode determines probable zero-sequence current in the motor, which may contribute to wastage of the inverter's current capacity and losses on electric resistance and electronic devices due to the current's heating effect. In independent mode, both sides' inverters are working and have on-state losses and switching losses. In conclusion, in terms of losses in electronic devices, star mode has the least and independent mode has the most, with triangle mode in the middle. In order to reduce those losses in electronic devices, the general winding modes switching strategy might be that if working conditions are appropriate, star mode is the first choice, then triangle mode; try not to use independent mode.

Under most circumstances, the differences between both sides' DC bus voltage are small and the highest rotary speeds of the motor in the non-flux-weakening region, triangle and independent mode increase in the order $n_{mY} < n_{m\triangle} < n_{mD}$. In accordance with this inequality, we define switching from star mode to triangle mode, from triangle mode to independent mode, and from star mode to independent mode as upswitching; switching from independent mode to triangle mode, from triangle mode to star mode, and from independent mode to star mode are downswitching.

We will begin with the upswitching strategy. During motor working period, parameters such as magnet flux and inductance fluctuate because of rising temperature, followed by fluctuations of working ranges of each winding mode. To avoid frequent switching when the working point of the motor fluctuates, this paper proposes an algorithm to judge the saturation state of torque. When judged as positive, upswitching is triggered to ensure the accuracy of the switching boundary.

This algorithm functions during the process of torque increasing from zero to the maximum. It calculates when the expected torque is at a maximum, the time period t_{int} during which torque increases from zero to the maximum at this specific rotary speed in the current winding mode (star or triangle) as integration time and integration of absolute torque error ΔT (the difference between actual torque T_l and expected torque T_e^*) as threshold I_{th} in this process. Then, it calculates integration I of absolute actual torque error ΔT from t_{int} to the present and compares it with integration threshold I_{th}. If $I \geq I_{th}$, then the motor's torque is almost saturated and does not match the expected torque. This algorithm is based on whether the output torque agrees with the expected torque in a particular period to eliminate uncertainties in the winding modes' working range because of fluctuations in the motor's parameters.

Voltage equations of motor in dq coordination are expressed in Equations (6) and (7):

$$u_d = R_s i_d + L_d \frac{di_d}{dt} - \omega_r L_q i_q \tag{6}$$

$$u_q = R_s i_q + L_q \frac{di_q}{dt} + \omega_r (L_d i_d + \psi_f). \tag{7}$$

Electromagnetic torque is shown in Equation (8):

$$T_e = p_0 i_q [\psi_f + (L_d - L_q) i_d]. \tag{8}$$

From the above equations, we can tell that electromagnetic torque T_e is mainly determined by axis q's current i_q; the smaller the inductance difference between axis d and q, the bigger the share that i_q determines. When $L_d = L_q = L_s$ the motor is surface-mounted; MTPA control is $i_d = 0$ control. For the sake of simplifying calculations, the integration time of torque t_{int} is derived as $i_d = 0$ control, and when the difference between L_d and L_q is small, MTPA control is still accurate using t_{int}. In addition, because $L_q \geq L_d$, the inertia of i_q is higher and the change rate of i_q is smaller if applied with the same voltage, which makes the calculated integration time t_{int} is bigger and the threshold of torque saturation is higher.

When using $i_d = 0$ control, due to the current of axis d being 0, voltage equations and torque expression of motor in dq coordination have changed to:

$$u_d = -\omega_r L_q i_q \tag{9}$$

$$u_q = R_s i_q + L_q \frac{di_q}{dt} + \omega_r \psi_f \tag{10}$$

$$T_e = p_0 \psi_f i_q. \tag{11}$$

Voltage vectors have the relationship:

$$u_s^2 = u_d^2 + u_q^2 = (-\omega_r L_q i_q)^2 + \left(R_s i_q + L_q \frac{di_q}{dt} + \omega_r \psi_f\right)^2. \tag{12}$$

When the voltage vector is at its maximum u_{smax}, we can get the change rate of axis q's current from Equation (13):

$$\frac{di_q}{dt} = \frac{\sqrt{u_{smax}^2 - (\omega_r L_q i_q)^2} - \omega_r \psi_f - R_s i_q}{L_q}. \tag{13}$$

In Equation (13), because of the mechanical inertia being greater than the electronic inertia, we can assume that ω_r is constant during torque integration time t_{int}, but during this time i_q increases from 0 to the maximum of stator current vector amplitude, i_{smax}. In order to simplify the torque following process into a linear process, di_q/dt must be a constant. By maximizing i_q to i_{smax}, we can get the minimum of di_q/dt for an increased integration time and thus a higher torque saturation threshold. If we neglect stator electric resistance R_s, then we get di_q/dt as:

$$\frac{di_q}{dt} = \frac{\sqrt{u_{smax}^2 - (\omega_r L_q i_{smax})^2} - \omega_r \psi_f}{L_q}. \tag{14}$$

It is obvious that with motor rotary velocity ω_r increasing, the partial voltage of electromotive force increases, voltage of stator gets almost saturated, and the voltage allowance used to control motor cuttent decreases, which in turn results in diminishing di_q/dt. When ω_r approaches the right boundary of the non-flux-weakening region's external characteristics in current winding mode, di_q/dt slips to 0, implying that the drive system is losing control of the motor's current. In order to guarantee a certain amount of voltage margin used to control the motor's current, we, in accordance with base speed ratio, can get the motor angular velocity ω_{rs} used to calculate di_q/dt in each winding mode by Equation (15):

$$\omega_{rs} = k_s \omega_{rb}. \tag{15}$$

In Equation (15), ω_{rb} is the rotator's base angular velocity of the motor in current winding mode. When ignoring R_s in $i_d = 0$ control strategy, we can get:

$$\omega_{rb} = \frac{u_{smax}}{\psi_s} = \frac{u_{smax}}{\sqrt{\psi_f^2 + (L_q i_{smax})^2}}. \tag{16}$$

Because the approximation of di_q/dt is constant, the integration time t_{int} of this process is:

$$t_{int} = \frac{\Delta i_q}{di_q/dt} = \frac{L_q i_{smax}}{\sqrt{u_{smax}^2 - (\omega_{rs} L_q i_{smax})^2} - \omega_{rs} \psi_f} = \frac{L_q i_{smax}}{\sqrt{u_{smax}^2 - (k_s \omega_{rb} L_q i_{smax})^2} - k_s \omega_{rb} \psi_f}. \tag{17}$$

In Equation (17), k_s is the rotator speed sensitivity coefficient and $k_s \in (0,1)$; the bigger k_s is, the closer the motor's angular velocity $k_s \omega_{rb}$ used to calculate integration time t_{int} is to the current winding mode's base angular velocity, and the smaller voltage margin and di_q/dt are, which leads to bigger t_{int} and vice versa. k_s reflects tolerance of torque following speed. The bigger k_s is, the lower the requirements for torque following speed, which leads to a slower response of the torque saturation judgmental algorithm. In the above two expressions, i_{smax} is the maximum of the stator current vector in the present winding mode and u_{smax} is the maximum of any angle voltage vector in the present winding mode.

Because the approximation of di_q/dt is constant, the torque following process is simplified to a linear process. The integration of torque error of this process is:

$$I_{lim} = \int_0^{t_{int}} |\Delta T dt| = \int_0^{t_{int}} |T_l - T_e^*| dt = \frac{1}{2} T_{emax} t_{int}. \tag{18}$$

For practical application, we define an integration threshold sensitivity coefficient so that the actual integration threshold is:

$$I_{th} = k_I I_{lim} = \frac{1}{2} k_I T_{emax} t_{int} = \frac{1}{2} k_I p_0 \psi_f i_{smax} t_{int}. \tag{19}$$

In Equation (19), T_{emax} is the maximum electromagnetic torque in the current winding mode. For integration threshold sensitivity coefficient $k_I \in (0,1)$, the bigger k_I is, the bigger integration threshold I_{th} is and the less sensitive the judgment of torque saturation, which can reduce the possibility of improper switching but will also increase delays in winding mode switching; the smaller k_I is, the smaller integration threshold I_{th} is and the more sensitive the judgment of torque saturation, which can make torque switching swift but may lead to improper switching when working conditions change unexpectedly. By changing k_s and k_I, we can modulate the sensitivity and stability of the torque saturation judgmental algorithm.

It is important to note that if the current mode is triangle mode, upswitching's goal is only independent mode; when the torque saturation threshold is reached, the current mode is switched to independent mode. However, if the current mode is star mode, there are two options: triangle and independent mode; when the torque saturation threshold is reached, terminals are decided by the degree of saturation of stator current vector. When the amplitude of stator current vector $i_s \geq i_{smax\triangle}$, it indicates that the amplitude of the stator current vector in the current mode exceeds the capacity of adjustment of triangle mode and upswitching's goal is independent mode; when $i_s < i_{smax\triangle}$, the goal is triangle mode.

Downswitching is determined by the rotary speed threshold, which is the base speed of each winding mode. When the motor's rotary speed decreases to the star or triangle mode's base speed, downswitching is triggered. This strategy staggers the working points of upswitching and downswitching to avoid frequent switching. Switching is not only decided by the motor's speed, like upswitching, but also by the degree of saturation of the stator current vector. Whether the current mode is triangle or independent, as long as the motor's speed has the relationship $n_r < n_{bY}$ the motor's speed is lower than the star mode's base speed, and the current mode is switched to star mode. However, if the current mode is independent and the motor's speed has the relationship $n_{bY} < n_r < n_{b\triangle}$, it has to be decided whether to switch to triangle mode or not. When the amplitude of stator current vector $i_s \geq i_{smax\triangle}$, this indicates that the amplitude of the stator current vector has exceeded the capacity of adjustment of triangle mode and current winding should stay in independent mode. If $i_s < i_{smax\triangle}$, the current mode is switched to triangle mode.

Switching principles among the three winding modes are summarized in Table 2.

Table 2. Switching algorithm of each winding mode.

Switching Goal Starting Mode	Star Mode	Triangle Mode	Independent Mode
Star mode	NA	Positive torque saturation decision and $i_s < i_{smax\triangle}$	Positive torque saturation decision and $i_s \geq i_{smax\triangle}$
Triangle mode	$n_r < n_{bY}$	NA	Positive torque saturation decision
Independent mode	$n_r < n_{bY}$	$n_{bY} \leq n_r < n_{b\triangle}$ and $i_s < i_{smax\triangle}$	NA

3. Dual Inverters' Current Modulation Method

This paper adopts a hysteresis control-based current control method. In single-sourced star and triangle modes, inverters could only provide two potentials for each winding. Thus, the traditional hysteresis current control method was adopted. In dual-sourced independent mode, dual inverters, when controlled coordinately, could provide three or four potentials. Hence, a multi-level hysteresis current control method was employed.

3.1. Current Modulation Method in Star and Triangle Modes

Star mode required ends of three-phase stator windings connected to one point, which could be accomplished by making all lower arms of the idle inverter's bridges turned on at the same time in dual inverters configuration while the other inverter was working to control the three-phase current. Based on expected torque T_e^*, we could get control variables i_d^* and i_q^* through the abovementioned MTPA and flux-weakening strategy. Then, with dq0 to ABC coordination transformation, expected phase currents i_A^*, i_B^* and i_C^* were obtained; meanwhile, actual phase currents i_A, i_B, and i_C were acquired via the current sensor. Phase current error represented the difference between each phase expected current and the actual current, which is demonstrated by following formula:

$$\Delta i_A = i_A - i_A^*, \Delta i_B = i_B - i_B^*, \Delta i_C = i_C - i_C^*. \tag{20}$$

From each phase current's error, we could control the three-phase current. Given the half-width of the current's hysteresis band, the acceptable current error is h. When one phase current error $\Delta i \geq h$, the upper arm of this phase's inverter bridge was turned off and the lower arm was turned on, making the midpoint voltage of this phase $-V_{dc}/2$, a lower voltage, and the phase current error Δi of this phase decreasing. When phase current error $\Delta i \leq -h$, the upper arm of this phase's inverter bridge was turned on and the lower arm was turned off, making the midpoint voltage of this phase $V_{dc}/2$, a lower voltage, and the phase current error Δi of this phase increasing.

Triangle mode requires the heads and tails of three-phase stator windings to be connected in a circle, which is completed by shutting down one power source and turning on triangle circuit switch. The other side's inverter was working to control the motor's current. What was different from star mode was that triangle mode, instead of controlling the motor's phase current i_A, i_B, and i_C directly, controlled motor's line current i_1, i_2, and i_3 to control motor's phase current indirectly. The relationship of triangle mode's phase current and line current is shown in Equation (3). With Equation (3), we can transform phase current's expected value i_A^*, i_B^*, and i_C^* to line current's expected value i_1^*, i_2^*, and i_3^*, then determine line current's error Δi_1, Δi_2, and Δi_3 to have each line current hysteresis controlled respectively.

If ignoring zero-sequence current in triangle circuit, we got $i_A + i_B + i_C = 0$, and then we could get the phase current expression, represented by the line current as Equation (21) via Equation (3):

$$i_A = (i_1 - i_2)/3, i_B = (i_2 - i_3)/3, i_C = (i_3 - i_1)/3. \tag{21}$$

If the acceptable line current error was h_l, the half-width of the hysteresis band of i_1, i_2 was h_l, and the acceptable error of $i_1 - i_2$ was $2h_l$, then we could calculate the acceptable error of phase current $i_A = (i_1 - i_2)/3$ as $2h_l/3$. Thus, it was clear that if the acceptable error of phase current stayed at h, the line current hysteresis comparator's acceptable current error was required to be set at $3h/2$, so we had:

$$h_l = 3h/2. \tag{22}$$

At this point, if ignoring the zero-sequence current, the tracking error of the phase current in triangle mode was the same as that in star mode. However, note that the topology of triangle mode is different from that of star mode and DC bus voltage was directly loaded on the phase current. In every combination of inverter switching states, there are five possible phase voltages in star mode: $-2V_{dc}/3$, $-V_{dc}/3$, 0, $V_{dc}/3$, $2V_{dc}/3$. In triangle mode, there are three possible phase voltages with higher amplitude: $-V_{dc}$, 0, V_{dc}. Therefore, if the half-width of the hysteresis band was determined according to Equation (22), even though the acceptable tracking error of the current in triangle mode was the same as that in star mode, the current saw tooth fluctuation in triangle mode is more drastic than that in star mode and the switching frequency of inverters, along with the core loss of motor, would be greater than those in star mode.

3.2. Current Modulation Method in Independent Mode

Different from star and triangle modes, independent mode is powered by two sources and controlled by two inverters, which means each winding's electric potential is controlled by both of the two inverters' bridges and has four statuses: (10), (01), (11), and (00).

The first number of switching status stands for the electric potential of inverter 1's bridge. The second number of switching status stands for the electric potential of inverter 2's bridge. The status of the inverter bridge upper arm (on) and lower arm (off) obtained high potential, which was substituted with number "1". The status of the inverter bridge upper arm (off) and lower arm (on) obtained low potential, which was substituted with number "0". Because the two sources were insulated, the mid-point voltage difference of the two sources u_{nm} varied with different inverter switching statuses, which brought about diverse phase voltages in different switching statuses, especially when $V_{dc1} \neq V_{dc2}$. When the mid-point voltage of the two sources was $u_{nm} = 0$, the phase voltage was the mid-point phase voltage. When the switching statuses were (10), (01), (11), and (00), the corresponding mid-point phase voltages were $(V_{dc1} + V_{dc2})/2$, $-(V_{dc1} + V_{dc2})/2$, $(V_{dc1} - V_{dc2})/2$, and $-(V_{dc1} - V_{dc2})/2$. Hence, when $V_{dc1} \neq V_{dc2}$, we could get four mid-point voltages via dual inverters' control. When $V_{dc1} = V_{dc2}$, we could get three mid-point voltages and the mid-point voltages with switching statuses (11) and (00) were both zero.

The traditional hysteresis current modulation algorithm has two potentials and corresponding two trigger areas: $\Delta i \geq h$ and $\Delta i \leq -h$. When $V_{dc1} \neq V_{dc2}$, there are four potentials in independent mode of the hysteresis current modulation algorithm. Apart from two trigger areas used for activating switching statuses (01) and (10), two intermediate trigger lines were needed for activating switching statuses (00) and (11)—two intermediate potentials. Provided the two intermediate trigger lines were $\Delta i = d$ and $\Delta i = -d$, with d as the linear current deviation of the intermediate line, we could get Equation (23) based on the principle that mid-point voltage is proportional to current error Δi when triggered:

$$h : \frac{V_{dc1} + V_{dc2}}{2} = d : \frac{V_{dc1} - V_{dc2}}{2}. \tag{23}$$

Equation (23) could be transformed to:

$$d = \frac{V_{dc1} - V_{dc2}}{V_{dc1} + V_{dc2}} h. \tag{24}$$

It was obvious that when $V_{dc1} > V_{dc2}$, $d > 0$; when $V_{dc1} < V_{dc2}$, $d < 0$; when $V_{dc1} = V_{dc2}$, $d = 0$, and at this time, two intermediate trigger lines coincided at $\Delta i = 0$. It has to be pointed out that in each winding's four potentials in independent mode, two boundary potentials are triggered by Δi being in the corresponding area to ensure that Δi can depart from that triggered area swiftly in the relatively greater phase voltage generated by boundary potentials as long as Δi is in the corresponding area and restricted to hysteresis band $[-h, h]$. Two intermediate potentials were triggered by crossing the corresponding intermediate trigger lines without considering the crossing direction; this control logic mainly mattered in slowing fluctuations of Δi down to load smaller voltage. A rather small phase voltage would be loaded if Δi is close to the center line ($\Delta i = 0$) to avoid drastic fluctuations frequently hitting the hysteresis boundary if constantly loaded with large phase voltage. When applied with this modulation method, the changing rate would decrease after Δi crosses the control line $\Delta i = d$ or $\Delta i = -d$ until it hits the boundary again. Two additional potentials would effectively retard the fluctuation speed of Δi to make current changes milder and reduce the inverters' switching frequency.

Two improved multi-level hysteresis modulation methods were proposed on the foundation of the above multi-level hysteresis current modulation strategy. They were the low switching frequency method and high power difference method. These two methods were achieved by adding the trigger conditions of inverter bridges' switching statuses (00) and (11) in independent mode.

Due to independent mode being powered by two sources, power distribution between two sources was involved. Thus, a major power source was proposed and was expected to have greater power output than the other source in independent mode. Low switching frequency and high power difference methods could determine which one was the major power source and switch over it at any time to accomplish power distribution. When the load is low, the major power source can even charge another source. Table 3 indicates the relationships between inverter bridges' switching statuses, direction of phase current i, and current flow. Given that phase current flowing from left to right is positive in Figure 1, we could tell that energy flow between the two sources could only be accomplished when the inverter bridges' switching statuses were at two intermediate potentials (00) and (11).

Table 3. Relationship between inverter bridge's switching states and power flow direction.

Direction of Phase Current / Inverter Switching States		$i < 0$	$i > 0$
Boundary	(01)	Both power sources discharging	Both power sources charging
potentials	(10)	Both power sources charging	Both power sources discharging
Intermediate	(00)	Power source 1 discharging Power source 2 charging	Power source 1 charging Power source 2 discharging
potentials	(11)	Power source 1 charging Power source 2 discharging	Power source 1 discharging Power source 2 charging

The trigger regulation of inverter bridges' switching statuses in low switching frequency and the high power difference method is displayed in Table 4.

Table 4. Dual inverter trigger rules of two different current modulation methods.

Modulation Pattern / Inverter Switching States		Low Switching Frequency Method	High Power Difference Method
Boundary	(01)	Δi is in the area of $\Delta i \geq h$	Δi is in the area of $\Delta i \geq h$
potentials	(10)	Δi is in the area of $\Delta i \geq h$	Δi is in the area of $\Delta i \geq h$
Intermediate potentials	(00)	Δi crossed control line $\Delta i = d$ and switching state of inverter bridge on major power source's side is 0	Δi crossed control line $\Delta i = d$ and when power source 1 is major power source: phrase current $i < 0$; when power source 2 is major power source: phrase current $i > 0$
	(11)	Δi crossed control line $\Delta i = -d$ and switching state of inverter bridge on major power source's side is 1	Δi crossed control line $\Delta i = -d$ and when power source 1 is major power source: phrase current $i > 0$; when power source 2 is major power source: phrase current $i < 0$

These two improved methods added trigger conditions of two intermediate potentials, which made two potentials, instead of being triggered when Δi crossing control lines $\Delta i = \pm d$, triggered at other specific conditions. Normally only one intermediate potential was triggered in one hysteresis period. The low switching frequency method, needed to confirm the switching status of inverter bridge on major power source's side, remains unchanged after switching when Δi crossing control lines $\Delta i = \pm d$. In this case, the switching statuses of both two inverters' bridges would not be changed simultaneously when Δi crosses the control lines and the switching frequency of inverter devices could be lowered to a minimum. In the high power difference method, when Δi crosses the control lines $\Delta i = \pm d$, we need to decide whether to switch based on the present phase current i's direction to ensure the major power source could charge the other source when the switching status of inverter bridge is at two intermediate potentials. This method increases the difference between two sources' power outputs as much as possible.

4. Results of Simulations of OW-PMSM Drive System

We ran simulation models of the OW-PMSM drive system on the Matlab/Simulink platform. The basic parameters of OW-PMSM are shown in Table 1 and controller parameters are shown in Table 5. A PI speed controller was used to generate the expected torque to make the motor speed follow the preset value.

Table 5. Parameters of simulation setting and controller.

Module Affiliation	Item	Parameters
Model as a whole	Time step T_s/s	5×10^{-7}
Inverter devices	On-resistance R_{on}/Ω	0.01
	Forward voltage drop of IGBT V_f/V	0.8
	Forward voltage drop of diode V_{fd}/V	0.8
	Current fall time T_f/s	1×10^{-6}
	Current tailing time T_t/s	1.5×10^{-6}
Winding mode controller	Sampling time T_{s_MS}/s	1×10^{-4}
	Sensitivity coefficient of rotor speed $[k_{sY}, k_{s\Delta}]$	[0.9, 0.9]
	Sensitivity coefficient of integral threshold $[k_{IY}, k_{I\Delta}]$	[0.35, 0.75]
PI controller of motor speed	Sampling time T_{s_SC}/s	1×10^{-4}
	Proportionality coefficient P	0.4
	Integral coefficient I	4
MTPA and flux weakening controller	Sampling time T_{s_CC}/s	1×10^{-4}
	Voltage saturation coefficient k_u	0.95
Current hysteresis controller	Sampling time T_{s_CR}/s	1×10^{-5}
	Half width of hysteresis band h/A	3
	Maximum switching frequency of devices f_{max}/Hz	1×10^4

This simulation's duration was 0.9 s. In this process, the expected motor rotary speed linearly increased to 5500 r/min in 0–0.3 s and stayed at 5500 r/min till 0.6 s. Then speed linearly dropped to 0 in 0.6–0.9 s; loaded torque jumped from 0 to 50 N·m at 0.05 s and remained 50 N·m until the simulation finished. In order to monitor the results of power distribution, we shifted the current modulation method from low switching frequency to high power difference at 0.4 s and switched the major power source from power source 1 to power source 2 at 0.5 s. Another OW-PMSM drive system under traditional two-level current hysteresis modulation without winding mode switching function was also simulated for comparison.

Curves of expected torque T_e^*, electromagnetic torque T_e of proposed and contrast model, and rotary speed n_m are demonstrated in Figure 6a–c.

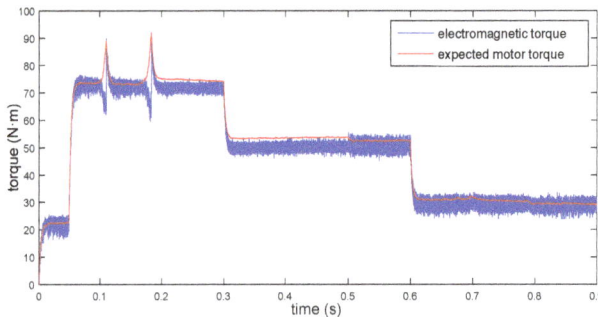

(**a**) Expected torque and electromagnetic torque of motor

Figure 6. *Cont.*

(b) Electromagnetic torque of motor in contrast model

(c) Rotary speed of motor

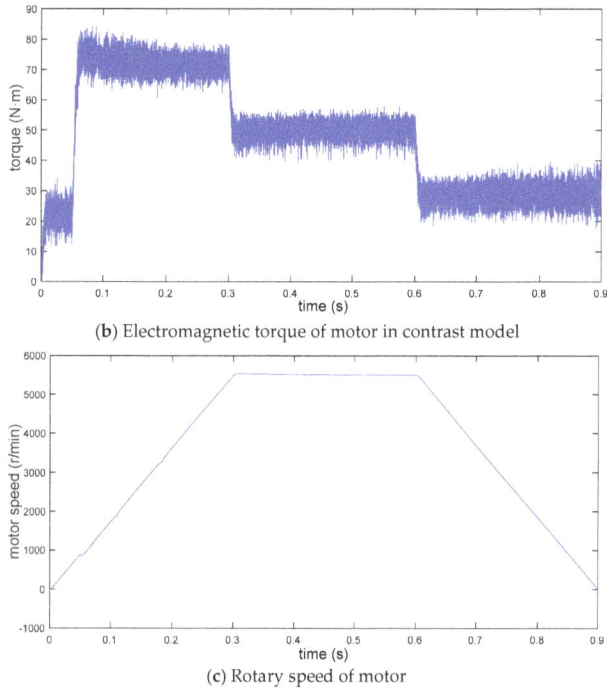

Figure 6. Curves of expected motor torque, electromagnetic torque, and motor speed.

We understand from Figure 6 that the motor rotary speed could smoothly and swiftly follow the preset value and only had a slight fluctuation at 0.05 s when the loaded torque jumped. Switching of modes had no impact on rotary speed. Electromagnetic torque T_e could also follow expected torque T_e^* well. In the present current following accuracy ($h = 3$ A), the amplitude of electromagnetic torque's fluctuation was limited within 5 N·m or so, but after switching to lower voltage power source 2 as major power source, the fluctuation was more drastic, which indicated that setting the higher voltage power source as the major power source is preferable. We also found that the motor under the proposed multi-level current hysteresis modulation had about 30% less torque fluctuation than the motor under traditional two-level current hysteresis modulation.

Winding mode, torque saturation judgmental threshold I_{th}, and actual integration I curves are displayed in Figure 7a,b.

In Figure 7a, the mode signals are demonstrated: 1 is star mode, 2 stands for triangle mode, 3 and 4 are the low switching frequency and high power difference methods, respectively, in independent mode, and 3.5 and 4.5 represented the flux-weakening region of the two methods, respectively, in independent mode. In Figure 6a,b, two peaks of expected torque curve at 0.12 s and 0.17 s indicate that with increasing speed, voltage is almost saturated and electromagnetic torque cannot follow expected torque, leading to the speed controller further increasing the expected torque; from the electromagnetic torque curve, it is evident that the actual electromagnetic torque will gradually decline because of voltage saturation before the expected torque peak; when torque error accumulates enough to satisfy torque saturation conditions, upswitching is triggered and electromagnetic torque rapidly follows the expected torque after switching; From electromagnetic torque being unable to follow to the upswitching point, it only took 0.02 s. From Figure 7b, we understand that 0.02 s before upswitching, with the voltage almost saturated, the switching integration of torque error rose sharply to the threshold and upswitching was triggered. After winding mode switching to triangle mode from star mode, the integration threshold and integration of torque error declined due to the integration time being smaller.

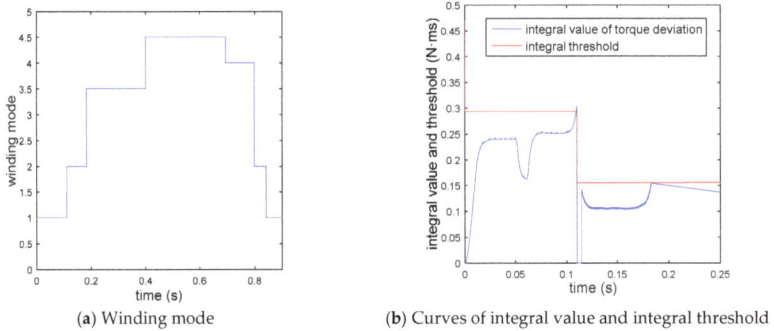

(a) Winding mode

(b) Curves of integral value and integral threshold

Figure 7. Curves of winding mode, integral value, and threshold of torque saturation decision.

Phase A's voltage curves at star and triangle, and triangle and independent mode switching points are shown in Figure 8a,b. Phase A's current and its local curves are shown in Figure 8c,d.

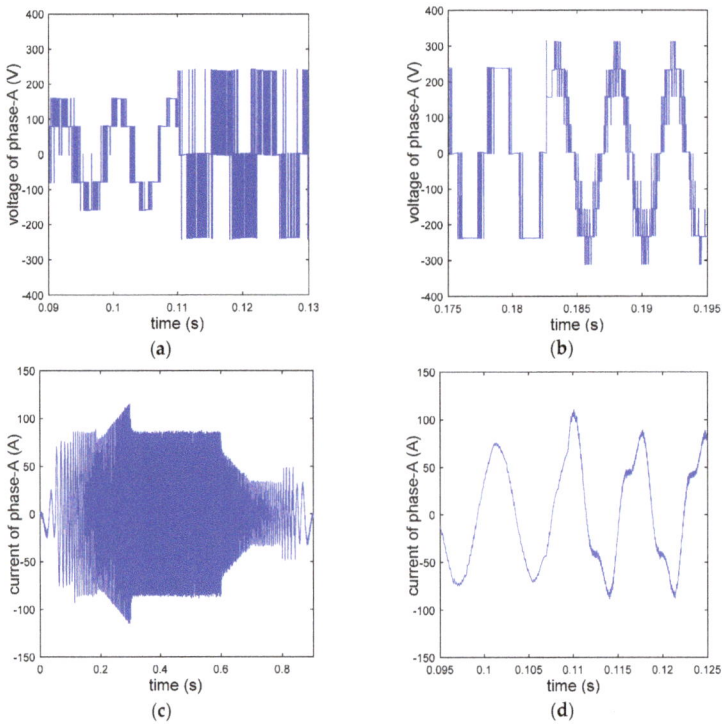

(a)

(b)

(c)

(d)

Figure 8. Curves and partial waves of phase A's voltage and current. (**a**) Partial wave of phase—A voltage at switching point 1; (**b**) partial wave of phase—A voltage at switching point 2; (**c**) curve of Phase—A current; (**d**) partial wave of phase—A current at switching point 1.

It is clear that the phase voltage of motor changes follows a sinusoidal waveform; amplitudes of phase voltage in star, triangle and independent modes increased in sequence. From the local curve of phase A's voltage, there were five phase voltage values, $-2V_{dc}/3$, $-V_{dc}/3$, 0, $V_{dc}/3$, and $2V_{dc}/3$, when the motor was in star mode. There were only three phase voltage values, $-V_{dc}$, 0, and V_{dc}, when the motor was in triangle mode; if we ignore the voltage difference between the two sides' power sources, there were nine phase voltage values, $-4V_{dc}/3$, $-V_{dc}$, $-2V_{dc}/3$, $-V_{dc}/3$, 0, $V_{dc}/3$, $2V_{dc}/3$, V_{dc}, and $4V_{dc}/3$ when the motor was in independent mode, which made the current control more smooth. The phase current of the motor changed according to the sinusoidal rule; because of the zero-sequence current effect, the amplitude of the phase current in triangle mode was slightly greater than those in star and independent modes with the same amplitude of stator's current. From Figure 8d, we can see that after switching from star mode to triangle mode, the waveform of phase current changed and was no longer a standard sinusoidal waveform due to the zero-sequence current, which led to occupation of the inverter's extra capacity and increasing switching frequency of devices. At the same time, the phase current was restricted in the hysteresis band and the current followed accordingly.

Total switching frequency of inverter devices (sum of all IGBT devices' switching frequencies) of both the proposed and the contrasting model is shown in Figure 9a. Inverters' loss is shown in Figure 9b.

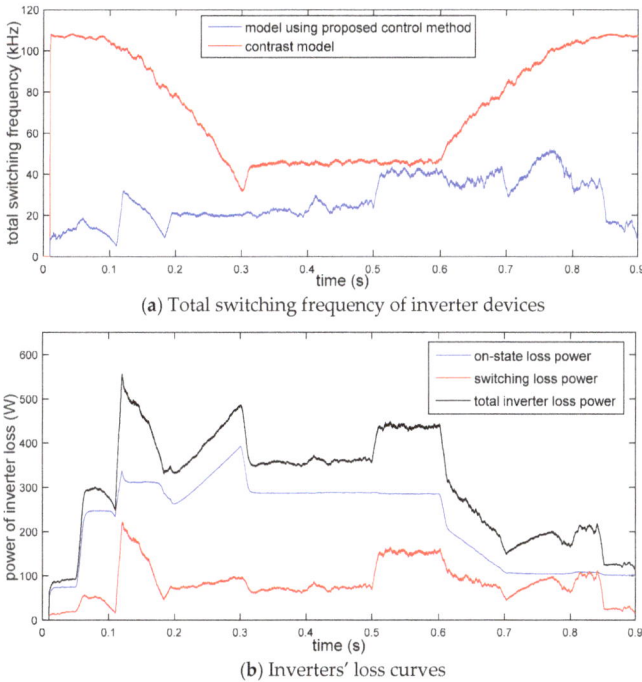

(a) Total switching frequency of inverter devices

(b) Inverters' loss curves

Figure 9. Curves of total switching frequency of inverter devices and inverter loss.

From Figure 9a, we see that the switching frequencies of inverter devices in triangle mode are higher than those in star mode because the phase voltage values in triangle mode are lower and their amplitudes were greater, leading to a higher changing rate of phase current and more frequently hitting the hysteresis boundary; the total switching frequency of inverter devices in independent mode is also high due to the two inverters working together in independent mode. However, there were more phase voltage values in independent mode and the control of the phase current could be more smooth and flexible, which might reduce the switching frequencies of inverter devices. The total switching frequency of devices increased after shifting from the low switching frequency method to the high power difference method at 0.4 s; the total switching frequency increased even more after switching the major power source from power source 1 to power source 2, a lower voltage power source, at 0.5 s, which indicated that setting a higher voltage power source as the major power source could reduce the switching frequencies of devices. In addition, the total switching frequency of devices in steady state was lower than that in transient state and it was higher when the motor speed was decreasing than when motor speed was increasing. Under the proposed control method, total switching frequency is far lower than in the contrasting model, especially in the period of star and triangle mode. In Figure 9b, the switching loss of inverters was proportional to the total switching frequency of devices and in most circumstances the share of switching losses of inverters was under 30%, which was ideal.

Input power curves of both inverters are shown in Figure 10.

It was evident that before 0.5 s, power source 1 was the major power source; in this period, star mode and triangle mode were all powered by power source 1 and all input power was generated by inverter 1; after 0.5 s, power source 2 was the major power source. In this period, star mode and triangle mode were supplied by power source 2 and all input power was generated by inverter 2. When in independent mode, both power sources provided power. After shifting from the low switching frequency method to the high power difference method at 0.4 s, the power difference between two inverters increased; after switching the major power source from power source 1 to power source 2 at 0.5 s, inverter 2's power was higher than inverter 1's, becoming the major power output inverter; nevertheless, because of the voltage of power source 2 being lower than that of power source 1, after switching the major power source to power source 2, the power difference between two inverters decreased, which indicated that the power difference of the inverters was affected by the voltage difference of the power sources.

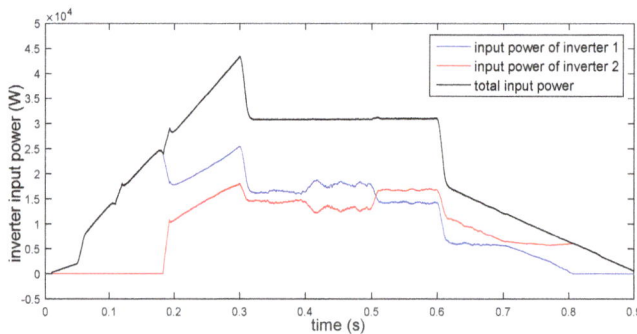

Figure 10. Curves of input power of both inverters.

An efficiency MAP of the drive system with switched winding modes and multi-level current hysteresis modulation is shown in Figure 11a; that of the contrasting drive system is shown in Figure 11b.

(a) Efficiency MAP under proposed control method

(b) Efficiency MAP for contrast

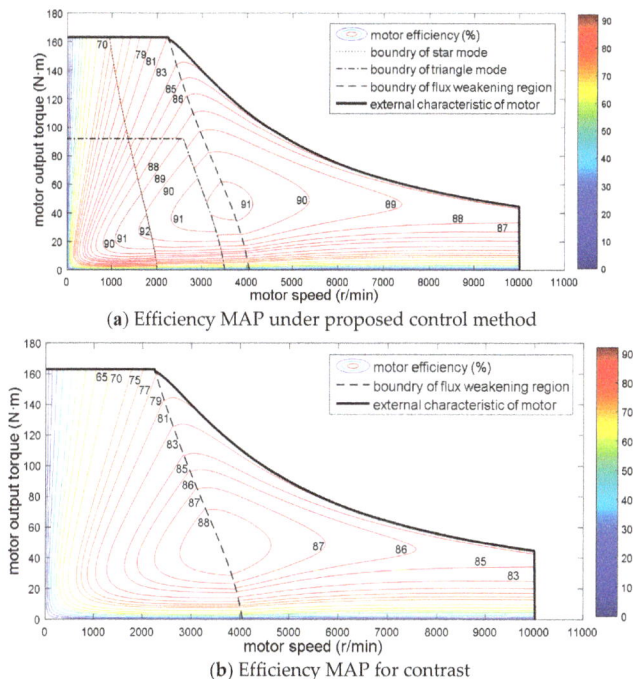

Figure 11. Efficiency MAP of drive system.

It is clear that the peak efficiency of the drive system under the proposed multi-level current hysteresis modulation is 3% higher than that of the contrasting drive system, because multi-level current hysteresis modulation reaches a lower switching frequency of inverter devices and then reduces inverter switching loss. We could also see that the high-efficiency area of the drive system is larger than in the contrasting system, extending to low speed and low torque areas, because in the region of star mode and triangle mode there is only one inverter working and producing switching loss. The iron loss caused by the stator voltage vector in the unsteady state is less than that in independent mode because the stator voltage amplitude in star and triangle mode is lower.

We also conducted a simulation of a small electric vehicle for economic performance. The drive system under the proposed control method was equipped, and the drive system under the traditional two-level current hysteresis modulation without winding mode switching function was also used for contrast. The basic parameters of the vehicle are shown in Table 6.

Table 6. Parameters of vehicle.

Items	Parameters
Vehicle weight m_0/kg	950
Drag coefficient C_d	0.30
Windward area A/m^2	2.11
Reduction gear ratio i_0	8.4
Rolling radius r_g/m	0.307
Rotational mass conversion factor δ	1.1
Rolling resistance coefficient f	0.015
Transmission efficiency η_m	0.95
Rate of braking energy regeneration R_b	0.6

We simulated four different type of driving cycles, NEDC (New European Driving Cycle), UDDS (Urban Dynamometer Driving Schedule), JC08 (Made by Japanese 2005 emission regulation), and HWFET (Highway Fuel Economy Test Cycle), representing a standard driving situation, an urban driving situation, a frequent acceleration and deceleration situation, and a highway driving situation respectively. The motor operating point distributions of the different driving cycles are shown in Figure 12.

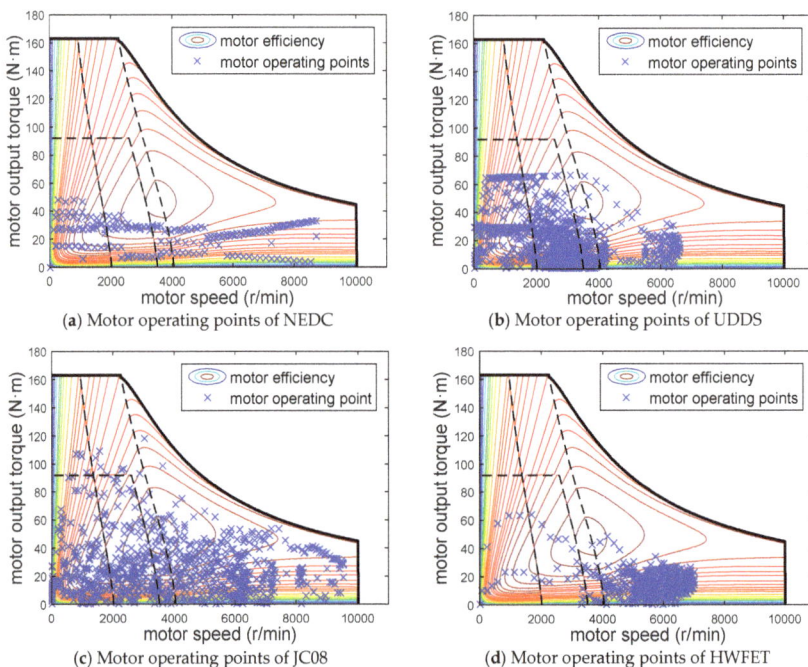

Figure 12. Motor operating points in four driving cycles.

The efficiency distribution and power consumption of the proposed drive system and the contrasting drive system are shown in Table 7. The average efficiency distributions of the four driving cycles are shown in Figure 13.

Table 7. Efficiency distribution and power consumption.

Driving Cycles		Efficiency Distribution/%					Power Consumption/ kWh 100 km^{-1}
		$\geq$0.85	0.8–0.85	0.7–0.8	0.5–0.7	<0.5	
Proposed drive system	NEDC	32.55	38.20	2.57	2.11	24.57	11.33
	UDDS	44.21	14.97	9.71	7.08	24.02	10.26
	JC08	46.62	9.71	6.31	5.52	31.84	15.17
	HWFET	28.41	42.52	19.33	5.36	4.38	12.52
	Average	37.95	26.35	9.48	5.02	21.20	12.32
Contrast drive system	NEDC	9.79	12.92	43.18	8.15	25.96	12.18
	UDDS	17.49	19.75	22.86	10.37	29.54	11.17
	JC08	24.66	17.14	13.74	9.55	34.91	16.23
	HWFET	6.14	24.30	52.51	10.45	6.60	13.27
	Average	14.52	18.53	33.07	9.63	24.25	13.21

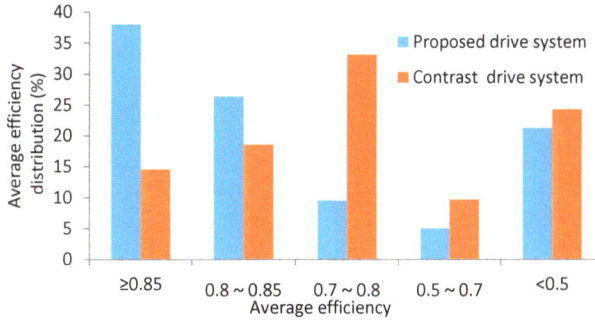

Figure 13. Average efficiency distribution of four driving cycles.

It is evident that the proportion of high-efficiency working points of the proposed drive system is larger than that of the contrasting system, thus the power consumption is on average 6.75% lower than that of the contrasting system.

5. Conclusions

This paper, aimed at an electric vehicle equipped with an OW-PMSM drive system with dual power sources and dual inverters; based on analyzing the external characteristics of each winding mode, we produced a winding mode switching strategy whose torque saturation judgmental algorithm, insensitive to a motor's parameters, could automatically realize upswitching of winding mode. The proposed multi-level current hysteresis modulation algorithm can set the major power source and switch it at any time in independent mode, accomplishing energy distribution between two power sources; its two control methods, the low switching frequency method and the high power difference method, could achieve different energy distribution effects. From the simulation results, compared with an OW-PMSM drive system with traditional two-level current hysteresis modulation, under the same conditions the proposed system has 30% lower torque ripple and a lower switching frequency of inverter devices. Thus the proposed system has 3% higher peak efficiency and a larger high-efficiency area than a traditional OW-PMSM system. By applying it to electric vehicles, the power consumption is 6.75% lower on average under the proposed control methods. Moreover, through its two different control methods in independent mode, energy distribution between the two power sources can be realized, thus the DC/DC converter between them can be cancelled.

Finally, the proposed winding mode switching strategy and multi-level current hysteresis modulation method take full advantage of each winding mode's working range, reduce inverter switching loss, increase system efficiency, and realize energy distribution between two power sources. It provides a theoretical basis and implementation scheme a for dual-power OW-PMSM drive system in electric vehicles. Future research will be directed towards finding an energy distribution method matched with this system for electric vehicles to maximize the overall efficiency and driving range. After solving the existing practical issues, an experimental verification of the proposed system will also be conducted in an electric vehicle.

Acknowledgments: This work was supported by the China Postdoctoral Science Foundation (2014M561290), the Energy Administration of Jilin Province [2016]35, and the Jilin Province Science and Technology Development Fund (20150520115JH).

Author Contributions: Nan Xu and Liang Chu conceived the control method and revised the full manuscript; Yi-fan Jia and Dong-sheng Chen performed the simulation and wrote the full manuscript; Yan-wei Wang and Xin Tang analyzed the data; Zhe Xu analyzed and evaluated the simulation results and gave valuable suggestions.

Conflicts of Interest: The authors declare no conflict of interest.

References

1. Chowdhury, S.; Wheeler, P.; Patel, C.; Gerada, C. A multilevel converter with a floating bridge for open-ended winding motor drive applications. *IEEE Trans. Ind. Electron.* **2016**, *63*, 5366–5375. [CrossRef]
2. Loncarski, J.; Leijon, M.; Srndovic, M.; Rossi, C.; Grandi, G. Comparison of output current ripple in single and dual three-phase inverters for electric vehicle motor drives. *Energies* **2015**, *8*, 3832–3848. [CrossRef]
3. Welchko, B.A.; Nagashima, J.M. A comparative evaluation of motor drive topologies for low-voltage, high-power EV/HEV propulsion systems. In Proceedings of the IEEE International Symposium on Industrial Electronics, Rio De Janeiro, Brazil, 9–11 June 2003; pp. 379–384.
4. An, Q.T.; Duan, M.H.; Sun, L.; Wang, G.L. SVPWM strategy of post-fault reconfigured dual inverter in open-end winding motor drive systems. *Electron. Lett.* **2014**, *50*, 1238–1240. [CrossRef]
5. Engelmann, G.; Kowal, M.; De Doncker, R.W. A highly integrated drive inverter using direct FETs and ceramic dc-link capacitors for open-end winding machines in electric vehicles. In Proceedings of the Applied Power Electronics Conference and Exposition, Charlotte, NC, USA, 15–19 March 2015; pp. 290–296.
6. Sandulescu, P.; Meinguet, F.; Kestelyn, X.; Semail, E.; Bruyere, E. Control strategies for open-end winding drives operating in the flux-weakening region. *Power Electron. IEEE Trans.* **2014**, *29*, 4829–4842. [CrossRef]
7. An, Q.; Liu, J.; Peng, Z.; Sun, L.; Sun, L. Dual-space vector control of open-end winding permanent magnet synchronous motor drive fed by dual inverter. *IEEE Trans. Power Electron.* **2016**, *31*, 8329–8342. [CrossRef]
8. An, Q.; Sun, L.; Sun, L. Research on novel open-end winding permanent magnet synchronous motor vector control systems. *Proc. CSEE* **2015**, *22*, 5891–5898.
9. Park, J.S.; Nam, K. Dual inverter strategy for high speed operation of HEV permanent magnet synchronous motor. In Proceedings of the Industry Applications Conference 2006, Ias Meeting, Tampa, FL, USA, 8–12 October 2006; pp. 488–494.
10. Kwak, M.S.; Sul, S.K. Flux weakening control of an open winding machine with isolated dual inverters. In Proceedings of the Industry Applications Conference 2007, Ias Meeting, New Orleans, LA, USA, 23–27 September 2007; pp. 251–255.
11. Rossi, C.; Grandi, G.; Corbelli, P.; Casadei, D. Generation system for series hybrid powertrain based on the dual two-level inverter. In Proceedings of the European Power Electronic Conferences, Barcelona, Spain, 8–10 September 2009; pp. 1–10.
12. Griva, G.; Oleschuk, V.; Profumo, F. Hybrid traction drive with symmetrical split-phase motor controlled by synchronized PWM. In Proceedings of the International Symposium on Power Electronics, Electrical Drives, Automation and Motion, Ischia, Italy, 11–13 June 2008; pp. 1033–1037.
13. Deng, Q.; Wei, J.; Zhou, B.; Han, C.; Chen, C. Research on control strategies of open-winding permanent magnetic generator. In Proceedings of the International Conference on Electrical Machines and Systems, Beijing, China, 20–23 August 2011; pp. 1–6.
14. Lee, Y.; Ha, J.I. Hybrid modulation of dual inverter for open-end permanent magnet synchronous motor. *IEEE Trans. Power Electron.* **2015**, *30*, 3286–3299. [CrossRef]
15. Casadei, D.; Grandi, G.; Lega, A.; Rossi, C. Multilevel operation and input power balancing for a dual two-level inverter with insulated DC sources. *IEEE Trans. Ind. Appl.* **2008**, *44*, 1815–1824. [CrossRef]
16. Welchko, B.A. A double-ended inverter system for the combined propulsion and energy management functions in hybrid vehicles with energy storage. In Proceedings of the Industrial Electronics Society 2005, IECON 2005, Raleigh, NC, USA, 6–10 November 2005.
17. Nipp, E. Permanent Magnet Motor Drives With Switched Stator Windings. Ph.D. Thesis, Royal Institute of Techonology, Stockholm, Sweden, May 1999.
18. Nguyen, N.K.; Semail, E.; Meinguet, F.; Sandulescu, P.; Kestelyn, X.; Aslan, B. Different virtual stator winding configurations of open-end winding five-phase PM machines for wide speed range without flux weakening operation. In Proceedings of the European Conference on Power Electronics and Applications, Lille, France, 2–6 September 2013; pp. 1–8.
19. Welchko, B.; Nagashima, J.M. The influence of topology selection on the design of EV/HEV propulsion systems. *IEEE Power Electron. Lett.* **2003**, *99*, 36–40. [CrossRef]
20. Nguyen, N.; Meinguet, F.; Semail, E.; Kestelyn, X. Fault-tolerant operation of an open-end winding five-phase PMSM drive with short-circuit inverter fault. *IEEE Trans. Ind. Electron.* **2014**, *63*, 595–605. [CrossRef]

21. Gu, C.; Zhao, W.; Zhang, B. Simplified minimum copper loss remedial control of a five-phase fault-tolerant permanent-magnet vernier machine under short-circuit fault. *Energies* **2016**, *9*, 860. [CrossRef]
22. Zhao, J.; Gao, X.; Li, B.; Liu, X.; Guan, X. Open-phase fault tolerance techniques of five-phase dual-rotor permanent magnet synchronous motor. *Energies* **2015**, *8*, 12810–12838. [CrossRef]
23. Zhan, H.; Zhu, Z.Q.; Odavic, M.; Li, Y.X. A novel zero sequence model based sensorless method for open-winding PMSM with common DC bus. *IEEE Trans. Ind. Electron.* **2016**, *63*, 6777–6789. [CrossRef]
24. Heng, N.; Zeng, H.; Zhou, Y. Zero sequence current suppression strategy for open winding permanent magnet synchronous motor with common DC bus. *Trans. China Electrotech. Soc.* **2015**, *30*, 40–48.
25. Sun, T.; Wang, J.; Chen, X. Maximum torque per ampere (MTPA) control for interior permanent magnet synchronous machine drives based on virtual signal injection. *IEEE Trans. Power Electron.* **2015**, *30*, 5036–5045. [CrossRef]
26. Zhang, Y.; Cao, W.; Mcloone, S.; Morrow, J. Design and flux-weakening control of an interior permanent magnet synchronous motor for electric vehicles. *IEEE Trans. Appl. Supercond.* **2016**, *26*, 1–6. [CrossRef]

energies

MDPI

Article

Design Methodology of a Power Split Type Plug-In Hybrid Electric Vehicle Considering Drivetrain Losses

Hanho Son, Kyusik Park, Sungho Hwang and Hyunsoo Kim *

School of Mechanical Engineering, Sungkyunkwan University, Suwon-si 16419, Korea;
hanho1014@naver.com (H.S.); rbtlr9010@naver.com (K.P.); hsh@me.skku.ac.kr (S.H.)
* Correspondence: hskim@me.skku.ac.kr; Tel.: +82-31-290-7911

Academic Editor: Jih-Sheng (Jason) Lai
Received: 11 January 2017; Accepted: 23 March 2017; Published: 25 March 2017

Abstract: This paper proposes a design methodology for a power split type plug-in hybrid electric vehicle (PHEV) by considering drivetrain losses. Selecting the input split type PHEV with a single planetary gear as the reference topology, the locations of the engine, motor and generators (MGs), on the speed lever were determined by using the mechanical point considering the system efficiency. Based on the reference topology, feasible candidates were selected by considering the operation conditions of the engine, MG1, and a redundant element. To evaluate the fuel economy of the selected candidates, the loss models of the power electronic system and drivetrain components were obtained from the mathematical governing equation and the experimental results. Based on the component loss model, a comparative analysis was performed using a dynamic programming approach under the presence or absence of the drivetrain losses. It was found that the selection of the operating mode and the operation time of each mode vary since the drivetrain loss affects the system efficiency. In addition, even if the additional modes provide the flexibility of selecting the operating mode that results in a higher system efficiency for the given driving condition, additional drivetrain elements for realizing the modes can deteriorate the fuel economy due to their various losses.

Keywords: design methodology; power split type; plug-in hybrid electric vehicle (PHEV); drivetrain losses; dynamic programming

1. Introduction

To meet the regulations for reducing CO_2 emissions and increasing the fuel economy, the development of electric drive vehicles such as the battery electric vehicle (BEV) and hybrid electric vehicle (HEV) are an inevitable necessity.

The plug-in hybrid electric vehicle (PHEV) has emerged as a viable solution to meet these regulations, while overcoming the disadvantage of the relatively short travel distance of the BEV [1]. PHEV can be driven only using electric energy until the battery state of charge (SOC) decreases to the lower limit, which is called the 'All Electric Range' (AER) or the 'charge depleting (CD) mode'. After AER, PHEV has to run using the internal combustion engine and motor(s) to sustain the battery SOC, which is called the 'charge sustaining (CS) mode'. In CS mode, various operating modes such as series, parallel, etc. are used depending on the PHEV configuration. When the PHEV is operated in CS mode, the fuel economy is directly related to the operating modes, which are determined from its configuration. The PHEV configuration can be classified into series, parallel, and power-split type configurations [2]. The Toyota hybrid system (THS) is a typical example of the input split type configuration in which the engine power is split at the input side. The THS enables the engine to operate on the optimal operating line (OOL) via the electrically continuous variable transmission function. In addition, the THS has a relatively simple structure without using the clutch and brake to change the mode [3]. However, the power circulation that occurs when the vehicle runs at high speed has been mentioned as a major drawback that deteriorates the fuel economy [4,5]. The THS uses two operating modes; the EV and HEV modes. The number of modes that the PHEV can realize is directly related to the fuel economy because, with a greater number of PHEV modes, more flexibility can be provided for the given driving conditions [6]. On the other hand, additional drivetrain elements, such as the clutch or brake, which are required to implement an additional mode, can cause parasitic loss in the drivetrain.

To develop the PHEV configuration with a high system efficiency, design methodologies or new system configurations have been investigated [7]. To reduce the power circulation, an optimal gear ratio design for the planetary gear and the final reduction gear was proposed [8], and the speed ratio control was investigated to drive the vehicle near the mechanical point [9]. A topology optimization was performed for HEV with transmission to improve CO_2 emission and fuel economy [10]. To implement multi-mode operation in the input split type PHEV, a design methodology to find a feasible design from the possible combinations, using the clutches and brakes, was investigated. For all possible candidates, the fuel economy and driving performance were evaluated using a backward simulator [8,11]. In addition, a comprehensive design methodology was suggested to find the optimal configuration of the input and output-split type HEV in terms of the fuel economy and driving performance, using the clutch topology and gear ratio [12,13].

In the aforementioned studies, there are some limitations; (1) parasitic losses of the drivetrain components, such as the clutch and brake, which are required to realize the new configuration, were neglected and (2) the unloaded loss of the power electronic systems, such as motor and generators (MGs), were not considered. For PHEVs that have relatively short AER, the above factors influence the fuel economy when the vehicle is operated in CS mode.

In this study, a design methodology for a new PHEV configuration was proposed using the input split type as a design reference. Based on the reference topology, feasible candidates were selected by considering the infeasible and redundant conditions. For each candidate, the improvement ratio of the fuel economy was evaluated by dynamic programming, and a comparative study was performed in terms of the positive aspects of the additional mode and negative aspects of the drivetrain loss.

2. Topology Design Based on the Input Power Split Type PHEV

Search for Feasible Topology to Realize the Multi-Mode PHEV

Since power split type PHEV has the advantage of both series and parallel type configurations [4], power split type was selected as a reference topology in design of a new PHEV configuration. In general, the power split type PHEV is designed by combining one engine, two motor-generators (MGs), and multiple planetary gears. In addition, clutches and/or brakes are used to realize the demanded operating modes [8,11]. When we design a new PHEV using multiple planetary gear sets, numerous configurations can be constructed by connecting the power source (engine, MGs) to each shaft of the planetary gears [14,15]. Since it is unrealistic to evaluate all the possible configurations, it is necessary to narrow the variety of choices. In this study, the following constraints were introduced:

(1) Single planetary gear system:

Even if multiple planetary gears can provide more freedom than a single planetary gear, additional drivetrain elements such as the clutch and brake are required to connect the planetary gears, and it is difficult to avoid a complicated system structure compared with the single planetary gear system. Also, we can easily deduce that the transmission efficiency of the multiple planetary gear system is lower than that of the single planetary gear system [16]. Furthermore, additional drivetrain elements such as the planetary gear, clutch, and brake may exacerbate the packaging problem, as well as incur increased costs. Therefore, a single planetary gear was selected.

(2)   Input split type:

In the power split transmission (PST) using the single planetary gear (PG), the power is split at the input side or output side of the PST depending on the location of PG [4,17]. The split power flows to the mechanical path and electrical path. Since the efficiency of the electrical path, which consists of the power electronic (PE) system including MG1, MG2, and the inverter, is much lower than that of the mechanical path, the PST efficiency increases as the power ratio of the electrical path decreases. When all the power flows through the mechanical path, the PST efficiency shows the highest value, and this is called the 'mechanical point' (MP) [18].

In the input split type PHEV, the PST efficiency decreases rapidly when the vehicle speed becomes higher than the speed of the mechanical point. This is because power circulation occurs along the electrical path, which causes decreased PST efficiency. In contrast, in the output split type, the power circulation occurs when the vehicle speed is lower than the mechanical point. Therefore, if the vehicle drives mostly in the city (urban dynamometer driving schedule cycle), in other words, at a low to medium speed, it is desirable to use the input split type that has a relatively high efficiency in the low to medium speed region without the power circulation when the mechanical point is positioned at high speed [18].

Considering the above two constraints, the input split configuration using a single planetary gear was selected as a reference topology in development of a new PHEV.

A power split structure can be represented as a generalized single lever model [19]. Figure 1 shows a lever model of the input split type, which involves one engine and two MGs using the single planetary gear. Since the single PG is used, there are three nodes in the lever. Now, we can find a feasible PST configuration as follows:

(a)   If we use MG2 as a main driving motor, it should be connected to the output node. This is because MG2 needs to be operated to propel the vehicle directly in the electric vehicle (EV) mode, as can be observed in most PHEVs.

(b)   In the single PG, the node of the speed lever should be positioned in the order of the sun (S), carrier (C), and ring gear (R), or vice versa from the lever analogy.

(c)   The mechanical point needs to be located at a high vehicle speed to achieve high PST efficiency in the low to medium speed range.

Therefore, we have two possible configurations, as shown in Figure 1. In Figure 1, i is the lever distance from the output to the engine and a and b are the lever distances from the output to MG1 and to MG2, respectively. It is seen from Figure 1 that b is equal to zero because MG2 needs to be located at the output from the above condition (a).

Since the electrical power becomes zero at the mechanical point, we can find the speed ratio of the mechanical point under the condition that the speed of MG1 or MG2 is zero. We define the speed ratio (SR) as,

$$SR = \frac{\omega_e}{\omega_{out}},$$

(1)

where ω_e is the engine speed and ω_{out} is the output speed (vehicle speed). From the speed lever in Figure 1, when the speed of MG1 or MG2 is zero, the speed ratio at the mechanical point is obtained as

$$SR = \frac{a-i}{a} = \frac{(a/i)-1}{(a/i)} \text{ at mechanical point.}$$

(2)

In Figure 2, the speed ratio at the mechanical point is shown with respect to a/i for an input split type transmission [20]. To meet condition (c) above, the speed ratio at the mechanical point should be the overdrive ratio (region F) for which the output speed of the PST is higher than the input (engine) speed. This implies that the lever distances a and i need to be in the same direction and that a/i is larger than one. It is found that lever A satisfies the aforementioned requirements, while lever

B does not meet the requirements. In addition, it is found that the engine needs to be located at the center node of lever A, i.e., at carrier C, since the order of the PG node should be S, C, R or R, C, S (from condition (b)).

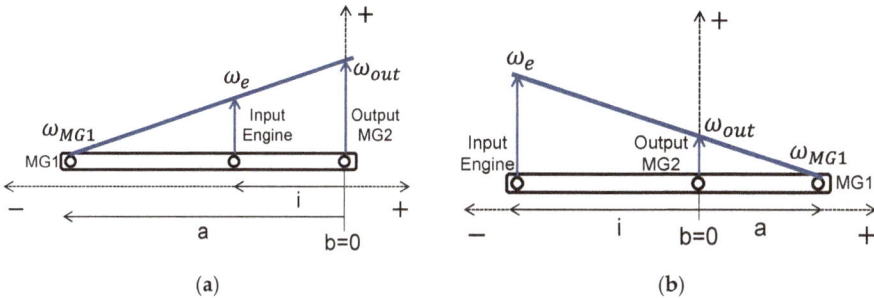

Figure 1. A generalized single lever model for the input split plug-in hybrid electric vehicle (PHEV). (a) Lever A; (b) Lever B.

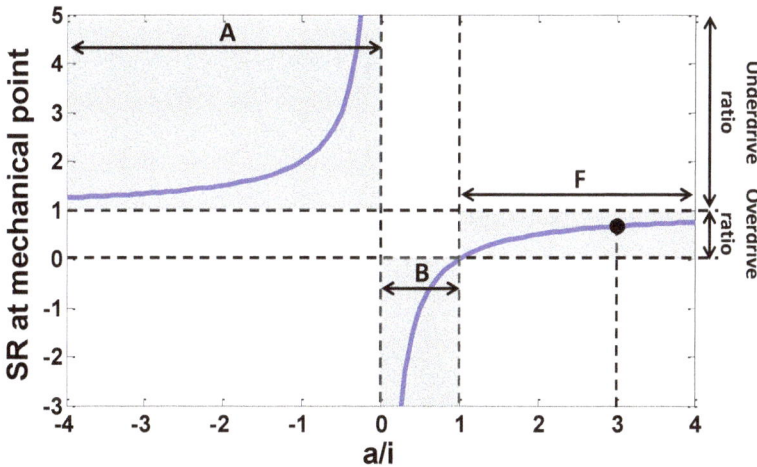

Figure 2. Speed ratio at the mechanical point [20].

Now, we determine the position of MG2 (output). MG2 can be located at the sun gear, S (Figure 3a), or the ring gear, R (Figure 3b). From the speed lever analogy, it is found that MG2 is positioned at the ring gear when a/i is greater than 2; meanwhile it is positioned at the sun gear for $1 < a/i < 2$.

From the lever analogy of the speed and torque, we can obtain the power of each node using a/i. Defining the power ratio PR as the ratio of the MG1 power (electrical power) to the engine (input) power, the power ratio is represented as

$$PR = \frac{P_{MG1}}{P_e} = \frac{\omega_{out}}{\omega_e} \left(\frac{1}{(a/i) - 1} \right) + 1 \tag{3}$$

In Figure 4, the power ratio is shown for a/i between 1 and 3.5 when the engine speed is 1500 rpm, the engine torque is 100 Nm, and the output speed is 2000 rpm. It is seen from Figure 4 that the power ratio decreases as a/i increases. Since the PST efficiency increases as the power ratio decreases, in other words, the power through MG1 (electrical path) decreases, we find that MG2 needs to be positioned at the ring gear.

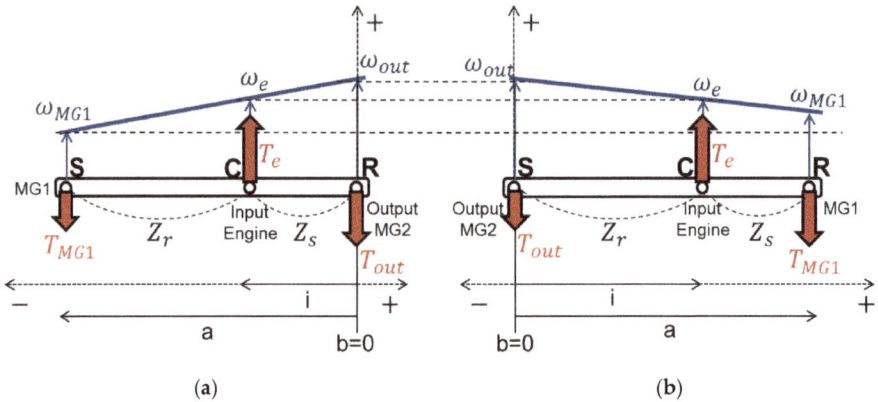

Figure 3. Lever analysis for the input split type by a/i. (**a**) a/i is larger than 2; (**b**) a/i is smaller than 2.

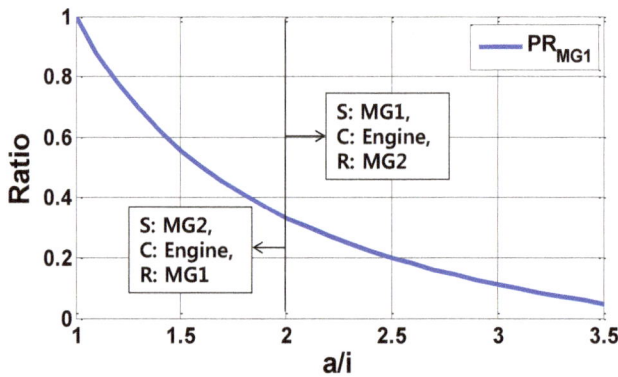

Figure 4. Speed, torque, and power ratio versus a/i when ω_e = 1500 rpm, T_e = 100 Nm, ω_{out} = 2000 rpm.

In Figure 5, a schematic diagram of the input split type PST using the single PG is shown, which is selected from the design procedure. The PST in Figure 5 consists of one engine at the carrier (C), MG2 at the ring gear (R), and MG1 at the sun gear (S). The PST in Figure 5a provides two operating modes when driving; (1) electric vehicle (EV) and (2) power split mode. In the EV mode, the vehicle is propelled by MG2 using electric energy (Figure 5b). In the power split mode, the engine power is split at the PG and transmitted to MG1 and to the output (Figure 5c). At this moment, the power may circulate through the closed path, depending on the vehicle speed.

Using the PST in Figure 5 as a reference topology, a design methodology of the PHEV that can realize more operating modes with high efficiency is investigated.

To realize more than two operating modes, the degree of freedom of the single PG needs to be changed using the clutch and brake. In Figure 6a, the possible positions where the clutch and brake can be added are shown. Since the PG has three nodes (sun, carrier, and ring gear), the clutches, CL1, CL2, CL3, and the brakes, BK1, BK2, and BK3, can be added at each node, as shown in Figure 6a [11,15]. Mathematically, the number of combinations that can be constructed from three clutches and three brakes is 2^6. However, infeasible combinations can be eliminated by considering the real driving environment. In the power split mode, in other words, in the hybrid electric vehicle (HEV) mode in which the engine and MG2 work together, the engine and MG1 should always operate together due to the following reasons:

(1) The engine needs to supply the power to propel the vehicle and to generate MG1.
(2) When the engine works, MG1 is required to control the engine operation on the OOL.

From the above constraints, it is seen that clutches CL2 and CL3 should always be engaged, which means that CL2 and CL3 are not necessary, as shown Figure 6b. Therefore, the number of combinations becomes $2^4(=16)$ by eliminating CL2 and CL3.

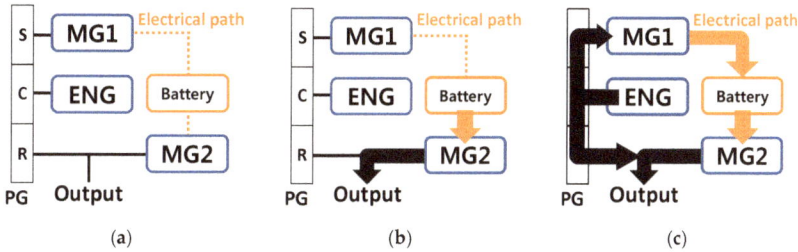

Figure 5. Reference power split transmission. (**a**) Input split power split transmission (PST); (**b**) electric vehicle (EV) mode; (**c**) Power split mode.

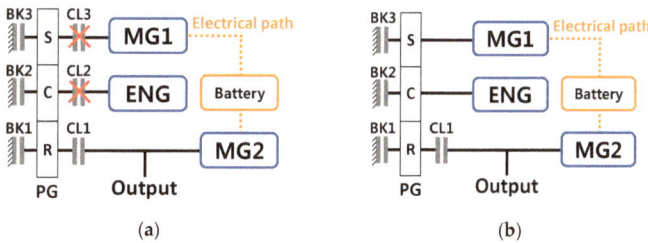

Figure 6. Schematic diagram for the reference topology. (**a**) All possible locations for adding the clutch and brake [11]; (**b**) Feasible locations for adding the clutch and brake.

By adding CL1, BK1, BK2, and BK3 to the reference topology, the following modes can be realized:

EV#1:	The vehicle is propelled only by MG2. MG1 and the engine are off. No clutch or brake is required.
Power split:	This mode can be implemented as explained in Figure 5c. No additional clutch or brake is needed from the reference topology.
EV#2:	MG2 and MG1 propel the vehicle together. Since the engine does not work, the reaction force acting on the carrier must exist to transmit the MG1 torque through the PG. Therefore, the carrier is fixed by activating BK2.
Parallel:	The engine and MG2 propel the vehicle using the parallel path. MG1 does not function, which requires BK3 to make the MG1 speed zero. The engine power is transmitted to the output through the mechanical path without using the electrical path. MG2 works using the battery energy.
Series:	To implement this mode, the engine needs to be separated from the output, which requires CL1 to disengage the power from the engine and BK1 to ground the ring gear. The engine only drives MG1 to generate the electrical power, which is transmitted to the output through the electrical path.

In Table 1, the operation of the clutch and brake, the lever analogy, and the speed and torque equation for each mode are shown. It is seen from Table 1 that EV#1 and power split mode are implemented as a basic operating mode without any additional elements. For the EV#2 mode, BK2 is required; meanwhile BK3 is required for the parallel mode. Finally, it is noted that BK1 and CL1 are needed for the series mode.

Table 1. Possible operating modes with the speed and torque equation.

Operating Mode	Lever Analogy	Speed and Torque Equation ($N_{PG} = Z_r/Z_s$)
EV#1		$\begin{bmatrix} \omega_{MG1} \\ \omega_{MG2} \end{bmatrix} = \begin{bmatrix} N_{PG} & 0 \\ 1 & 0 \end{bmatrix} \begin{bmatrix} \omega_{out} \\ \omega_e \end{bmatrix}$ $[T_{out}] = [\,0\ \ 0\ \ 1\,] \begin{bmatrix} T_{MG1} \\ T_e \\ T_{MG2} \end{bmatrix}$
Power split		$\begin{bmatrix} \omega_{MG1} \\ \omega_{MG2} \end{bmatrix} = \begin{bmatrix} -N_{PG} & 1+N_{PG} \\ 1 & 0 \end{bmatrix} \begin{bmatrix} \omega_{out} \\ \omega_e \end{bmatrix}$ $[T_{out}] = [\,1\ \ 1\ \ 1\,] \begin{bmatrix} T_{MG1} \\ T_e \\ T_{MG2} \end{bmatrix}$
EV#2		$\begin{bmatrix} \omega_{MG1} \\ \omega_{MG2} \end{bmatrix} = \begin{bmatrix} N_{PG} & 0 \\ 1 & 0 \end{bmatrix} \begin{bmatrix} \omega_{out} \\ \omega_e \end{bmatrix}$ $[T_{out}] = [\,N_{PG}\ \ 0\ \ 1\,] \begin{bmatrix} T_{MG1} \\ T_e \\ T_{MG2} \end{bmatrix}$
Parallel		$\begin{bmatrix} \omega_{MG1} \\ \omega_{MG2} \end{bmatrix} = \begin{bmatrix} 0 & 0 \\ 1 & 0 \end{bmatrix} \begin{bmatrix} \omega_{out} \\ \omega_e \end{bmatrix}$ $[T_{out}] = [\,0\ \ 1+N_{PG}\ \ 1\,] \begin{bmatrix} T_{MG1} \\ T_e \\ T_{MG2} \end{bmatrix}$
Series		$\begin{bmatrix} \omega_{MG1} \\ \omega_{MG2} \end{bmatrix} = \begin{bmatrix} 0 & 1+N_{PG} \\ 1 & 0 \end{bmatrix} \begin{bmatrix} \omega_{out} \\ \omega_e \end{bmatrix}$ $[T_{out}] = [\,0\ \ 0\ \ 1\,] \begin{bmatrix} T_{MG1} \\ T_e \\ T_{MG2} \end{bmatrix}$

In Table 2, sixteen candidates are shown with the operating mode and additional element(s). It is seen from Table 2 that some of the candidates have a redundant element when realizing the target operating modes. For example, candidate #14 can realize the EV#1, power split, EV#2, and parallel mode only by using BK2 and BK3 without CL1. Therefore, CL1 is a redundant element, and it is found that candidate #14 is equivalent to candidate #11. Similarly, candidate #15 is equivalent to #11, #2 and #3 are equivalent to #1, etc. After eliminating the candidates that have a redundant element (gray rows), eight candidates were selected, as shown in Figure 7.

Table 2. All feasible candidates to realize multi-mode PHEV.

Candidate No.	Additional Element	Operating Mode	Redundant Element	Equivalent Candidate
#1	Reference	Basic (EV#1, Power split)	-	-
#2	CL1	Basic	CL1	#1
#3	BK1	Basic	BK1	#1
#4	BK2	Basic + EV#2	-	-
#5	BK3	Basic + Parallel	-	-
#6	CL1, BK1	Basic + Series	-	-
#7	CL1, BK2	Basic + EV#2	CL1	#4
#8	CL1, BK3	Basic + Parallel	CL1	#5
#9	BK1, BK2	Basic + EV#2	BK1	#4
#10	BK1, BK3	Basic + Parallel	BK1	#5
#11	BK2, BK3	Basic + EV#2, Parallel	-	-
#12	CL1, BK1, BK2	Basic + EV#2, Series	-	-
#13	CL1, BK1, BK3	Basic + Parallel, Series	-	-
#14	CL1, BK2, BK3	Basic + EV#2, Parallel	CL1	#11
#15	BK1, BK2, BK3	Basic + EV#2, Parallel	BK1	#11
#16	CL1, BK1, BK2, BK3	Basic + EV#2, Parallel, Series	-	-

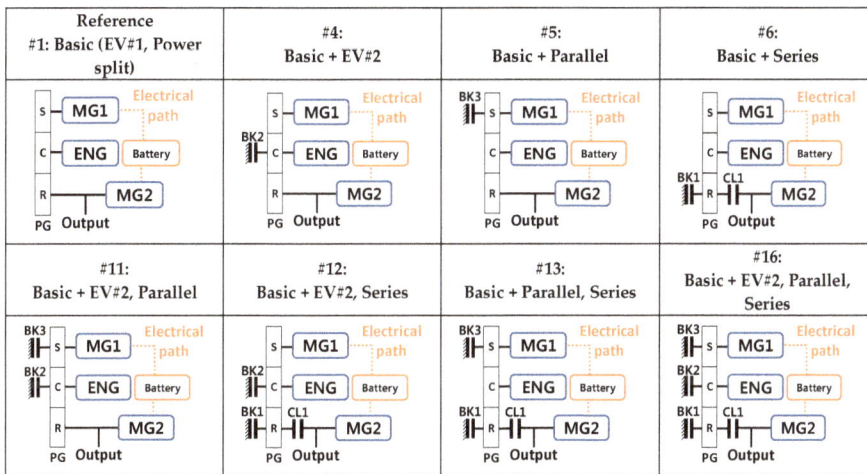

Figure 7. Eight candidates with operating modes.

3. Component Loss Model

As shown in Figure 7, eight candidates for new PHEV configurations were obtained using the additional clutch and brake. In general, it is expected that, as the number of the operating modes increases, the fuel economy increases, since the PHEV can be operated with high efficiency by selecting the proper operating mode for the given wheel torque and speed. However, additional drivetrain components such as the clutch and brake cause parasitic power loss. Therefore, the pros and cons of the multi-mode operation need to be evaluated when adding the operating mode.

In this study, to evaluate the effect of the power electronic (PE) loss and drivetrain loss on the fuel economy, the component loss models were obtained based on the mathematical governing equation and experimental results.

3.1. PE Loss

The MG1 and MG2 losses were calculated from the efficiency map. The efficiency of MG1 or MG2 can be determined from the operation point for the given torque and speed (Figure 8a,b). The high-voltage DC/DC converter (HDC) boosts the battery voltage in consideration of the operating conditions of MG1 or MG2. When the HDC boosts the battery voltage, a boost loss occurs. The HDC loss was determined from the HDC efficiency with respect to the battery power (Figure 8c). In addition, the battery loss was obtained from the charge (discharge) efficiency. In this study, the battery charge (discharge) efficiency was assumed to be 98.5% [21].

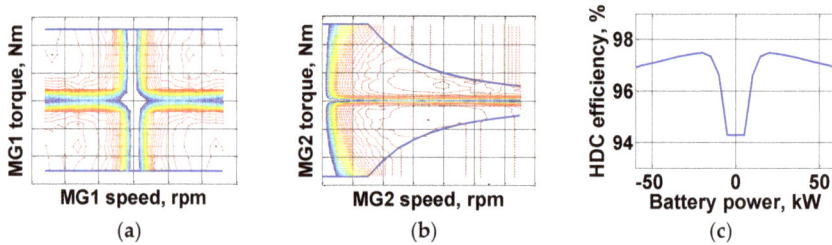

Figure 8. The efficiency map derived from the experimental result for: (**a**) motor generator 1 (MG1); (**b**) MG2; (**c**) high-voltage DC/DC converter (HDC).

3.2. Drivetrain Loss

The drivetrain losses have been described in detail in the literature [22]. A short summary of the drivetrain losses is as follows:

Gear loss: Gear loss was assumed to be 1% of the transmitted torque [23], which is widely used in the automotive industry.

Planetary gear loss: When the sun, pinion, and ring gear are meshing, the planetary gear loss occurs due to the gear teeth friction. The planetary gear loss is represented as [24,25]:

$$T_{loss\_PG} = \begin{cases} k_{PG} \times T_{in\_PG}, & \text{when the carrier is fixed} \\ k_{PG} \times \left(\frac{Z_r}{Z_r+Z_s}\right) \times T_{in\_PG}, & \text{when the ring gear is fixed} \\ k_{PG} \times \left(\frac{Z_s}{Z_r+Z_s}\right) \times T_{in\_PG}, & \text{when the sun gear is fixed,} \end{cases} \tag{4}$$

where T_{loss} is the torque loss, T_{in} is the input torque, and k is the coefficient of friction for PG.

Bearing loss: There are loaded and unloaded losses in the bearing. Loaded loss is proportional to the bearing load. The loaded loss is calculated as follows [26,27]:

$$T_{loss\_BRGload} = f_1 \times P_1{}^a \times d_m{}^b \times 10^{-3}, \tag{5}$$

where a, b, and f_1 are the coefficients according to the bearing type, P_1 is the equivalent bearing load and is determined from the axial and radial bearing reaction force, and d_m is the bearing mean diameter.

Bearing unloaded loss is caused by the slip between the rotating surface and the lubrication oil film. The bearing unloaded loss is represented as

$$T_{loss\_BRGunload} = \begin{cases} 1.6 \times 10^{-8} \times f_0 \times d_m{}^3, & \text{if } (v_{oil} \times n) < \frac{2000 \text{ mm}^2}{s \cdot min} \\ 10^{-10} \times f_0 \times (v_{oil} \times n)^{\frac{2}{3}} \times d_m{}^3, & \text{if } (v_{oil} \times n) \geq \frac{2000 \text{ mm}^2}{s \cdot min}, \end{cases} \tag{6}$$

where f_0 is the coefficient for the bearing unloaded loss, v_{oil} is the kinematic viscosity, and n is the bearing rotational speed.

Churning loss: To reduce friction between the gear teeth, the final reduction gear rotates in lubrication oil. During the gear rotation, the churning loss occurs in proportion to the rotational speed. Churning loss is calculated as follows [28]:

$$T_{loss\_churning} = \frac{1}{2}\rho\omega^2 R_p{}^3 S_m C_m,\qquad(7)$$

where ω is the rotational speed, ρ is the lubricant density, R_p is the gear pitch effective radius, S_m is the contact surface coefficient, and C_m is the dimensionless churning torque loss.

Brake and clutch loss: The brake and clutch unloaded losses are the drag losses between the friction surface and lubricant in the disengaged state. The brake and clutch unloaded losses were modeled using the experimental results (Figure 9a).

MG1 unloaded loss: MG1 unloaded loss is caused by the mechanical and electrical components when MG1 rotates freely [29]. For the input split type, MG1 does not produce the power to propel the vehicle in the EV#1 mode and is freely rotating because it is connected to the vehicle through the planetary gear. MG1 unloaded loss was modeled using the experimental results (Figure 9b).

Oil pump loss: The oil pump provides a flow rate for the lubrication, cooling, and the control of the clutch and brake. Since a mechanical oil pump, which is driven by the driveshaft via a gear was used in this study, the oil flow is supplied in proportion to the vehicle speed. While the clutch or brake is engaged, additional oil flow is needed to generate the control pressure. Therefore, the oil pump loss depends on the vehicle speed and control pressure. In this study, the oil pump loss was obtained from Prius THS experimental results, shown in Figure 9c [16].

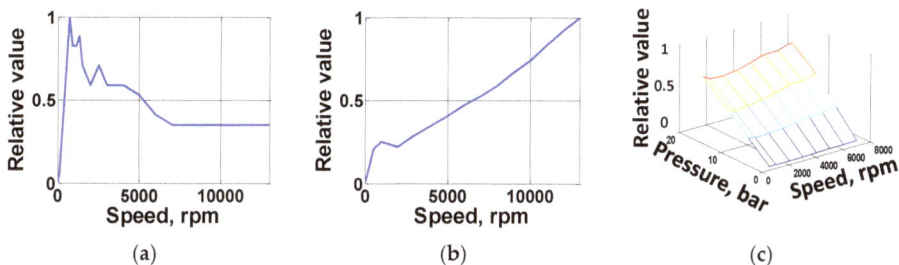

Figure 9. The torque loss map derived from the experimental result for: (**a**) Brake/clutch; (**b**) MG1 unloaded; (**c**) Oil pump.

To calculate the drivetrain loss, the design specification and location of each element are required. In Figure 10, a schematic diagram of the reference PHEV, using the design concept described in Figure 6b, is shown. The installation positions of the bearings were determined referring to the 3rd generation Toyota Prius. The bearing losses inside the MG1 and MG2 were considered in the motor efficiency map. MG2 was connected to the output through two reduction gears, G1-G2 and G3-G4. The friction face area of the brake and clutch and the number of clutch friction faces were determined by considering the transmitted torque [30].

The magnitude of the PE loss and drivetrain loss vary depending on the power flow path, which is determined by the operating mode. In Table 3, the components of the PE loss and drivetrain loss are shown for each operating mode.

In EV#1 and parallel mode, the PE losses occur from the battery charge/discharge, MG2, and HDC operation. The MG1 loss needs to be considered in the EV#2 mode, power split, and series mode, in addition to the battery, MG2, and HDC losses. The drivetrain losses come from the gear, planetary gear, and bearings. The unloaded losses occur due to the drag when the clutch or brake is freely rotating. The churning loss always exists at the final reduction gear. In addition, the oil pump loss needs to be considered when the clutch or brake is engaged.

	Vehicle specification	
Engine	Max power	115 kW
	Max torque	185 Nm
MG1	Max power	70 kW
	Max torque	50 Nm
MG2	Max power	90 kW
	Max torque	270 Nm
Battery	Max power	50 kW
	Capacity	25Ah
Vehicle	Mass	1800 kg
	Tire radius	0.32 m
Gear ratio	PG	2.6
	G1-G2	2.478
	G3-G4	1.0

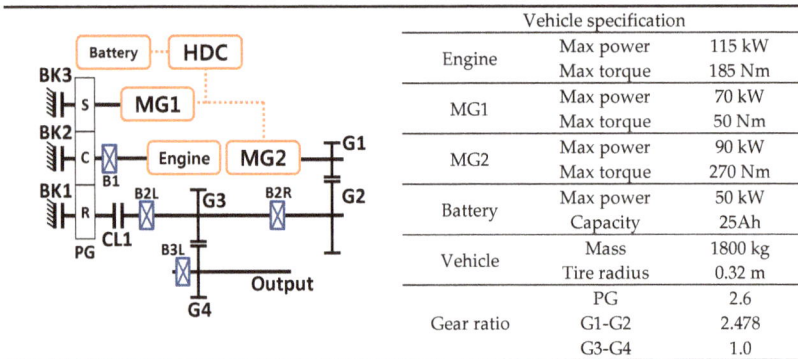

Figure 10. Reference topology and specifications with drivetrain elements and additional clutch and brakes. MG: motor and generator, CL#: clutch, PG#: planetary gear, B#L: bearing on left side, B#R: bearing on right side, HDC: high voltage DC/DC converter, BK#: brake, G#: gear.

Table 3. Power electronics (PE) and drivetrain loss for each operating mode.

Operating Mode		EV#1	EV#2	Power Split	Parallel	Series
PE loss		Battery, HDC, MG2	Battery, HDC, MG2, MG1	Battery, HDC, MG2, MG1	Battery, HDC, MG2	Battery, HDC, MG2, MG1
Drivetrain loss	Loaded	Gear, BRG	Gear, BRG, PG	Gear, BRG, PG	Gear, BRG, PG	Gear, BRG, PG
	Un-loaded	BRG, Churning, MG1 unloaded, BK#	BRG, Churning, BK#, Pump	BRG, Churning, BK#	BRG, Churning, BK#, Pump	BRG, Churning, BK#, CL1, Pump

4. Backward Simulator Using Dynamic Programming

To evaluate the maximum potential in the fuel economy of the eight candidates in Figure 7, a backward simulator was developed using dynamic programming (DP). Since DP is able to find the optimal SOC trajectory regardless of the control strategy, which guarantees minimum fuel consumption for the given PHEV configuration [31,32], it was used for the comparative analysis of the candidates for the presence or absence of the drivetrain losses.

The PHEV system in Figure 7 has two control variables, engine speed and torque, and one state variable, battery SOC [33]. For each time step, $k - 1$, the instantaneous optimal operating point of the engine is determined for the specific battery power. When the operating point of the engine is given, the PE loss and drivetrain loss are calculated [22]. Considering the component loss, the optimal fuel consumption rate, g_{k-1}, is obtained through the instantaneous optimization process.

After the process is completed, global optimization is performed to find the minimum fuel consumption over the whole driving cycle. The global optimization process can be represented as a recursive equation [33,34]. The recursive equation and constraint are represented as

$$\text{Recursive equation}: J_k^*(SOC_k) = \{g_{k-1}(\omega^e_{k-1}, T^e_{k-1}) + J_{k-1}^*(SOC_{k-1})\}$$
$$\text{Constraint}: SOC_{initial} - SOC_{final} = 0,$$
(8)

where k is the discrete time step, J_k^* is the minimum fuel consumption from 1 to k step, J_{k-1}^* is the minimum fuel consumption from 1 to $k - 1$ step, ω^e_{k-1} and T^e_{k-1} are respectively the speed and torque of the engine that has the minimum fuel consumption rate at the $k - 1$ step, and g_{k-1} is the fuel consumption rate at the $k - 1$ step.

Using the recursive equation, we can find the minimum fuel consumption, J_k^*; in other words, the maximum potential in the fuel economy over the whole driving schedule [35,36].

5. Comparative Analysis Using Dynamic Programming

PE Loss and Drivetrain Loss

To evaluate the influence of the drivetrain losses on the fuel economy, simulation was performed for candidate #16 using the backward simulator, in which the target PHEV undergoes the highway fuel economy test (HWFET) cycle. In the simulation, the final battery SOC was assumed to be equal to the initial SOC. As described in Section 4, the operation points of the engine, MG1, and MG2 were determined, which guarantee the minimum fuel consumption by the dynamic programming results. In addition, the operating mode that provides the minimum fuel consumption was selected.

In Figure 11, the simulation results are shown. The engine torque (b) and speed (c) were almost maintained at the optimal operating point at 80 Nm and 1600 rpm, where the engine has the highest thermal efficiency for the demanded engine power. The MG2 torque (b) showed a negative value when regenerative braking was performed. The battery SOC (d) decreased from the initial value when the vehicle accelerated and increased during the regenerative braking. The vehicle started in the EV#1 mode and used the EV#2, power split, parallel, and regenerative braking modes (e) during the driving. It is seen that the parallel mode was mostly used for the HWFET cycle, while the series mode was never used.

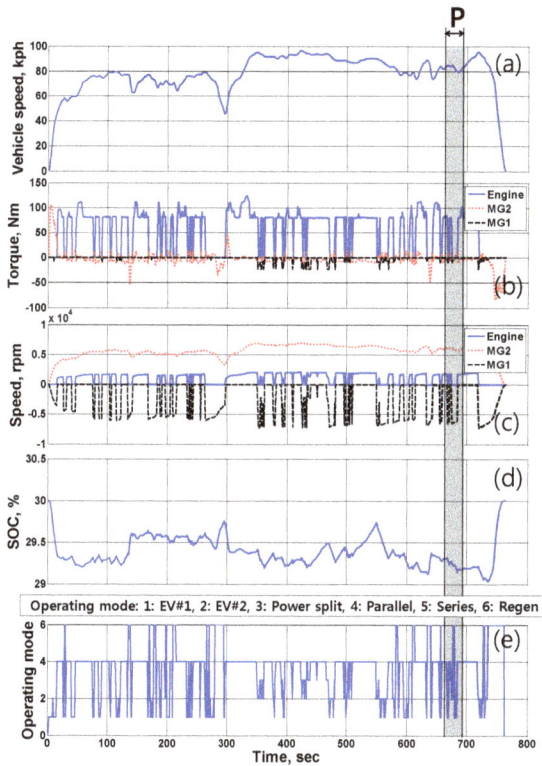

Figure 11. Backward simulation result for candidate #16 considering PE and drivetrain loss (highway fuel economy test (HWFET) cycle). (**a**) Vehicle speed; (**b**) Torque; (**c**) Speed; (**d**) Battery SOC; (**e**) Operating mode.

In Figure 12, the drivetrain loss and PE loss in region P (*t* = 665–695 s) of Figure 11 are shown. The gear (9–150 W), bearing (150–200 W), and PG losses (3–210 W) (Figure 12a) always occur when the vehicle drives. The gear and PG losses showed an almost zero value at an instant when the transmitted torque is very small during regenerative braking. An MG1 unloaded loss of 400 W occurred when MG1 was freely rotating, such as in the EV#1 and regenerative braking mode. When MG1 works, the MG1 unloaded loss does not appear. Instead, the PE loss of MG1 occurs (c). The drag losses of BK1, BK2, and BK3 (b) were 20–45 W when they were disengaged. In region P, since the series mode was not used (e), BK1 was disengaged and CL1 was engaged, causing a drag loss of BK1 of 30 W and a zero drag loss of CL1. The BK2 loss was zero in the EV#2 mode when it was engaged but showed a 20 W loss in the parallel mode when it was disengaged. It is seen that a drag loss of 45 W occurred in BK3 when BK3 was disengaged. In addition, a pump loss (b) of 45 W occurred, supplying the pressure for the engagement of BK and CL1. In Figure 12c, the PE losses are shown. The MG2 loss of 400–900 W occurred because MG2 always works when the vehicle drives. The MG1 loss (500–700 W) occurred in the EV#2 mode. The HDC loss showed a range of 0–50 W. In Figure 12d, the total drivetrain loss is shown with the PE loss. It is seen that the total drivetrain loss (300–900 W) was almost the same as the PE loss (385–1260 W), which demonstrates that drivetrain loss should be considered when evaluating the fuel economy for a new PHEV configurations.

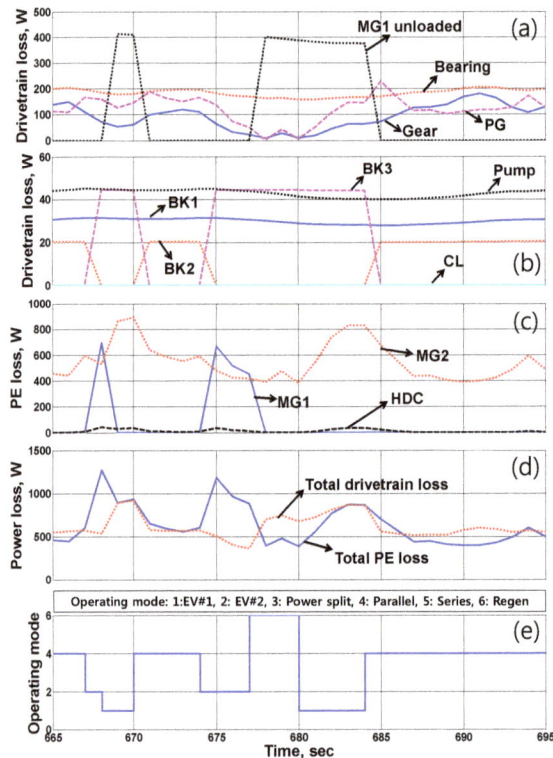

Figure 12. PE and drivetrain loss for candidate # 16 when driving the HWFET cycle (665–695 s). (**a**,**b**) Drivetrain loss; (**c**) PE loss; (**d**) Power loss; (**e**) Operating mode.

In Figure 13a,b, the operating modes are plotted, which provide the minimum fuel consumption for the demanded wheel power and vehicle speed. The operating mode was selected from the global optimization using dynamic programming. It is seen from the dynamic programming results that the

PHEV (candidate #16) does not use the EV#2 and series mode without consideration of the drivetrain loss, even if it can implement the EV#2 and series mode. This is because these two operating modes cannot provide the minimum fuel consumption for the given driving condition. Instead, the EV#1, power split and parallel mode were used with the operation times of 283 s, 27 s, and 365 s, respectively, across the total driving time of 765 s (Figure 13c). On the other hand, when the drivetrain loss is considered, as can be seen from Figure 13b,c, the EV#2 mode is used at low wheel power (less than 8 kW) and high vehicle speed (72–95 kph). The operation time of the EV#2 mode is 79 s, while the EV#1, power split, and parallel modes are used for 110 s, 30 s, and 456 s, respectively. It is noted that the series mode was not used, even when drivetrain loss was considered.

From the comparative analysis, it was found that the mode selection and operation time of each mode varies depending on the presence or absence of drivetrain loss. This is because the drivetrain loss affects the system efficiency for the given driving conditions. As a result, a different operating mode was selected and a different mode operation time was obtained, which leads to a different fuel economy. When the drivetrain loss is not considered, the fuel economy was obtained as 28.76 km/L, and when the drivetrain loss is taken into consideration, the fuel economy is decreased by as much as 8.1% to 26.43 km/L. It is seen from the comparative analysis for candidate #16 that the drivetrain loss has a significant impact on the fuel economy, which demonstrates that the drivetrain loss should be considered in fuel economy evaluations.

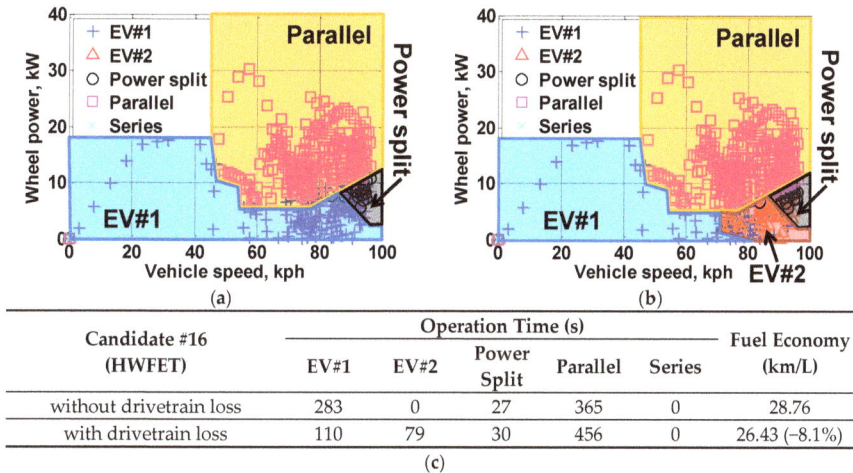

| Candidate #16 | Operation Time (s) | | | | | Fuel Economy |
(HWFET)	EV#1	EV#2	Power Split	Parallel	Series	(km/L)
without drivetrain loss	283	0	27	365	0	28.76
with drivetrain loss	110	79	30	456	0	26.43 (−8.1%)

(c)

Figure 13. Comparison of optimal operating mode for candidate #16 in the presence or absence of drivetrain losses (HWFET). (**a**) Without drivetrain loss; (**b**) With drivetrain loss; (**c**) Operation time and fuel economy.

Now, considering the drivetrain loss, the fuel economies of the eight candidates in Figure 7 were evaluated. In Table 4, simulation results of the operation time, fuel economy, PE loss, and drivetrain loss are compared for the HWFET cycle when the vehicle is operated in CS mode.

It is seen from Table 4 that the fuel economies of candidates #6 and #12 were decreased compared with the reference (candidate #1) in spite of the additional modes. This is because brake drag and pump loss occurred from the additional elements, BK1, BK2, and CL1. The fuel economies of candidates #5, #11, #13, and #16 were improved by 3.65%–4.04%, and we found that all these candidates have the parallel mode in common. It is seen that the operation time of the parallel mode is 437–456 s, which replaced most of the power split mode operation. The reason why the fuel economy was improved when adding the parallel mode can be explained by the reduced PE loss. In the parallel mode, MG1 is turned off and there is no power flow through the electrical path, which provides a smaller PE loss than that of the power split mode. As a result, candidates #5, #11, #13, and #16, which can implement the parallel mode, have smaller PE loss (524.3–540.1 kJ) than the other candidates (935–950 kJ). It is also noted that the fuel economy improvement (3.77%) of candidate #16 is less than that of candidate #11 (4.04%), even if it can implement one more operating mode than candidate #11; this is because the drivetrain loss was increased due to the additional elements, BK1 and CL1.

From Table 4, we can select candidate #11 as a new PHEV configuration that provides the best fuel economy using the EV#1, EV#2, power split, and parallel modes.

As shown in the design procedure, which considers the speed and torque lever analogy, drivetrain loss, and PE loss, it is seen that the design methodology proposed in this study can be used effectively for the development of a new PHEV configuration and that the drivetrain losses must be included in the fuel economy evaluation.

Table 4. Fuel economy and component losses of eight candidates for HWFET when the vehicle is operated in charge sustaining (CS) mode.

Candidate		#1 Basic (EV#1, Power Split)	#4 Basic + EV#2	#5 Basic + Parallel	#6 Basic +Series	#11 Basic + EV#2, Parallel	#12 Basic + EV#2, Series	#13 Basic + Parallel Series	#16 Basic + EV#2, Parallel, Series
Additional element		-	BK2	BK3	BK1, CL1	BK2, BK3	BK1, BK2, CL1	BK1, BK3, CL1	BK1, BK2, BK3, CL1
No. of modes		2	3	3	3	4	4	4	5
Operation time (s)	EV#1	89	64	169	89	118	60	172	110
	EV#2	X	45	X	X	69	47	X	79
	Power split	586	566	69	586	33	568	61	30
	Parallel	X	X	437	X	455	X	442	456
	Series	X	X	X	0	X	0	0	0
FE (km/L)		25.47	25.47	26.47 (+3.92%)	25.37 (−0.39%)	26.5 (+4.04%)	25.36 (−0.43%)	26.4 (+3.65%)	26.43 (+3.77%)
PE loss (kJ)	MG2	658.2	645.5	486.1	658.1	463	645.1	485.7	460.3
	MG1	264.1	290.4	15.2	265.9	49.4	292.7	13.3	55
	HDC	12.6	11.5	25.3	12.3	25.2	11.7	25.3	24.8
Total PE loss (kJ)		934.9	947.4	526.6	936.3	537.6	949.5	524.3	540.1
Drivetrain loss (kJ)	MG1-unload	49.7	41.1	80.6	50	60.4	39.7	81.5	57.3
	PG	182.6	175.7	129	182.1	122.4	175.4	128.1	121.6
	Gear	182.3	182.3	182.3	182.3	182.3	182.3	182.3	182.3
	Bearing	133.1	133.1	133.1	133.1	133.1	133.1	133.1	133.1
	CL1 & BK#	0	10.6	25.1	20.4	34.7	31	33.3	42.5
	Pump	18.3	19	25.5	30.9	27	30.9	30.9	30.9
Total drivetrain loss (kJ)		566	561.8	575.6	598.8	559.9	592.4	589.2	567.7

6. Conclusions

A design methodology of a power split type PHEV was proposed by considering the drivetrain losses. As a design reference, an input split type PHEV using a single planetary gear was selected. First, to determine the engine position on the speed lever of the single planetary gear, the mechanical point (MP) at which the power split transmission (PST) has the highest efficiency was investigated with respect to the speed ratio, and it was found that the engine should be located at the carrier to have the MP at high speeds, which provided a higher PST efficiency in the main driving range. In addition, the positions of MG1 and MG2 on the speed lever were determined, which provides a better PST efficiency by reducing the power that flows through the electrical path. Based on the reference topology, feasible locations of the additional clutch and brake were investigated to realize the multi-mode in addition to the basic operating mode of the reference PST. Among the mathematically possible combinations of 2^6 candidates, sixteen candidates were selected by considering the operation condition of the engine and MG1 in a real driving environment, and, finally, eight candidates were obtained by eliminating the candidates that had a redundant element. To evaluate the fuel economy of the selected candidates, the loss models of the power electronic system (MG1, MG2, HDC) and drivetrain components (gear, planetary gear, clutch, brake, bearing, MG1 unloaded loss, etc.) were obtained based on the mathematical governing equation and experimental results. Based on the component loss model, a backward simulator was developed using dynamic programming to find the maximum potential for the fuel economy of the PHEV candidates for the given driving duty cycle. Using the backward simulation, a comparative analysis was performed under the presence or absence of the drivetrain losses, and it was found that the selection of the operating mode and the operation time of each mode vary, since the drivetrain losses affect the system efficiency. The fuel economy also decreased by as much as 8.1% for the HWFET cycle.

In addition, it was found from the comparative analysis that, even if the additional modes result in flexibility when selecting the operating mode, thus providing a higher system efficiency for the given wheel power and vehicle speed, additional drivetrain elements to realize the modes can deteriorate the fuel economy due to the losses of the additional elements. It is also noted that the series mode was never used due to its low system efficiency. On the other hand, the parallel mode can improve the system efficiency since the PE loss is reduced compared with the other modes.

It is expected that the design methodology proposed in this study, which considers the drivetrain losses, can be used in development of new PHEV configurations.

Acknowledgments: This material is based upon work supported by the Ministry of Trade, Industry, and Energy (MOTIE, Korea) under Industrial Technology Innovation Program. No. 10062742.

Author Contributions: Hanho Son developed the design methodology of a power split type PHEV considering drivetrain losses and wrote the paper. Kyusik Park developed the individual models of the target PHEV powertrain and drivetrain losses. Hyunsoo Kim and Sungho Hwang supervised this research and wrote the paper.

Conflicts of Interest: The authors declare no conflict of interest.

References

1. U.S. DOE. Chapter 8: Advanced Clean Transportation and Vehicle Systems and Techonologies, Quadrennial Technology Review 2015. Available online: https://energy.gov/sites/prod/files/2016/01/f28/QTR2015-8E-Plug-in-Electric-Vehicles.pdf (accessed on 11 January 2017).
2. Ehsani, M.; Gao, Y.; Gay, S.; Emadi, A. *Modern Electric, Hybrid Electric, and Fuel Cell Vehicles*, 2nd ed.; CRC Press: Boca Raton, FL, USA, 2009.
3. Matsubara, T.; Yaguchi, H.; Takaoka, T.; Jinno, K. *Development of New Hybrid System for Compact Class Vehicles*; SAE Technical Paper: Detroit, MI, USA, 2009. [CrossRef]
4. Brendan, C. *Comparative Analysis of Single and Combined Hybrid Electrically Variable Transmission Operating Modes*; SAE Technical Paper: Detroit, MI, USA, 2005. [CrossRef]
5. Schulz, M. Circulating mechanical power in a power-split hybrid electric vehicle transmission. *Proc. Inst. Mech. Eng. Part D J. Automob. Eng.* **2004**, *218*, 1419–1425. [CrossRef]
6. Ma, C.; Kang, J.; Choi, W.; Song, M.; Ji, J.; Kim, H. A comparative study on the power characteristics and control strategies for plug-in hybrid electric vehicles. *Int. J. Automot. Technol.* **2012**, *13*, 505–516. [CrossRef]
7. Silvas, E.; Hofman, T.; Murgovski, N.; Pascal Etman, L.F.; Steinbuch, M. Review of optimization strategies for system-level design in hybrid electric vehicles. *IEEE Trans. Veh. Technol.* **2017**, *66*, 57–70. [CrossRef]
8. Zhang, X.; Peng, H.; Sun, J. A near-optimal power management strategy for rapid component sizing of power split hybrid vehicles with multiple operating mode. In Proceedings of the American Control Conference (ACC), Washington, DC, USA, 17–19 June 2013.
9. Kim, J.; Kim, N.; Hwang, S.; Hori, Y.; Kim, H. Motor control of input-split hybrid electric vehicles. *Int. J. Automot. Technol.* **2009**, *10*, 733–742. [CrossRef]
10. Hofman, T.; Ebbesen, S.; Guzzella, L. Topology optimization for hybrid electric vehicles with automated transmissions. *IEEE Trans. Veh. Technol.* **2012**, *61*, 2442–2451. [CrossRef]

11. Zhang, X.; Li, C.; Kum, D.; Peng, H. Prius+ and Volt-: Configuration analysis of power-split hybrid vehicles with a single planetary gear. *IEEE Trans. Veh. Technol.* **2012**, *6*, 857–865.
12. Kim, H.; Kum, D. Comprehensive design methodology of compound-split hybrid electric vehicles: In search of optimal configurations. *IEEE Trans. Mechatron.* **2016**, *21*, 2912–2923. [CrossRef]
13. Zhuang, W.; Zhang, X.; Peng, H.; Wang, L. Simultaneous optimization of topology and component sizes for double planetary gear hybrid powertrains. *Energies* **2016**, *9*, 411. [CrossRef]
14. Mashadi, B.; Emadi, S. Dual-mode power-split transmission for hybrid electric vehicle. *IEEE Trans. Veh. Technol.* **2010**, *59*, 3223–3232. [CrossRef]
15. Zhang, X.; Li, S.; Peng, H.; Sun, J. Efficiency exhaustive search of power-split hybrid powertrains with multiple planetary gears and clutches. *J. Dyn. Syst. Meas. Control* **2015**, *137*. [CrossRef]
16. Kang, J.; Choi, W.; Kim, H. Development of a control strategy based on the transmission efficiency with mechanical loss for a dual mode power split-type hybrid electric vehicle. *Int. J. Automot. Technol.* **2012**, *13*, 825–833. [CrossRef]
17. Yang, H.; Cho, S.; Kim, N.; Lim, W.; Cha, S. Analysis of planetary gear hybrid powertrain system part 1: Input split system. *Int. J. Automot. Technol.* **2007**, *8*, 771–780.
18. Kim, N.; Kim, J.; Kim, H. Control strategy for dual-mode electromechanical, infinitely variable transmission for a hybrid electric vehicles. *Proc. Inst. Mech. Eng.* **2008**, *222*, 1587–1601. [CrossRef]
19. Benford, H.; Leising, M. *The Lever Analogy: A New Tool Transmission Analysis*; SAE Paper: Detroit, MI, USA, No. 810102; 1981. [CrossRef]
20. Kim, J. Design and Control of Power Split Transmission for a Hybrid Electric Vehicle. Ph.D. Thesis, Sungkyunkwan University, Seoul, Korea, 2009.
21. Rydh, C.J.; Sanden, B.A. Energy analysis of batteries in photovoltaic systems. Part II: Energy return factors and overall battery efficiencies. *Energy Convers. Manag.* **2005**, *46*, 1980–2000. [CrossRef]
22. Son, H.; Kim, H. Development of near optimal rule-based control for plug-in hybrid electric vehicles taking into account drivetrain component losses. *Energies* **2016**, *9*, 420. [CrossRef]
23. Kim, I.; Kim, H. Configuration Analysis of plug-in hybrid systems using global optimization. In Proceedings of the Electric Vehicle Symposium and Exhibition (EVS27), Barcelona, Spain, 17–20 November 2013.
24. Haka, R.J. Determination of efficiency (torque related losses) in planetary gearsets-generalized theory for simple and compound gearsets. In Proceedings of the ASME 2003 International Design Engineering Technical Conferences and Computers and Information in Engineering Conference, Chicago, IL, USA, 2–6 September 2003; pp. 1085–1097.
25. Chen, J.Y.; Borgerson, J.B. Analytical and test evaluation of planetary gear train efficiency (torque related losses) with multiple power flow arrangements. In Proceedings of the ASME 2003 International Design Engineering Technical Conferences and Computers and Information in Engineering Conference, Chicago, IL, USA, 2–6 September 2003; pp. 1057–1065.
26. *Gears—Thermal Capacity—Part 2: Thermal Load-Carrying Capacity*; ISO/TR 14179-2:2001; International Organization for Standardization: Geneva, Switzerland, 2001.
27. SKF General Catalogue. Available online: http://www.worldcat.org/title/skf-general-catalogue/oclc/225258275 (accessed on 13 January 2016).
28. Changenet, C.; Velex, P. A model for the prediction of churning losses in geared transmissions-preliminary results. *J. Mech. Des.* **2006**, *129*, 128–133. [CrossRef]
29. Yamazaki, K.; Watari, S. Loss analysis of permanent-magnet motor considering carrier harmonics of pwm inverter using combination of 2-D and 3-D finite-element method. *IEEE Trans. Magn.* **2005**, *41*, 1980–1983. [CrossRef]
30. Yuan, Y.; Liu, E.A.; Hill, J.; Zou, Q. An improved hydrodynamic model for open wet transmission clutches. *J. Fluids Eng.* **2007**, *129*, 333–337. [CrossRef]
31. Kirk, D.E. *Optimal Control Theory: An Introduction*; Dover Publications, Inc.: Mineola, NY, USA, 1970.
32. Wang, X.; He, H.; Sun, F.; Zhang, J. Application study on the dynamic programming algorithm for energy management of plug-in hybrid electric vehicles. *Energies* **2015**, *8*, 3225–3244. [CrossRef]
33. Kim, N. Energy Management Strategy for Hybrid Electric Vehicles Based on Pontryagin's Minimum Prinsiple. Ph.D. Thesis, Seoul National University, Seoul, Korea, 2009.
34. Vinot, E.; Trigui, R.; Cheng, Y.; Espanet, C.; Bouscayro, A. Improvement of an EVT-based HEV using dynamic programming. *IEEE Trans. Veh. Technol.* **2014**, *63*, 40–50. [CrossRef]

35. Karbaschian, M.A.; Söffker, D. Review and comparison of power management approaches for hybrid vehicles with focus on hydraulic drives. *Energies* **2014**, *7*, 3512–3536. [CrossRef]
36. Li, Y.; Kar, N.C. Advanced design approach of power split device of plug-in hybrid electric vehicles using dynamic programming. In Proceedings of the Vehicle Power and Propulsion Conference (VPPC), Chicago, IL, USA, 6–9 September 2011.

Article

Optimization of Key Parameters of Energy Management Strategy for Hybrid Electric Vehicle Using DIRECT Algorithm

Jingxian Hao [1,2], Zhuoping Yu [1], Zhiguo Zhao [1,*], Peihong Shen [1] and Xiaowen Zhan [1]

1 School of Automotive Studies, Tongji University, Shanghai 201804, China; haojingxian@saicmotor.com (J.H.); yuzhuoping@tongji.edu.cn (Z.Y.); shenpeihong@tongji.edu.cn (P.S.); xiaowenzhan1@163.com (X.Z.)
2 SAIC Motor Commercial Vehicle Technical Center, Shanghai 200438, China
* Correspondence: zhiguozhao@tongji.edu.cn; Tel.: +86-21-6958-9117

Academic Editors: Michael Gerard Pecht and Chunhua Liu
Received: 22 August 2016; Accepted: 22 November 2016; Published: 26 November 2016

Abstract: The rule-based logic threshold control strategy has been frequently used in energy management strategies for hybrid electric vehicles (HEVs) owing to its convenience in adjusting parameters, real-time performance, stability, and robustness. However, the logic threshold control parameters cannot usually ensure the best vehicle performance at different driving cycles and conditions. For this reason, the optimization of key parameters is important to improve the fuel economy, dynamic performance, and drivability. In principle, this is a multiparameter nonlinear optimization problem. The logic threshold energy management strategy for an all-wheel-drive HEV is comprehensively analyzed and developed in this study. Seven key parameters to be optimized are extracted. The optimization model of key parameters is proposed from the perspective of fuel economy. The global optimization method, DIRECT algorithm, which has good real-time performance, low computational burden, rapid convergence, is selected to optimize the extracted key parameters globally. The results show that with the optimized parameters, the engine operates more at the high efficiency range resulting into a fuel savings of 7% compared with non-optimized parameters. The proposed method can provide guidance for calibrating the parameters of the vehicle energy management strategy from the perspective of fuel economy.

Keywords: fuel economy; hybrid electric vehicle; energy management strategy; logic threshold value; DIRECT; parameters optimization

1. Introduction

The main factors affecting the fuel consumption and emission performance of a hybrid electric vehicle (HEV) include the performance parameters of various powertrain components and vehicle control strategy parameters. Optimizing the parameters of the powertrain and control strategy will not only result in a reasonable match for the powertrain, but also reduce the vehicle fuel consumption and emissions.

At this stage, the energy management strategy based on logic threshold is mainly used in HEVs [1,2]. The focus is to predetermine a number of threshold parameters that make the engine and battery work in the high efficiency area. The battery charging and discharging efficiency are also considered in order to properly distribute the driver's required torque to the engine and motor, thereby attaining a good vehicle fuel economy and emission performance.

In vehicle tests, the predefined parameter values of the logic threshold control strategy are usually obtained by trial and error based on engineering experience. This method requires considerable debugging time to acquire satisfactory results both in simulation and vehicle transfer hub test for the defined typical driving cycle. In addition, this method cannot ensure optimal global parameters. Therefore, it is necessary to adopt an optimization method that can automatically search the globally optimized threshold parameters for the energy management strategy.

The optimization of key parameters for logic threshold energy management strategies is a mathematical nonlinear problem with many variable constraints. The genetic algorithm was applied in optimizing various governing parameters of hybrid electric vehicles (HEVs) and the fuel economy was improved significantly [3–6]. The multi-objective genetic algorithm was adopted to optimize the control parameters of the HEV for improving the fuel economy and emission performance [7,8]. Li et al. utilized a modified non-dominated sorting genetic algorithm-II to effectively optimize the logic threshold control strategy parameters of the HEV to minimize the equivalent fuel consumption [9]. Li et al. applied a hybrid genetic algorithm (HGA), which combines an enhanced genetic algorithm with simulated annealing, in optimizing the powertrain and control parameters of plug-in hybrid electric bus simultaneously. Simulation results show that HGA has a better convergence speed and global searching ability [10]. The particle swarm optimization (PSO) algorithm was applied to search the optimal value of the power system and control parameters of HEV to improve the fuel economy [11–13]. In order to achieve a better fuel economy and emission performance, Deng et al. presented an optimization method for logic threshold control strategy parameters of a parallel HEV using the simulated annealing particle swarm optimization [14]. Wang et al. utilized evolutionary algorithm in conjunction with an instantaneous optimal energy management strategy to optimize the propulsion system parameters as well as the energy control parameters for plug-in HEV [15]. Zhang et al. used differential evolution algorithm to globally optimize the plug-in HEV control parameters [16]. Long et al. optimized the key component and control parameters by using the bees algorithm [17]. Chris et al. showed that the DIvided RECTangle (DIRECT) algorithm has a better optimal effect compared with the genetic algorithm, simulated annealing, PSO and other algorithms by test, because it can cover the global space for parameter optimization without missing any optimization value [18].

The DIRECT algorithm [19] does not require a clear expression of the objective function equation as well as derivative information, but decides on the next searching area based on the estimated value of the function at the sampling points of each iteration and the division of a hyper-rectangle. Thus, it is ideal for simulation of the black-box function optimization [20]. However, it requires a large number of samples in the region to ensure the final global optimum. Besides, the number of estimated function is relatively larger than that of the gradient-based optimization method. In practical engineering optimization, the meta-model optimization is often very complex and the simulation time is relatively long [21]. Instead of the complex meta-model, an approximate model built by the sampling points of each DIRECT algorithm iteration is utilized, thereby reducing the number of simulations, improving the convergence speed, shortening the optimization time, and saving computing resources.

As mentioned above, the advantage of DIRECT algorithm is to obtain the global optimization result compared with other optimization algorithm. Besides, it also has low computational burden, rapid convergence. So it is meaningful to utilize the DIRECT algorithm to acquire the global optimized value of the parameters of HEV energy management strategy. However, a few works have been found to optimize the control strategy parameters of hybrid electric vehicle utilizing the DIRECT algorithm. Rousseau et al. and He et al. established a power component parameter optimization model for a HEV to minimize the fuel consumption. The constraints are the dynamic design specifications and variables, namely the engine power, battery power, battery capacity, battery bus voltage, etc. The DIRECT algorithm is utilized to optimize these parameters. Whereas, the logic threshold control strategy parameters have not been analyzed and optimized [22,23]. Panday et al. utilized DIRECT algorithm to optimize partial parameters of HEV control strategy, such as state of charge in the battery, engine idle speed, engine on duration and power demand [24].

The general comparison of different algorithms is presented in Table 1. The research works which optimize the parameters of HEV energy management strategy utilizing GA, PSO, etc. may lead to local optimization. The DIRECT algorithm based optimization method can ensure the global optimized parameters of HEV energy management strategy. However, the literatures which optimized hybrid electric vehicle parameters based on DIRECT algorithm merely optimized the parameters of powertrain configuration or partial parameters of control strategy to improve energy efficiency. Few research works have comprehensively analyzed and optimized the influencing parameters of logic threshold control strategy for HEV, especially for all-wheel-drive HEV. The logic threshold control strategy of all-wheel-drive HEV is more complicated, for the all-wheel-drive HEV has more freedom of power sources. And the optimization of the logic threshold control strategy parameters for all-wheel-drive HEV comprehensively is more challenging and more meaningful to improve the fuel economy.

Table 1. General comparison of different algorithms.

Algorithm	Convergence	Computation Burden	Global Optimization
Genetic algorithm	good	general	bad
Hybrid genetic algorithm	good	general	general
Particle swarm optimization	good	good	bad
Simulated annealing	bad	good	general
Bees algorithm	good	bad	general
DIRECT algorithm	good	general	good

In this paper, the logic threshold parameters of the all-wheel-drive HEV energy management strategy are comprehensively analyzed, and the seven energy efficiency influencing parameters to be optimized are extracted. Then, the minimized equivalent fuel consumption per 100 km is set as the target, and the DIRECT algorithm is implemented to optimize the proposed parameters globally. Finally, the effectiveness of the algorithm to solve the problem is analyzed by simulation.

2. HEV Powertrain Model

The research object in this study is an all-wheel-drive full HEV, and its powertrain structure is shown in Figure 1 [25]. Its front axle adopts the driving structure including the engine, integrated starter and generator (ISG) motor, and automated mechanical transmission (AMT) gearbox. Its rear axle is driven by two in-wheel motors. The ISG motor shares the same axle with the engine; therefore, it can function as a cranking motor to start the engine quickly. Besides, the output torque of the in-wheel motor can be directly transmitted to the wheel and is capable of driving the vehicle alone at low speed. At the same time, the ISG and in-wheel motors can be used both as driving motors for the vehicle and function as generators to regenerate the excess kinetic energy.

Figure 1 shows that the key parts of the HEV powertrain include the engine, ISG motor, AMT gearbox, in-wheel motor, and power battery. The main technical parameters of these components are shown in Tables 2–7. In consideration of the complexity for acquiring the model parameters and control accuracy, the static model with dynamic correction for main power components is adopted.

Table 2. Vehicle parameters of four-wheel drive HEV.

Vehicle Curb Mass	Tire Radius	Frontal Area	Correction Coefficient of Rotating Mass	Coefficient of Air Resistance	Coefficient of Rolling Resistance
1650 kg	0.308 m	2.095 m^2	1.05	0.293	0.0137

Table 3. Engine parameters of four-wheel drive HEV.

Engine Capacity	Peak Torque	Peak Power	Maximum Rotational Speed
1.8 L	250 N·m	150 kW	6500 rpm

Table 4. ISG motor parameters of four-wheel drive HEV.

Rated Power	Peak Torque	Peak Power	Maximum Rotational Speed
10 kW	72 N·m	15 kW	6500 rpm

Table 5. In-wheel motor parameters of four-wheel drive HEV.

Rated Power	Peak Torque	Peak Power	Maximum Rotational Speed
6 kW	200 N·m	16 kW	2000 rpm

Table 6. Gear ratio of AMT gearbox.

Gear	1st	2nd	3rd	4th	5th	6th
Gear Ratio of AMT gearbox	3.615	2.042	1.257	0.909	0.902	0.0137

Table 7. Transmission parameters of four-wheel drive HEV.

Gear Ratio of Main Reducer	Reverse Gear Ratio	Transmission Efficiency
3.894	4.293	0.92

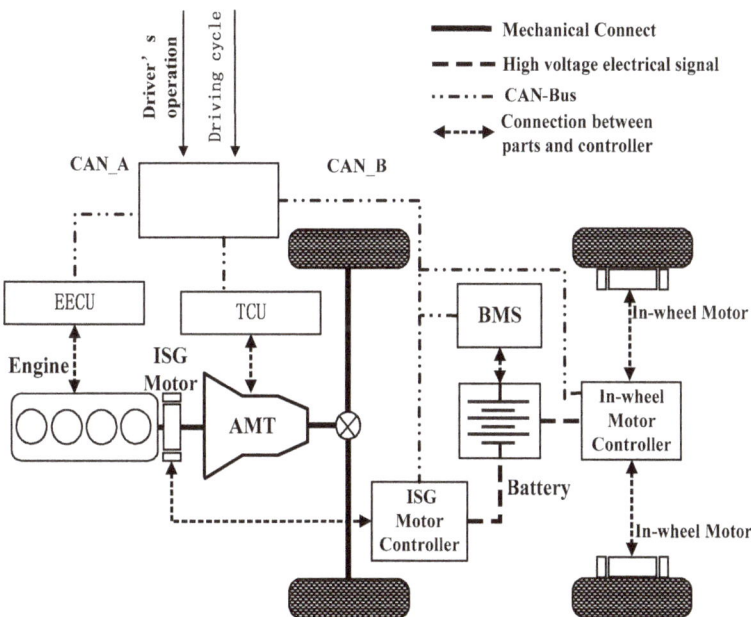

Figure 1. Powertrain structure of hybrid electric vehicle (HEV).

2.1. Engine Model

Since the engine output torque in steady-state condition is a function of its speed and throttle opening, the numerical model of the engine can be established by polynomial fitting based on the engine test data. Considering that the throttle opening changes quickly under the condition of starting or speed-changing, a dynamic process is needed for the engine to be steady. For this reason, a first-order inertia link is used to amend the engine torque representing its dynamic property [26].

$$T_e = \frac{1}{\tau_e s + 1} f(\omega_e, \alpha), \tag{1}$$

where T_e is the engine output torque; ω_e is the engine speed; α is the throttle opening; τ_e is the time constant of the engine torque response; f is a function of the engine torque characteristic.

The data regarding the external characteristic, fuel consumption, and emissions are acquired from the existing engine test. Then, the working characteristic of the engine is obtained by a lookup table or fitting. The current maximum torque and fuel consumption rate of the engine can be acquired by a lookup table according to the current engine speed and torque in the Simulink model. Figure 2 shows the engine fuel consumption rate curve and external characteristic curve used in the model.

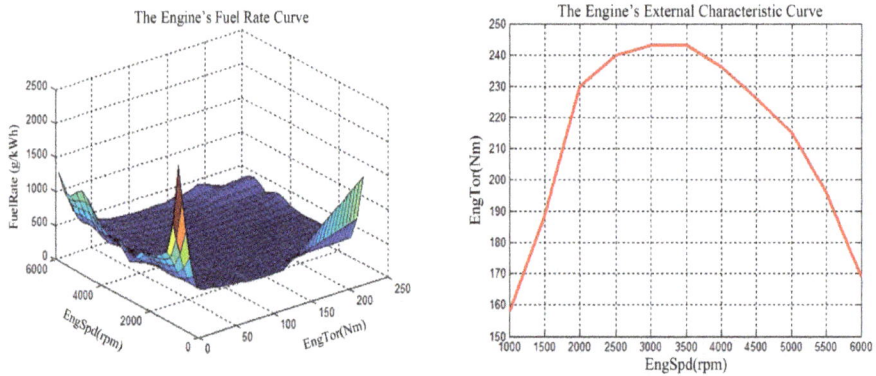

Figure 2. Engine fuel consumption rate curve and external characteristic curve.

2.2. Motor Model

Both the ISG and in-wheel motors are permanent magnet synchronous motors. The motor model is established according to the data of motor efficiency. The dynamic correction is done via a first-order inertia link [26].

$$T_m = \begin{cases} \frac{1}{\tau_m s + 1} \min(T_{mr}, T_{dismax}(\omega_e)) & T_{mr} > 0 \\ \frac{1}{\tau_m s + 1} \max(T_{mr}, T_{chmax}(\omega_e)) & T_{mr} < 0 \end{cases}, \tag{2}$$

where T_m is the output torque of the motor; T_{mr} is the required torque of the motor; τ_m is the time constant of the motor; $T_{dismax}(\omega_e)$ is the maximum output torque of the motor at the speed ω_e when discharging; $T_{chmax}(\omega_e)$ is the minimum output torque of the motor at the speed ω_e when charging.

The motor power can be calculated based on the following equation:

$$P_m = I_m U_m = \begin{cases} \frac{T_m \omega_e}{9550 \eta_m}, & T_m \geq 0 \\ \frac{T_m \omega_e \eta_m}{9550}, & T_m < 0 \end{cases}, \tag{3}$$

where I_m is the motor current; U_m is the bus voltage; η_m refers to the motor efficiency; P_m is the motor power.

The working efficiency and external characteristic curves of the ISG and in-wheel motors, which are shown below in Figures 3 and 4, respectively, are obtained from the experiment.

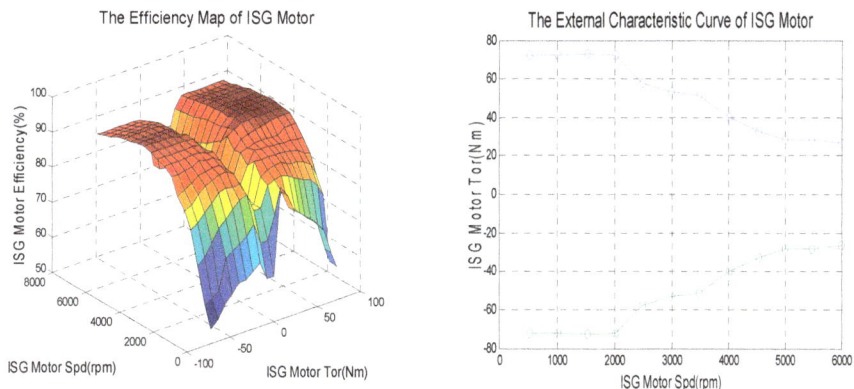

Figure 3. Working efficiency curve and external characteristic curve of integrated starter generator (ISG) motor.

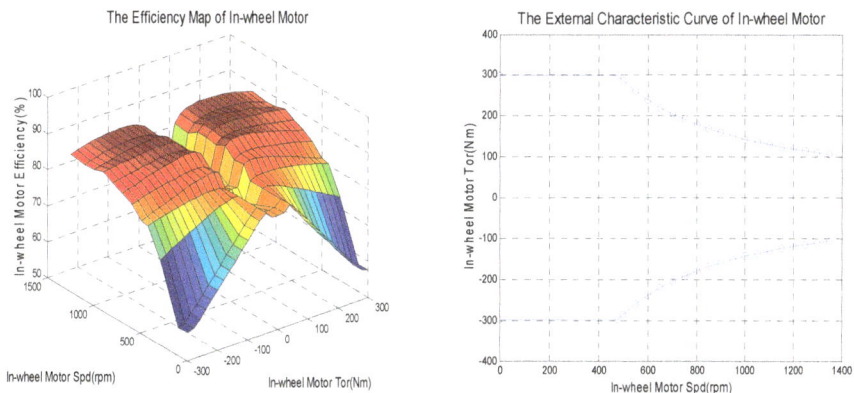

Figure 4. Working efficiency curve and external characteristic curve of in-wheel motor.

2.3. Powertrain Model

The powertrain components in this research include the clutch, AMT gearbox, and main reducer. The powertrain model diagram is shown in Figure 5. The combination and separation of the clutch is determined by the transmission control unit. The vehicle management system controls the clutch state only in the starting process. The transmission and reducer output the powertrain speed and torque according to the throttle percentage, current gear, state of clutch, output torque of clutch, and vehicle velocity.

Considering the practical needs of the control strategy when modeling, we simplify the clutch model using 0 and 1 to represent the complete separation and combination of the clutch. The function of the AMT gearbox and main reducer is to slow down the speed and increase the torque. In particular, when the AMT gearbox is running, the gear changes with shifting control strategy, which directly influences the dynamic performance, fuel economy, and comfort in the vehicle. Therefore, it is necessary to introduce a gear shifting control strategy in the model. Figure 6 shows the gear shift curve calculation module and gear shift control module in the Simulink/Stateflow. The gear ratio is determined by the engine throttle position and vehicle velocity, which can be attained by the lookup table based on the current gear.

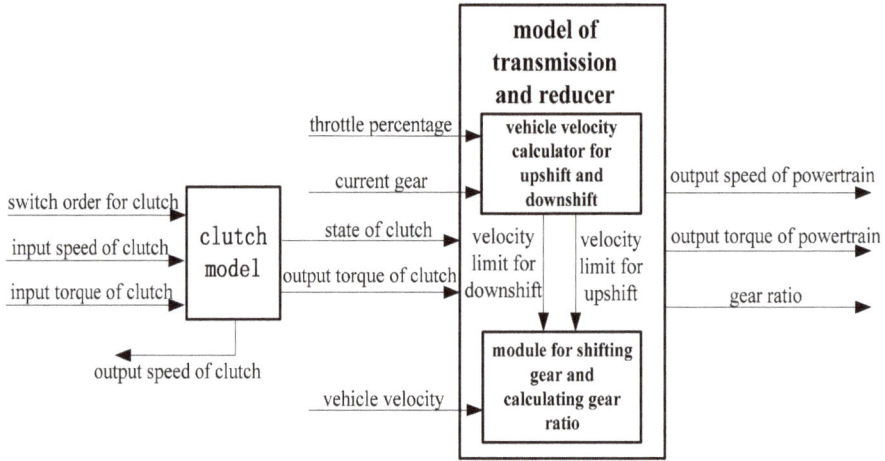

Figure 5. The powertrain model diagram.

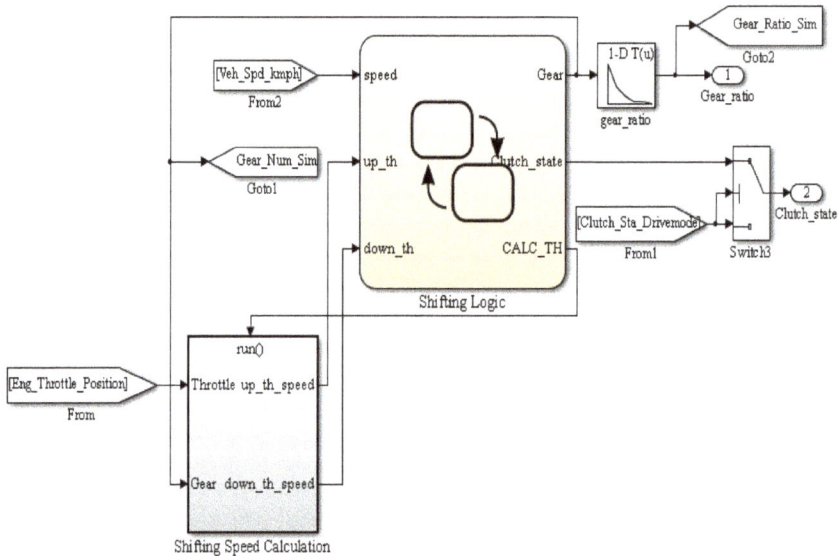

Figure 6. Gear shift model of the automated mechanical transmission (AMT) gearbox.

2.4. Battery Model

The battery functions as an auxiliary power supply for HEVs to provide energy for the motor as well as recycle the excess kinetic energy when braking. The recovery of braking energy is important for improving the vehicle efficiency and saving energy. Although the chemical reaction inside the battery is relatively complex, only the external characteristic of the battery is used in the model. The commonly used battery models include the equivalent circuit model and neural network model. The equivalent circuit model can accurately reflect the battery characteristics using the circuit components such as resistors, capacitors, and voltage to simulate the dynamic performance of the battery. The equivalent circuit models mainly used are the Rint, Thevenin, and PNGV models [27]. The Rint model, which is shown in Figure 7, is selected in this study [28].

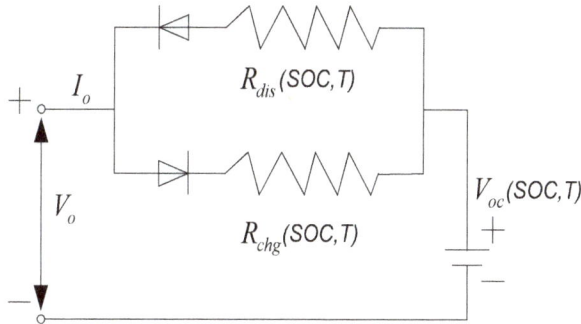

Figure 7. Battery equivalent circuit of the Rint model.

In the diagram, R_{dis} and R_{chg} are the internal resistance of the battery when charging and discharging, and V_{oc} is the open circuit voltage of the battery. R_{dis}, R_{chg}, and V_{oc} are functions of the battery state of charge (*SOC*) and temperature T. When the battery power demand is known, the terminal voltage of the battery, V_o, and current, I_o, can be calculated based on the mathematical models described below.

$$P_{cmd} = V_o I_o, \tag{4}$$

$$V_o = V_{oc} - R_{int} I_o, \tag{5}$$

By combining Equations (4) and (5), Equation (6) can be obtained.

$$I_o = \frac{V_{oc}}{2R_{int}} - \sqrt{\left(\frac{V_{oc}}{2R_{int}}\right)^2 - \frac{P_{cmd}}{R_{int}}}, \tag{6}$$

Here, R_{int} is the internal resistance of the battery when charging or discharging; and P_{cmd} is the battery demand power when charging or discharging.

The current battery *SOC* value can be calculated by the ampere hour algorithm [26] method with the specific formula shown as follows:

$$SOC(t) = SOC_{start} - \frac{\int_0^t I_o dt}{Q_{max}}, \tag{7}$$

where SOC_{start} is the initial battery *SOC* value, and Q_{max} is the maximum battery capacity.

The working state and related parameters of the battery are related to the temperature, so the change of the internal battery temperature should be considered. The specific calculation formula is presented as follows:

$$mc_p \frac{dT(t)}{dt} = R_{int} I_o^2(t) - h_c A[T(t) - T_a], \tag{8}$$

where A is the total heat dissipation area of the battery, c_p is the specific heat capacity of the battery, h_c is the heat transfer coefficient of the battery cooling system, and T_a refers to the ambient temperature.

Based on the above equations, the battery model is established in Simulink. The input and output signals, as well as the calculation module of battery model, are shown in Figure 8. The battery model mainly includes the current, internal resistance, terminal voltage, output power, *SOC*, and temperature calculation modules. The main characteristic parameters of the battery are shown in Table 8.

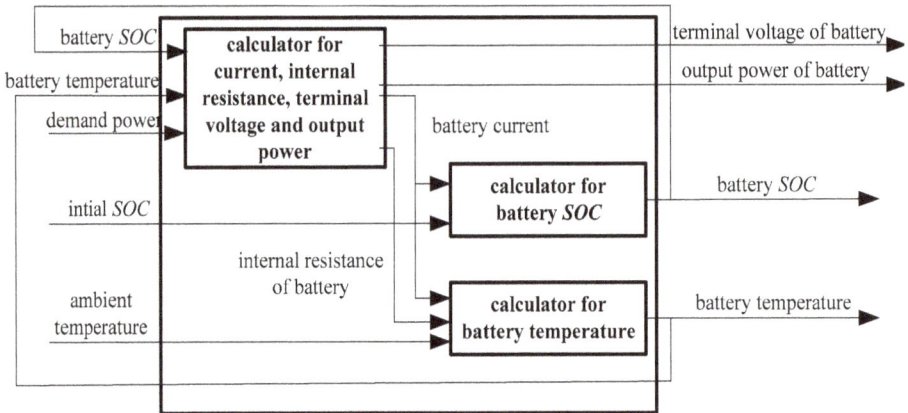

Figure 8. The battery model diagram.

Table 8. Main characteristic parameters of the battery.

Characteristic Parameter	Value	Characteristic Parameter	Value
Capacity	8 Ah	Total heat dissipation area	1.6 m^2
Voltage	312 V	Specific heat capacity	800 J/(kg·K)
Mass	70 kg	Heat transfer coefficient	24 W/(m^2·K)

2.5. Longitudinal Dynamics Model of Vehicle

This research focuses on the fuel economy of the HEV, thus only the longitudinal dynamics is considered in the vehicle model, regardless of the vertical vibration and handling stability. According to the vehicle kinematics equation [29], we have:

$$F_t = mgf\cos\alpha + \frac{C_D A}{21.15}u^2 + mg\sin\alpha + \delta mu, \tag{9}$$

where F_t is the driving force, m is the total mass of the vehicle, f is the rolling resistance coefficient, α is the slope angle, A is the vehicle frontal area, C_D is the air drag coefficient, u is the vehicle speed, and δ refers to the correction coefficient of rotating mass.

2.6. Driver Model

The function of the driver model is to simulate the real driver's controllability. And the driver model diagram is shown in Figure 9. The driver controls the accelerator or the braking pedal opening based on the difference between the real velocity and driving cycle velocity. The proportional–integral (PI) controller is selected for the driver model [30].

The driver model can be described as follows:

$$\alpha = k_p(v_{target} - v) + k_i \int (v_{target} - v)dt, \tag{10}$$

where α is the pedal opening, with positive and negative values representing the accelerator pedal opening and brake pedal opening, respectively; v_{target} is the target velocity; v is the current actual speed; k_p and k_i refer to the proportional and integral coefficients of the PI controller, respectively.

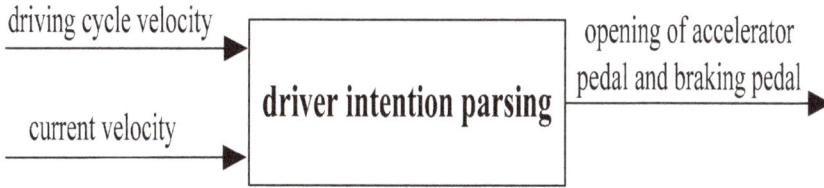

Figure 9. Driver model diagram.

3. Parameters to Be Optimized for HEV Logic Threshold Energy Management Strategy

3.1. Hybrid Electric Vehicle Energy Management Strategy

The all-wheel drive full-HEV has two power sources—the engine and power battery. According to the steady-state efficiency map diagram, the working efficiency differs at different working sections. Based on the power battery charging and discharging resistance characteristics, its internal resistance under various charged states is different, thus the working efficiency is also different. The energy management strategy based on the logic threshold aims to make the engine operate in the high efficiency range and keep the battery *SOC* within a specific range [1]. The working area of the engine is shown in Figure 10. The engine is set to work within the area between the upper limit and lower limit. The working area of motor is presented in Figure 11. The motor works when the battery *SOC* is between 0.3 and 0.8. When the *SOC* is high, the motor provides driving torque. On the contrary, the motor works as a generator. The parsing of the driver's intention and torque distribution are the main focus of the energy management strategy.

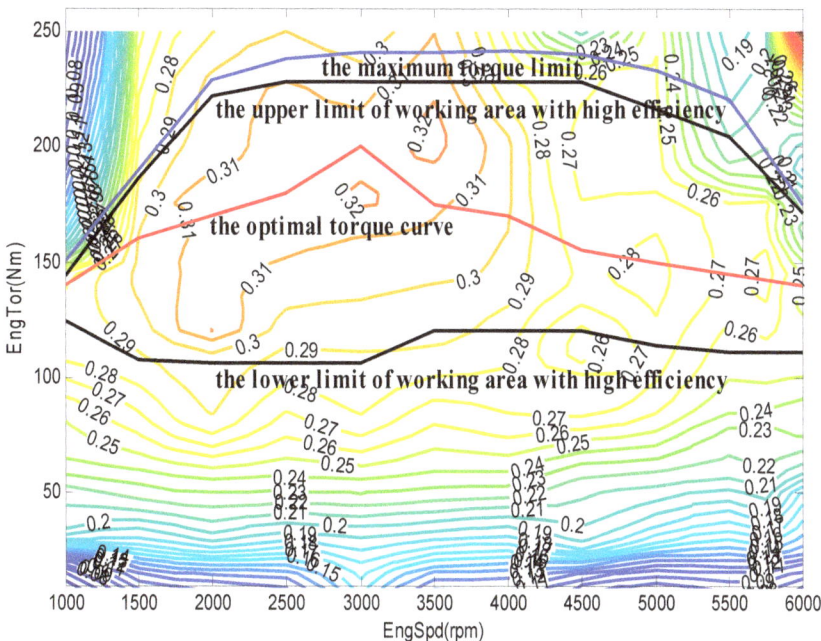

Figure 10. Engine working area division.

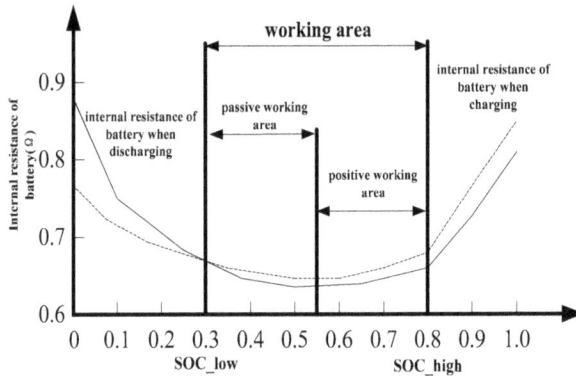

Figure 11. Working area division for nickel-metal hydride battery pack.

3.1.1. Parsing of Driver's Intention

In the process of driving, the driver shows his or her acceleration or deceleration demand by manipulating the accelerator pedal or braking pedal. Therefore, to parse the driver's demand, it is necessary to transfer the pedal signal into vehicle demand torques, which can be divided into driving demand torque and braking demand torque. In order to identify the demand torque, the pedal signal is normalized, and is expressed as α, within $(-1, 1)$. The positive value represents the accelerator pedal opening while the negative value is for the braking pedal opening. In order to maintain the driver's reaction when operating the pedal and obtain the same dynamic properties as the prototype, we consider the engine external characteristic curve as the driver's maximum demand torque curve for HEVs.

Therefore, when the driver controls the accelerator pedal, the demand driving torque can be represented as

$$T_{cmd} = \alpha T_{emax}, \tag{11}$$

where T_{cmd} is the demand driving torque, α is the accelerator pedal opening, and T_{emax} is the engine maximum torque.

When the driver controls the braking pedal, the demand braking torque can be represented as

$$T_{cmd} = \alpha(T_{MaxGen} + T_{MaxMechBrake}), \tag{12}$$

where T_{cmd} is the demand braking torque, α is the braking pedal opening, T_{MaxGen} is the maximum regenerative braking torque, and $T_{MaxMechBrake}$ refers to the highest mechanical braking torque.

For the all-wheel drive full-HEV, the energy management strategy distributes the driver's demand braking torque to the engine, ISG motor, in-wheel motor, and mechanical braking system.

The demand driving torque is as follows:

$$T_{cmd} = T_e + T_{ISG} + T_{hubmotor}/i_o i_g, \tag{13}$$

The demand braking torque is presented as follows:

$$T_{cmd} = T_{ISGGen}i_o i_g + T_{hubmotorGen} + T_{MechBrake}, \tag{14}$$

where T_e, T_{ISG}, and $T_{hubmotor}$ are the driving torques provided by the engine, ISG motor, and in-wheel motor, respectively; T_{ISGGen}, $T_{hubmotorGen}$, and $T_{MechBrake}$ refer to the braking torques provided by ISG motor, in-wheel motor, and mechanical braking system, respectively; i_o and i_g are the gear ratios of the main reducer and AMT gearbox, respectively.

3.1.2. Torque Distribution Strategy for Different Driving Modes

When the vehicle is running, the engine working points can be controlled to stay in the high efficiency area, and the battery can be controlled to work in the range within the low internal resistance by adjusting the output torques of the ISG and in-wheel motors. Many driving modes can be obtained by various combinations of the engine, ISG motor, and in-wheel motor.

(1) Pure electric driving mode

When starting, if the vehicle is at low load condition and the battery capacity is sufficient, the vehicle is driven only by the in-wheel motor, in order to avoid the engine from working in the low efficiency area. When the maximum speed of pure electric driving is exceeded, the vehicle switches into pure engine driving mode.

(2) Pure engine driving mode

When the vehicle works at medium load condition, the demand torque is provided by the engine only if it can work in the area of high efficiency. If the demand torque is greater than the maximum torque at the high efficiency area or the battery needs charging, the ISG motor or in-wheel motor starts to work, and the vehicle switches into hybrid driving mode.

(3) Hybrid driving mode

When the vehicle is at high load conditions such as climbing or accelerating, and the engine output maximum torque cannot meet the demand torque, the ISG and in-wheel motors provide power, if the battery capacity is sufficient. Then, the vehicle switches into hybrid driving mode. The hybrid driving mode can be subdivided into engine driving + ISG motor charging mode, engine driving + in-wheel motor driving mode, and engine driving + ISG motor driving + in-wheel motor driving mode.

The judgment condition and control logic for the different driving modes mentioned above can be displayed in Figure 12.

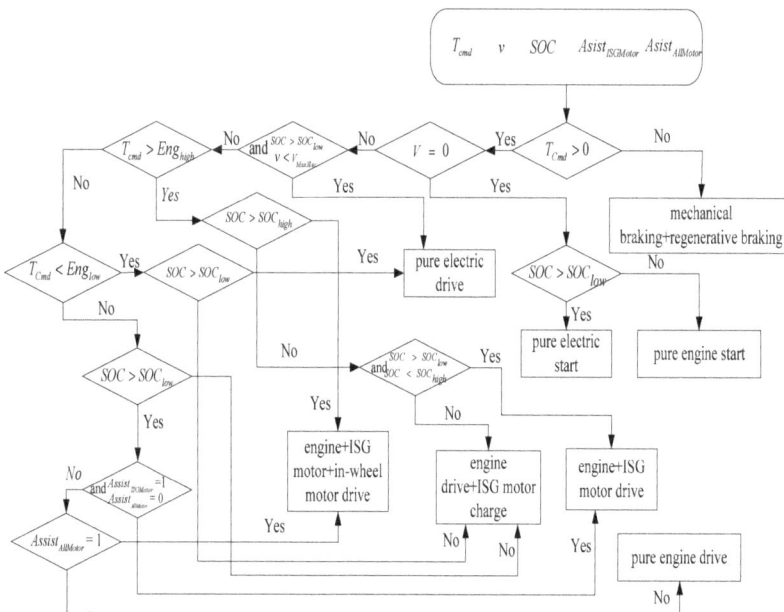

Figure 12. Control logic flow for various driving modes.

In the diagram, T_{cmd} is the driver demand torque; $Assist_{ISGMotor}$ is the power-assisting mark of ISG motor; $Assist_{AllMotor}$ is the power-assisting mark of all motors; v is the vehicle velocity; $v_{MaxElec}$ is the maximum speed of pure electric driving; SOC_{low} and SOC_{high} refer to the lower and upper limit of the battery working SOC, respectively; Eng_{low} and Eng_{high} refer to the lower and upper limit of the engine working area, respectively.

3.1.3. Torque Distribution Strategy in Braking Mode

For HEVs, there are two sets of braking system: the conventional mechanical braking system, and motor regenerative braking system. When braking, the mechanical braking system should coordinate with the motors to provide the demanded braking torque for braking safety, recover the excess kinetic energy, and improve the vehicle efficiency as well. The braking control strategy, which is shown in Figure 13, is formulated based on the battery SOC and the braking pedal opening.

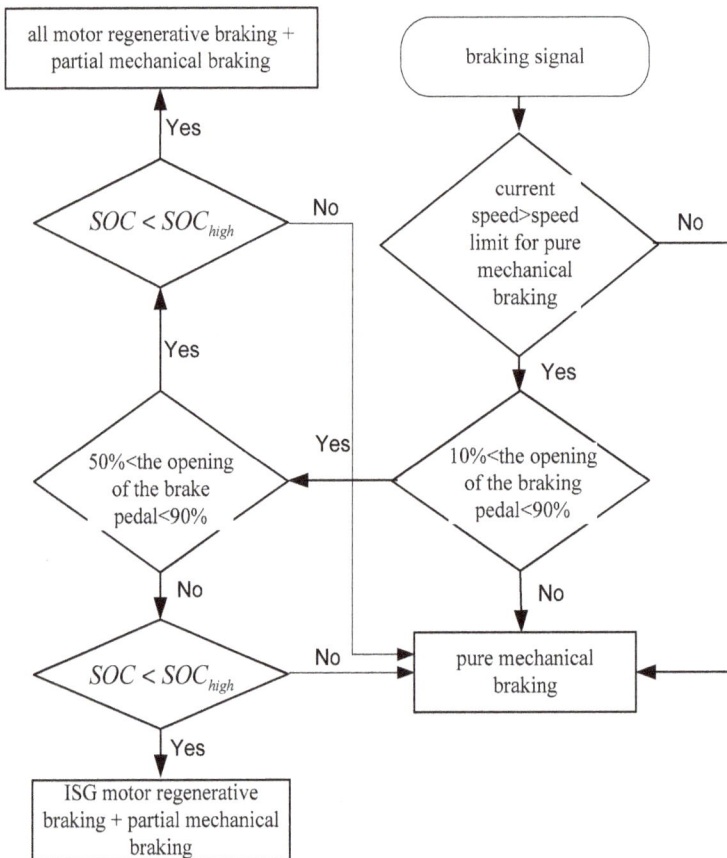

Figure 13. Control logic flow diagram of braking mode.

3.2. *Key Parameters of HEV Logic Threshold Energy Management Strategy*

According to the optimal working area division of the engine and motor, the driver's intention parsing, and the torque distribution strategy in different working modes, the energy management strategy based on a logic threshold is established in the MATLAB/Simulink. After studying the energy control strategy, the influencing parameters, which are presented in Table 9, can be obtained. These parameters, which have a direct effect on the vehicle performance especially the fuel economy, are the focus of the control strategy.

Table 9. Key parameters of logic threshold energy management strategy.

Name	Definition	Parameter	Simplified Version
Maximum torque coefficient of the engine high-efficiency area	The ratio of the maximum torque and engine peak torque for the engine high-efficiency area	Eng_{high}	x_1
Minimum torque coefficient of the engine high-efficiency area	The ratio of the minimum torque and engine peak torque for the engine high-efficiency area	Eng_{low}	x_2
Power battery *SOC* upper limit	The upper limit of the working *SOC* for power battery	SOC_{high}	x_3
Power battery *SOC* lower limit	The lower limit of the working *SOC* for power battery	SOC_{low}	x_4
Throttle opening threshold for ISG motor assisting	The minimum throttle opening when the ISG motor assists	$Assist_{ISGMotor}$	x_5
Throttle opening threshold for all motors assisting	The minimum throttle opening when all motors assist	$Assist_{AllMotor}$	x_6
Maximum speed in pure electric driving mode	The maximum speed in pure electric driving mode	$v_{MaxElec}$	x_7

4. Validation and Analysis of the Simulation Model of HEV and Energy Management Strategy

In order to verify the accuracy of the HEV simulation model and validity of the energy management strategy, a comparative analysis between the output data of the components from offline simulation and the real drum bench experiment was performed. It was ensured that the parameters of the offline simulation model were consistent with those of the real vehicle. In addition, the control strategy and parameters threshold were the same for both the simulation and real vehicle. The tests are carried out in the new European driving cycle (NEDC). The comparison between the offline simulation and drum test results during the driving cycle of NEDC are shown in Figures 14 and 15. As observed, there is a small difference between the output results of components from the simulation and bench test. The transient characteristic has not been accurately reflected in modeling, which contributed to the difference. However, the difference is within the acceptable range.

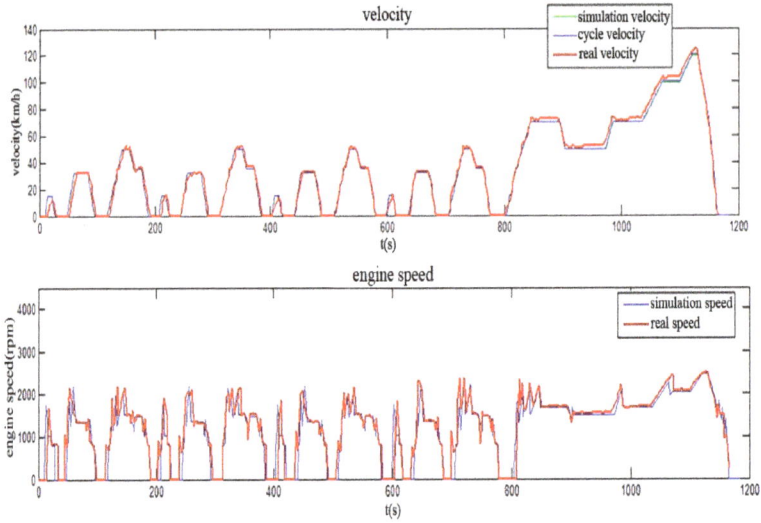

Figure 14. Driving cycle tracking and engine speed comparison of offline simulation and drum bench test in NEDC.

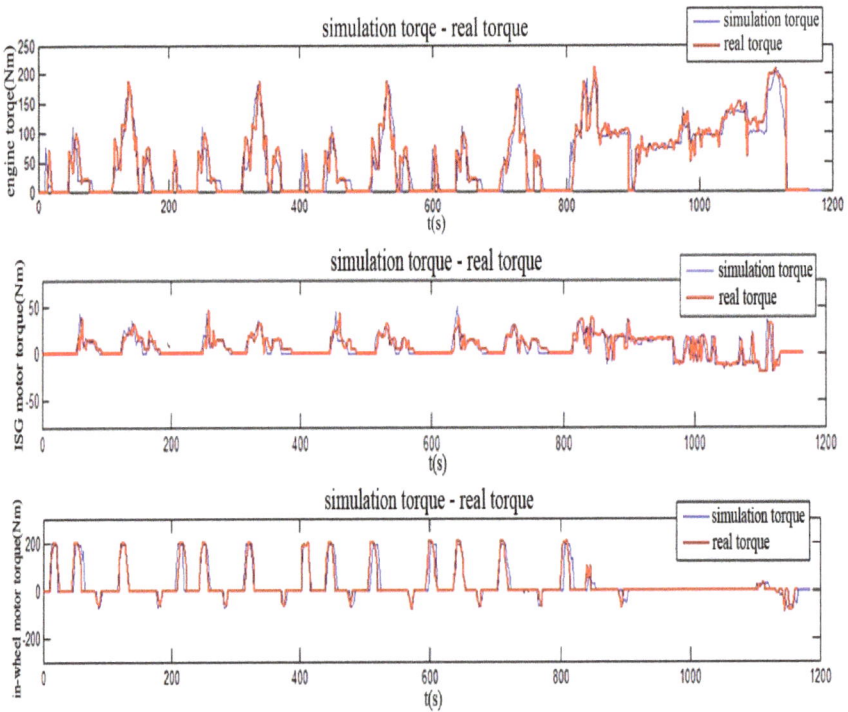

Figure 15. Output torques of components from offline simulation and drum bench test in NEDC.

This study focused on the HEV fuel economy. Table 10 shows the equivalent fuel consumption per 100 km and *SOC* values of the offline simulation and drum bench test. Comparing the values of the simulation and bench test, the error of the equivalent fuel consumption is 4.5%, which is within the generally allowed range of 5%. It is necessary to verify the accuracy of the vehicle simulation model and the effectiveness of the energy management strategy, which can provide assurance of the key parameters optimization for the HEV energy management strategy based on offline simulation.

Table 10. Comparison of results from offline simulation and drum bench test.

Parameter	Offline Result	Drum Bench Test Result	Error
SOC	0.7~0.52	0.7~0.56	—
Equivalent fuel consumption	7.69 L/100 km	7.36 L/100 km	4.5%

5. Optimization of HEV Control Parameters Using DIRECT Algorithm Based on Fuel Economy

5.1. Implementation of DIRECT Algorithm

The DIRECT algorithm is a deterministic global optimization algorithm proposed by Jones et al. in 1993 [19]. It is especially suitable for optimizing a multivariable function with specific variables and space [31]. Take the optimization problem with three-dimensional space as an example; it supposes that c is the center point of the hypercube and calculates the value of a function $f(x)$ at point $c \pm \delta e_i$ with δ equal to 1/3 of the hypercube length; e_i is a unit vector. The parameter, w_i is defined as follows:

$$w_i = \min(f(c + \delta e_i), f(c - \delta e_i)), \tag{15}$$

It splits the hypercube in the order of w_i. First, it cuts the hypercube in the direction perpendicular to the minimum value of w_i. Second, it cuts the hypercube in the direction perpendicular to weak minimal value of w_i. Then, it repeats the above steps until the hypercube is cut in all directions. Figure 16 shows an example of the hypercube division.

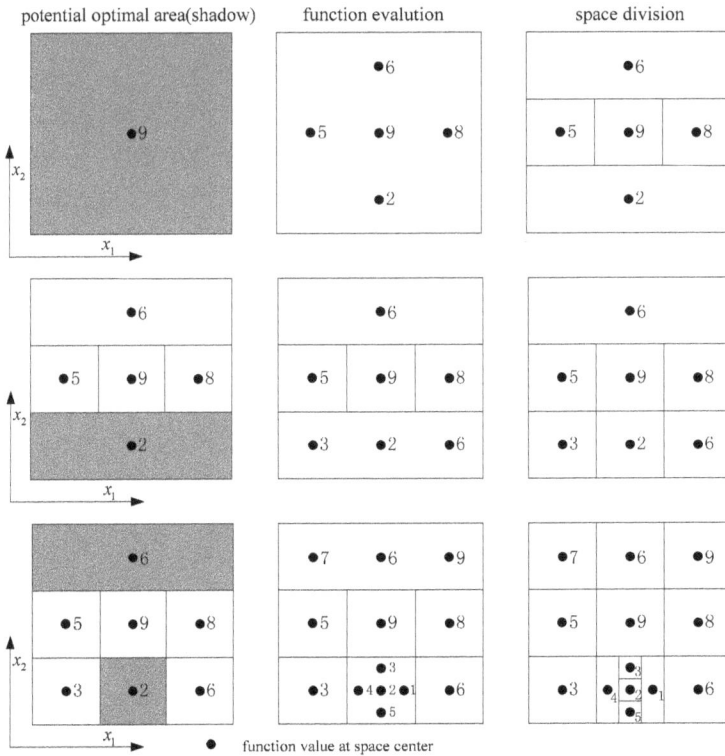

Figure 16. Division and selection of potential optimal hyper-rectangle in DIRECT algorithm.

It is supposed that $\omega_1 = \min(5,8) = 5$, and $\omega_2 = \min(6,2) = 2$. First, the hypercube is cut in the direction perpendicular to x_2. Then, the hypercube is cut in the direction perpendicular to x_1. The method of cutting the longest side of the hypercube can ensure decreasing the length of the longest side. In DIRECT algorithm, the side length of the hypercube is at most two values after each division. The hypercube, with a function value at the center point equal to 2 in Figure 16, is the potential optimal hyper-rectangle after two divisions.

Based on the above result, Figure 17 can be obtained by setting the function value at the center point as y-axis, and the distance between the center point and vertex as x-axis. In the figure, (1)–(3) are the potential optimal hyper-rectangle from the first to third iteration, respectively. In the first iteration, there is only one list of selectable points. The hypercube, with a function value at the center point equal to 2, is the potential optimal hyper-rectangle. We need to divide the cube further. Similarly, in the second iteration, we can get two hyper-cubes with the smallest value, to be divided in the third iteration. The process continues until it satisfies the stopping condition.

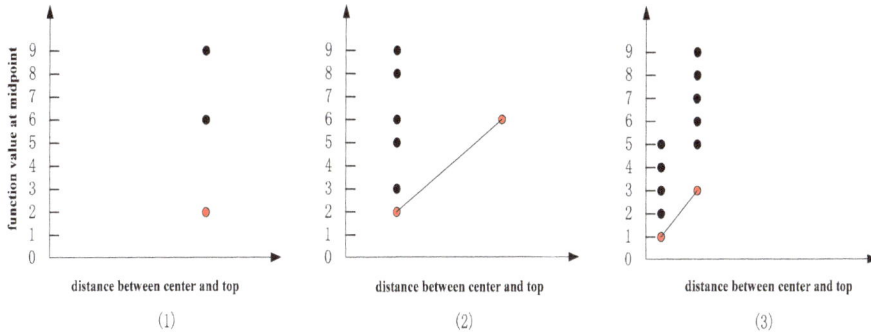

Figure 17. The potential optimal hyper-rectangle of each iteration.

For multidimensional space optimization problems, the DIRECT algorithm takes similar steps to select the best potential optimal hyper-rectangle.

5.2. Optimization of Key Parameters for Logic Threshold Energy Management Strategy Using DIRECT Algorithm Based on Fuel Economy

Based on the discussions in the third section, the key parameters of the logic threshold energy management strategy for the HEV are presented in Table 3. In this research, the purpose of the energy management strategy is to achieve the best fuel economy for a given driving cycle. Therefore, the target function is

$$FC = \min f(x), \tag{16}$$

where $f(x)$ is the equivalent fuel consumption per 100 km, which includes the engine fuel consumption and equivalent fuel consumption of the electric energy from the power battery. The unit is L/100 km. The calculation for $f(x)$ is shown as below.

$$f(x) = 100\frac{\frac{\int k_1 U I dt}{q}}{\rho \int v dt} + 100\frac{\int k_2 f_r(T_e, \omega_e)\frac{T_e \omega_e}{9550} dt}{\rho \int v dt} n_e \tag{17}$$

where ρ is the gasoline density in g/L; $f_r(T_e, \omega_e)$ is the current engine fuel consumption rate, which is a lookup function of the engine torque and speed, with the unit g/kWh; T_e and ω_e are the current engine torque and speed, with the units N·m and rpm, respectively; k_1 and k_2 are the gasoline–electric conversion constant coefficients; U and I are the present battery voltage and current, with the units V and A, respectively; q refers to the gasoline calorific value in J/kg; v is the current speed in km/h.

The engine torque and speed, battery voltage and current, and average speed are related to the seven parameters to be optimized as shown in Table 3.

Therefore, the optimization of key parameters for the HEV energy management strategy is converted to the optimization of seven dimensional parameters. The DIRECT algorithm is selected to solve this problem. The process is shown in Figure 18. First, we normalized n-dimensional space into n-dimensional unit hyper-cube and calculate the equivalent fuel consumption per 100 km at the center point as the initial minimum fuel consumption. The hyper-cube is the potential optimal hyper-rectangle when iteration starts. Then, we choose a potential optimal hyper-rectangle and divide it. Afterwards, we calculate the equivalent fuel consumption per 100 km at the center point of each rectangle. After that, we compare it with the minimal value collected in the last iteration. If this value is smaller than the previous minimum fuel consumption, we update and store the minimum fuel consumption. In addition, we update the potential optimal hyper-rectangle. The optimization of DIRECT algorithm will stop until the defined maximum number of iterations or the potential optimal hyper-rectangle is empty.

Energies **2016**, *9*, 997

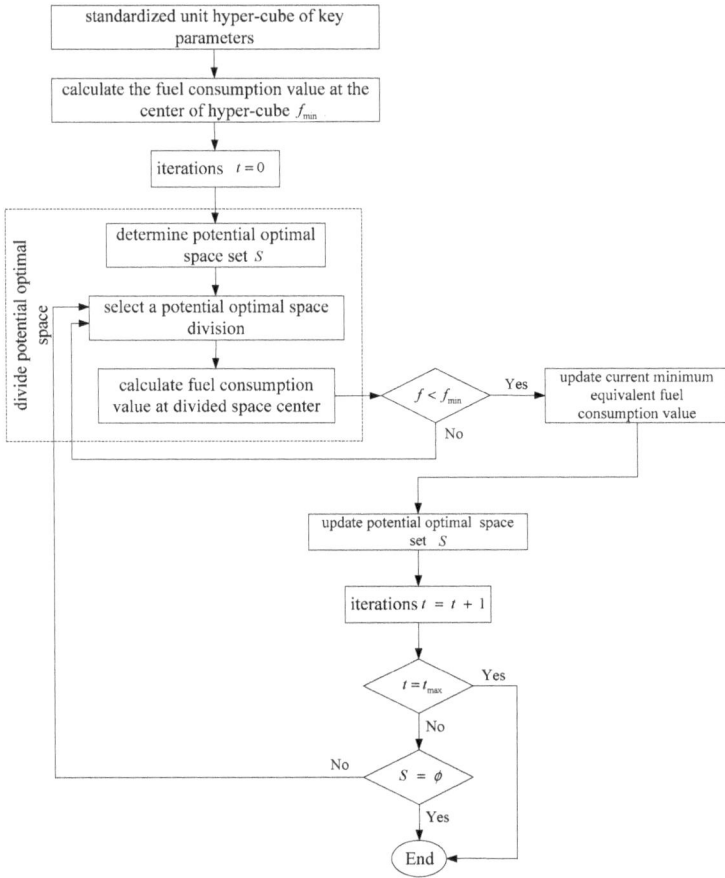

Figure 18. Optimization procedure of key parameters for HEV energy management using DIRECT algorithm.

5.3. Optimization Result and Analysis

Figure 19 shows the optimization model for the HEV energy management key parameters, which includes the previously established HEV closed-loop simulation model, target to minimize the fuel consumption per 100 km, and code of the DIRECT algorithm. The optimization model is established in MATLAB/Simulink. Besides, the constraint of vehicle's dynamic performance should also be taken into account. The 0 to 60 time for the all-wheel drive full-HEV studied should not be more than 10 s. The driving cycle of NEDC is selected, and the initial *SOC* is set to 55%. The key parameters of the DIRECT algorithm are set as shown in Table 11.

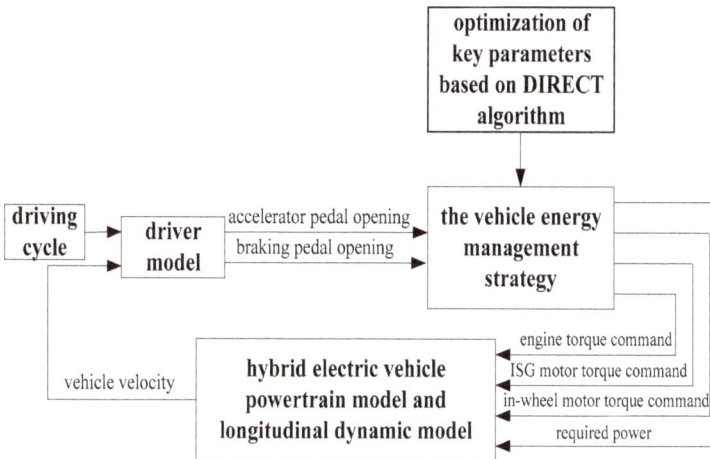

Figure 19. Optimization model for the HEV energy management key parameters.

Table 11. Key parameters of DIRECT algorithm.

Key Parameters	Value
Maximum number of iterations	20
Maximum number of function calculation	1000
Maximum division time per side of hyper-rectangle	100
Global/local weighting coefficient	0.0001
Relative error	0.01%

x_1, x_2, x_3, x_4, x_5, x_6, x_7 are the parameters that need to be optimized. The meaning of these parameters are described in Table 3. The initial value, range, and optimized value of these parameters are shown in Table 12.

Table 12. Optimized result of key parameters based on DIRECT algorithm.

Parameters	Initial Value	Lower Limit	Upper Limit	Optimized Value
x_1	0.4	0.2	0.6	0.3889
x_2	0.3	0.1	0.5	0.3556
x_3	70	50	70	53.333
x_4	40	30	50	47.963
x_5	20	15	25	22.222
x_6	50	40	60	50.667
x_7	15	10	20	18.333

Changes to the parameters to be optimized have a big impact on the fuel economy. In the optimization process of DIRECT algorithm, the equivalent fuel consumption per 100 km for different iteration function evaluations is shown in Figures 20 and 21, respectively.

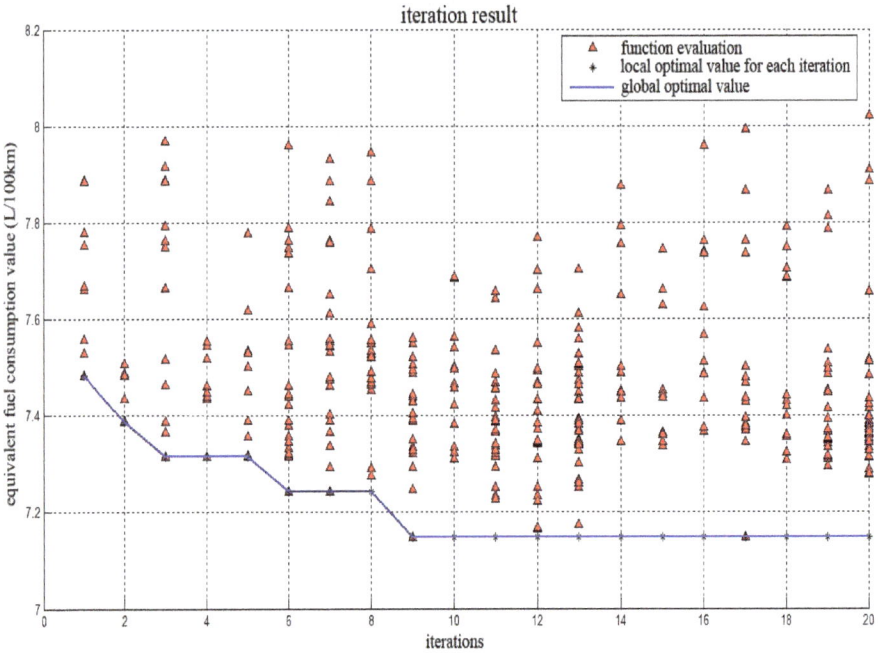

Figure 20. Equivalent fuel consumption of different iterations using DIRECT algorithm.

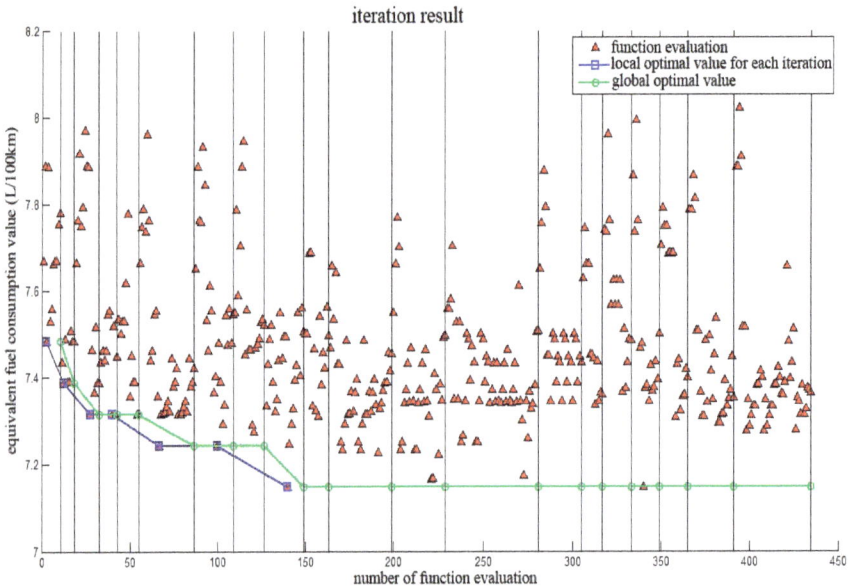

Figure 21. Equivalent fuel consumption per 100 km with different function evaluations using DIRECT algorithm.

As can be observed from the beginning, the equivalent fuel consumption per 100 km decreases rapidly with the increase of iterations and function evaluations in the driving cycle of NEDC. Then, it stabilizes after 9 iterations or 150 function evaluations. This stabilized value is the minimum equivalent fuel consumption per 100 km. The corresponding parameters are the globally optimized results using the DIRECT algorithm, which meet the minimum equivalent fuel consumption. It shows that the proposed optimization method using the DIRECT algorithm has good convergence and effectiveness to optimize the key parameters of energy management strategy from the perspective of fuel economy.

The simulation of the all-wheel-drive HEV is implemented in the driving cycle of NEDC, utilizing the optimized logic threshold control strategy parameters. The simulation results are shown in Figure 22. As can be seen, frequently gear change is avoided. And in the driving condition, the in-wheel motors output more torques than engine and ISG motor. During the braking cycle, both in-wheel motors and ISG motor work as generators to charge the battery. And the battery *SOC* changes from the initial value 0.55 to the terminal value 0.7. The instantaneous fuel consumption is controlled within low range in most of the time.

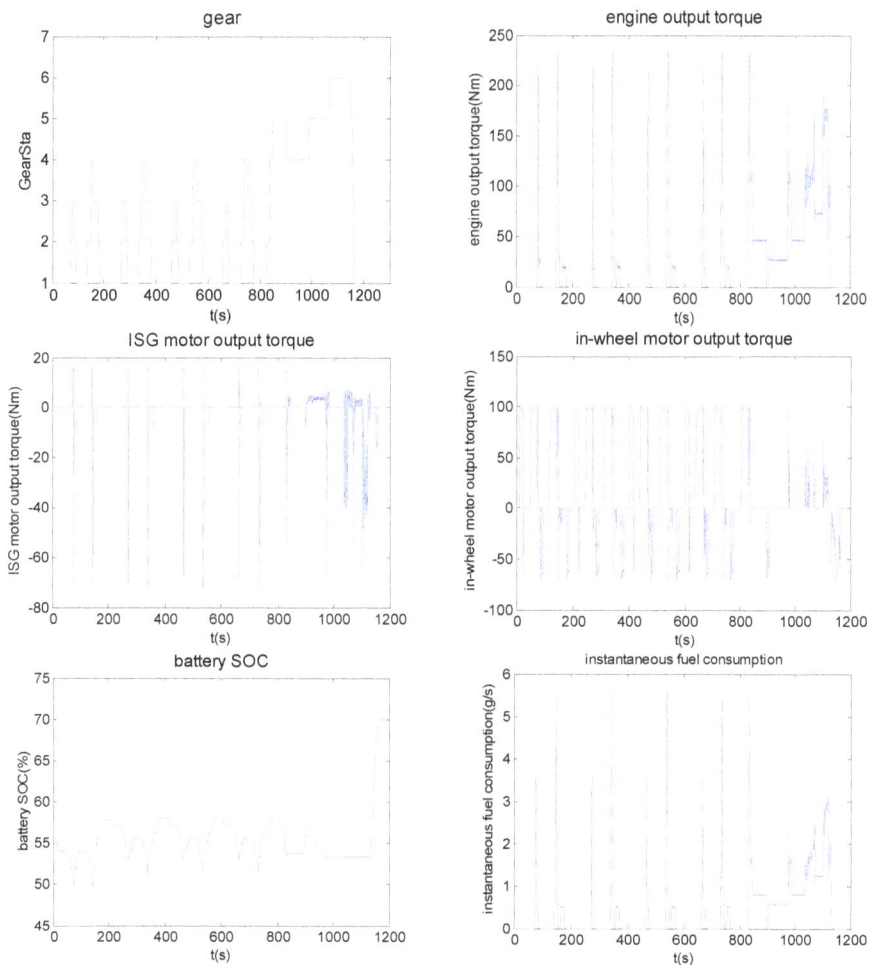

Figure 22. Simulation results of all-wheel drive HEV model utilizing the optimized parameters.

303

In the driving cycle of NEDC, the equivalent fuel consumption per 100 km decreases from 7.691 L/ 100 km using the initial parameters to 7.148 L/100 km utilizing the optimized parameters. Therefore, the equivalent fuel consumption using the optimized results decreases by 7.06% compared to the previous fuel consumption. Figure 23 shows the engine operating points before and after optimization.

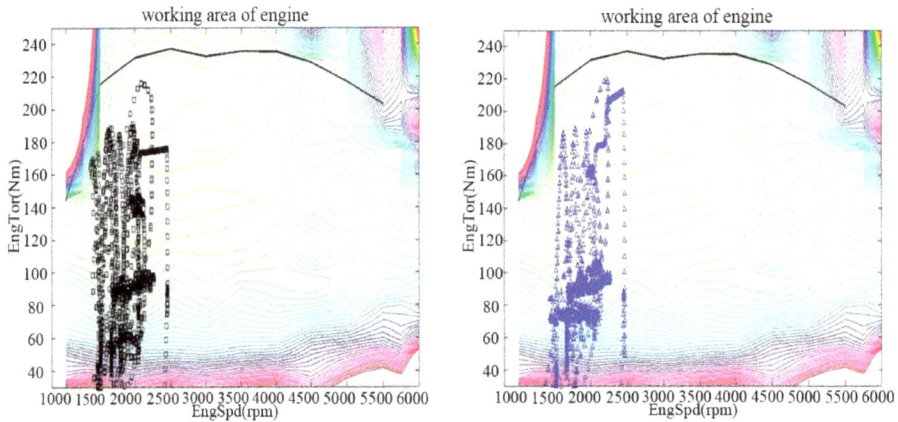

Figure 23. Engine working points distribution before and after optimization.

As shown in Figure 23, the engine operating points scattering in the area with low torque (0–60 N·m) and high fuel consumption have decreased. Furthermore, the engine operating points scattering in the zone with low speed (1000–1500 rpm) and high fuel consumption have also decreased. Hence, the energy management strategy using the optimized parameters makes the engine operate more in the area with high torque and low fuel consumption.

By comparing the equivalent fuel consumption per 100 km and the distribution of engine operating points before and after optimization, it can be concluded that the DIRECT algorithm can be applied to optimize the key parameters of the energy management strategy for the HEV with a positive effect. The optimized results obtained by the offline simulation can provide a reference for debugging the real vehicle.

6. Conclusions

(1) In this study, the closed-loop simulation model of the all-wheel–drive HEV powertrain is built in Matlab/Simulink with the power component model established based on the experimental test. The logic threshold energy management strategy is comprehensively analyzed and formulated. On this basis, the seven key parameters that influence the fuel economy of the HEV, which need to be optimized, are extracted. The accuracy of the simulation model and validity of the proposed logic threshold energy management strategy are verified by comparing the simulation test and real drum bench experiment.

(2) The optimization model of the key parameters based on the fuel economy is proposed. The implementation of the DIRECT algorithm is analyzed. Then, it is applied to solve this nonlinear multiparameter optimization problem with the objective of minimizing the equivalent fuel consumption.

(3) The optimization result shows that the logic threshold energy management strategy using the optimized parameters reduces the equivalent fuel consumption per 100 km by 7% and makes engine operate more in the high efficiency area. The simulation result validates the effectiveness of the DIRECT algorithm in solving the multiparameter energy consumption optimization problem. It will play a guiding role in calibrating the control strategy parameters for a real vehicle. Next,

we will verify the optimization method of the key parameters for the HEV energy management strategy based on logical threshold by testing a real vehicle.

Acknowledgments: This work was financially supported by China Automobile Industry Innovation and Development joint fund (Project No. U1564208), National Natural Science Foundation of China (Project No. 51275355) and National Key Scientific Instrument and Equipment Development Project (No. 2012YQ15025603).

Author Contributions: Peihong Shen and Xiaowen Zhan built the simulation model of HEV. Jingxian Hao and Zhuoping Yu formulated the rule-based logic threshold control strategy and analyzed its key parameters. Zhiguo Zhao established the optimization model of key parameters using DIRECT algorithm. All authors carried out data analysis. Jingxian Hao wrote the paper.

Conflicts of Interest: The authors declare no conflict of interest.

References

1. Peng, J.; Fan, H.; He, H.; Pan, D. A Rule-Based Energy Management Strategy for a Plug-in Hybrid School Bus Based on a Controller Area Network Bus. *Energies* **2015**, *8*, 5122–5142. [CrossRef]
2. Wirasingha, S.; Emadi, A. Classification and Review of Control Strategies for Plug-in Hybrid Electric Vehicles. *IEEE Trans. Veh. Technol.* **2011**, *60*, 907–914. [CrossRef]
3. Panday, A.; Bansal, H.O. Energy Management Strategy for Hybrid Electric Vehicles Using Genetic Algorithm. *J. Renew. Sustain. Energy* **2016**, *8*, 741–746. [CrossRef]
4. Jun, W.; Wang, Q.; Wang, P.; Han, B. The Optimization of Control Parameters for Hybrid Electric Vehicles Based on Genetic Algorithm. In Proceedings of the SAE World Congress, Detroit, MI, USA, 8–10 April 2014.
5. Varesi, K.; Radan, A. A Novel GA Based Technique for Optimizing Both the Design and Control Parameters in Parallel Passenger Hybrid Cars. *Int. Rev. Electr. Eng.* **2011**, *63*, 1279–1286.
6. Montazeri-Gh, M.; Poursamad, A.; Ghalichi, B. Application of Genetic Algorithm for Optimization of Control Strategy in Parallel Hybrid Electric Vehicles. *J. Frankl. Inst.* **2006**, *343*, 420–435. [CrossRef]
7. Fang, L.; Qin, S.; Xu, G.; Li, T.; Zhu, K. Simultaneous Optimization for Hybrid Electric Vehicle Parameters Based on Multi-Objective Genetic Algorithms. *Energies* **2011**, *4*, 532–544. [CrossRef]
8. Huang, B.; Wang, Z.C.; Xu, Y.S. Multi-Objective Genetic Algorithm for Hybrid Electric Vehicle Parameter Optimization. In Proceedings of the IEEE/RSJ International Conference on Intelligent Robots & Systems, Beijing, China, 9–15 October 2006.
9. Li, Y.; Lu, X.; Kar, N. Rule-Based Control Strategy with Novel Parameters Optimization Using NSGA-II for Power-Split PHEV Operation Cost Minimization. *IEEE Trans. Veh. Technol.* **2014**, *63*, 3051–3061. [CrossRef]
10. Li, L.; Zhang, Y.H.; Chao, Y.G.; Jiao, X.H.; Zhang, L.P.; Song, J. Hybrid Genetic Algorithm-based Optimization of Powertrain and Control Parameters of Plug-in Hybrid Electric Bus. *J. Frankl. Inst.* **2014**, *352*, 776–801. [CrossRef]
11. Chen, Z.; Xiong, R.; Wang, K.; Jiao, B. Optimal Energy Management Strategy of a Plug-in Hybrid Electric Vehicle Based on a Particle Swarm Optimization Algorithm. *Energies* **2015**, *8*, 3661–3678. [CrossRef]
12. Xiong, R.; He, H.W.; Sun, F.C. Methodology for Optimal Sizing of Hybrid Power System Using Particle Swarm Optimization and Dynamic Programming. *Energy Procedia* **2015**, *75*, 1895–1900. [CrossRef]
13. Wu, J.; Zhang, C.H.; Cui, N.X. PSO Algorithm-based Parameter Optimization for HEV Powertrain and Its Control Strategy. *Int. J. Automot. Technol.* **2008**, *91*, 53–59. [CrossRef]
14. Deng, Y.W.; Chen, K.L. Simulated Annealing Particle Swarm Algorithm Based Parameters Optimization for Hybrid Electric Vehicles. *Aut. Eng.* **2012**, *34*, 580–584.
15. Wang, Q.; Frank, A.A. Plug-in HEV with CVT: Configuration, Control, and Its Concurrent Multi-objective Optimization by Evolutionary Algorithm. *Int. J. Automot. Technol.* **2014**, *15*, 103–115. [CrossRef]
16. Zhang, L.P.; Lin, C.; Niu, X. Optimization of Control Strategy for Plug-in Hybrid Electric Vehicle Based on Differential Evolution Algorithm. In Proceedings of the Asia-Pacific Power and Energy Engineering Conference, Wuhan, China, 28–31 May 2009.
17. Long, V.T.; Nhan, N.V. Bees-algorithm-based Optimization of Component Size and Control Strategy Parameters for Parallel Hybrid Electric Vehicles. *Int. J. Automot. Technol.* **2012**, *13*, 1177–1183. [CrossRef]
18. Chris, M.; Abul, M.; David, G. *Hybrid Electric Vehicles: Principles and Applications with Practical Perspectives*; John Wiley & Sons: Hoboken, NJ, USA, 2011.

19. Jones, D.R.; Perttunen, C.D.; Stuckman, B.E. Lipschitz Optimization without Lipschitz Constant. *J. Optim. Theory Appl.* **1993**, *79*, 157–181. [CrossRef]
20. Finkel, D.E.; Kelley, C.T. *Convergence Analysis of the DIRECT Algorithm*; North Carolina State University Center for Research in Scientific Computation Technology: Raleigh, NC, USA, 2010.
21. Gu, L.A. Comparison of Polynomial Based Regression Models in Vehicle Safety Analysis. In Proceedings of the 2001 ASME Design Engineering Technical Conferences, Pittsburgh, PA, USA, 9–12 September 2001.
22. Rousseau, A.; Pagerit, S.; Gao, D. Plug-in Hybrid Electric Vehicle Control Strategy Parameter Optimization. *JAEV* **2007**, *6*, 1125–1133. [CrossRef]
23. He, H.W.; Huang, X.G.; Zhang, Y.M.; Zhang, X.W. DIRECT Algorithm Based Parameters Optimization Research for Hybrid Electric Vehicles. *Aut. Eng.* **2011**, *1*, 35–41.
24. Panday, A.; Bansal, H.O. Fuel Efficiency Optimization of Input-split Hybrid Electric Vehicle Using DIRECT Algorithm. In Proceedings of the IEEE International Conference on Industrial and Information System, Gwalior, India, 15–17 December 2014.
25. Zhao, Z.G.; Chen, H.J.; Yang, J.; Wu, X.W.; Yu, Z.P. Estimation of the Vehicle Speed in the Driving Mode for a Hybrid Electric Car Based on an Unscented Kalman Filter. *Proc. Inst. Mech. Eng. Part D J. Automob. Eng.* **2014**, *229*, 437–456. [CrossRef]
26. Zhao, Z.G.; He, N.; Zhu, Y.; Yu, Z.P. Mode Transition Switch Control for Four-wheel Drive Hybrid Electric Vehicle. *J. Mech. Eng.* **2011**, *47*, 100–109.
27. Chen, Q.S.; Lin, C.T. Review for Battery Performance Model Research of Electric Vehicles. *Aut. Technol.* **2005**, *1*, 1–5.
28. Johnson, V.H. Battery Performance Models in ADVISOR. *J. Power Sources* **2002**, *110*, 321–329. [CrossRef]
29. Yu, Z.S.; Xia, Q.S. *Theories of Automotive*; China Machine Press: Beijing, China, 2000. (In Chinese)
30. Shen, P.H.; Sun, Z.C.; Zeng, Y.J.; Wang, X.J.; Dai, H.F. Study on Power Ratio Between the Front Motor and Rear Motor of Distributed Drive Electric Vehicle Based on Energy Efficiency Optimization. In Proceedings of the 2016 SAE World Congress and Exhibition, Detroit, MI, USA, 4–6 April 2016.
31. Steven, E. *Multifidelity Global Design Optimization including Parallelization Potential*; University of Florid: Gainesville, FL, USA, 2002.

energies

MDPI

Article

Design and Control of a 3 kW Wireless Power Transfer System for Electric Vehicles

Zhenshi Wang [1,2], **Xuezhe Wei** [2,*] **and Haifeng Dai** [1,2]

[1] Clean Energy Automotive Engineering Center, Tongji University, No. 4800, Caoan Road, Shanghai 201804, China; 1022wangzhenshi@tongji.edu.cn
[2] School of Automotive Studies, Tongji University, No. 4800, Caoan Road, Shanghai 201804, China; tongjidai@tongji.edu.cn
* Correspondence: weixzh@tongji.edu.cn; Tel.: +86-135-0184-8129

Academic Editor: K. T. Chau
Received: 13 November 2015; Accepted: 17 December 2015; Published: 24 December 2015

Abstract: This paper aims to study a 3 kW wireless power transfer system for electric vehicles. First, the LCL-LCL topology and LC-LC series topology are analyzed, and their transfer efficiencies under the same transfer power are compared. The LC-LC series topology is validated to be more efficient than the LCL-LCL topology and thus is more suitable for the system design. Then a novel q-Zsource-based online power regulation method which employs a unique impedance network (two pairs of inductors and capacitors) to couple the cascaded H Bridge to the power source is proposed. By controlling the shoot-through state of the H Bridge, the charging current can be adjusted, and hence, transfer power. Finally, a prototype is implemented, which can transfer 3 kW wirelessly with ~95% efficiency over a 20 cm transfer distance.

Keywords: wireless power transfer (WPT); topology analysis; power regulation; electric vehicle

1. Introduction

Research on wireless power transfer began soon thereafter the famous Tesla coils were invented by Nikola Tesla in 1889 [1,2], and many good results have been achieved [3–6]. In 2007, researchers at MIT proposed strongly coupled magnetic resonances (SCMR), by which they were able to transfer 60 watts wirelessly with ~40% efficiency over distances in excess of 2 m [7]. Various research hot spots, including system architectures, optimization design, frequency splitting, impedance matching and special applications, have been investigated [8–14]. Wireless power transfer is very suitable for charging electric vehicles [15–17], as it can avoid the troublesome plug-in process, provide an inherent electrical isolation and adapt to harsh environments. However, SCMR is not appropriate for automotive applications, as its operating frequency is very high, which goes beyond the limitation of SAE J2954 (work in progress). As another kind of wireless power transfer techniques, inductive power transfer (IPT) has developed for more than twenty years [5], and it mainly focuses on the high power level applications, where the issues of concern normally include power conversion and control [18,19], magnetic structure design [20], control algorithm and strategy [21,22] as well as circuit topology [23]. Basically, both SCMR and IPT conforms to Faraday's and Ampere's laws, and their differences primarily include the design approaches, system architectures, parameter selection and transfer characteristics [6,24].

This paper aims to study a 3 kW vehicle-mounted wireless power transfer system, on which two key parts, the resonant topology analysis and comparison, and the online power regulation, are elaborated. Many resonant topologies are available for wireless power transfer system, but the most basic ones are only series-series, series-parallel, parallel-series and parallel-parallel [23], and the others are all derived from these ones. A wireless power transfer system for electric vehicles requires a resonant topology that should have a unity-power-factor and a current source characteristic [18], no matter whether the distance or angle between the chassis and the ground changes or not. In this paper the characteristics of the LC-LC series and LCL-LCL topologies are analyzed first, and we prove that they both have the required unity-power-factor and current source characteristics. Then the transfer efficiencies of the LC-LC series and LCL-LCL topologies are compared under the same transfer power conditions, and the comparison results validate that the LC-LC series topology is more suitable for the system design due to its higher transfer efficiency. In practice, the distance or angle between vehicle chassis and ground often changes [25], as drivers cannot park their cars at the same location every time, and naturally, online power regulation is indispensable in the battery charging process. Traditional power regulation methods include cascade DC/DC, dead time modulation and phase shifting control [26]. However, the cascade DC/DC may increase the number of devices, decrease the power density and lower the transfer efficiency, while dead time modulation may distort the output waves produced by the H Bridge, and the phase shifting control cannot boost the transfer power, so a novel q-Zsource-based power regulation method is proposed in this paper, which employs two pairs of inductors and capacitors as a unique boost network. The power regulation is realized by controlling the shoot-through state of the H Bridge, and there are no extra power switch devices. By combining the phase shifting control and shoot-through state control, the square-wave voltage produced by the H Bridge can be adjusted arbitrarily, and hence, the transfer power.

2. Comparative Analysis of Resonant Topologies

2.1. LC-LC Series Topology

Wireless power transfer systems normally consist of a power source, a H Bridge, a resonant topology, a rectifier as well as a load. With the classical frequency-domain analysis, we can easily get the amplitude-frequency characteristics of resonant topologies, which are steep spikes, and the maximal point just emerges at the resonant frequency. For example, the LC-LC series topology has a four order transfer function, and the LCL-LCL topology has a six order transfer function. These two resonant topologies only allow the fundamental components of the square-wave voltages produced by the H Bridge to pass through, thus the resonant topology input can be substituted by a quasi-sine voltage source. As the power batteries have strong voltage source characteristics, the rectifier and battery pack can be also simplified into a quasi-sine voltage source, the amplitude of which depends on the battery pack voltage multiplied by $4/\pi$. The simplified LC-LC topology [27,28] is shown in Figure 1.

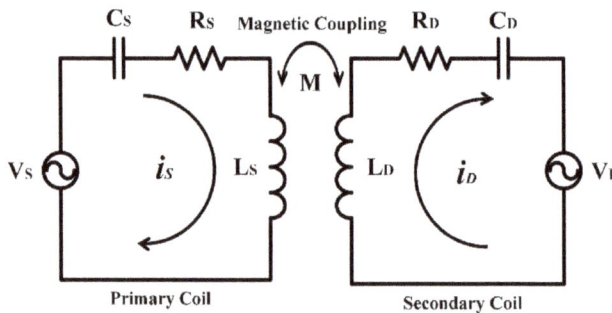

Figure 1. LC-LC series topology.

In Figure 1, L_S and L_D represent the magnetic coils, C_S and C_D represent the resonant capacitors, R_S and R_D represent the parasitic resistances, and M is the mutual inductance between L_S and L_D. When the system works in the resonant state, one has:

$$\begin{cases} V_S = i_S R_S - j\omega_r M i_D \\ j\omega_r M i_S = i_D R_D + jV_B \end{cases} \tag{1}$$

where i_S and i_D are the resonant currents in the primary and secondary coils, and ω_r is the system resonant frequency. By solving Equation (1), we have:

$$\begin{cases} |i_S| = \dfrac{V_S + \dfrac{\omega_r M V_B}{R_D}}{R_S + \dfrac{(\omega_r M)^2}{R_D}} \\ |i_D| = \dfrac{\dfrac{\omega_r M V_S}{R_S} - V_B}{R_D + \dfrac{(\omega_r M)^2}{R_S}} \end{cases} \tag{2}$$

Equation (2) shows that the charging current i_D depends on ω_r, M, R_S, R_D, V_S and V_B. Normally, V_B may change while charging, as it increases with the state of charge (SOC) of the battery. Because $\omega_r = 2\pi f_r$ is up to 100 kHz, M is from 20 μH to 100 μH, R_S is in the milliohm level, and V_S is usually larger than V_B, we have:

$$\frac{\omega_r M V_S}{R_S} \gg V_B \tag{3}$$

By substituting Equation (3) into Equation (2), we can find that i_D remains unchanged during the whole charging process, which realizes a constant-current charging function. As the product of R_S and R_D is smaller than either of them, Equation (2) is further simplified into:

$$\begin{cases} |i_S| = \dfrac{V_B}{\omega_r M} \\ |i_D| = \dfrac{V_S}{\omega_r M} \end{cases} \tag{4}$$

It is worth mentioning that the current source characteristic of LC-LC series topology is tenable only when the charging objects are batteries or some other capacitive load. Based on Equation (2), the transfer power can be expressed as:

$$P_{LC-LC} = |i_D| \cdot V_B = \frac{\omega_r M V_S - V_B R_S}{R_D R_S + \omega_r^2 M^2} \cdot V_B \tag{5}$$

2.2. LCL-LCL Topology

Similarly, the simplified LCL-LCL topology [18,29–31] is shown in Figure 2.

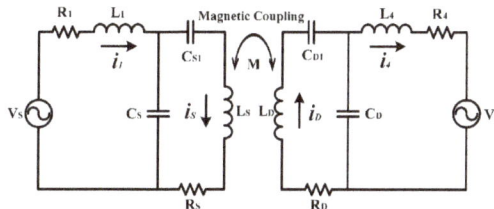

Figure 2. LCL-LCL topology.

In this figure, L_1 and L_4 are the matching inductors of L_S and L_D, R_1 and R_4 are their parasitic resistances. For high power applications, L_S and L_D are normally bulky, which make L_1 and L_4 bulky, and this is not beneficial for the objectives of miniaturization and lightness. Thus the compensating capacitors C_{S1} and C_{D1} are introduced to decrease L_1 and L_4. Still, they should satisfy the following equations:

$$C_{S1} = \frac{1}{\omega_r^2(L_S - L_1)}, \quad C_{D1} = \frac{1}{\omega_r^2(L_D - L_4)} \tag{6}$$

$$\omega_r = \frac{1}{\sqrt{L_1 C_S}} = \frac{1}{\sqrt{L_4 C_D}} \tag{7}$$

Actually, LCL is a transformation of the LC parallel topology [18]. It is well-known that the reflected impedance of the LC parallel topology contains an imaginary part [23], especially when the mutual inductance and the load change online, which makes the tuning process very cumbersome. The additional inductor of the LCL topology can just eliminate this imaginary part whether the mutual inductance and the load change or not. Using a method similar to that in Section 2.1, we can conclude that the LCL-LCL topology also has constant-current charging characteristics. The parasitic resistances are usually small due to the use of Litz wires, thus neglecting the parasitic resistances will not affect the system efficiency sharply, and in practice that loss is very small compared with the loss caused by the H Bridge and rectifier, so we can get the simplified calculation formulas of the LCL-LCL topology as follows:

$$i_S = \frac{V_S}{j\omega_r L_1}, \quad i_4 = j\frac{MV_S}{\omega_r L_1 L_4}, \quad i_D = \frac{V_B}{j\omega_r L_4}, \quad i_1 = j\frac{MV_B}{\omega_r L_1 L_4} \tag{8}$$

The transfer power can be written as:

$$P_{LCL-LCL} = \frac{MV_S V_B}{2\omega_r L_1 L_4} \tag{9}$$

Equation (9) shows that the charging power can be adjusted by V_S. Unlike Equation (5), there are two additional power regulation freedoms L_1 and L_4.

2.3. Comparison between the LC-LC Series Topology and LCL-LCL Topology

The LC-LC series topology and LCL-LCL topology are widely used in practice, as both can realize the constant-current charging characteristics, the unity-power-factor characteristics and even bidirectional power transfer characteristics. Their transfer power characteristics are however different, for instance, the transfer power of the LC-LC series topology increases with the increasing transfer distances according to Equation (5), and the transfer power of the LCL-LCL topology decreases with the increasing transfer distance according to Equation (9). However, their transfer efficiency characteristics have not been compared before, thus this section aims to compare them to provide some suggestions for practical engineering design. The charging power and magnetic coils of the two topologies must be identical, as only then can the efficiency comparison be meaningful. The charging power of the LC-LC series topology equals to that of LCL-LCL topology, if their charging currents are designed to be the same, as they both have the constant-current characteristic. Based on Equations (4) and (8), we can write:

$$\frac{V_S}{\omega_r M} = \frac{MV_S}{\omega_r L_1 L_4} \tag{10}$$

From Equation (10), one has $M^2 = L_1 L_4$. This means that the charging power of the two topologies are the same if the product of two compensating inductors in the LCL-LCL topology equals the mutual inductance M. When the transfer distances are 10, 15, 20, 25 and 30 cm, the measured mutual inductances between two magnetic coils (L_S and L_D) are 107.155 μH, 66.66 μH, 42.538 μH, 28.125 μH, 18.888 μH, respectively. Normally, the distance between the chassis and ground is around 20 cm, thus the corresponding mutual inductance M is around 42.538 μH. Assuming that L_1 equals L_4, one has

$L_1 = L_4 = 42.538$ μH. Because the magnetic coils of the LCL-LCL topology are the same as those of the LC-LC series topology, the electric parameters of the LC-LC series topology and LCL-LCL topology can be summarized as shown in Table 1.

Table 1. Detailed parameters of the LC-LC series topology and LCL-LCL topology.

LC-LC Series Topology		LCL-LCL Topology	
Electric parameters	Value	Electric parameters	Value
Primary inductor L_S	290.18 μH	Compensating inductor L_1	43.2 μH
Parasitic resistance R_S	193 mΩ	Parasitic resistance R_1	53 mΩ
Primary capacitor C_S	34.83 nF	Primary resonant capacitor C_S	234.75 nF
Secondary inductor L_D	329.4 μH	Primary inductor L_S	290.18 μH
Parasitic resistance R_D	213 mΩ	Parasitic resistance R_S	193 mΩ
Secondary capacitor C_D	30.89 nF	Compensating capacitor C_{S1}	40.68 nF
		Secondary inductor L_D	329.4 μH
		Parasitic resistance R_D	213 mΩ
		Compensating capacitor C_{D1}	35.07 nF
		Compensating inductor L_4	42.5 μH
		Parasitic resistance R_4	42.5 mΩ
		Secondary resonant capacitor C_D	238.63 nF

The detailed efficiency expressions of LC-LC topology and LCL-LCL topology are given by references [27,30–32], and can be also deduced using Maple. Then we substitute the data of Table 1 into the power and efficiency expressions of the LC-LC series and LCL-LCL topologies, and their resulting transfer characteristics are as shown in Figure 3.

Figure 3a shows that the charging power of these two topologies are the same, despite the different battery voltages, and Figure 3b shows that the efficiency of the LC-LC series topology is higher than that of the LCL-LCL topology. Note that the theoretical results ignore the losses caused by the H Bridge and rectifier, so the efficiency losses are mainly due to the parasitic resistances of the inductors and capacitors. The parasitic resistances of compensating capacitors are usually ignored, for they are far smaller than those of magnetic coils. Compared with the LC-LC series topology, the LCL-LCL topology has another two compensating inductors, the parasitic resistances of which further cause a drop in the transfer efficiency. In order to verify the correctness of theoretical calculations, the experiments are performed, and the results are shown in Figure 4.

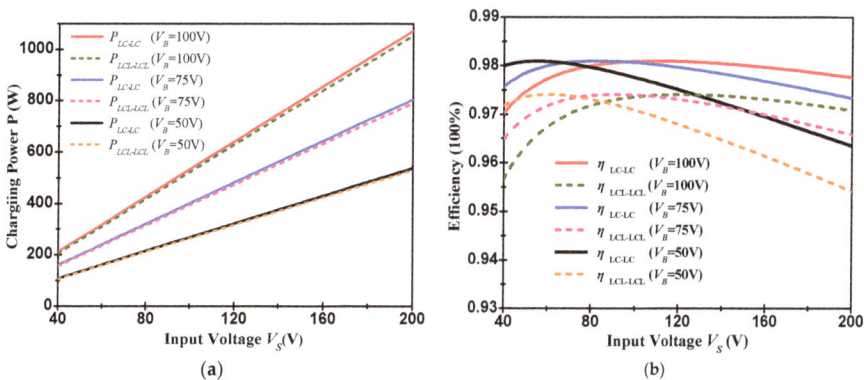

Figure 3. Theoretical transfer power (**a**) and efficiencies (**b**) of the LC-LC series and LCL-LCL topologies.

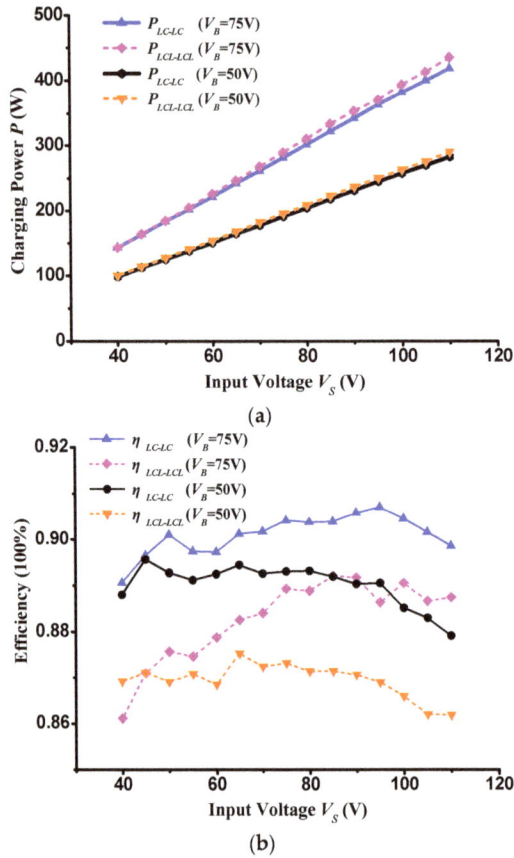

Figure 4. Experimental transfer power (**a**) and efficiencies (**b**) of the LC-LC series and LCL-LCL topologies.

The results in Figure 4a show a good agreement with those in Figure 3a. However, the experimental transfer efficiencies decline sharply compared with those in Figure 3b, which is mainly because the H Bridge and rectifier introduce additional losses. Still, it is obvious that the efficiency of the LC-LC series topology is superior to that of the LCL-LCL topology, so this paper adopts the LC-LC series topology as the power transfer carrier.

3. Online Power Regulation

3.1. Principles of q-Zsource

Z-source is a unique impedance network with two pairs of inductors and capacitors connected in an X shape [33] as shown in Figure 5a, and it was initially used for stabilizing the widely changing voltage produced by fuel cell stacks. Compared with the Z-source, the q-Zsource shown in Figure 5b has better performance [34]. The most obvious two virtues are as follows: first, there is an input inductor L_1, which enables the input current to be continuous and thus limits the transient peak loss. Second, the withstand voltage for C_1 is lowered, for C_1 always combines with power source to charge or discharge L_1. This allows the volume and weight of C_1 to be reduced, and improves system power density.

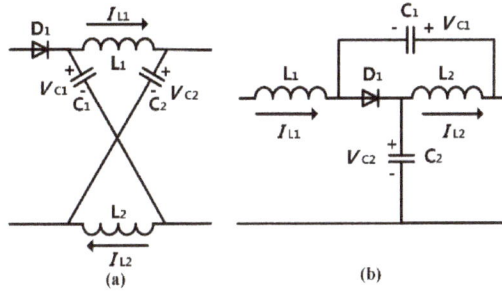

Figure 5. Topologies of Z-source (**a**) and q-Zsource (**b**).

This paper introduces the q-Zsource, not for stabilizing the output, but to produce a changeable output voltage, which can adjust the charging current, and hence, the charging power. The overall system schematic is depicted in Figure 6.

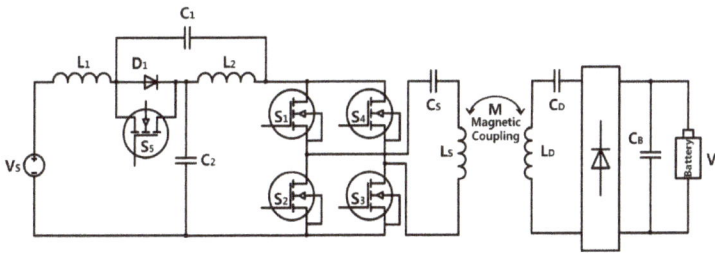

Figure 6. Overall schematic of the wireless power transfer system.

It clearly shows that the whole system consists of the power source, the q-Zsource, the H Bridge, the primary and secondary resonators, the rectifier, the filter and the battery pack. Unlike other system structures, the q-Zsource between the power source and H Bridge is first employed to boost the wireless charging power. By controlling the shoot-through state of the H Bridge, the input voltage to the H Bridge can be boosted through the q-Zsource, so the transmitter current can be adjusted, and hence, the charging current and power for the battery pack. Compared with some typical primary unit current control methods [35], the proposed method has two major merits: first, the digital control of the q-Zsource can be integrated into the primary microprocessor, and there are no active power switches in the proposed method, which can lower the system cost. Secondly, all the current or power regulation devices are included in the primary unit, and this design is beneficial to the vehicle miniaturization and lightness, as the secondary unit can be small and light. There is an additional MOSFET, S_5, connected in parallel with D_1 in Figure 6, and it is used to avoid discontinuous operation conditions when the load is light. To further demonstrate it, assume that there is a light load and the q-Zsource only consists of D_1. The diode D_1 will turn off when the current flowing through it decreases to zero, thus the connection between the q-Zsource and power source is disconnected, and the relationship between them will not be tenable. This abnormal state makes the output voltage of q-Zsource change freely, and further induces a decline of the system transfer characteristics. This unwanted phenomenon will disappear if S_5 is turned on or off actively. S_5 can be also removed if the system always works at the rated state. The q-Zsource has two typical operating states, and the equivalent circuits are depicted in Figure 7.

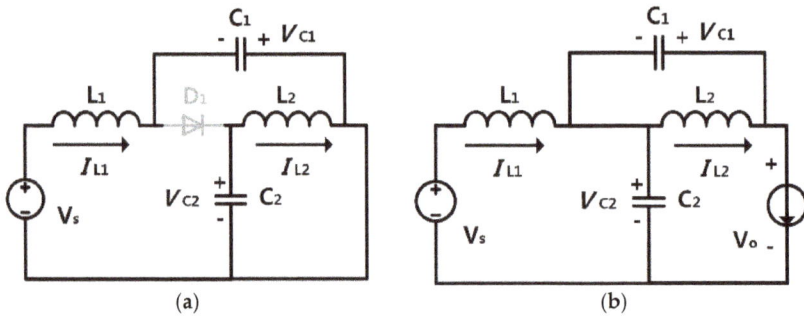

Figure 7. Equivalent circuits of q-Zsource in shoot-through state (**a**) and non-shoot-through state (**b**).

The operating mode of the q-Zsource is controlled by the cascade H Bridge. When the H Bridge works in the traditional phase shifting state, no voltage boosting phenomenon happens, but when the system charging power needs to be increased, making either of the bridge arms to be shoot-through, the voltage boosting phenomenon happens. The output of the q-Zsource is shorted by the H Bridge in the shoot-through state, and L_2 is charged by C_2, L_1 is charged by C_1 and V_S, whereas the H Bridge becomes an equivalent current source in the non-shoot-through state as shown in Figure 7b, C_2 and L_2 provide the output voltage together, C_1, L_1 and V_S provide the output voltage together. This also explains why the withstand voltage across C_1 reduces. By controlling the ratio between shoot-through time and non-shoot-through time, the output voltage of the q-Zsource can be adjusted. From Figure 7a, we have:

$$V_S + V_{C1} = V_{L1}, \; V_{C2} = V_{L2} \tag{11}$$

Similarly, from Figure 7b, we have:

$$V_S + V_{L1} + V_{C1} = V_O, \; V_S + V_{L1} = V_{C2}, \; V_{C1} = V_{L2}, \; V_{C1} + V_{C2} = V_O \tag{12}$$

Normally, L_1 equals L_2, and the currents flowing through them are the same, thus one has:

$$V_{L1} = V_{L2} \tag{13}$$

Since the volt-seconds of the inductor should be identical in the steady state, we can get:

$$(V_{C1} + V_S) \cdot T_S = (V_O - V_{C1} - V_S) \cdot T_N \tag{14}$$

T_S represents the shoot-through time and T_N represents the non-shoot-through time, the sum of those is the whole cycle time T. Then substituting Equations (11)–(13) into Equation (14), we have:

$$V_O = \frac{T}{T_N - T_S} \cdot V_S = \frac{1}{1 - 2D} \cdot V_S = BF \cdot V_S \tag{15}$$

$D = T_S/T$ is the duty cycle, BF is the boost factor produced by the shoot-through state, and it is always greater than or equal to 1. Additionally, we can also obtain the voltages across the capacitor C_1 and C_2:

$$V_{C1} = \frac{V_O - V_S}{2} = \frac{T_S}{T_N - T_S} \cdot V_S = \frac{D}{1 - 2D} \cdot V_S \tag{16}$$

$$V_{C2} = \frac{V_O + V_S}{2} = \frac{T_N}{T_N - T_S} \cdot V_S = \frac{1 - D}{1 - 2D} \cdot V_S \tag{17}$$

3.2. Control Method

Unlike the traditional buck or boost converter, the duty cycle D of the q-Zsource cannot reach 50% according to Equation (15). The voltage gain curve of the q-Zsource is shown in Figure 8, and it clearly shows that there are two operating regions.

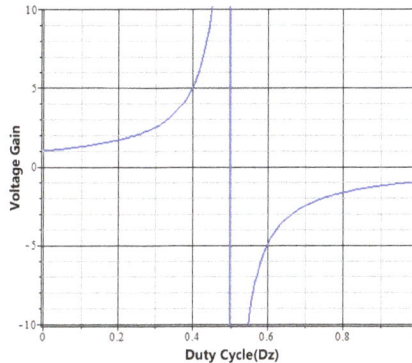

Figure 8. Voltage gain of the q-Zsource.

When D is greater than 0.5, the q-Zsource enters the negative gain region, and produces a negative output voltage, which is hardly used in practice. When D is less than 0.5, it produces a positive output voltage, thus we should limit the duty cycle D to below 0.5. All the traditional control strategies [26] can be used to control the q-Zsource and their theoretical input-output relationships still hold, the only difference is that the shoot-through time is added. The traditional phase shifting control is widely used to produce the square-wave voltages and realize the soft-switching conditions. However, it will not be discussed in this paper, as this control method has already been explained before [36]. It is worth mentioning that the q-Zsource has no effect on the output waves of the H Bridge in this mode, and only acts as a kind of filter.

When the charging power needs to be boosted, the H Bridge enters the new operating mode shown in Figure 9, which supplies the shoot-through state for the q-Zsource to boost the output voltage. Unlike the traditional phase shifting control, an additional shoot-through time $T_{shoot-through}$ is added into the control sequences, The dead time T_{dead}, shifting time $T_{shifting}$ and shoot-through time $T_{shoot-through}$ influence the output waves together. The shoot-through state in Figure 9 is realized by turning on S_3 and S_4 simultaneously, or it can be also realized by simultaneously turning on S_1 and S_2, which depends on the practical situations. The interval between t_0 and t_7 is the whole control cycle, as it is symmetrical, only the operating mode among $t_0 \sim t_3$ needs to be demonstrated. S_1 is turned off at t_0, while I_H is still positive, thus it is forced to flow through the free-wheeling diode of S_2. Before I_H changes, S_2 should be turned on at t_1, which can realize its soft switching. These two steps are similar to the control of a phase-shift-full-bridge, but not exactly the same, as the cascade loads are different. Before S_3 is turned off, S_4 is turned on at t_2, and this state is forbidden in the traditional control. However, precisely because of that, the shoot-through state is provided, which allows the q-Zsource to boost voltage, and different boost factors can be achieved by adjusting the interval between t_2 and t_3. It is noticeable here that the equivalent switching frequency viewed from the q-Zsource is two times the operating frequency of the H Bridge, which greatly reduces the volume and weight of the inductors and capacitors existed in q-Zsource. In addition, the lagging leg (S_3, S_4) is turned off with soft switching, but turned on with hard switching, which lowers the efficiency and needs to be further studied.

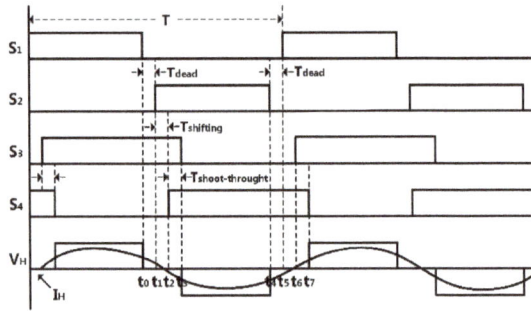

Figure 9. q-Zsource based voltage boost control sequences.

The transmitter and receiver in Figure 6 are high order resonant filters, which only allow the power signals at resonant frequency to pass through. Thus we should evaluate the fundamental components of the waves produced by the H Bridge. The FFT series of square waves depicted in Figure 10 is given as:

$$V = \sum_{n=1,2,3...}^{\infty} \frac{2A}{n\pi} \left[\cos\frac{n\pi T_{non-effective}}{T} - \cos n\pi (1 - \frac{T_{non-effecitve}}{T}) \right] \sin n\omega_r t \qquad (18)$$

where A, n, $T_{non-effective}$ and T represent the amplitude of the square wave, harmonic order, non-effective time and cycle time, respectively, and $\omega_r = 2\pi/T$ is the system angular frequency. According to Equation (15), A is actually determined by BF, and $T_{non-effective}$ is determined by T_{dead}, $T_{shifting}$ and $T_{shoot-through}$. Thus the output voltage produced by the H Bridge can be regulated by controlling these parameters appropriately, whereas the voltage stress across the power switches needs to be considered before designing the BF parameter.

The degree of approximation between the square wave shown in Figure 10 and a quasi-sinusoidal wave can be quantified by THD, and the lower the THD, the less the harmonic loss. The THD of the square wave consisted of different non-effective times can be calculated according to Equation (19), where V_n represents the different harmonic amplitudes:

$$THD = \frac{\sqrt{\sum_{n=2,3,4...}^{\infty} V_n^2}}{V_1} \qquad (19)$$

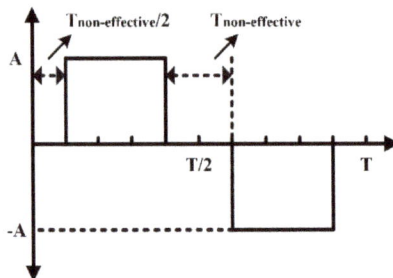

Figure 10. Typical square wave produced by the H Bridge.

The corresponding THD calculation results at different frequencies are shown in Figure 11, which clearly indicates that the lowest THD happens around 2 μs~3 μs non-effective times at 50 kHz and 1 μs~2 μs non-effective times at 80 kHz, rather than 0 μs non-effective time. The optimal non-effective time is where THD has the smallest decrease with the increasing frequencies. Because we adopt 80 kHz as the system operating frequency, the non-effective time should be designed around 1 μs~2 μs to reduce THD as well as harmonic loss.

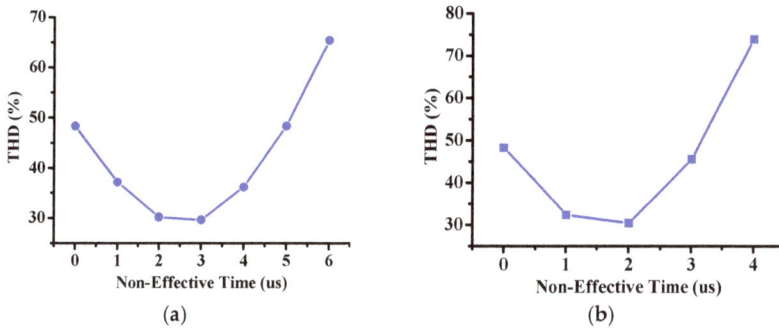

(a)   (b)

Figure 11. THDs of square waves with different non-effective times at 50 kHz (**a**) and 80 kHz (**b**).

4. Experiments

A prototype was implemented to validate our research results, as shown in Figure 12, and the magnetic coils are designed based on the nested three-layer optimization method, which will be discussed in our other papers. The prototype is fabricated according to the schematic shown in Figure 6, where the power source adopts a 62100H-600 high-voltage DC power supply (Chroma, Taoyuan, Taiwan), the H Bridge employs four SPW47N60C3 MOSFETs (Infineon, Neubiberg, Germany), the resonant capacitors adopt B32672L thin-film series (TDK-EPCOS, Tokyo, Japan) ,the rectifier consists of four IDW30E65D1 fast recovery diodes (Infineon, Neubiberg, Germany) and the battery pack consists of 24 lead-acid battery units.

Figure 13 shows the transfer characteristics of the wireless power transfer system at 20 cm transfer distance, and it is worth mentioning that the q-Zsource does not work, and is only present as a filter. In Figure 13a, the transfer power increases with rising input voltages or battery pack voltages. In practice, the power factor correction (PFC) with 400 V output voltage is employed to enhance AC power quality, and the 300 V battery pack is widely used for many production-ready vehicles, like Toyota Prius, Chevrolet Volt, Mitsubishi i-MiEV as well as Nissan Leaf, thus we define this situation as the system rated operating state. In the rated state, the charging power is 3220 W as shown in Figure 13a, which is a little bigger than 3 kW, since the battery pack voltage increases from 300 V to 309.7 V when the charging current (RMS value is 10.4 A as shown in Figure 14) flows through the battery resistance, causing an extra voltage drop. Figure 13b shows that the transfer efficiencies are nearly unchanged despite the increasing input voltages or battery pack voltages, and the rated efficiency where the input voltage equals 400 V and the battery pack voltage equals 300 V is around 95%.

(a)   (b)

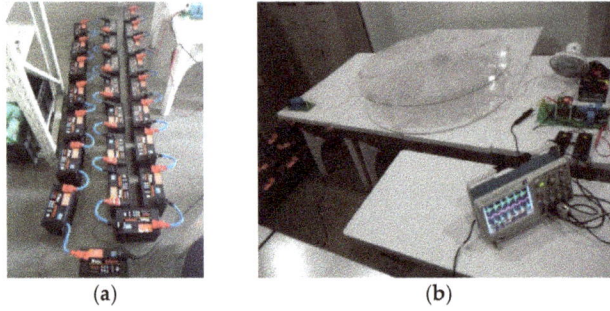

Figure 12. Prototype of wireless power transfer system. (**a**) 300V lead-acid battery pack. (**b**) Transmitter and receiver of wireless power transfer system. The magnetic coil is placed on the top of a perspex plate and fixed by eight perspex bars.

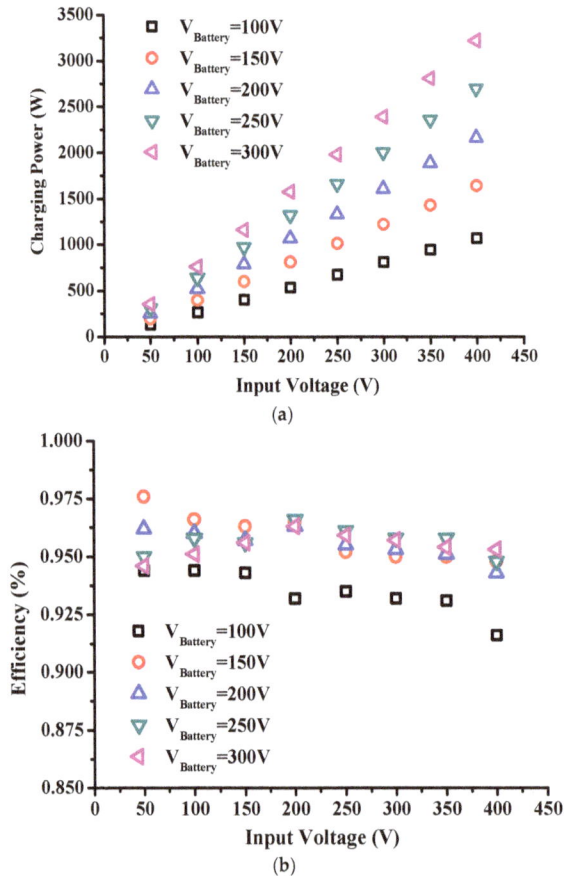

(a)

(b)

Figure 13. Wireless transfer power (**a**) and efficiency (**b**) at different input voltages and battery pack voltages at 20 cm transfer distance.

Figure 14 shows some critical waveforms of the wireless power transfer system, where the cyan curve shows the output voltage produced by the H Bridge, the green curve shows the voltage across the resonant capacitor in the transmitter and the purple curve shows the charging current for the battery pack. Because the reduction scale of high voltage probe is 200, the measured voltages need to be multiplied by 200. The amplitude of the square-wave voltage produced by the H Bridge is 400 V. The RMS value of the charging current is 10.4 A, and there are some ripple waves, the amplitude of which depends on the filter capacitors. The bigger the filter capacitors, the smaller the ripple waves.

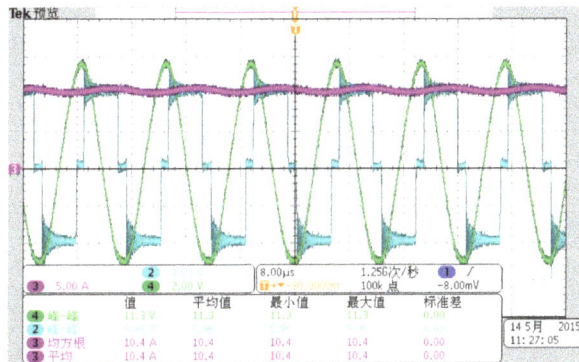

Figure 14. Purple curve: the charging current for the battery pack; cyan curve: the output voltage produced by the H Bridge; green curve: the voltage across the resonant capacitor in the transmitter. They are measured when the system works in the rated state (V_{in} = 400 V, V_B = 309.7 V, $P_{charging}$ = 3220 W, $D_{transfer}$ = 20 cm).

The following experimental results are measured when the q-Zsource works, and two different shoot-through times are shown to clearly demonstrate the q-Zsource principle. Figure 15 indicates that different shoot-through times determine different boost factors. Although the input voltages (green curve) of both Figure 15a,b are identical (200 V), their output voltages (cyan curve) are different. When the shoot-through time is 1 μs, the boost factor is around 1.5, thus the output voltage is around 300 V, and the charging current is 8.26 A. When the shoot-through time is 1.5 μs, the boost factor is around 2, thus the output voltage is around 400 V, and the charging current is 10.2 A. Summarily, the charging currents can be adjusted by controlling the shoot-through times.

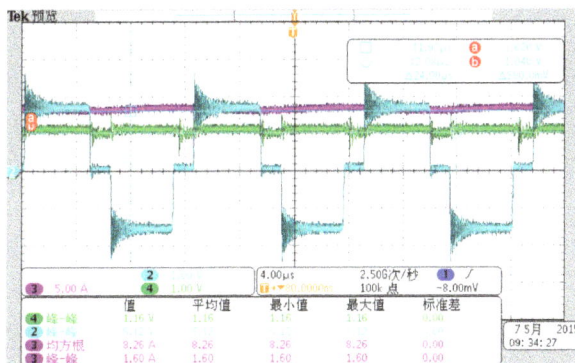

(a)

Figure 15. *Cont.*

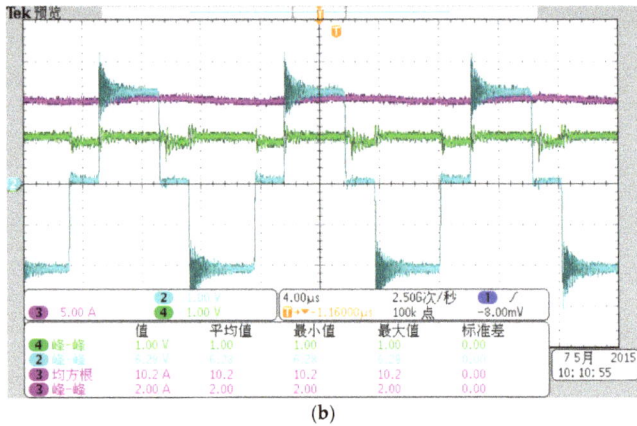

(b)

Figure 15. Purple curve: the charging currents for the battery pack; green curve: the input voltage of the q-Zsource; cyan curve: the output voltage produced by the H Bridge measured (**a**) at the 1 μs shoot-through time; and (**b**) at the 1.5 μs shoot-through time.

In Figure 16a,b the q-Zsource input voltages (green curve) are the same, but their output voltages (cyan curve) are different due to the different boost factors. Unlike the square-wave voltages produced by the H Bridge, the q-Zsource output voltages are pulsatile and always positive. In the shoot-through state, where the q-Zsource output voltage equals zero, the q-Zsource input current (purple curve) increases with a positive slope. However, it decreases with a negative slope when the q-Zsource enters into the non-shoot-through state.

Figure 17 shows the output currents (purple curve) and voltages (cyan curve) produced by the H Bridge, which validate that the system is basically in the resonant state. However, the quasi-resonant state is not beneficial for soft-switching, and hence, efficiency improvement. In practice, the current produced by the H Bridge should lag the voltage to a certain degree, thus an additional 1 nF capacitor is added into the transmitter capacitor array in this paper. The green curve represents the voltage across the q-Zsource capacitor (C_2 in Figure 6), and it is smaller than the output voltage produced by the H Bridge, but it is bigger than the input voltage of the q-Zsource shown in Figure 16.

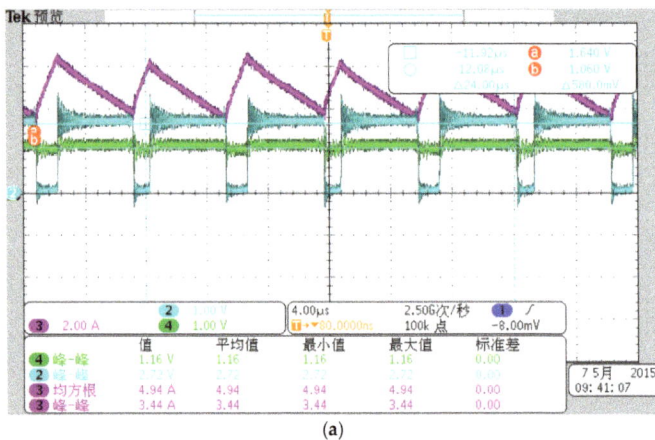

(a)

Figure 16. *Cont.*

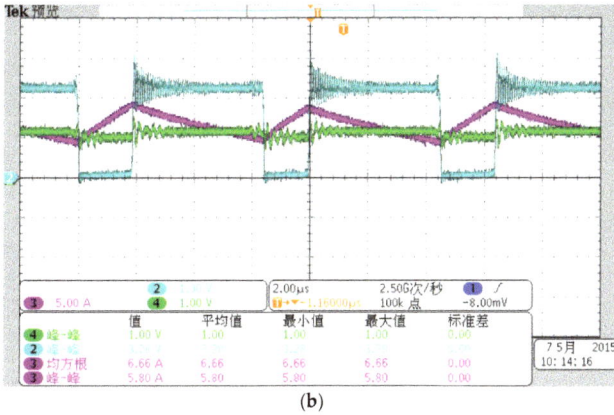

(b)

Figure 16. Purple curve: the input current of q-Zsource, green curve shows the input voltage of q-Zsource, cyan curve shows the output voltage of q-Zsource measured (**a**) at the 1 µs shoot-through time; and (**b**) at the 1.5 µs shoot-through time.

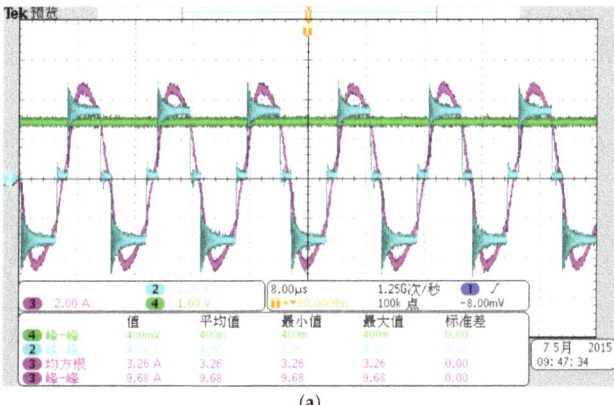

(a)

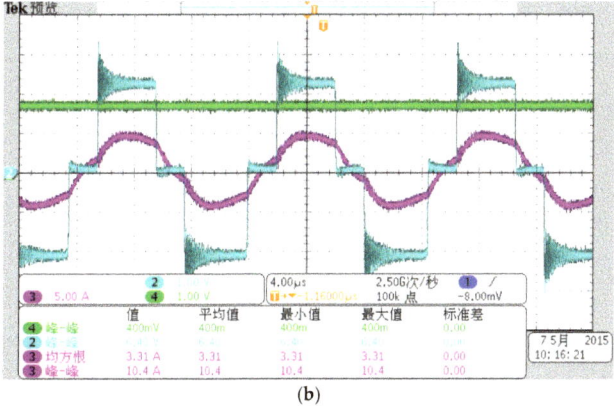

(b)

Figure 17. Purple curve: the currents produced by the H Bridge; cyan curve: the output voltages produced by the H Bridge; green curve: the capacitor voltages of the q-Zsource measured (**a**) at the 1 µs shoot-through time; and (**b**) at the 1.5 µs shoot-through time.

Figure 18a indicates that the charging power or current can be adjusted by changing the shoot-through times. Take the 200 V input voltage as an example, the charging current is 5.26 A when there is no shoot-through time, and it is increased to 8.26 A with 1 μs shoot-through time, and it is further increased to 10.2 A when the shoot-through time is 1.5 μs. Because the boost factor is close to 2 at 1.5 μs shoot-through time, the input voltages above 200 V are not allowed, which may damage the resonate capacitors. In Figure 18b, the efficiencies decline with the increasing shoot-through times, for the shoot-through state makes the MOSFETs of latter bridge arm lost their soft-switching conditions. Additionally, the operating frequency (160 kHz) of q-Zsource doubles that (80 kHz) of wireless power transfer system, which further causes the decline in the transfer efficiency. This phenomenon can be suppressed by reducing the operating frequency of q-Zsouce. If it is decreased to 40 kHz, the loss can be theoretically reduced as much as four times. However, the reduction of the frequency requires bigger inductors and capacitors than before, thus the q-Zsource parameters need to be re-optimized, which is our future work.

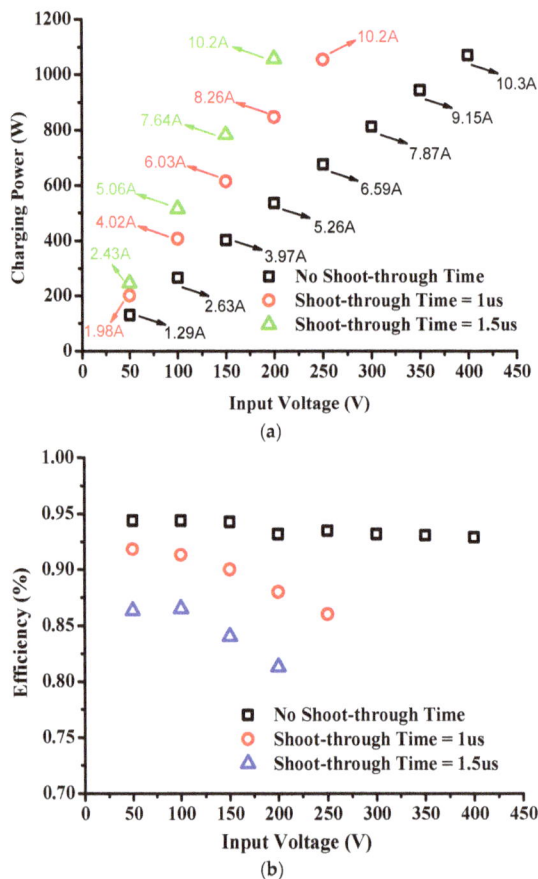

Figure 18. Wireless transfer power (**a**) and efficiency (**b**) characteristics based on the q-Zsource voltage boosting method at 20 cm transfer distance.

5. Conclusions

This paper studies a 3 kW vehicle-mounted wireless power transfer system. First, the efficiency of the LC-LC series topology is verified to be higher than that of the LCL-LCL topology when their transfer power are the same. Then a q-Zsource-based power regulation method is proposed to adjust the charging current online. At last, a 3 kW prototype with ~95% efficiency over a 20 cm transfer distance is implemented to validate our research results. Different shoot-through time durations determine different charging currents despite the same input voltage. When the input voltage is set to be 200 V, a 1 µs shoot-through time can boost the charging current from 5.26 A to 8.26 A, and a 1.5 µs shoot-through time can boost the charging current from 5.26 A to 10.2 A. We hope the work presented in this paper is beneficial to the development of wireless power transfer systems.

Acknowledgments: Acknowledgments: This work is financially supported by the Major State Basic Research Development Program of China (973 Program, Grant No. 2011CB711202) and the National Natural Science Foundation of China (NSFC, Grant No. 51576142).

Author Contributions: Author Contributions: Zhenshi Wang designed the system and analyzed the results, Xuezhe Wei and Haifeng Dai provided guidance and key suggestions.

Conflicts of Interest: Conflicts of Interest: The authors declare no conflict of interest.

References

1. Brown, W.C. The history of power transmission by radio waves. *IEEE Trans. Microwave Theory Tech.* **1984**, *32*, 1230–1242. [CrossRef]
2. Tucker, C.A.; Warwick, K.; Holderbaum, W. A contribution to the wireless transmission of power. *Int. J. Electr. Power Energy Syst.* **2013**, *47*, 235–242. [CrossRef]
3. Brown, W.C. Status of the microwave power transmission components for the solar power satellite (SPS). *IEEE MTT-S Int. Microwave Symp. Dig.* **1981**, *81*, 270–272.
4. Glaser, P.E. Power from the Sun: Its future. *Science* **1968**, *162*, 857–861. [CrossRef] [PubMed]
5. Covic, G.A.; Boys, J.T. Modern trends in inductive power transfer for transportation applications. *IEEE J. Emerg. Sel. Top. Power Electr.* **2013**, *1*, 28–41. [CrossRef]
6. Wei, X.Z.; Wang, Z.S.; Dai, H.F. A critical review of wireless power transfer via strongly coupled magnetic resonances. *Energies* **2014**, *7*, 4316–4341. [CrossRef]
7. Kurs, A.; Karalis, A.; Moffatt, R.; Joannopoulos, J.D.; Fisher, P.; Soljacic, M. Wireless power transfer via strongly coupled magnetic resonances. *Science* **2007**, *317*, 83–86. [CrossRef] [PubMed]
8. Sample, A.P.; Meyer, D.A.; Smith, J.R. Analysis, experimental results, and range adaptation of magnetically coupled resonators for wireless power transfer. *IEEE Trans. Ind. Electron.* **2011**, *58*, 544–554. [CrossRef]
9. Li, X.H.; Zhang, H.R.; Peng, F.; Li, Y.; Yang, T.Y.; Wang, B.; Fang, D. A wireless magnetic resonance energy transfer system for micro implantable medical sensors. *Sensors* **2012**, *12*, 10292–10308. [CrossRef] [PubMed]
10. Puccetti, G.; Stevens, C.J.; Reggiani, U.; Sandrolini, L. Experimental and numerical investigation of termination impedance effects in wireless power transfer via metamaterial. *Energies* **2015**, *8*, 1882–1895. [CrossRef]
11. Sun, L.; Tang, H.; Zhang, Y. Determining the frequency for load-independent output current in three-coil wireless power transfer system. *Energies* **2015**, *8*, 9719–9730. [CrossRef]
12. Sanghoon, C.; Yong-Hae, K.; Kang, S.-Y.; Myung-Lae, L.; Jong-Moo, L.; Zyung, T. Circuit-model-based analysis of a wireless energy-transfer system via coupled magnetic resonances. *IEEE Trans. Ind. Electron.* **2011**, *58*, 2906–2914.
13. Kiani, M.; Uei-Ming, J.; Ghovanloo, M. Design and optimization of a 3-coil inductive link for efficient wireless power transmission. *IEEE Trans. Biomed. Circuits Syst.* **2011**, *5*, 579–591. [CrossRef] [PubMed]
14. Dukju, A.; Songcheol, H. A study on magnetic field repeater in wireless power transfer. *IEEE Trans. Ind. Electron.* **2013**, *60*, 360–371.
15. Musavi, F.; Eberle, W. Overview of wireless power transfer technologies for electric vehicle battery charging. *IET Power Electron.* **2014**, *7*, 60–66. [CrossRef]
16. Del Toro, T.G.X.; Vázquez, J.; Roncero-Sanchez, P. Design, implementation issues and performance of an inductive power transfer system for electric vehicle chargers with series-series compensation. *IET Power Electron.* **2015**, *8*, 1920–1930.
17. Siqi, L.; Mi, C.C. Wireless power transfer for electric vehicle applications. *IEEE J. Emerg. Sel. Top. Power Electron.* **2015**, *3*, 4–17. [CrossRef]

18. Keeling, N.A.; Covic, G.A.; Boys, J.T. A unity-power-factor IPT pickup for high-power applications. *IEEE Trans. Ind. Electron.* **2010**, *57*, 744–751. [CrossRef]
19. Hwang, S.-H.; Kang, C.G.; Son, Y.-H.; Jang, B.-J. Software-based wireless power transfer platform for various power control experiments. *Energies* **2015**, *8*, 7677–7689. [CrossRef]
20. Zaheer, A.; Covic, G.A.; Kacprzak, D. A bipolar pad in a 10-kHz 300-W distributed IPT system for AGV applications. *IEEE Trans. Ind. Electron.* **2014**, *61*, 3288–3301. [CrossRef]
21. Madawala, U.K.; Thrimawithana, D.J. New technique for inductive power transfer using a single controller. *IET Power Electron.* **2012**, *5*, 248–256. [CrossRef]
22. Gao, Y.; Farley, K.; Tse, Z. A uniform voltage gain control for alignment robustness in wireless EV charging. *Energies* **2015**, *8*, 8355–8370. [CrossRef]
23. Chwei-Sen, W.; Covic, G.A.; Stielau, O.H. Power transfer capability and bifurcation phenomena of loosely coupled inductive power transfer systems. *IEEE Trans. Ind. Electron.* **2004**, *51*, 148–157.
24. Ricketts, D.S.; Chabalko, M.J.; Hillenius, A. Experimental demonstration of the equivalence of inductive and strongly coupled magnetic resonance wireless power transfer. *Appl. Phys. Lett.* **2013**, *102*, 053904. [CrossRef]
25. Birrell, S.A.; Wilson, D.; Yang, C.P.; Dhadyalla, G.; Jennings, P. How driver behaviour and parking alignment affects inductive charging systems for electric vehicles. *Transp. Res. C Emerg. Technol.* **2015**, *58*, 721–731. [CrossRef]
26. Choi, W.P.; Ho, W.C.; Liu, X.; Hui, S.Y.R. Comparative study on power conversion methods for wireless battery charging platform. In Proceedings of the 2010 14th International Power Electronics and Motion Control Conference (EPE/PEMC), Ohrid, Macedonia, 6–8 September 2010; pp. S15:9–S15:16.
27. Xuan, N.B.; Vilathgamuwa, D.M.; Foo, G.H.B.; Peng, W.; Ong, A.; Madawala, U.K.; Trong, D.N. An Efficiency Optimization Scheme for Bidirectional Inductive Power Transfer Systems. *IEEE Trans. Power Electron.* **2015**, *30*, 6310–6319.
28. Jun-Young, L.; Byung-Moon, H. A Bidirectional Wireless Power Transfer EV Charger Using Self-Resonant PWM. *IEEE Trans. Power Electron.* **2015**, *30*, 1784–1787.
29. Madawala, U.K.; Thrimawithana, D.J. A Bidirectional Inductive Power Interface for Electric Vehicles in V2G Systems. *IEEE Trans. Ind. Electron.* **2011**, *58*, 4789–4796. [CrossRef]
30. Madawala, U.K.; Thrimawithana, D.J. Modular-based inductive power transfer system for high-power applications. *IET Power Electron.* **2012**, *5*, 1119–1126. [CrossRef]
31. Xuan, N.B.; Foo, G.; Ong, A.; Vilathgamuwa, D.M.; Madawala, U.K. Efficiency optimization for bidirectional IPT system. In Proceedings of the 2014 IEEE Transportation Electrification Conference and Expo (ITEC), Dearborn, MI, USA, 15–18 June 2014; pp. 1–5.
32. Madawala, U.K.; Thrimawithana, D.J. Current sourced bi-directional inductive power transfer system. *IET Power Electron.* **2011**, *4*, 471–480. [CrossRef]
33. Zheng, P.F. Z-source inverter. *IEEE Trans.Ind. Appl.* **2003**, *39*, 504–510. [CrossRef]
34. Peng, F.X.; Guang, W.X.; Qiao, C.Z. A single-phase AC power supply based on modified Quasi-Z-Source Inverter. *IEEE Trans. Appl. Supercond.* **2014**, *24*, 1–5. [CrossRef]
35. Boys, J.T.; Covic, G.A. Decoupling Circuits. U.S. 7279850B2, 9 October 2005.
36. Hothongkham, P.; Kongkachat, S.; Thodsaporn, N. Performance comparison of PWM and phase-shifted PWM inverter fed high-voltage high-frequency ozone generator. In Proceedings of the 2011 IEEE Region 10 Conference on Convergent Technologies for the Asia-Pacific Region, Bali, Indonesia, 21–24 November 2011; pp. 976–980.

Article

An Autonomous Coil Alignment System for the Dynamic Wireless Charging of Electric Vehicles to Minimize Lateral Misalignment

Karam Hwang [1], Jaeyong Cho [1], Dongwook Kim [1], Jaehyoung Park [1], Jong Hwa Kwon [2], Sang Il Kwak [2], Hyun Ho Park [3] and Seungyoung Ahn [1,*]

[1] Graduate School for Green Transportation, KAIST, Daejeon 34141, Korea; hwang8@kaist.ac.kr (K.H.); fable@kaist.ac.kr (J.C.); dwkim88@kaist.ac.kr (D.K.); jaehyoung.park@kaist.ac.kr (J.P.)
[2] Electromagnetic Environment Research, ETRI, Daejeon 34129, Korea; hjkwon@etri.re.kr (J.H.K.); sikwak@etri.re.kr (S.I.K.)
[3] Department of Electronic Engineering, University of Suwon, Hwaseong 18123, Korea; hhpark@suwon.ac.kr
* Correspondence: sahn@kaist.ac.kr; Tel.: +82-42-350-1263

Academic Editor: Sheldon S. Williamson
Received: 2 January 2017; Accepted: 2 March 2017; Published: 7 March 2017

Abstract: This paper proposes an autonomous coil alignment system (ACAS) for electric vehicles (EVs) with dynamic wireless charging (DWC) to mitigate the reduction in received power caused by lateral misalignment between the source and load coils. The key component of the ACAS is a novel sensor coil design, which can detect the load coil's left or right position relative to the source coil by observing the change in voltage phase. This allows the lateral misalignment to be estimated through the wireless power transfer (WPT) system alone, which is a novel tracking method for vehicular applications. Once misalignment is detected, the vehicle's lateral position is self-adjusted by an autonomous steering function. The feasibility of the overall operation of the ACAS was verified through simulation and experiments. In addition, an analysis based on experimental results was conducted, demonstrating that 26% more energy can be transferred during DWC with the ACAS, just by keeping the vehicle's load coil aligned with the source coil.

Keywords: electric vehicle (EV); dynamic wireless charging (DWC); wireless power transfer (WPT); power degradation; coil misalignment; magnetic sensing

1. Introduction

With the realization of electric vehicles (EVs) to reduce greenhouse gases, significant research and development directed towards improving EVs' feaatures has been conducted during recent years. However, one main disadvantage of EVs is their battery technology due to its high cost and limited driving range. To overcome the mentioned issues and minimize dependency on battery systems, many have looked into the utilization of infrastructure such as vehicle to grid (V2G) [1], or residential distribution grid systems [2]. Dynamic wireless charging (DWC) for EVs is another promising method of utilizing infrastructure to overcome battery issues in EVs. DWC is a wireless power transfer (WPT) system that allows EVs to be charged wirelessly while in motion. With DWC, EVs are less dependent on battery systems because they receive power from the road. As a result, EVs with DWC offer several advantages over conventional EVs such as lower vehicle cost and reduced charging time due to their smaller battery systems [3–5]. In light of the many benefits it can provide, extensive research on WPT and DWC has been conducted at various institutions around the globe. Further benefits can be realized when DWC is implemented as an EV charging lane on highways [6]. If a DWC lane is installed at strategic locations along the highway, the EV driving range can be extended, allowing owners to drive longer distances without the anxiety of having to worry about EV range [7].

However, the one main disadvantage of this DWC scheme is the reduction in received power and efficiency that occurs when there is a lateral misalignment of the vehicle's load coil and the source coil embedded under the road. The vehicle must be aligned within a certain range of the source coil to achieve high power in the load coil [8], which also maximizes the driving distance of the vehicle. Keeping the load coil of a vehicle aligned at all times with the source coil while in motion is very difficult, even for an experienced driver. In addition, attempting to keep the vehicle aligned with the source coil may distract the driver from oncoming traffic or obstacles and eventually lead to serious traffic accidents.

To minimize the problem with transferring power to the load coil due to lateral misalignment in WPT, many methods have been proposed to maximize lateral misalignment tolerance. Some proposed methods include changing the geometry of the coil [9], placing multiple coils in an orthogonal configuration [10] or an overlapping configuration [11], and even combining multiple coils of different geometry into one unit [12]. Another popular method is the utilization of ferromagnetic materials, where E-shape or U-shape ferrite cores [13,14] are utilized at the load/source coils. Other methods also include active coil resonance frequency tuning circuits to maximize lateral tolerance [15,16]. All the proposed methods described above are constructive, but their implementation in vehicular applications can be very limited, due to the vehicle's limited installation space, weight constraints, as well as the dynamic driving environment. Even assuming that the proposed methods were implemented in the vehicle, misalignment would still be unavoidable as long as a person is controlling the vehicle. In addition, even a DWC-equipped EV with a high lateral misalignment tolerance (45 cm) will still have regions on a highway lane without wireless power delivery, as shown in Figure 1. And within the tolerance range, loss of power will still inevitably occur whenever the lateral misalignment increases.

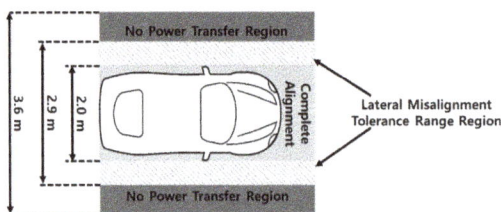

Figure 1. Power transfer coverage area on a 3.6 m width highway lane (based on FHWA standards [17]) for a WPT system with a 45 cm lateral misalignment tolerance. Equipped on a standard sized vehicle with a 2 m track width.

A vehicle tracking and autonomous guidance system using magnetic sensing can also be applied to reduce the DWC power transfer problem. The system concept is shown in Figure 2.

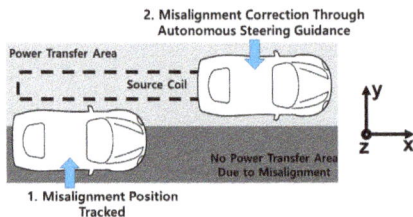

Figure 2. Concept of autonomous vehicle tracking and guidance to reduce degraded power transfer in a DWC system.

The misalignment between the coils is detected using sensors and the vehicle's position is then adjusted by appropriate autonomous steering until the degraded power transfer in the DWC is restored to an optimum level.

Many vehicle tracking methods have been previously proposed where one popular method is using global positioning system (GPS) sensors and/or magnetic markers [18]. Other tracking methods specific for DWC applications are also proposed in [19–21], where a radio-frequency identification (RFID) tag and reader were implemented. However, one disadvantage of magnetic markers or RFID tags is the rapid decay in the strength of the magnetic field with distance, as the detector moves away from the marker/tag range. Therefore, the magnetic markers or RFID tags have to be placed close to each other in order to achieve high tracking resolution, which increases overall construction costs. To reduce construction costs, magnetic sensing hardware with a wider detection range has been proposed as shown in [22] or using a Gaussian function-based algorithm to have higher detection accuracy [23]. However, this leads to bulkier sensor hardware which is not desirable, especially when vehicle space is very limited with the DWC system installed.

Another method of magnetic tracking is the autonomous coil alignment system (ACAS), which has been proposed by the authors of this paper in [24]. ACAS is a novel method of tracking a vehicle's misalignment position by only measuring the voltage in the vehicle's load coil. Since this method only utilizes the existing DWC system, the use of external magnetic or RFID markers is eliminated, which leads to a significant reduction in implementation costs for the magnetic tracking functions. One main challenge of this approach is how to determine whether the load coil was misaligned to the left or the right side relative to the center of the source coil. This is because the voltage readings are nearly identical on both sides. The proposed two-sensor coil unit design in [24] detected the left/right side position by calculating the difference in the voltage readings of the two sensor coils. Even though the operational feasibility of the ACAS was verified, it was only compatible with a specific DWC system, and its algorithm was very complex. If any significant changes were made to the DWC system, the accuracy of position tracking could be affected.

The ACAS proposed in this paper is a significant improvement over the previous design described in [24]. It is a single-sensor coil unit, and detects change in voltage phase to identify the left/right side position rather than the voltage difference between two sensor coils. With the new ACAS design, the complexity of the algorithm and hardware can be significantly reduced, providing a more reliable system. This also allows wider application compatibility with other DWC systems with varying specifications.

The paper is organized as follows: an analysis of power loss due to lateral misalignment, as well as the reasons for proposing the ACAS system, are discussed in Section 2. In Section 3, the main components of the ACAS are discussed and its concept feasibility is verified by simulation. The operational feasibility of the ACAS is further verified by experiments in Section 4. Conclusions are presented in Section 5.

2. Analysis of Power Loss Due to Lateral Misalignment in Dynamic Wireless Charging (DWC) Systems

DWC can be viewed as a general WPT system circuit model, shown in Figure 3, which consists of a source coil and a load coil section. The efficiency of the WPT system is defined by the ratio of the power delivered from the source coil, P_S, and the load coil, P_L, as follows:

$$\eta = \frac{P_L}{P_S} \tag{1}$$

From Equation (1), P_S and P_L can be defined as:

$$P_S = I_1 V_S = \left(\frac{R_L + \frac{1}{j\omega C_L} + j\omega L_L}{j\omega M} I_2 \right) V_S \tag{2}$$

$$P_L = I_2 V_L = I_2^2 R_L = \left(\frac{j\omega M}{R_L + \frac{1}{j\omega C_L} + j\omega L_L} I_1 \right)^2 R_L \tag{3}$$

where R_S, C_S, L_S represent the resistor, capacitor, or inductor components of the source coil, respectively, and R_L, C_L, L_L represent the resistor, capacitor, or inductor components of the load coil, respectively. ω represents the frequency, and I_1, I_2 represent the currents flowing in the source coil and load coil, respectively. M is the mutual inductance between the source coil and load coil, which is expressed as follows:

$$M = k\sqrt{L_S L_L} \tag{4}$$

where k is a coupling coefficient.

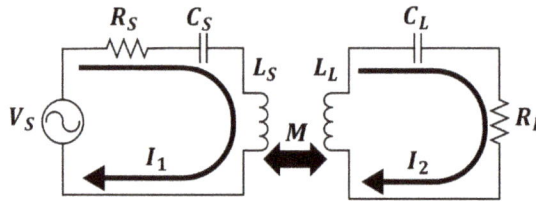

Figure 3. General circuit model for wireless power transfer (WPT).

The mutual inductance between the source coil and the load coil greatly affects the received power at the load coil as well as the overall efficiency of the WPT system, as noted in previous works [10,25]. Therefore, it is important for any WPT system to maintain the highest mutual inductance to maximize the received power and efficiency to the load coil. Based on the Neumann formula, the mutual inductance shown in (4) can also be described as the number of flux linkages in the load coil resulting from the current flowing from the source coil [10]:

$$M = \frac{\mu_0}{4\pi} \oint \oint \frac{dl_S dl_L}{|r_{SL}|} \tag{5}$$

where μ_0, dl_S and dl_L represent the permeability of free-space, and the infinitesimal segments of the source coil and load coil, respectively, while $|r_{SL}|$ represents the distance between the source coil and the load coil segments, which can also be seen as misalignment. Assuming that a constant current is applied, the only variable that will affect M is $|r_{SL}|$. Therefore, from (1) to (5), it can be determined that increased misalignment reduces M, thus reducing the magnitude of received power at the load coil and the overall WPT efficiency.

Based on (5), there are several factors that will change the magnitude of $|r_{SL}|$: vertical, lateral, and angular misalignment. However, not all of these factors are likely to be observed in practice, for the following reasons: because the load coil is fixed to the underbody of a vehicle, the misalignment between the source coil and load coil is highly dependent on vehicle movement. When a vehicle is operating on highways or on roads that are flat, it can be assumed that the vertical misalignment and angular misalignment relative to the x and y axes will have small variations.

The two factors that remain are the lateral, and angular misalignment relative to the z axis. Between these two factors, the dominant factor that will affect M is the lateral misalignment. To prove this statement, a 3-D electromagnetic (3-D EM) simulation was conducted. A general rectangular shaped load coil placed on top of the source coil lines was designed to replicate the DWC system shown in Figure 4. Its electrical parameters are presented in Table 1 as well.

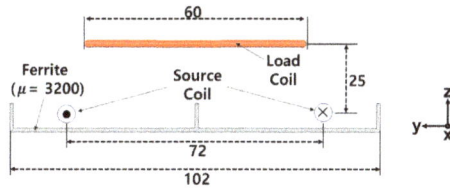

Figure 4. 3-D EM simulation model of the source coil and load coil and their dimensional parameters (in cm).

Table 1. Parameter setup for the 3-D EM simulation.

Parameter	Source Coil	Load Coil	Sensor Coil
# of turns	8	50	10
Inductance	116 µH	2.35 mH	732 nH
Operating frequency		20 kHz	
Current through source coil		200 A	

The simulation was conducted to observe the changes in M, k and the load coil's received power resulting from a lateral misalignment of −50 cm to 50 cm, and also an angular misalignment of 0, 10, and 20 degrees relative to the z axis. The simulation only considered an angular misalignment up to 20 degrees, because an angular misalignment greater than the stated value can be considered an extreme steering angle for a vehicle moving at high speeds.

The changes in M, and k from the lateral and angular misalignments are shown in Figure 5. There are slight variations with increased angular misalignment, but it can be seen that lateral misalignment is the dominant factor in reducing M. Therefore, the proposed ACAS was designed to minimize lateral misalignment to allow higher M so that higher power will be received at the load coil, increasing overall efficiency.

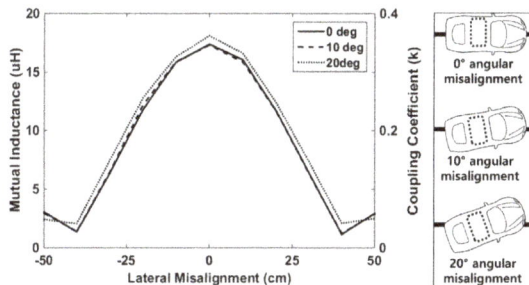

Figure 5. 3-D EM simulation result showing the change in the mutual inductance and coupling coefficient due to lateral misalignment and angular misalignment.

3. Concept of the Autonomous Coil Alignment System (ACAS)

Figure 6 shows a general block diagram of the proposed ACAS. It is divided into three parts: the ACAS sensor coil unit, the lateral position detection unit, and the fuzzy steering controller. The lateral misalignment position is estimated by the sensor coil unit and further processed by the lateral misalignment detection unit. Based on the detected lateral misalignment, the fuzzy steering controller will send steering commands to the electronic power steering system (EPS), after which the vehicle is steered autonomously to correct the misalignment. EPS is already used in most EVs and is being equipped in most newly manufactured commercial vehicles as well. In case an obstacle is present which obstructs the path of maximum charging, the user must have the ability to interrupt the operation of the ACAS. Therefore, the ACAS is designed so that the user can enable or disable the operation of the ACAS at any time, as shown in Figure 6.

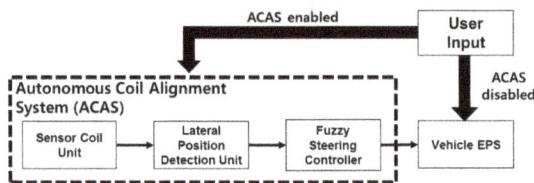

Figure 6. Block diagram of the ACAS.

3.1. ACAS Sensor Coil Unit

In the ACAS, the range of lateral misalignment is estimated based on voltage readings alone. Based on (3), the voltage on the load coil can be expressed as follows:

$$V_L = \frac{j\omega M}{R_L + \frac{1}{j\omega C_L} + j\omega L_L} I_1 R_L \tag{6}$$

When a constant current source is being supplied to the source coil, the dominant parameter that will change the load coil voltage is M. To prove this statement, a 3-D FEM simulation was conducted again using the same model shown in Figure 4 and Table 1, to measure the induced voltage in the load coil. The results are shown in Figure 7. It can be seen in the graphed trends that the peaks of the mutual inductance and the induced voltage in the load coil are nearly identical; thus, it shows that the lateral misalignment can be identified through the load coil voltage alone.

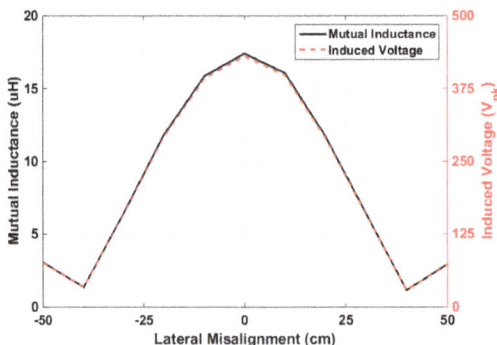

Figure 7. 3-D FEM simulation result showing a waveform comparison of the mutual inductance and the induced voltage on the load coil.

As mentioned in the introduction, the main challenge is to determine whether the load coil is misaligned to the left or the right side. To solve this problem, a sensor coil unit is proposed, which is composed of a single coil unit wound around the middle of the leading section of the load coil, as illustrated in Figure 8.

With this configuration, the load coil's left/right position can be determined based on the source coil's reference. The sensor coil does not occupy much space, and the number of turns for the coil is sufficient to induce a signal strong enough for the ACAS main controller to recognize, while small enough to not disrupt the power transfer between the source coil and the load coil.

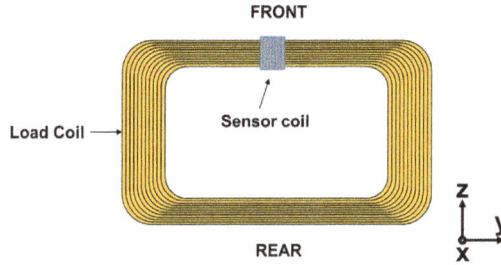

Figure 8. Configuration of the sensor coil within the load coil. Shown from a top view perspective.

Figure 9a shows how the load coil/sensor coil unit is implemented under the vehicle body when the load coil and source coil are in perfect alignment. In addition, it shows a cross section along the x-y plane of the source coil as well.

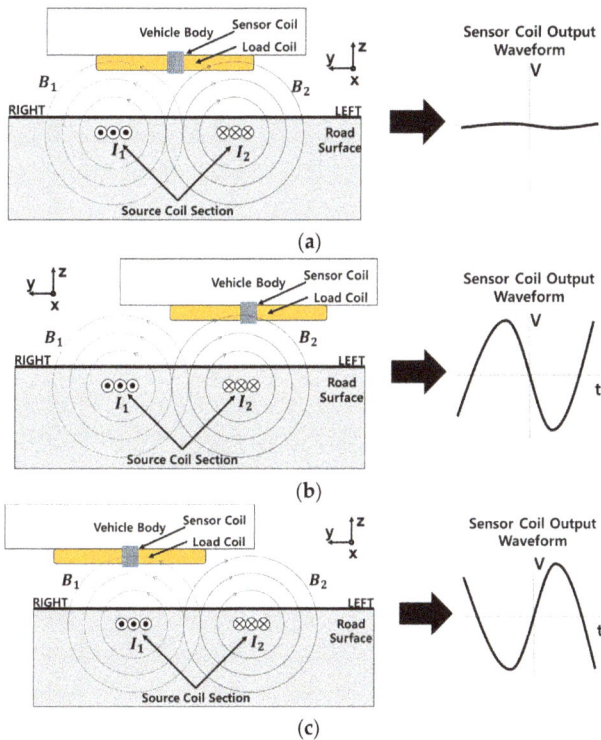

Figure 9. The resulting voltage waveform of the sensor coil in different positions (**a**) when the load coil is aligned with the source coil (origin point); (**b**) when the load coil is misaligned towards the left, and (**c**) when the load coil is misaligned to the right.

As shown in Figures 8 and 9, the sensor coil is placed in a specific location, where the loop of the sensor coil is facing in the y-axis direction. Viewing the source coil in Figure 9, the magnetic field generated on the source coil, $\mathbf{B_S}$, can be expressed in vector form as follows:

$$\mathbf{B_S} = \mathbf{B_1} + \mathbf{B_2} \tag{7}$$

where $\mathbf{B_1}$ and $\mathbf{B_2}$ are the right and left magnetic fields, respectively, as shown in Figure 9. When viewed from the *y*-axis direction, from the direction the loop of the sensor coil is facing, $\mathbf{B_S}$ can be expressed in vector form as follows:

$$\mathbf{B_y^s} = \mathbf{B_y^1} + \mathbf{B_y^2} = -\frac{\mu_0 I_1}{2\pi r_1}\sin\theta_1 + \frac{\mu_0 I_2}{2\pi r_2}\sin\theta_2 \tag{8}$$

In (8), I_1, I_2, θ_1 and θ_2 are the current flow and the angle needed to generate $\mathbf{B_1}$ and $\mathbf{B_2}$, respectively. It should be noted that I_1 and I_2 have the same magnitudes but flow in opposite directions. μ_0 is the free-space permeability and r_1, r_2 are the reference point distances of $\mathbf{B_1}$ and $\mathbf{B_2}$, respectively. In Figure 9a, the magnetic field sensed by the sensor coil is near zero because $\mathbf{B_1}$ and $\mathbf{B_2}$ cancel each other out. As the sensor coil moves towards the left, as shown in Figure 9b, the sensor coil is relatively more exposed to the $\mathbf{B_2}$ magnetic field, thus the magnitude of $\mathbf{B_2}$ will be more dominant than $\mathbf{B_1}$. The opposite phenomenon will occur when the sensor coil moves towards the right, where the magnitude of $\mathbf{B_1}$ will be more dominant than $\mathbf{B_2}$, as shown in Figure 9c.

The induced voltage in the sensor coil, V_{se}, can be expressed as follows:

$$V_{se} = -\frac{d\Phi}{dt} = -\frac{d\mathbf{B_S} A}{dt} \tag{9}$$

where the equation is based on Faraday's law. Φ, t and A represent the magnetic flux, time, and cross sectional area of the sensor coil, respectively. When the load coil is misaligned to the left or right, the phase angle difference between the sensor coil will always have a ±90 degree difference, respectively. The phase difference will only change when the load coil is shifted from the right region to the left region, or vice versa. Under these conditions, the difference in phase when the coil is misaligned to the left (shown in Figure 9b) and right (shown in Figure 9c) will be around 180 degrees.

Verification of the ACAS Sensor Coil Unit through Simulation

3-D EM simulations were conducted to verify that the placement of the sensor coil does not have much influence on the power transfer between the source coil and load coil, and also to verify the concept shown in Figure 9. The sensor coil was added to the simulation model shown in Figure 4 to match the configuration shown in Figure 8. The sensor coil parameters were as listed in Table 1. Figures 10 and 11 show comparisons of the induced voltage and magnetic flux density of the load coil with the sensor coil, and the load coil without the sensor coil, respectively, when a lateral misalignment occurs from −50 cm to 50 cm.

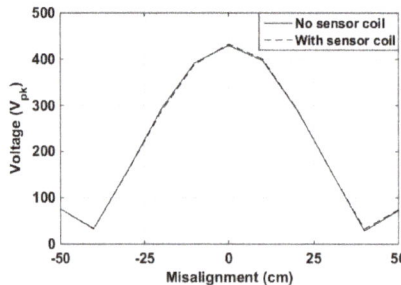

Figure 10. Simulation results showing induced voltage vs. lateral misalignment. A comparison of load coils with and without the sensor coil.

Based on Figure 10, the two output waveforms are nearly identical, and the magnetic flux density comparison in Figure 11 shows that the existence of the sensor coil unit creates almost no interference with the magnetic flow between the load coil and sensor coil. This verifies that the placement of the sensor coil has little effect on the performance of the overall WPT system.

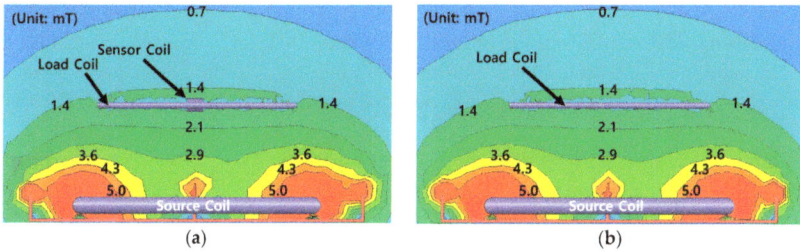

(a)   (b)

Figure 11. Simulation results showing the magnetic flux density for the load coil (**a**) with the sensor coil unit and (**b**) without the sensor coil unit.

The output voltage waveforms generated from the sensor coil when it is misaligned to the left or right side are shown in Figure 12. The results show that the resulting voltage for the left side and right side maintains a 90 degree phase difference, thus validating the theory shown in Figure 9, and clearly distinguishing the left/right side regions regardless of position. It can also be observed that the relationship between the sensor coil's induced voltage and lateral misalignment is non-linear. However, this information can be neglected because the left/right directions can be clearly identified by the sensor coil. With the left/right information, the exact lateral misalignment position can be identified by measuring the voltage through the load coil, which is further processed in the ACAS lateral position detection unit.

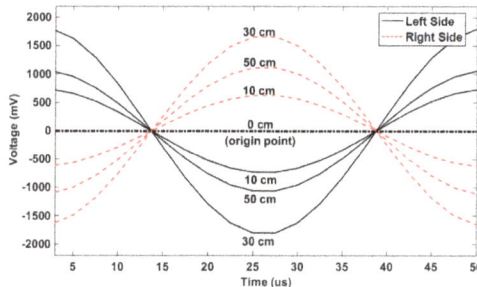

Figure 12. Simulated sensor coil voltage output vs. time, showing the left side and right side positions can be clearly identified.

3.2. ACAS Lateral Position Detection Unit

As mentioned in Section 3.1, the left and right sides can be detected based on the phase difference. The ACAS lateral position detection unit is responsible for converting those values into information that can be processed by the ACAS fuzzy steering controller. The block diagram of the ACAS lateral position detection unit is shown in Figure 13.

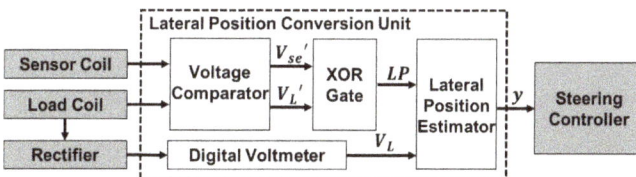

Figure 13. Block diagram showing the ACAS lateral position detection unit.

It consists of a voltage comparator, exclusive OR (XOR) gate, and digital voltmeter. The load coil waveform and the sensor coil waveform are each converted into logic square waveforms through the voltage comparators as shown in the block diagram in Figure 13. From here, the converted sensor coil logic signal, V_{se}', and the converted load coil logic signal, V_L' are sent into the XOR gate, where the two signals are compared with each other to determine the left or right lateral position. Since the phase of the load coil remains unchanged, the load coil voltage (in AC form) can be used as a reference to determine the phase change in the sensor coil, as shown in Figure 9. The logic flow of the XOR gate output used to determine the left and right position is shown in Figure 14. Two input signals V_{se}', and V_L' are sent as input into the XOR gate. Based on the nature of operation of the XOR gate, the output at the XOR gate, expressed as LP, is either a 5 V output (logic HI) or a 0 V (logic LO) signal to indicate left or right, respectively.

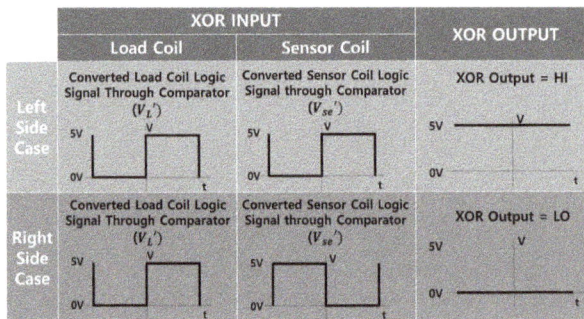

Figure 14. The XOR gate output from the converted load coil and the sensor coil logic signal inputs based on left or right side lateral misalignment.

With the left or right side detected, the location of the lateral misalignment, y, can be determined as follows:

$$y = \begin{cases} -f(V_L) \ if \ LP = HI \\ f(V_L) \ if \ LP = LO \end{cases} \tag{10}$$

where $f(V_L)$ is a function that represents the relationship between the load coil voltage and the lateral position, y. With the known left/right position, the lateral misalignment y is estimated and directly fed as input into the ACAS fuzzy steering controller.

3.3. ACAS Fuzzy Steering Controller

The lateral misalignment is estimated based on the voltage measurement alone, but irregularities can occur in the reading of the angular misalignment of the load coil, or even because of the non-linear characteristics of the DWC system. In addition, the source coil used in DWC systems is typically installed in segments to maximize efficiency, as described in [12,26]. Thus, it can be expected that the voltages generated in each segment will not be entirely identical. To tolerate such characteristics, a fuzzy logic based steering control method is used in the ACAS, because it can provide better dynamic response compared to other conventional controllers [27,28]. The fuzzy logic steering controller used for the ACAS in this paper was previously introduced by the authors. In the present study, only experiments to verify the performance of the fuzzy steering controller were conducted. The design process and simulation results can be obtained in [24].

4. Experimental Validation

To observe and validate the overall performance of the ACAS, a smaller scale DWC system and experimental vehicle were constructed, and two experiments were conducted. In the first experiment, the main purpose was to validate the power loss due to lateral misalignment, and also to validate the operating concept of the proposed sensor coil. Because the first experiment operated near a 90 W power level, a static load (10 Ω resistor) was used instead of the experimental vehicle because the stated power level exceeded the recommended operation settings used in the experimental vehicle. For the second experiment, the main purpose was to validate the overall operating concept of the ACAS. The same DWC system from the first experiment were used, but the power level was reduced to 10 W, so that the experimental vehicle could be used. The operation of the experimental vehicle can be viewed in the video shown in S1.

4.1. Experimental Setup

A DWC system was constructed in laboratory scale to validate the operation of the ACAS system. The source coil unit and the load coil/sensor coil unit are shown in Figure 15. The electrical and dimension parameters of the coil units are shown in Table 2.

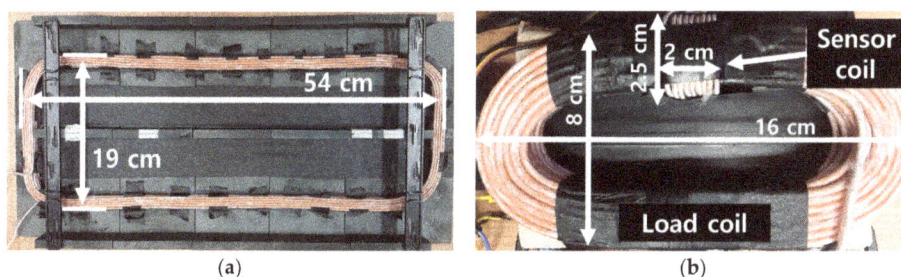

Figure 15. Coil construction of the (**a**) source coil unit and (**b**) load coil/sensor coil unit.

Table 2. Dimensions and electrical parameters of the coils used in the experiment.

Parameter	Source Coil	Load Coil	Sensor Coil
Dimensions (W × L × H)	19.0 cm × 54.0 cm × 1.5 cm	8.0 cm × 16.0 cm × 2.0 cm	2.0 cm × 2.5 cm × 2.5 cm
# of turns	20	42	10
inductance	590.00 µH	186.15 µH	2.45 µH

The source coil unit and the load coil each consist of ferrite core assemblies as shown in Figure 16. In case of the source coil ferrite core assembly, it consists of ferrite blocks of two different types, which has been identified as ferrite block type A and type B, respectively. The difference between the two ferrite blocks are its dimensions, while the other specifications are identical, as shown in Table 3.

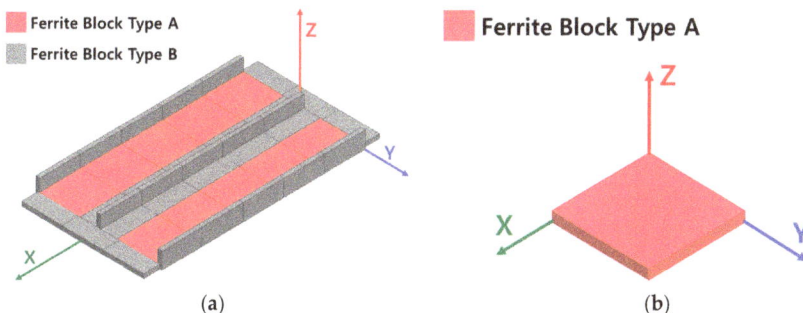

Figure 16. Ferrite core assemblies for the (**a**) source coil and (**b**) load coil unit.

Table 3. Material characteristics of ferrite blocks types A and B identified in Figure 16.

Parameter	Ferrite Block Type A	Ferrite Block Type B
Dimensions (W × L × H)	10.0 cm × 10.0 cm × 1.0 cm	10.0 cm × 4.0 cm × 1.0 cm
Material	Manganese-Zinc (Mn-Zn)	
Permeability (μ)	3200	
Saturation flux density ($\mathbf{B_S}$)	520 mT	

The laboratory scale vehicle is shown in Figure 17. The lateral position detection unit as well as the fuzzy steering controller unit described in [24] were programmed into the microcontroller as shown in Figure 17a, and the constructed load coil/sensor coil unit was mounted underneath the vehicle as shown in Figure 17b. The complete experimental setup is shown in Figure 18, where a 20 kHz inverter was used as the main source for delivering power wirelessly from the source coil to the load coil. It can also be observed that an acrylic plate has been placed between the source coil and load coil. The mentioned acrylic plate was moved laterally left and right to generate misalignment between the source coil and load coil.

(a)   (b)

Figure 17. Experimental vehicle. (**a**) Top view, showing the microcontroller unit; (**b**) Bottom view, showing the load coil/sensor coil unit.

Figure 18. Experimental setup of the experimental vehicle with the ACAS unit.

Figure 19 shows the schematic of the overall experiment, and the values of parameters shown are listed in Table 4. The V_S is the main voltage source generated from the 20 kHz inverter unit, and it was adjusted to generate 90 W and 10 W at the load for the first experiment and second experiment, respectively. During the construction phase of the experimental vehicle, it was observed that a lot of noise was generated which affected the sensor coil's measurement accuracy. Therefore, an active non-inverting (NI) low-pass filter (LPF) was constructed using an instrumentation amplifier, and placed between the sensor coil and the voltage comparator as shown in Figure 19. The R, C values for the LPF, assigned as R_{LPF}, and C_{LPF}, respectively, was assigned to match the cut-off frequency of 20 kHz as close as possible, which is also listed in Table 4.

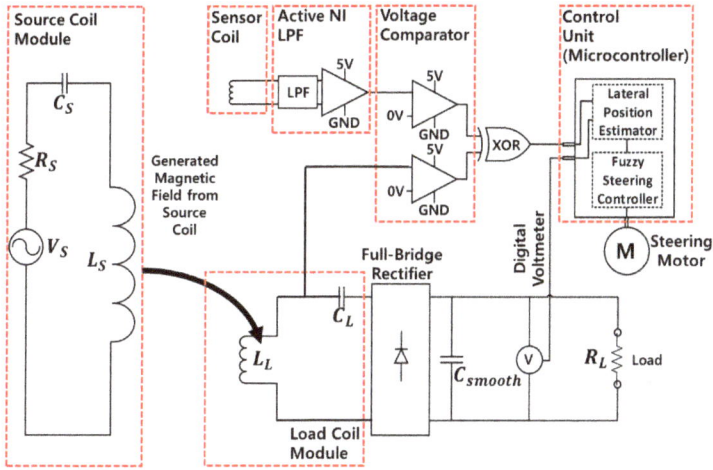

Figure 19. Overall circuit schematic for the experiment.

Table 4. Measured value of each components for the circuit schematic shown in Figure 19.

Component	Symbol	Value
Source/load coil components to match resonance @ 20 kHz	L_S	590.00 µH
	R_S	170 mΩ
	C_S	107.33 nF
	L_L	186.15 µH
	R_L	103 mΩ
	C_L	328.11 nF
Rectifier smoothing capacitor	C_{smooth}	3200 µF
Low-pass filter (LPF) components	R_{LPF}	919 Ω
	C_{LPF}	9.8 nF
Load resistance for first experiment (static load)	R_L	10 Ω
Load resistance for second experiment (experimental vehicle load)	R_L	2~3 Ω (varying load due to motor operation)

In the first experiment (static load), the load coil was moved from −8 cm (left) to 8 cm (right) in 1 cm increments. The change in voltage and current was monitored through the voltmeter and current meter, respectively, and recorded. Also, the changes in the load coil and sensor coil voltage waveforms were monitored through the oscilloscope as well. Because the power was reduced to 10W in the second experiment (experimental vehicle load), the load coil was moved less; −5 cm (left) to 5 cm (right) in 1 cm increments, unlike ±8 cm in the first experiment. The sensor coil voltage waveform as well as its converted logic signals were monitored through the oscilloscope, and the detected voltage and steering position data were recorded through the microcontroller in the experimental vehicle.

4.2. Experimental Results from the First Experiment (Static Load: 10 Ohm Resistor)

Figure 20 shows the measured voltage and current as the load coil was moved from −8 cm (left) to 8 cm (right) in 1 cm increments. Based on Figure 20, it can be seen that the power is significantly reduced when the lateral misalignment increases, both from the right side and left side. Half of the load coil's power was lost at ±4 cm, and nearly all of its power was lost at ±8 cm. The results in Figure 20 verify the theory explained in Section 2, and thus, justify the need for the proposed ACAS system.

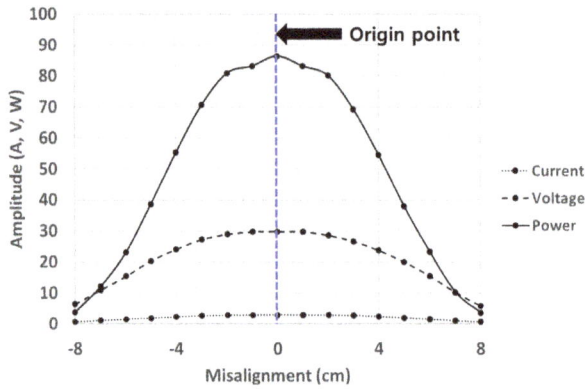

Figure 20. Voltage and current measured at the load, and its calculated power (first experiment: static load).

Figure 21 shows the voltage waveforms of the load coil and the sensor coil measured from the oscilloscope. Figure 21a,b represent when the load coil has a left misalignment, and a right misalignment, respectively. The captured results indicate that significant noise is present in the sensor coil waveform. These are filtered out through a low-pass filter before being converted into a logic signal as shown in the circuit schematic in Figure 19. Based on the results shown in Figure 21a, the phase difference between the load coil and sensor coil is 118.9 degrees (left misalignment), while the phase difference in Figure 21b is −57.3 degrees (right misalignment). The difference between the two recorded phase values is 176.2 degrees, which is near the 180 phase angle difference described in the last paragraph in Section 3.1.

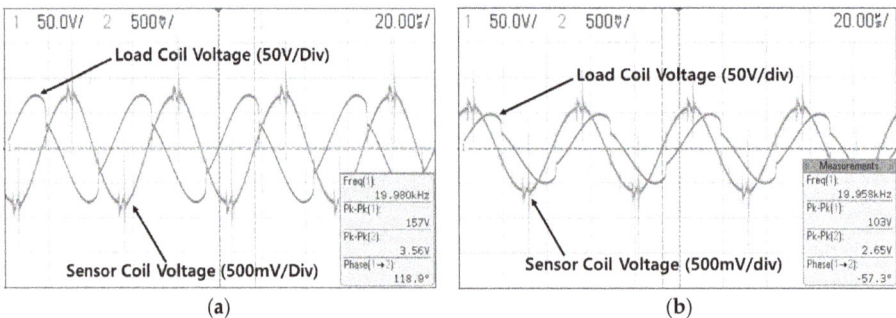

Figure 21. Oscilloscope readings from the first experiment showing the voltage waveform of the load coil and the sensor coil unit at (**a**) left misalignment; and (**b**) right misalignment.

4.3. Experimental Results from the Second Experiment (Load: Experimental Vehicle)

The recorded load coil and sensor coil waveforms for the second experiment are shown in Figure 22. These results show characteristics similar to the results shown in Figure 21, except there is a lower voltage amplitude, because the source coil was operating at a 10 W power level. Based on the recorded data, the phase of the sensor coil relative to the load coil for the left side (Figure 22a) and the right side (Figure 22b) are 139.5 degrees and −58.7 degrees, respectively, which makes the phase difference between left and right side 198.2 degrees. This slightly exceeds the theoretical 180 degree phase difference, but the difference between the left region and right region can still be clearly distinguished.

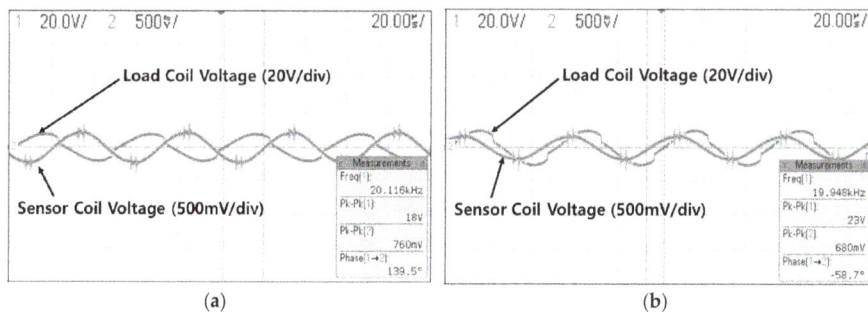

Figure 22. Oscilloscope readings from the second experiment showing the voltage waveform of the load coil and sensor coil unit at (**a**) left misalignment; and (**b**) right misalignment.

The waveforms (1) and (2) shown in Figure 23 show the converted square waveforms of the load coil and sensor coil outputs from Figure 22, respectively. These converted square waveforms are fed into the XOR gate, and the output signals are shown as waveform (3) in Figure 23. Figure 23a,b represent the left misalignment case and the right misalignment case, respectively. As can be seen from Figure 21, a phase difference between the load coil and sensor coil exists, thus the output generated by the XOR gate is an "unclean" HIGH or LOW logic output. However, it still verifies the concept illustrated in Figure 14.

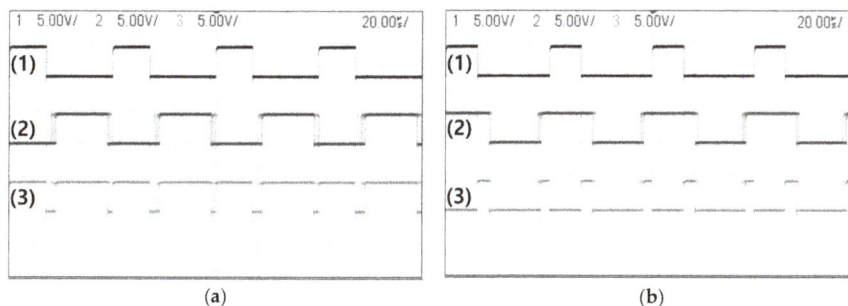

Figure 23. Oscilloscope readings from the second experiment showing the converted logic waveform of the (**1**) load coil; (**2**) sensor coil; and (**3**) the XOR gate output for (**a**) left misalignment; and (**b**) right misalignment.

To compensate the "unclean" signal from the two signals shown in waveform (3) in Figure 21a,b, a moving average filter was implemented in the main controller as follows:

$$LP_{total} = \frac{1}{N} \sum_{j=0}^{N-1} LP[i+j] \tag{11}$$

where LP_{total} is the filtered lateral position output based on the number of samples, N. From this, the LP described in Equation (10) can finally be identified, as follows:

$$LP = \begin{cases} left, & if \ LP_{total} > threshold \\ right, & if \ LP_{total} < threshold \end{cases} \tag{12}$$

where *threshold* is a certain fixed value used as a reference to compare with LP_{total}. The left or right position of the vehicle can be clearly distinguished when LP_{total} is greater or lower than the *threshold*, respectively. Figure 24 illustrates the recorded DC voltage and current values as the vehicle was moved from left to right. These values have a trend similar to the data shown in Figure 20, where power decreased significantly with increasing misalignment. The data from Figure 24 were used to determine the relationship between voltage and misalignment distance, $f(V_L)$, shown in Equation (10), which came out as $y = -0.1834x^2 - 0.055x + 6.7756$. In the equation, x is the measured voltage, and y is the determined misalignment range output.

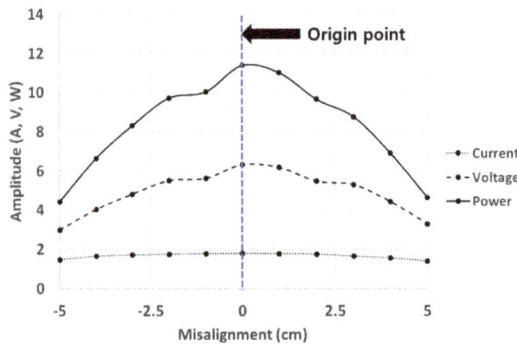

Figure 24. Voltage and current measured at the load, and its calculated power, and its calculated power (second experiment: experimental vehicle load).

Figure 25 shows the data acquired from the experimental vehicle from the second experiment, indicating the estimated lateral misalignment based on voltage readings from the digital voltmeter, and the determined left/right region from the sensor coil. The negative values and positive values on the misalignment range axis are the left and right lateral misalignment location values, respectively. It can be observed that the vehicle's position is always above or below the origin point (0 cm). This is because a tolerance range has been assigned in the fuzzy steering controller, which is ±1.7 cm at the origin point. If the desired position was set at the origin point (0 cm), it can be expected that continuous steering oscillations will occur, which is undesirable for both the vehicle's safety and comfort.

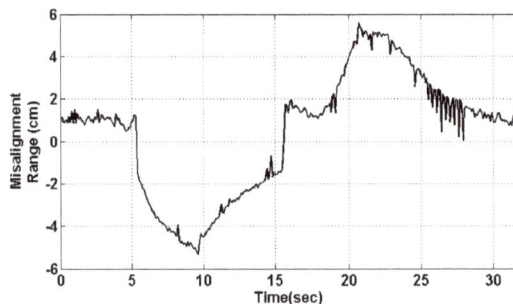

Figure 25. Estimated misalignment range from the ACAS controller based on voltage readings and the left/right position given by the sensor coil.

Based on the estimated lateral misalignment values, the steering can be controlled, as shown in the experimental vehicle's recorded data in Figure 26. It should be noted that the experimental setup differs greatly from a real driving environment, because the vehicle's lateral position was forcibly changed using the acrylic plate. However, the test demonstrates that the fuzzy controller designed in [24] outputs a steering value corresponding to the lateral misalignment position, thus verifying the feasibility of the operating concept of the autonomous steering control.

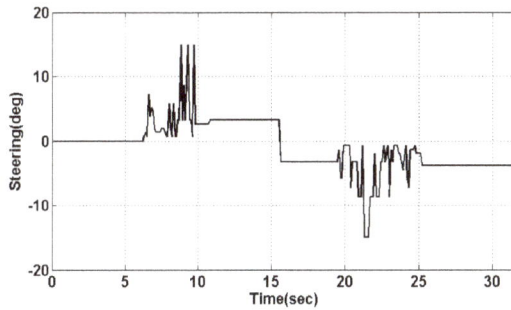

Figure 26. Steering angle output generated by the Fuzzy steering controller designed in [24] based on the estimated lateral misalignment shown in Figure 22.

Figure 27 shows the estimated power received in the load coil of the experimental vehicle. The estimated power is the product of voltage and current. The voltage was measured by a digital voltmeter installed in the vehicle. The current can be calculated by finding the relationship function between voltage and current based on Figure 24. The figure shows that when the vehicle's load coil is aligned with the source coil, roughly 11 W of power is received, and when the vehicle is at a ±5 cm lateral misalignment, the received power falls to roughly 4 W. This indicates that the received power is reduced by more than two thirds of its maximum level, which can result in a significant loss in the vehicle's driving range.

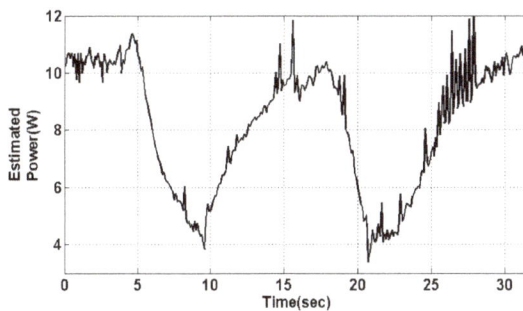

Figure 27. Estimated power received by the load coil of the experimental vehicle during the second experiment.

To show the significance of the ACAS, an analysis comparing the generated energy of a vehicle with ACAS and without ACAS was conducted. Here, two assumptions are made:

1. For the vehicle without ACAS, the estimated power shown in Figure 27 was assumed to be the level of power that would actually be received by the DWC vehicle while in operation.
2. For the vehicle with ACAS, it was assumed that a constant 11W of power is generated, as the vehicle's load coil is kept in constant alignment with the source coil on the road.

Based on these two assumptions, an analysis comparing the energy accumulated by the vehicle with ACAS and the vehicle without ACAS was conducted, and the results are shown in Figure 28. The recorded accumulated energy at the 30 second mark (shown in Figure 28) is 0.067 Wh for the vehicle without ACAS, and 0.091 Wh for the vehicle with ACAS, respectively. Assuming that the experiment continued at this rate for an hour, the total accumulated energies would be 8.04 Wh and 10.92 Wh for the vehicle without ACAS and the vehicle with ACAS, respectively. Based on these values, the vehicle with the ACAS received 26% more accumulated energy than the vehicle without the ACAS.

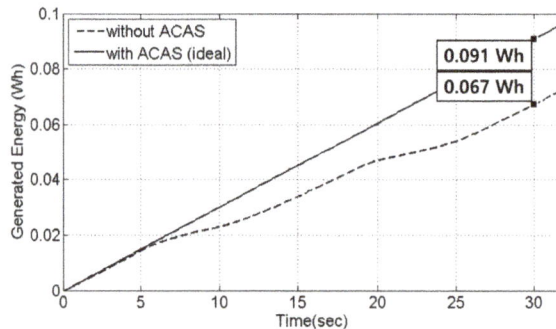

Figure 28. Comparison of estimated energy accumulated by a vehicle with ACAS and a vehicle without ACAS (based on generated power in the load coil shown in Figure 24).

5. Conclusions

This paper has proposed an ACAS for EVs with DWC to maximize the power delivered to the load coil, by keeping the vehicle aligned with the source coil as much as possible. The proposed ACAS system was verified through simulations and experiments. Two separate experiments were conducted at different power levels. One was conducted at ~90 W, and the other at ~10 W. The experiments verified the operating concept of the sensor coil, which is the key component used to determine the left/right side misalignment. In addition, an analysis based on the results of the second experiment demonstrated the advantage of the proposed ACAS in the vehicle, where 26% more energy could be accumulated by keeping the vehicle's load coil aligned with the source coil. By incorporating an improved steering controller which has a faster response to the vehicle's characteristics, it can be expected that more energy could be saved. The ACAS provides a solution to resolve the reduction in power received in the load coil during DWC due to misalignment, and ultimately provide higher efficiency and longer driving ranges for EVs with DWC.

Supplementary Materials: The following are available online at www.mdpi.com/1996-1073/10/3/315/s1, Video S1: video showing the overall operation of the autonomous coil alignment system (ACAS). It can be seen that the vehicle's steering position is autonomously adjusted once significant misalignment is detected.

Acknowledgments: This research was supported by the MSIP (Ministry of Science, ICT and Future Planning), Korea, under the ITRC (Information Technology Research Center) support program (IITP-2016-R2718-16-0012) supervised by the IITP (National IT Industry Promotion Agency) and the IT R&D program of MSIP/IITP. (B0138-16-1002, Study on the EMF exposure control in smart society).

Author Contributions: Karam Hwang conceived and designed the experiments; Karam Hwang performed the experiments; Karam Hwang, Jaeyong Cho, Dongwook Kim, and Jaehyoung Park, and Seungyoung Ahn analyzed the data; Jong Hwa Kwon, Sang Il Kwak, Hyun Ho Park, and Seungyoung Ahn contributed reagents/materials/analysis tools; Karam Hwang and Seungyoung Ahn wrote the paper.

Conflicts of Interest: The authors declare no conflict of interest.

References

1. Aziz, M.; Oda, T.; Mitani, T.; Watanabe, Y.; Kashiwagi, T. Utilization of Electric Vehicles and Their Used Batteries for Peak-Load Shifting. *Energies* **2015**, *8*, 3720–3738. [CrossRef]
2. Cao, C.; Wang, L.; Chen, B. Mitigation of the Impact of High Plug-in Electric Vehicle Penetration on Residential Distribution Grid Using Smart Charging Strategies. *Energies* **2016**, *9*, 1024. [CrossRef]
3. Tan, L.; Liu, H.; Liu, Z.; Guo, J.; Yan, C.; Wang, W.; Huang, X. Power Stabilization Strategy of Random Access Loads in Electric Vehicles Wireless Charging System at Traffic Lights. *Energies* **2016**, *9*, 811. [CrossRef]
4. Choi, S.Y.; Gu, B.W.; Jeong, S.Y.; Rim, C.T. Advances in wireless power transfer systems for roadway-powered electric vehicles. *IEEE J. Sel. Top. Power Electron.* **2015**, *3*, 18–36. [CrossRef]
5. Shekhar, A.; Prasanth, V.; Bauer, P.; Bolech, M. Economic Viability Study of an On-Road Wireless Charging System with a Generic Driving Range Estimation Method. *Energies* **2016**, *9*, 76. [CrossRef]
6. Nagendra, G.R.; Chen, L.; Covic, G.A.; Boys, J.T. Detection of EVs on IPT highways. *IEEE J. Sel. Top. Power Electron.* **2014**, *2*, 584–597. [CrossRef]
7. Jang, Y.J.; Jeong, S.; Lee, M.S. Initial Energy Logistics Cost Analysis for Stationary, Quasi-Dynamic, and Dynamic Wireless Charging Public Transportation Systems. *Energies* **2016**, *9*, 483. [CrossRef]
8. Chen, Z.; Jing, W.; Huang, X.; Tan, L.; Chen, C.; Wang, W. A Promoted Design for Primary Coil in Roadway-Powered System. *IEEE Trans. Magn.* **2015**, *51*, 8402004. [CrossRef]
9. Chen, W.; Liu, C.; Lee, C.H.; Shan, Z. Cost-Effectiveness Comparison of Coupler Designs of Wireless Power Transfer for Electric Vehicle Dynamic Charging. *Energies* **2016**, *9*, 906. [CrossRef]
10. Chow, J.P.; Chen, N.; Chung, S.H.; Chan, L.L. An investigation into the use of orthogonal winding in loosely coupled link for improving power transfer efficiency under coil misalignment. *IEEE Trans. Power Electron.* **2015**, *30*, 5632–5649. [CrossRef]
11. Choi, S.Y.; Jeong, S.Y.; Lee, E.S.; Gu, B.W.; Lee, S.W.; Rim, C.T. Generalized models on self-decoupled dual pick-up coils for large lateral tolerance. *IEEE Trans. Power Electron.* **2015**, *30*, 6434–6445. [CrossRef]
12. Kalwar, K.A.; Mekhilef, S.; Seyedmahmoudian, M.; Horan, B. Coil Design for High Misalignment Tolerant Inductive Power Transfer System for EV Charging. *Energies* **2016**, *9*, 937. [CrossRef]
13. Shin, J.; Shin, S.; Kim, Y.; Ahn, S.; Lee, S.; Jung, G.; Jeon, S.; Cho, D. Design and implementation of shaped magnetic-resonance-based wireless power transfer system for roadway-powered moving electric vehicles. *IEEE Trans. Ind. Electron.* **2014**, *61*, 1179–1192. [CrossRef]
14. Kim, J.; Kim, J.; Kong, S.; Kim, H.; Suh, I.; Suh, N.; Cho, D.; Kim, J.; Ahn, S. Coil design and shielding methods for a magnetic resonant wireless power transfer system. *Proc. IEEE* **2013**, *101*, 1332–1342. [CrossRef]
15. Gao, Y.; Farley, K.B.; Tse, Z.T.H. A Uniform Voltage Gain Control for Alignment Robustness in Wireless EV Charging. *Energies* **2015**, *8*, 8355–8370. [CrossRef]
16. Hu, P.; Ren, J.; Li, W. Frequency-Splitting-Free Synchronous Tuning of Close-Coupling Self-Oscillating Wireless Power Transfer. *Energies* **2016**, *9*, 491. [CrossRef]
17. Federal Highway Administration. Available online: http://safety.fhwa.dot.gov/geometric/pubs/mitigationstrategies/chapter3/3_lanewidth.cfm (accessed on 15 October 2014).
18. Hernandez, J.I.; Kuo, C. Steering control of automated vehicles using absolute positioning GPS and magnetic markers. *IEEE Trans. Veh. Technol.* **2003**, *52*, 150–161. [CrossRef]
19. Ryu, H.; Har, D. Wireless power transfer for high-precision position detection of railroad. In Proceedings of the IEEE Power, Communication and Information Technology Conference (PCITC), Bhubaneswar, India, 15–17 October 2015; pp. 605–608.
20. Chen, S.; Liao, C.; Wang, L. Research on positioning technique of wireless power transfer system for electric vehicles. In Proceedings of the IEEE Conference and Expo Transportation Electrification Asia-Pacific (ITEC Asia-Pacific), Beijing, China, 31 August–3 September 2014; pp. 1–4.
21. Choi, Y.; Kang, D.; Lee, S.; Kim, Y. The autonomous platoon driving system of the on line electric vehicle. In Proceedings of the ICROS-SICE International Joint Conference, Fukuoka, Japan, 18–21 August 2009; pp. 3423–3426.
22. Xu, H.; Yang, M.; Wang, C.; Yang, R. Magnetic sensing system design for intelligent vehicle guidance. *IEEE/ASME Trans. Mechatron.* **2010**, *15*, 652–656.
23. Byun, Y.-S.; Jeong, R.-G.; Kang, S.-W. Vehicle Position Estimation Based on Magnetic Markers: Enhanced Accuracy by Compensation of Time Delays. *Sensors* **2015**, *15*, 28807–28825. [CrossRef] [PubMed]

24. Hwang, K.; Park, J.; Kim, D.; Park, H.H.; Kwon, J.H.; Kwak, S.I.; Ahn, S. Autonomous Coil Alignment System Using Fuzzy Steering Control for Electric Vehicles with Dynamic Wireless Charging. *Math. Probl. Eng.* **2015**, *2015*, 205285. [CrossRef]

25. Kim, J.; Lee, B.; Lee, S.; Park, C.; Jung, S.; Lee, S.; Yi, K.; Baek, J. Development of 1-MW inductive power transfer system for a high-speed train. *IEEE Trans. Power Electron.* **2015**, *62*, 6242–6249. [CrossRef]

26. Ko, Y.; Jang, Y. The optimal system design of the online electric vehicle utilizing wireless power transmission technology. *IEEE Trans. Intell. Transp. Syst.* **2013**, *14*, 1255–1265. [CrossRef]

27. Cheng, C.-H. Implementation of a Small Type DC Microgrid Based on Fuzzy Control and Dynamic Programming. *Energies* **2016**, *9*, 781. [CrossRef]

28. Sayed, K.; Gabbar, H.A. Electric Vehicle to Power Grid Integration Using Three-Phase Three-Level AC/DC Converter and PI-Fuzzy Controller. *Energies* **2016**, *9*, 532. [CrossRef]

Review

Progress in Heat Pump Air Conditioning Systems for Electric Vehicles—A Review

Qinghong Peng [1,2] and Qungui Du [1,*]

[1] School of Mechanical and Automotive Engineering, South China University of Technology,
     Guangzhou 510641, China; pqh0720@sina.com
[2] Department of Mechanical and Electrical Engineering, Shunde Polytechnic, Foshan 528333, China
[*] Correspondence: ctqgdu@scut.edu.cn; Tel./Fax: +86-20-8711-2553

Academic Editor: Vincent Lemort
Received: 24 October 2015; Accepted: 17 March 2016; Published: 25 March 2016

Abstract: Electric vehicles have become increasingly popular in recent years due to our limited natural resources. As a result, interest in climate control systems for electric vehicles is rising rapidly. According to a variety of research sources, the heat pump air conditioning system seems to be a potential climate control system for electric vehicles. In this paper, an extensive literature review has been performed on the progress in heat pump air conditioning systems for electric vehicles. First, a review of applications of alternative environmentally friendly refrigerants in electric vehicles is introduced. This is followed by a review of other advanced technologies, such as the inverter technology, innovative components and the system structure of the heat pump air conditioning system for electric vehicles. Lastly, recent developments in multiple source heat pump systems are presented. The use of these advanced technologies can provide not only sufficient refrigerating capacity for the electric vehicle, but also higher climate control system efficiency. We believe that ideal practical air conditioning for electric vehicles can be attained in the near future as the mentioned technical problems are gradually resolved.

Keywords: air conditioning; heat pump; electric vehicle; heat source

---

1. Introduction

Due to pollution reduction and greenhouse gas emission reduction policies, fully electric vehicles (EVs) are being strongly promoted. In both EVs and internal combustion engine vehicles, a comfortable cabin environment is essential for passengers. However, in consideration of the absence of heat from the engine coolant in EVs, an innovative air conditioning (AC) system design must be provided. In recent years, solutions for the AC system in fully EVs have been studied extensively.

Some authors have presented the thermoelectric AC, whereby the vehicle cabin can be cooled and heated by thermoelectric modules, which have the advantage of having no moving parts, no noise, long life, small size and precise temperature control [1,2], but this technology has not been accepted due to poor efficiency. Currently, it is used only in seat heating and cooling in some luxury cars. In addition, this technology is applied to short-distance small EVs in view of limited resources and the low figure of merit of thermoelectric materials [3].

The simplest solution is to use an electric compressor instead of a mechanical compressor for cooling, meanwhile a positive temperature coefficient (PTC) heater is adapted to provide heating in place of the engine coolant heater core [4–6]. A 42 V electric AC system was proposed. The system consisted of a compressor, a blower, an integrated PTC heater, an inverter, pipes and some heat exchangers [5]. The cabin temperature would initially decline quickly and then change more consistently. The results showed the 42 V electric AC system could maintain a stable and comfortable interior environment under hot weather conditions. Moreover, it could achieve a relatively better thermal environment than the AC system used in conventional vehicles under very cold weather conditions. Although the PTC heater could provide sufficient heat energy to warm up the cabin, its energy was derived from battery electricity. It resulted in 24% losses of the driving range for fully EVs due to the low energy efficiency of PTC heaters [7]. A fuel fired heater was another option proposed for heating without electricity consumption, but it did not meet environmental demands.

Much research has focused on the heat pump AC system. It is based on the vapor compression cycle, which provides both cooling and heating capacity by adopting a 4-way valve that reverses the direction of refrigerant flow. Lee [8] declared that the power consumption of a heat pump system was about one third that of the electric PTC heating system for the same heating capacity. Moreover, the coefficient of performance (COP) of a heat pump AC system is larger than 1, so the heat pump AC system seems to be a more reasonable solution than other climate control systems proposed for EVs [9].

Various studies have been performed to enhance the heat pump AC system efficiency, especially the heating performance when faced with low outdoor temperatures. Besides the single air source heat pump AC system, multiple source heat pump AC systems have been developed for EVs. These systems can supply sufficient cooling or heating capacity while minimizing the influence of the AC system on driving ranges.

It is the intent of this paper to review the most recent progress concerning heat pump AC technologies for EVs. This review is broadly divided into two key categories and will be systematically organized. First, single source heat pump AC systems for EV applications are introduced. In this section, several advanced technologies and strategies concerning single air source heat pump AC systems are comprehensively reviewed. Second, multiple source heat pump AC systems are analyzed. These systems are applied for all possible heat sources in EVs to enhance the heating capacity under very low outdoor temperature conditions, as well as to achieve high energy efficiency. Finally, conclusions are drawn based on the various reviews and analyses.

2. Single Source Heat Pump AC Systems

Considering its convenient replacement, low cost and easy maintenance, the single source heat pump AC system is still dominant in EVs, especially in mild climate areas. However, the heat pump AC system, which only involves the necessary modifications based on conventional vehicle AC systems, has low system efficiency [10]. Therefore many scholars have presented innovative technologies in various aspects.

2.1. Alternative Refrigerants

At present, the refrigerant R134a, which has a global warming potential (GWP) of 1300, is still dominant in automotive AC systems, but for future environmental considerations, the Kyoto and Montreal protocols have banned or limited the use of chemical refrigerants [11]. Similarly, the European Union has passed regulations to restrict the use of refrigerants with a GWP higher than 150 in mobile AC systems [12]. A directive for the gradual phase-out of high GWP refrigerants in mobile AC systems was ratified in 2007 and went into effect at the beginning of 2008 [13]. In light of this situation, automotive AC systems using other potential substitute refrigerants have been studied [14]. CO_2 is one of the most studied options since it has adequate thermophysical properties with no ozone depletion potential (ODP) and a GWP = 1 [15,16]. As a result, more and more authors are devoted to investigating automotive AC systems using CO_2 as a refrigerant. Prototype CO_2 automotive AC systems were presented in [17,18]. They concluded that, in the heat pump mode, high capacity and COP can also be achieved at low ambient temperature and with high air supply temperature to the passenger compartment [19]. Furthermore, the system performance was equal or superior to that of the current R134a system [20]. The cooling COP ratio to R134a system was 1, while the heating/dehumidifying COP ratio was 1.31. Kim *et al.* [21] studied the effects of operating parameters on the performance of a CO_2 AC system for vehicles with various operating conditions, which include different gas cooler inlet pressures, compressor speeds and frontal air temperatures/flow rates passing through the evaporator and the gas cooler. They also proposed an algorithm for optimum high pressure control for the transcritical CO_2 cycle to achieve a maximum COP.

Although CO_2 AC systems have special benefits, they have some disadvantages, such as low critical temperature and high operating pressure [22]. Other problems in defrosting of exterior heat exchangers and performance deterioration under cold ambient temperature conditions exist in CO_2 automotive heat pump systems as well as R134a heat pump systems. By re-arranging the radiator and outdoor heat exchangers of the CO_2 heat pump system in electric cars, the heating capacity and COP were increased by 54% and 22%, respectively [23]. To enhance cooling performance, Lee *et al.* [24] presented an electrical AC system with CO_2 that used an inverter-driven compressor. The cooling capacity and COP of this tested system were increased by 36.8% up to 6.4 kW and by 30.3% up to 2.5 kW, respectively. Ma *et al.* [25] conducted a thorough review of the CO_2 heat pump and refrigeration cycle. They concluded that some modifications, such as using an internal heat exchanger, two-stage compression, and expansion work recovery as well as enhancing heat transfer, could improve the CO_2 transcritical cycle performance to a level similar to that of a conventional heat pump system.

In addition to using CO_2, other possible refrigerants in the heat pump AC system include R1234yf, R152a, R290, R245fa and water [26–29]. The performance of the R1234yf "drop-in" automotive AC system was analyzed and compared with that of systems with CO_2 and R32 by experimentation and simulation [30]. The COP and capacity of R1234yf system were up to 2.7% and 4.0% lower than those of R134a system, respectively, while the compressor discharge temperature and amount of refrigerant charge were 6.5 °C and 10% lower than those of R134a system [31]. Consequently, the R1234yf "drop-in" AC system was the most feasible candidate for automobiles from the standpoint of system performance and operating conditions. However, more work must be completed before the R1234yf system can be widely accepted in EVs [31–33]. Ghodbane [34,35] presented the secondary loop system using R152a as the working fluid in mobile climate control systems. This system showed a very good cooling and heating performance, but had a slower response to load changes, complex system connections and a high cost.

2.2. Application of Inverter Technology

Frequency variation technology, a common way to save energy, is also widely used in AC systems. An R134a automotive AC system capable of operating as an air-to-air heat pump is described in Figure 1 [36].

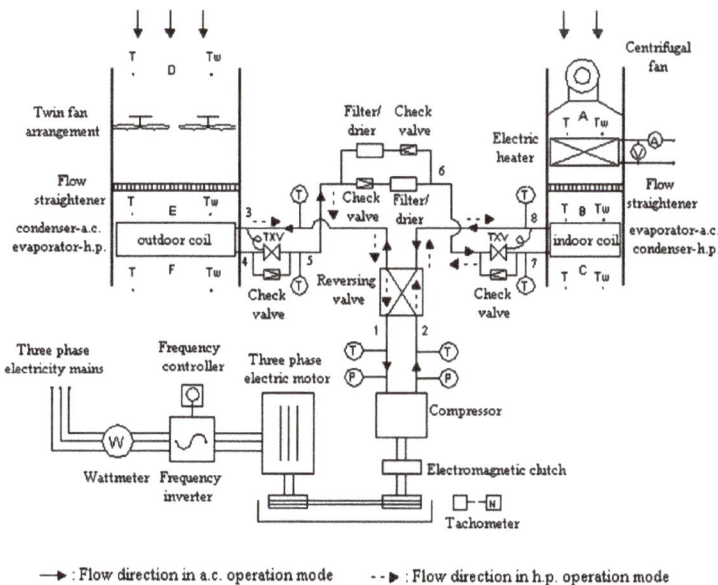

Figure 1. Schematic diagram of the experimental automotive air conditioning/heat pump system [36].

347

This system was tested by varying the compressor speed. The conclusions showed that both the heating and the cooling capacities of the system increased with the rise of compressor speed, whereas the COPs in both cases decreased. Jabardo *et al.* [37] also reported an automotive AC system equipped with a variable capacity compressor run by an electric motor controlled by a frequency converter. The impact on system performance of operational parameters such as compressor speed, return air in the evaporator and condensing air temperatures, was experimentally evaluated and simulated by means of the developed model. To better develop the inverter AC system, all factors which influence the performance of the variable frequency AC system have been discussed in [38]. Effects of compressor frequency on the performance and parameters are shown in Figure 2. At a fixed ambient temperature and heat transfer area of heat exchangers, the cooling capacity and power consumption increased as the compressor frequency increased. In contract, the energy efficiency ratio (EER) initially increased but subsequently decreased, so the compressor frequency should be increased in order to improve the cooling capacity, while it should be decreased in order to reduce power consumption. These results are very useful for optimizing the design, and automatically controlling and diagnosing exceptions in the operation of variable frequency AC systems.

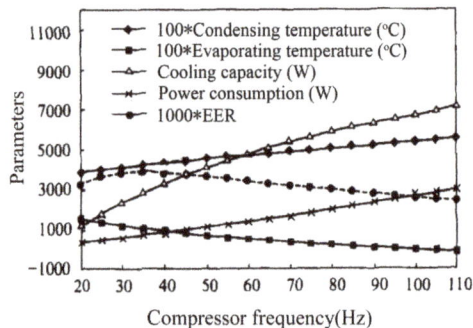

Figure 2. Effects of compressor frequency on the performance and parameters [38].

In addition to variable-speed compressors, adjusting the fan speed can lead to further performance and efficiency improvements. Lee *et al.* [39] focused on the effect of outdoor coil fan speed on the performance variation of the heat pump system adopting the hot gas bypass method. The integrated heating capacity with hot gas bypass was highest at 60% (780 rpm) fan speed. This value was 4.4% higher than that of the constant speed fan. On the other hand, the averaged COP of the heat pump in this case was higher by 2.8% than the constant speed fan. As mentioned in [40], three frequency converters were equipped to control the speeds of the compressor, the evaporator fans and the condenser fans, respectively. The analysis showed that the three speeds could be adjusted simultaneously according to both actual working conditions and operation mode so that the AC system could operate in an optimum state, but in practical applications only one fan speed and compressor speed need to be modulated in real-time, considering hardware costs and system complexity.

Advanced control algorithms have also been studied by many authors. Yeh *et al.* [41] proposed two control algorithms. The first algorithm, which modulated the outdoor fan speed, could enhance the steady-state power efficiency. The second algorithm, which added one more degree of freedom to control by modulating the indoor fan speed, could improve the transient response. The performance of the AC system could be improved if both algorithms were simultaneously implemented in the way that the second algorithm was responsible for the control action during the startup/transient phase of operation, and the first algorithm took over at steady state. Shi *et al.* [42] described three control algorithms for the inverter AC system: the matrix control, the system-relative commands control and the fuzzy control. Moreover, to realize comfort and system energy efficiency, they explained the mechanism to regulate the running speed of the compressor, indoor and outdoor fans, and the opening of the expansion valve. As shown in Figure 3, an artificial neural network-based controller was developed to simultaneously control the indoor air temperature and humidity by varying the compressor speed and supply fan speed [43]. The controllability tests including command following test and disturbance rejection test showed that the artificial neural network-based was able to track changes in setpoints and to resist disturbances with an adequate control accuracy. Although these control algorithms are mainly applied in building AC systems, in theory, they are also suitable for the AC systems of EVs.

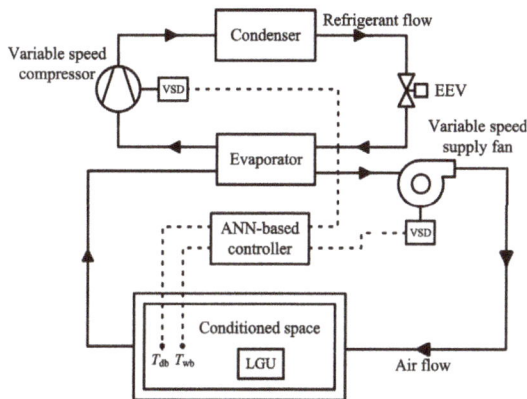

Figure 3. Schematic diagram of the artificial neural network-based controller arrangement [43].

2.3. Novel Components

Many scholars have focused on improving the performance of diverse components and parts of the heat pump AC, especially the compressor and heat exchangers. The electric scroll compressor is the most common type used in the AC systems of EVs. It has been widely accepted automakers like Toyota, BYD, Denso and so on [44–47]. Makino *et al.* [48] developed an electric compressor with various technologies as shown in Figure 4. This compressor had high reliability, low vibration and noise, small size and light weight, and high efficiency. At the same time, it showed superior comfort and cooling performance, equivalent to that of current engine-driven compressors. To further enhance the electric compressor performance and efficiency, the drive motor was included in [49–51]. Besides of the electric scroll compressor, other types of compressors were also proposed for the EV heat pump AC system. Wei *et al.* [52] presented experimental investigations of an EV heat pump AC system separately integrated with a swash plate variable displacement compressor, a scroll compressor and an electric scroll compressor. For the ambient temperature of $-10\,^{\circ}$C, the average vehicle cabin temperatures were 12 °C, 10 °C and 5 °C, respectively. The conclusions showed that, when the ambient temperature was below $-10\,^{\circ}$C, the average vehicle cabin temperature using the swash plate based system was higher than using the other two options. That is to say, the swash plate variable displacement compressor is a good choice for an EV heat pump AC system in low temperature environments. A type of vane compressor with double working cavities that was driven by a frequency modulated electric motor was designed in [53]. There was no obvious difference in performance between the new compressor and the electric scroll compressor, while the former compressor was distinctive in simple structure, manufacture and assembly. Therefore, it can be applied to EV heat pump AC systems instead of the electric scroll compressor. In [54], a miniature electrically driven turbocompressor was presented. The measurements showed that the heating, ventilation, and AC system with this turbocompressor had an ultra-compact size and high efficiency. Sakai *et al.* [55] developed a 2-way driven compressor, but this compressor can only be used in hybrid EVs. Although these novel compressors were developed, there are more tasks that must be completed before they can be widely applied in EVs, except for the electric scroll compressor.

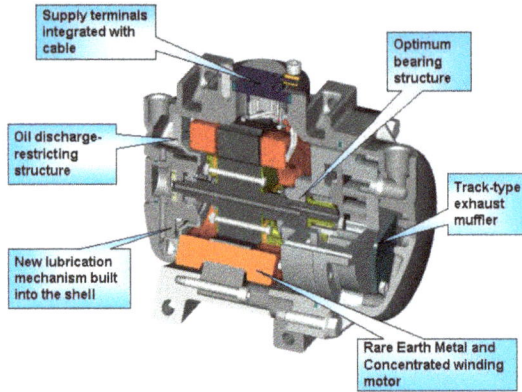

Figure 4. Main element technologies of the electric compressor [48].

In addition to compressors, heat exchangers also have research highlights. Cummings *et al.* [56] performed a comprehensive review of testing of AC heat exchangers in vehicles. They evaluated the actual performance of condensers and evaporators of AC systems through wind tunnel testing and road tests. Huang *et al.* [57] investigated the frosting characteristics of an air-source heat pump by varying the fin type of the outdoor heat exchanger. Under frosting conditions, the decreasing orders of both the average and the maximum values of the heating capacity, COP and input power were flat, wavy and wavy/slit fins. The average values of heating capacity, COP and input power for the wavy/slit fins, compared with the flat fins, were decreased by 14.57%, 8.26% and 7.11%, respectively. This conclusion provides a basis for selecting the fin type for outdoor heat exchangers. AC systems with micro-channel heat exchangers were proposed in [58–60]. The representative micro-channel evaporator is shown in Figure 5 [60].

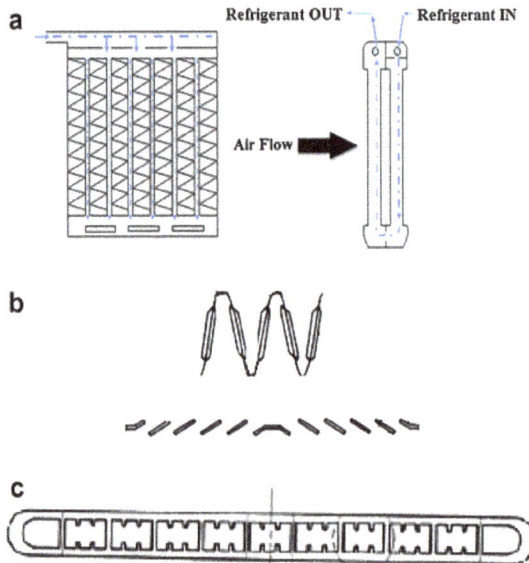

Figure 5. Schematic diagram of the microchannel evaporator: (**a**) Front view (left) and side view (right) of the microchannel evaporator; (**b**) Louver fin used in the microchannel evaporator; (**c**) Microchannel tube [60].

Compared with the use of fin-tube heat exchangers, the cooling and heating efficiencies of heat pump AC systems were increased by at least 20% with the use of micro-channel heat exchangers over a constant heat transfer area [58]. Besides higher-efficiency, the micro-channel heat exchangers have other advantages such as a neeed for less refrigerant charge, compactness and low cost. The heat pump AC system (using micro-channel heat exchangers) was applied to EVs by Wu *et al.* [61]. They concluded that the size of the indoor and outdoor heat exchangers decreased by 57.6% and 62.5%, respectively, so the AC weight was effectively reduced, which contributed to an increase in the mileage of the EV. At the same time, this system could cut the refrigerant charge by 26.5%, which reduced the greenhouse effect. The disadvantages of this system were also presented in this paper. The AC system frequently worked on the defrosting cycle in cold weather conditions, which immensely affected the heating capacity and the heating performance coefficient. Denso developed an ejector integrated evaporator, as shown in Figure 6, to reduce the power consumption of vehicle cabin AC systems [62]. The ejector system was introduced into the market May 2009. However, the noise and the temperature distribution were two main challenges in developing an evaporator with integrated ejector.

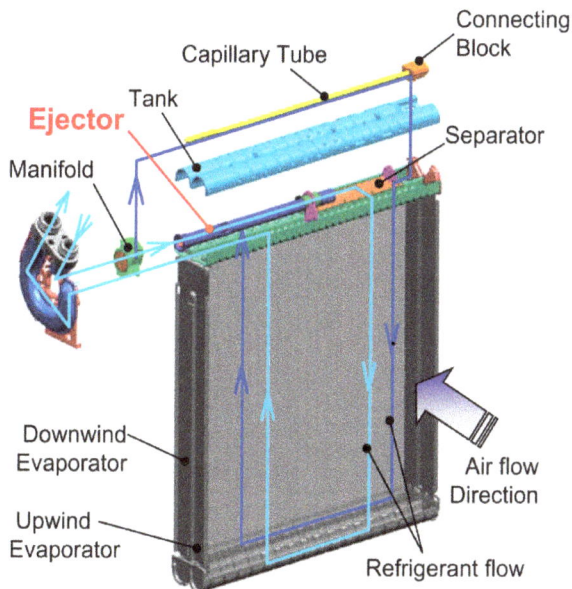

Figure 6. Ejector evaporator structure [62].

2.4. Innovative System Structure

Reforming the integral structure of the heat pump AC system for the EV is also a widely popular approach. Wang *et al.* [63] adopted three heat exchangers instead of a four-way valve to achieve cooling and heating for an EV cabin. They concluded that the heat pump AC system with three heat exchangers had advantages in demisting and dehumidifying, but the capacity and COP of this system were slightly lower than that of the heat pump AC system with a four-way valve. Suzuki *et al.* [47] proposed a representative system structure of the AC system for EVs. The construction and mechanism are shown in Figure 7.

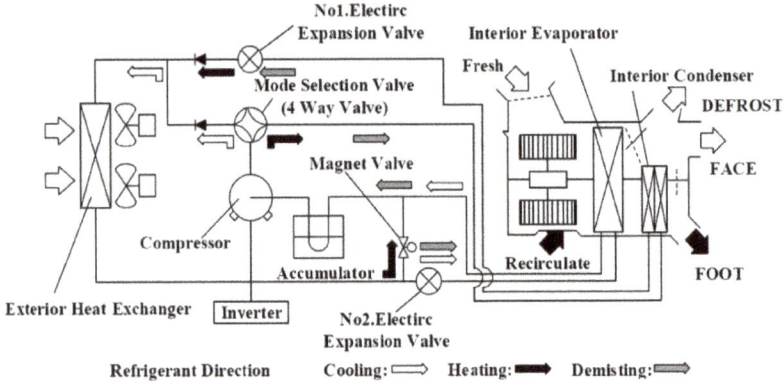

Figure 7. Air conditioning system structure and operation for electric vehicles [47].

With the two heat exchangers in the interior unit, the system could not only provide cooling, heating and demisting/dehumidifying, but also ensure safe driving when the operation mode was switched from cooling to heating. Subsequently, Xie and Min *et al.* measured the performance of this representative heat pump AC system [64,65]. In [66], an internal heat exchanger was installed in an automobile AC system to improve system performance. In [67,68], a suction line heat exchanger was added to a car AC system. The results showed that both the capacity and the COP could be improved by up to 25%, while the compressor discharge temperatures were also increased. Furthermore, Ahn *et al.* [69] compared the performance of the AC, heat pump and dual-evaporator heat pump systems, which were all combined with a heater. The experimental results showed that the dual-evaporator heat pump system as shown in Figure 8, had a superior performance in the dehumidifying and heating operation compared with the other two systems. The specific moisture extraction rate and COP of the dual-evaporator heat pump system were 53% and 62% higher, respectively, than those of the heat pump system at the indoor air wet bulb temperature of 13 °C. Moreover, the specific moisture extraction rate and COP of the heat pump system were 154% and 180% higher, respectively, than the AC system at the indoor air wet bulb temperature of 15 °C.

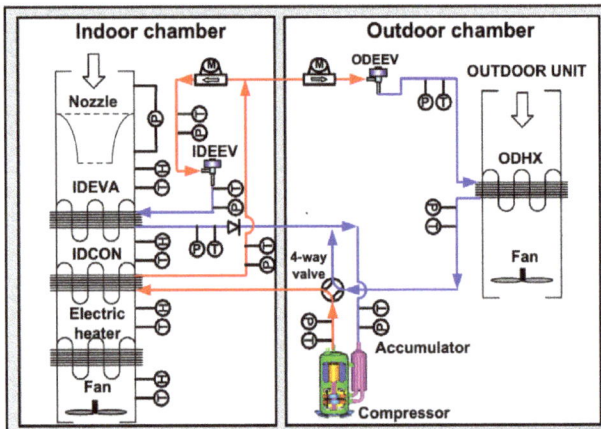

Figure 8. Schematic diagram of the dual-evaporator heat pump system for electric vehicles [69].

A dual-loop cooling and heating system for automotive applications was designed and fabricated [70,71]. The structure and flow diagram of the dual-loop system in cooling mode are shown in Figure 9 [70]. In addition to the main refrigeration loop, the system had two separate secondary fluid loops using a 50% glycol-water mixture to exchange energy with the refrigeration loop. The experimental results showed that the COP of this system varied from 0.9 to 1.8 in cooling mode, while for the heating mode it varied from 2 to 5, depending on the outdoor air conditions. A heat pump cycle with an economizer and a modified reciprocating was introduced [72]. For mobile application, the heat pump with Voorhees economizer demonstrated better performance compared to the conventional heat pump without economizer when the evaporating temperature is lower than −20 °C. It could increase the capacity at low ambient temperatures of more than two times. Wang [73] took the two-stage cycle technology and applied it in a rail vehicle AC. Li *et al.* [74,75] presented a low temperature heat pump AC system for fully EVs based on the two-stage compression refrigeration cycle as shown in Figure 10. They studied the characteristics of this system by simulation and experimentation. The results revealed that, when the environment temperature was −20 °C, the system could still run normally with a COP of 1.5. At the same time, it also possessed good performance under standard cooling and heating condition [75]. That is to say, this system could steadily and efficiently extend its operating range.

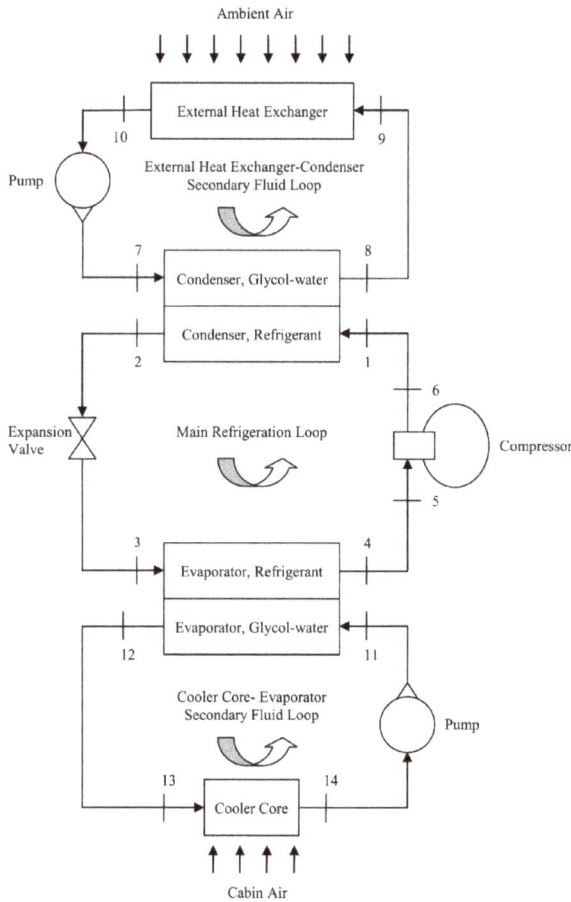

Figure 9. Ejector evaporator structure [70].

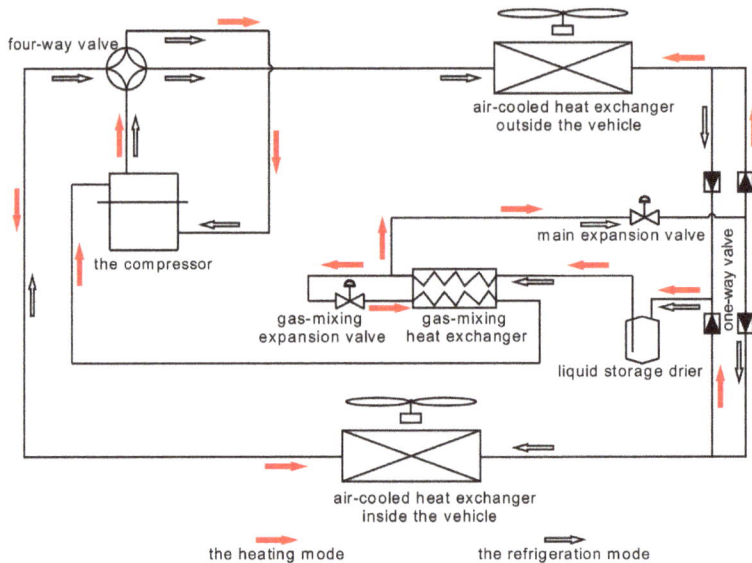

Figure 10. Schematic of gas-mixing heat pump system for fully EVs [74].

The improvement of defrosting method in addition to using the two-stage cycle technology to increase the heating performance, was another research hotspot. As described in [76,77], defrosting the heat supply and energy consumption adversely impacted the heating performance. To decrease this influence, Qu *et al.* [78] reported two control strategies for the electronic expansion valve. They investigated the two control strategies effects on reverse-cycle defrosting performance of an air source heat pump. Zhang *et al.* [79] compared the three defrosting methods, *i.e.*, reverse cycle defrosting, hot gas bypass defrosting and phase change thermal energy storage defrosting. The experimental results indicated that the defrosting method, which used phase change thermal energy storage, could shorten defrosting time and reduce energy consumption [80].

3. Multiple Source Heat Pump AC Systems

The single source heat pump AC system can be qualified to heat and cool for EVs in most weather conditions by the abovementioned methods. However the heating capacity and heating COP drop sharply with decreasing outdoor temperature. The heating capacity is insufficient in extremely cold weather. To solve these problems, many authors have proposed multiple source heat pump systems.

3.1. Additional Waste Heat Source

Waste heat discharged from electric devices, such as motors, batteries and inverters of fully EVs, is available. However, it is greatly insufficient for heating the cabin directly. To maximize the use of the waste heat, an integrated climate control system was developed by Groupe Enerstat Inc. [81]. The test results showed that the use of a dual source heat pump, which uses both air and waste heat, was one of the best methods for EVs. Promme [82] proposed a reversible heat pump system with an additional heat source which could utilize the waste heat of the battery, driven electric motor and electronic control unit. Ahn and Woo [83,84] investigated a dual source heat pump (using both air and waste heat) in EVs, which is shown in Figure 11.

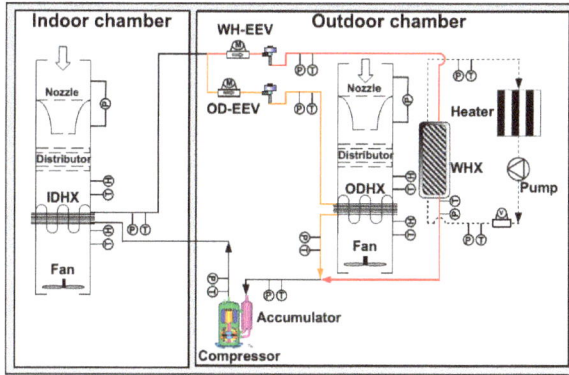

Figure 11. Schematic diagram of the dual heat source heat pump [83].

They compared the heating performance of the dual source heat pump in various operation modes: air source-only, waste heat-only and dual heat sources. The experimental results indicated that the heating capacity and COP in the dual source heat pump were increased by 20.9% and 8.6%, respectively, compared to those of the air source-only heat pump system, while it became very close to the waste heat-only mode at low outdoor temperature. Cho *et al.* [85] measured the heating performance of a coolant source heat pump, which used waste heat from electric devices on an electric bus. As shown in Figure 12, the test setup was composed of a refrigerant loop, an air circulation loop, and a coolant loop. Both an evaporator and a condenser, with plate heat exchangers, were installed for the purpose of exchanging heat between the refrigerant and the coolant source using the waste heat from the electric devices. The same heat transfer mechanism was adopted in [86,87]. Besides of the feasibility of integrating a heat pump into the AC system of the EV, both cooling and heating performances under various experimental conditions, including variations in outdoor and indoor temperatures, the water flow rate for the condenser and the evaporator sides, were investigated [86]. The system was also optimized by varying the refrigerant charge and the compressor frequency as well as using a control algorithm for operational energy management. The proposed heat pump AC system could meet target power capacities which had been set as 28 kW of cooling and 26 kW of heating with COPs of more than 1.6 for cooling and 2.6 for heating, which were required for system energy efficiency and customer comfort [87]. The target energy consumption by cooling and heating had been met at less than 20% and 25% of the total electrical energy consumption of the electric bus, respectively. Zou *et al.* [88] also presented a heat pump AC system coupled with the battery cooling system. The authors declared that the battery dissipation heat was not only a useful heat source for the heat pump AC system, especially at low outdoor temperature, but also an additional cooling load since the ambient temperature was too high to dissipate the battery heat to the ambient air directly.

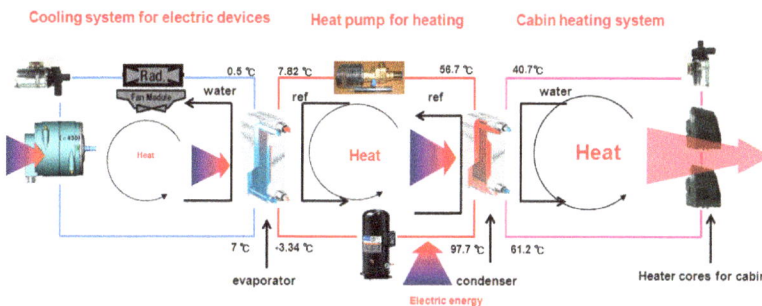

Figure 12. Heat transfer mechanism of the heat pump system using waste heat from electric devices [85].

Kim and Lee focused on heating performance enhancement of the CO_2 heat pump system using waste heat from the stacks in fuel cell EVs [89,90]. In [89], a heater core that used stack coolant was placed upstream of a cabin heater to preheat incoming air into the cabin heater. The performance of this heat pump system with heater core was compared with that of a conventional heating system with heater core and that of a heat pump system without heater core. The heating capacity of the heat pump system with heater core, which used recovered heat from the stack coolant, was improved by 100% over the heat pump system without heater core and by 70% over the conventional heating system with heater core. Furthermore, the coolant to the air heat pump system with heater core showed a significantly better performance than the air to air heat pump system with heater core. Lee *et al.* [90] concluded that when the waste heat from the stack coolant was used as the evaporating heat source, the heat pump using R744 could provide sufficient heating capacity and heating COP under cold weather conditions.

3.2. Supplemental External Heat Source

In addition to the waste heat discharged from interior electric devices in EVs, as an assisted heat source, there are other heat pump combined systems with external heat. Solar-assisted heat pump AC systems are one interesting option for EV climate control systems since solar cells can not only provide a heat insulating layer, but also recharge the battery. This concept was also applied to motor train units by Yin [91]. In [92], a solar controller and an AC controller were designed. The solar controller managed the battery fatures, such as charging and over-discharging protection, and communicated with the vehicle control system. The experimental results showed that the solar-assisted heat pump AC system could operate stably in the heating mode as well as in the cooling mode. In [93], solar cells covered the roof of a compact car, and could generate about 225 W of power. The solar-assisted heat pump AC system could improve the refrigerating capacity by about 8%, which significantly reduced the peak cooling load and the driving mileage losses of the EV. Zhao [94] declared that two hours of generating electricity capacity by a solar panel could keep the solar-assisted system running for half an hour.

In [95], a heat pump system with integrated thermoelectric modules was proposed. The authors discussed the application of this technique to provide supplemental heating for EVs. At an ambient temperature of −17.8 °C, the integrated automotive AC system could achieve an additional 2 kW of heating capacity with almost the same COP compared to the heat pump system without thermoelectric modules. The test results showed that this integrated system could increase the heating capacity in an energy-efficient way, especially for cold climate operation.

The PTC heater is another additional heat source. Kim *et al.* [96] investigated a combined system consisting of a heat pump and a PTC heater as a heating unit in EVs. Compared to the standard of the PTC heater at an indoor temperature of 20 °C, the heating capacity was increased by 59% for the combined system, and the COP was increased by 100% for the heat pump system. The conclusions showed that the heat pump cycle should be always operated for better efficiency, and the PTC heater should be controlled for better performance. Therefore the PTC heater and heat pump combined system is an optional AC system for EVs, especially in extremely cold weather conditions.

4. Conclusions

The heat pump AC system seems to be the most reasonable solution to control the climate of EVs, although there are currently many other solutions used in EVs. In this paper, an extensive literature review has been performed on the progress of heat pump AC systems for the EV.

Not only single air source heat pump systems, which have been widely considered by many researchers, have been comprehensively analyzed, but also multiple source heat pump systems have been included. In single air source heat pump AC systems, many advanced technologies and strategies were described, such as alternative refrigerants, the inverter technology, novel components, as well as innovative system structures. These advanced technologies can improve the AC system efficiency and vehicle mileage as verified in the cited reports. Furthermore, the multiple source heat pump system is also a useful method to enhance the cooling/heating capacity and the COP of the AC system in EV, especially in cold weather conditions. The heat sources include air, waste heat, solar heat, water and so on.

Considering the tremendous development in the heat pump AC system field, it is worthwhile to be mindful that there is no one single technology that can obtain ideal results to control the climate in an EV all year round. A combination of several of these technologies is still necessary more often than not. Although some of the innovative technologies described in this paper are still part of on-going research, we believe the real practical application in EVs is imminent. The integrated heat pump AC system based on the two-stage CO_2 cycle, which is equipped with a variable capacity compressor and uses the waste heat from electric devices as an additional heat source, should be considered for EVs in the coming research.

Acknowledgments: This work was supported by the Industry-University-Research Institute Collaboration Fund of Guangdong Province (No. 2011B090400572), Strategic Emerging Industry Development Fund of Guangdong Province (No. [2011]891), Natural Science Foundation of Guangdong Province (No. S2012010010199), Science and Technology Innovation Fund Project of Foshan City (No. 2013AG100063).

Conflicts of Interest: The authors declare no conflict of interest.

References

1. Mei, V.C.; Chen, F.C.; Mathiprakasam, B.; Heenan, P. Study of solar-assisted thermoelectric technology for automobile air conditioning. *J. Sol. Energy* **1993**, *115*, 200–205. [CrossRef]
2. Ma, G.Y. Study on thermoelectric air conditioning for electric vehicles. *Refrig. Air Cond.* **1998**, *14*, 5–10.
3. Cao, Z.Y. Solution to air conditioning on EVs. *Auto Electric. Parts* **2008**, *47*, 1–4.
4. Zhang, J.Z. Structural features of fully electric air conditioning system. *Automob. Maint.* **2006**, *12*, 4–5.
5. Guyonvarch, G.; Aloup, C.; Petitjean, C. *Savasse ADMD. 42V Electric Air Conditioning Systems (E-A/CS) for Low Emissions, Architecture, Comfort and Safety of Next Generation Vehicles*; SAE Technical Paper No. 2001-01-2500; SAE International: Warrendale, PA, USA, 2001.
6. Randall, B. Blowing Hot and Cold. Available online: https://www.teslamotors.com/blog/blowing-hot-and-cold (accessed on 20 December 2006).
7. Torregrosa, B.; Payá, J.; Corberán, J.M. Modeling of mobile air conditioning systems for electric vehicles. In Proceedings of the 4th European Workshop—Mobile Air Conditioning and Vehicle Thermal Systems, Torino, Italy, 1–2 December 2011.
8. Lee, D. Experimental study on the heat pump system using R134a refrigerant for zero-emission vehicles. *Int. J. Automot. Technol.* **2015**, *16*, 923–928. [CrossRef]
9. Qi, Z.G. Advances on air conditioning and heat pump system in electric vehicles—A review. *Renew. Sustain. Energy Rev.* **2014**, *38*, 754–764. [CrossRef]
10. Shi, B.X.; Ma, G.Y.; Chen, G.S. Research on heat pump systems for electric vehicle air conditioning. *Fluid Mach.* **2002**, *30*, 48–50.
11. Billiard, F. Refrigeration and air conditioning: What's new at regulatory level. In Proceedings of the Ninth European Conference on Technological Innovations in Refrigeration, Air Conditioning and in the Food Industry, Milano, Italy, 29–30 June 2001.
12. European Parliament (EUROPA). Regulation (EC) No. 842/2006 of the European Parliament and of the Council of 17 May 2006 on certain fluorinated greenhouse gases. *Off. J. Eur. Union* **2006**, *L161*, 1–11.
13. European Parliament (EUROPA). Directive 2006/40/EC of the European Parliament and of the Council of 17 May 2006 relating to emissions from air-conditioning systems in motor vehicles and amending council directive 70/156/EEC. *Off. J. Eur. Union* **2006**, *L161*, 12–18.
14. Pettersen, J.; Lorentzen, G. *A New, Efficient and Environmentally Benign System for Automobile Air Conditioning*; SAE Technical Paper No. 931129; SAE International: Warrendale, PA, USA, 1993.
15. McEnaney, R.; Hrnjak, P. *Clutch Cycling Mode of Compressor Capacity Control of Transcritical R744 Systems Compared to R134a Systems*; SAE Technical Paper No. 2005-01-2033; SAE International: Warrendale, PA, USA, 2005.

16. Petersen, M.; Bowers, C.; Elbel, S.; Hrnjak, P. Development of high-efficiency carbon dioxide commercial heat pump water heater. *HVAC R Res.* **2013**, *19*, 823–835.

17. Hafner, A. Experimental Study on Heat Pump Operation of Prototype CO_2 Mobile Air Conditioning System. In Proceedings of the 4th IIR-Gustav Lorentzen Conference on Natural Working Fluids; International Institute of Refrigeration (IRR): Paris, France, 2000.

18. Hafner, A.; Pettersen, J.; Skaugen, G.; Nekså, P. An Automobile HVAC System with CO_2 as the Refrigerant. IIR. In Proceedings of the Gustav Lorentzen Conference Natural Working Fluids; International Institute of Refrigeration (IRR): Paris, France, 1998.

19. Hafner, A.; Jakobsen, A.; Nekså, P.; Pettersen, J. Life Cycle Climate Performance (LCCP) of Mobile Air-conditioning Systems. In Proceedings of the Verband der Automobilindustrie (VDA) Alternate Refrigerant Wintermeeting, Saafelden, Austria, 18–19 February 2004.

20. Tamura, T.; Yakumaru, Y.; Nishiwaki, F. Experimental study on automotive cooling and heating air conditioning system using CO_2 as a refrigerant. *Int. J. Refrig.* **2005**, *28*, 1302–1307. [CrossRef]

21. Kim, S.C.; Won, J.P.; Kim, M.S. Effects of operating parameters on the performance of a CO_2 air conditioning system for vehicles. *Appl. Therm. Eng.* **2009**, *29*, 2408–2416. [CrossRef]

22. Kim, M.H.; Pettersen, J.; Bullard, C.W. Fundamental process and system design issues in CO_2 vapor compression systems. *Prog. Energy Combust.* **2004**, *30*, 119–174. [CrossRef]

23. Kim, S.C.; Kim, M.S.; Hwang, I.C.; Lim, T.W. Performance evaluation of a CO_2 heat pump system for fuel cell vehicles considering the heat exchanger arrangements. *Int. J. Refrig.* **2007**, *30*, 1195–1206. [CrossRef]

24. Lee, M.Y.; Lee, H.S.; Won, H.P. Characteristic evaluation on the cooling performance of an electrical air conditioning system using R744 for a fuel cell electric vehicle. *Energies* **2012**, *5*, 1371–1383. [CrossRef]

25. Ma, Y.T.; Liu, Z.Y.; Tian, H. A review of transcritical carbon dioxide heat pump and refrigeration cycles. *Energy* **2013**, *55*, 156–172. [CrossRef]

26. Zou, Y.; Hrnjak, P. Effects of fluid properties on two-phase flow and refrigerant distribution in the vertical header of a reversible microchannel heat exchanger-Comparing R245fa, R134a, R410A and R32. *Appl. Therm. Eng.* **2014**, *70*, 966–976. [CrossRef]

27. Wongwises, S.; Kamboon, A.; Orachon, B. Experimental investigation of hydrocarbon mixtures to replace HFC-134a in an automotive air conditioning system. *Energy Convers. Manag.* **2006**, *47*, 1644–1559. [CrossRef]

28. Hoffmann, G.; Plehn, W. *Natural Refrigerants for Mobile Air-Conditioning in Passenger Cars*; German Federal Environment Agency Press Office: Dessau, Germany, 2010; pp. 1–10.

29. Chamoun, M.; Rulliere, R.; Haberschill, P.; Berail, J.F. Dynamic model of an industrial heat pump using water as refrigerant. *Int. J. Refrig.* **2012**, *35*, 1080–1091. [CrossRef]

30. Zhao, Y. Study on the Performance of Automotive Air Conditioning Systems with R1234yf. Ph.D. Thesis, Shanghai Jiao Tong University, Shanghai, China, 2012.

31. Lee, Y.H.; Jung, D.S. A brief performance comparison of R1234yf and R134a in a bench tester for automobile applications. *Appl. Therm. Eng.* **2012**, *35*, 240–242. [CrossRef]

32. Seybold, L.; Hill, W.; Zimmer, C. *Internal Heat Exchanger Design Performance Criteria for R134a and HFO-1234yf*; SAE Technical Paper No. 2010-01-1210; SAE International: Warrendale, PA, USA, 2010.

33. Qi, Z.G. Experimental study on evaporator performance in mobile air conditioning system using HFO-1234yf as working fluid. *Appl. Therm. Eng.* **2013**, *53*, 124–130. [CrossRef]

34. Ghodbane, M. *An Investigation of R152a and Hydrocarbon Refrigerants in Mobile Air Conditioning*; SAE Technical Paper No. 1999-01-0874; SAE International: Warrendale, PA, USA, 1999.

35. Ghodbane, M.; Baker, J.A.; Kadle, P.S. *Potential Applications of R-152a Refrigerant in Vehicle Climate Control Part II*; SAE Technical Paper No. 2004-01-0918; SAE International: Warrendale, PA, USA, 2004.

36. Hosoz, M.; Direk, M. Performance evaluation of an integrated automotive air conditioning and heat pump system. *Energy Convers. Manag.* **2006**, *47*, 545–559. [CrossRef]

37. Saiz Jabardo, J.M.; Gonzales Mamani, W.; Ianella, M.R. Modeling and experimental evaluation of an automotive air conditioning system with a variable capacity compressor. *Int. J. Refrig.* **2002**, *25*, 1157–1172. [CrossRef]

38. Shao, S.Q.; Shi, W.X.; Li, X.T.; Yan, Q.S. Study on the adjusting performance of variable frequency air conditioning system. *Refrig. Air Cond.* **2001**, *1*, 17–20.

39. Lee, J.; Byun, J.S. Experiment on the performance improvement of air-to-air heat pump adopting the hot gas bypass method by outdoor fan speed variation. *J. Mech. Sci. Technol.* **2009**, *23*, 3407–3415. [CrossRef]

40. Peng, Q.H.; Du, Q.G. Performance evaluation of a variable frequency heat pump air conditioning system for electric bus. *IJFMS* **2015**, *8*, 13–22. [CrossRef]

41. Yeh, T.J.; Chen, Y.J.; Hwang, W.Y.; Lin, J.L. Incorporating fan control into air-conditioning systems to improve energy efficiency and transient response. *Appl. Therm. Eng.* **2009**, *29*, 1955–1964. [CrossRef]

42. Shi, W.X.; Shi, B.H.; Yan, Q.S. Three control algorithms for air conditioners with frequency control. *Heat Vent. Air Cond.* **2000**, *30*, 16–19.

43. Li, N.; Xia, L.; Deng, S.M.; Xu, X.G.; Chan, M.Y. Dynamic modeling and control of a direct expansion air conditioning system using artificial neural network. *Appl. Energy* **2012**, *91*, 290–300. [CrossRef]

44. Zhang, J.Z. Structure characteristics of electric air-conditioning system in Prius car. *Automob. Maint.* **2006**, *8*, 4–5.

45. Li, S. BYD: Active research on air-conditioning system for green car. *Refrig. Air Cond.* **2015**, *15*, 85–86.

46. Li, X.H. Development and application of frequency conversion air conditioner with electric heat pump for electric bus. *Bus Technol. Res.* **2009**, *31*, 23–25.

47. Suzuki, T.; Ishii, K. *Air Conditioning System for Electric Vehicle*; SAE Technical Paper No. 960688; SAE International: Warrendale, PA, USA, 1996.

48. Makino, M.; Ogawa, N.; Abe, Y.; Fujiwara, Y. *Automotive Air-Conditioning Electrically Driven Compressor*; SAE Technical Paper No. 2003-01-0734; SAE International: Warrendale, PA, USA, 2003.

49. Naidu, M.; Nehl, T.W.; Gopalakrishnan, S.; Wurth, L. *Electric Compressor Drive with Integrated Electronics for 42 V Automotive HVAC Systems*; SAE Technical Paper No. 2005-01-1318; SAE International: Warrendale, PA, USA, 2005.

50. Hwang, K.Y.; Park, G.B.; Cho, H.S. *Design of IPMSM for the Electrical Compressor in EV*; SAE Technical Paper No. 2011-28-0063; SAE International: Warrendale, PA, USA, 2011.

51. Baumgart, R.; Aurich, J.; Ackermann, J.; Danzer, C. *Comparison and Evaluation of a New Innovative Drive Concept for the Air Conditioning Compressor of Electric Vehicles*; SAE Technical Paper No. 2015-26-0045; SAE International: Warrendale, PA, USA, 2015.

52. Wei, M.S.; Huang, H.S.; Song, P.P.; Peng, S.Z.; Wang, Z.X.; Zhang, H. Experimental investigations of different compressors based electric vehicle heat pump air-conditioning systems in low temperature environment. In Proceedings of the IEEE Transportation Electrification Conference and Expo, ITEC Asia-Pacific, Beijing, China, 31 August–3 September 2014.

53. Chen, G.S.; Shi, B.X.; Ma, G.Y. The simulating test of vane compressor with double working cavity. *J. Guangdong Univ. Technol.* **2000**, *17*, 11–14.

54. Krahenbuhl, D.; Zwyssig, C.; Weser, H.; Kolar, J.W. A miniature 500000-r/min electrically driven turbo compressor. *IEEE Trans. Ind. Appl.* **2010**, *46*, 2459–2466. [CrossRef]

55. Sakai, T.; Ueda, M. *2-Way Driven Compressor for Hybrid Vehicle Climate Control System*; SAE Technical Paper No. 2004-01-0906; SAE International: Warrendale, PA, USA, 2004.

56. Cummings, R.W.; Shah, R.K. *Experimental Performance Evaluation of Automotive Air-Conditioning Heat Exchangers as Components and in Vehicle Systems*; SAE Technical Paper No. 2005-01-2003; SAE International: Warrendale, PA, USA, 2005.

57. Huang, D.; Liu, X.Y.; Wang, Y.L. Effect of fin type on frosting characteristics of an air-source heat pump. *J. Refrig.* **2012**, *33*, 12–17.

58. Liu, N.; Li, J.M.; Li, H.Q. Performance research on heat pump air conditioner using micro-channel heat exchangers. *Refrig. Air Cond.* **2011**, *11*, 96–99.

59. Zhang, L. Analysis of performance of air conditioners with micro-channel heat exchangers. *Refrig. Technol.* **2010**, *30*, 33–36.

60. Qi, Z.G.; Zhao, Y.; Chen, J.P. Performance enhancement study of mobile air conditioning system using microchannel heat exchangers. *Int. J. Refrig.* **2010**, *33*, 301–312. [CrossRef]

61. Wu, J.H.; Xie, F.; Liu, C.P.; Ouyang, G. Adaptability Research on micro-channel heat exchanger applied to heat pump air conditioning system for electrical vehicle. *J. Mech. Eng.* **2012**, *48*, 141–147. [CrossRef]

62. Brodie, B.R.; Takano, Y.; Gocho, M. *Evaporator with Integrated Ejector for Automotive Cabin Cooling*; SAE Technical Paper No. 2012-01-1048; SAE International: Warrendale, PA, USA, 2012.

63. Wang, Y.; Shi, J.Y.; Chen, J.P.; Wang, X.N.; Kang, Z.J. Comparative study of two kinds of automotive air conditioning system with three heat exchangers and four-way valve. *J. Refrig.* **2014**, *35*, 71–76.

64. Xie, Z. Research of Electrical Vehicle Heat Pump Air Conditioner and Its Automatic Control System. Master's Thesis, Shanghai Jiao Tong University, Shanghai, China, 2006.

65. Min, H.T.; Wang, X.D.; Zeng, X.H.; Li, S. Parameter design and computation study for air conditioning system of electric vehicle. *Automob. Technol.* **2009**, *40*, 19–22.

66. Cho, H.; Lee, H.; Park, C. Performance characteristics of an automobile air conditioning system with internal heat exchanger using refrigerant R1234yf. *Appl. Therm. Eng.* **2013**, *61*, 563–569. [CrossRef]

67. Boewe, D.; Yin, J.; Park, Y.C.; Bullard, C.W.; Hrnjak, P.S. *The Role of Suction Line Heat Exchanger in Transcritical R744 Mobile A/C Systems*; SAE Technical Paper No. 1999-01-0583; SAE International: Warrendale, PA, USA, 1999.

68. Li, G.Q.; Yuan, X.L.; Xu, H.T.; Li, M.X. Application of sution line heat exchanger for R134a automotive air-conditioning system. *J. Refrig.* **2002**, *2*, 56–59.

69. Ahn, J.H.; Kang, H.; Lee, H.S.; Kim, Y. Performance characteristics of a dual-evaporator heat pump system for effective dehumidifying and heating of a cabin in electric vehicles. *Appl. Energy* **2015**, *146*, 29–37. [CrossRef]

70. Jokar, A.; Hosni, M.H.; Eckels, S.J. *New Generation Integrated Automotive Thermal System*; SAE Technical Paper No. 2005-01-3476; SAE International: Warrendale, PA, USA, 2005.

71. Jokar, A.; Hosni, M.H.; Eckels, S.J. A heat pump for automotive application. In Proceedings of the 8th International Energy Agency Heat Pump Conference, Las Vegas, NY, USA, 30 May–2 June 2005.

72. Zha, S.; Hafner, A.; Neksa, P. Investigation of R-744 Voorhees transcritical heat pump system. *Int. J. Refrig.* **2008**, *31*, 16–22. [CrossRef]

73. Wang, X.J. Application of low-temperature heat pump technology in rail vehicle air conditioning. *Mech. Electr. Eng. Technol.* **2011**, *40*, 165–168.

74. Li, H.J.; Zhou, G.H.; Li, A.G.; Li, X.G.; Chen, J. Simulation study on characteristics of ultra-low temperature heat pump air conditioning system for pure electric vehicles. *Appl. Mech. Mater.* **2014**, *580–583*, 2475–2479. [CrossRef]

75. Li, H.J. Study on Performance of Low Temperature Heat Pump Air-Conditioning System for Pure Electric Vehicle. Ph.D. Thesis, Xi'An University of Architecture and Technology, Xian, China, 2015.

76. Dong, J.K.; Deng, S.M.; Jiang, Y.Q.; Xia, L.; Yao, Y. An experimental study on defrosting heat supplies and energy consumptions during a reverse cycle defrost operation for an air source heat pump. *Appl. Therm. Eng.* **2012**, *37*, 380–387. [CrossRef]

77. Lee, M.Y.; Kim, Y.C.; Lee, D.Y. Experimental study on frost height of round plate fin-tube heat exchangers for mobile heat pumps. *Energies* **2012**, *5*, 3479–3491. [CrossRef]

78. Qu, M.L.; Xia, L.; Deng, S.M.; Jiang, Y.Q. An experimental investigation on reverse-cycle defrosting performance for an air source heat pump using an electronic expansion valve. *Appl. Energy* **2012**, *97*, 327–333. [CrossRef]

79. Zhang, J.; Lan, J.; Du, R.H.; Gao, G.F. The performance comparison of several defrosting modes for air-source heat pump. *J. Refrig.* **2012**, *33*, 47–49.

80. Dong, J.K.; Jiang, Y.Q.; Yao, Y.; Hu, W.J. Experimental study of the characteristic of defrosting for air source heat pump with phase change energy storage. *J. Hunan Univ.* **2011**, *38*, 18–22.

81. Bilodeau, S. *High Performance Climate Control for Alternative Fuel Vehicle*; SAE Technical Paper No. 2001-01-1719; SAE International: Warrendale, PA, USA, 2001.

82. Pomme, V. *Reversible Heat Pump System for an Electrical Vehicle*; SAE Technical Paper No. 971772; SAE International: Warrendale, PA, USA, 2001.

83. Ahn, J.H.; Kang, H.; Lee, H.S.; Jung, H.W.; Baek, C.; Kim, Y. Heating performance characteristics of a dual source heat pump using air and waste heat in electric vehicles. *Appl. Energy* **2014**, *119*, 1–9. [CrossRef]

84. Woo, H.S.; Ahn, J.H.; Oh, M.S.; Kang, H.; Kim, Y.C. Study on the heating performance characteristics of a heat pump system utilizing air and waste heat source for electric vehicles. *Air Cond. Refrig. Eng.* **2013**, *25*, 180–186. [CrossRef]

85. Cho, C.W.; Lee, H.S.; Won, J.P.; Lee, M.Y. Measurement and evaluation of heating performance of heat pump systems using wasted heat from electric devices for an electric bus. *Energies* **2012**, *5*, 658–669. [CrossRef]

86. Lee, D.Y.; Cho, C.W.; Won, J.P.; Park, Y.C.; Lee, M.Y. Performance characteristics of mobile heat pump for a large passenger electric vehicle. *Appl. Therm. Eng.* **2013**, *50*, 660–669. [CrossRef]

87. Suh, I.S.; Lee, M.; Kim, J.; Oh, S.T.; Won, J.P. Design and experimental analysis of an efficient HVAC (heating, ventilation, air-conditioning) system on an electric bus with dynamic on-road wireless charging. *Energy* **2015**, *81*, 262–273. [CrossRef]
88. Zou, H.M.; Jiang, B.; Wang, Q.; Tian, C.Q.; Yan, Y.Y. Performance analysis of a heat pump air conditioning system coupling with battery cooling for electric vehicles. *Energy Procedia* **2014**, *61*, 891–894. [CrossRef]
89. Kim, S.C.; Kim, M.S.; Hwang, I.C.; Lim, T.W. Heating performance enhancement of a CO_2 heat pump system recovering stack exhaust thermal energy in fuel cell vehicles. *Int. J. Refrig.* **2007**, *30*, 1215–1226. [CrossRef]
90. Lee, H.S.; Won, J.P.; Cho, C.W.; Kim, Y.C.; Lee, M.Y. Heating performance characteristics of stack coolant source heat pump using R744 for fuel cell electric vehicles. *J. Mech. Sci. Technol.* **2012**, *26*, 2065–2071. [CrossRef]
91. Yin, J. Theoretic Study of Motor Train Unit of Solar Air Conditioning System. Master's Thesis, Southwest Jiaotong University, Chengdu, China, 2012.
92. Sun, H. Research of Automatic Control System of the Solar-Assisted Air-Conditioning for Pure Electric Vehicle. Master's Thesis, Nanchang University, Nanchang, China, 2012.
93. Ma, G.Y.; Shi, B.X.; Wu, L.Z.; Chen, G.S. Study on solar-assisted heat pump system for electric vehicle air conditioning. *Acta Energiae Sol. Sin.* **2001**, *22*, 176–180.
94. Zhao, C. Soar-Assisted System Study of Automobile Air Conditioner Based on Parallel Technology. Master's Thesis, Heifei University of Technology, Hefei, China, 2012.
95. Okuma, T.; Radermacher, R.; Hwang, Y. A novel application of thermoelectric modules in an HVAC system under cold climate operation. *J. Electron. Mater.* **2012**, *41*, 1749–1758. [CrossRef]
96. Kim, K.Y.; Kim, S.C.; Kim, M.S. Experimental studies on the heating performance of the PTC heater and heat pump combined system in fuel cells and electric vehicles. *Int. J. Automot. Technol.* **2012**, *13*, 971–977. [CrossRef]

MDPI

St. Alban-Anlage 66

4052 Basel, Switzerland

Tel. +41 61 683 77 34

Fax +41 61 302 89 18

http://www.mdpi.com

MDPI Books Editorial Office,

Email: books@mdpi.com,

http://www.mdpi.com/books

www.ingramcontent.com/pod-product-compliance
Lightning Source LLC
Chambersburg PA
CBHW051710210326
41597CB00032B/5427